ALL THE HELP, **RESOURCES**, AND PERSONAL **SUPPORT** YOU AND YOUR STUDENTS NEED!

2-Minute Tutorials and all of the resources you & your students need to get started
www.wileyplus.com/firstday

Student support from an experienced student user. Ask your local representative for details!

Collaborate with your colleagues, find a mentor, attend virtual and live events, and view resources
www.WhereFacultyConnect.com

Pre-loaded, ready-to-use assignments and presentations
www.wiley.com/college/quickstart

Technical Support 24/7 FAQs, online chat, and phone support
www.wileyplus.com/support

Your *WileyPLUS* Account Manager. Training and implementation support
www.wileyplus.com/accountmanager

www.w001

MAKE IT YOURS!

9TH EDITION

FUNDAMENTALS OF PHYSICS

9TH EDITION

HALLIDAY & RESNICK

FUNDAMENTALS OF PHYSICS

Jearl Walker

Cleveland State University

WILEY

John Wiley & Sons, Inc.

SPONSORING EDITOR   Geraldine Osnato
EXECUTIVE EDITOR   Stuart Johnson
ASSISTANT EDITOR   Aly Rentrop
ASSOCIATE MARKETING DIRECTOR   Christine Kushner
SENIOR PRODUCTION EDITOR   Elizabeth Swain
TEXT DESIGNER   Madelyn Lesure
COVER DESIGNER   M77 Design
DUMMY DESIGNER   Lee Goldstein
PHOTO EDITOR   Hilary Newman
EXECUTIVE MEDIA EDITOR   Thomas Kulesa
COVER IMAGE   ©Eric Heller/Photo Researchers, Inc.

This book was set in 10/12 Times Ten by Prepare and was printed and bound by
R.R.Donnelley/Jefferson City. The cover was printed by R.R.Donnelley/Jefferson City.

This book is printed on acid free paper.

Library of Congress Cataloging-in-Publication Data

Halliday, David
 Fundamentals of physics / David Halliday, Robert Resnick, Jearl Walker.—9th ed.
    p. cm.
 Includes index.
Part 1: ISBN 978-0-470-54791-5 (pbk.)
Also catalogued as
Extended version: ISBN 978-0-470-46908-8
 1. Physics—Textbooks. I. Resnick, Robert II. Walker, Jearl III. Title.
   QC21.3.H35 2011
   530—dc22
                                        2009033774

Printed in the United States of America

10 9 8 7 6 5 4 3 2 1

CONTENTS

APPENDICES

ANSWERS

INDEX I-1

WHY I WROTE THIS BOOK

Fun with a big challenge. That is how I have regarded physics since the day when Sharon, one of the students in a class I taught as a graduate student, suddenly demanded of me, "What has any of this got to do with my life?" Of course I immediately responded, "Sharon, this has everything to do with your life—this is physics."

She asked me for an example. I thought and thought but could not come up with a single one. That night I began writing the book *The Flying Circus of Physics* (John Wiley & Sons Inc., 1975) for Sharon but also for me because I realized her complaint was mine. I had spent six years slugging my way through many dozens of physics textbooks that were carefully written with the best of pedagogical plans, but there was something missing. Physics is the most interesting subject in the world because it is about how the world works, and yet the textbooks had been thoroughly wrung of any connection with the real world. The fun was missing.

I have packed a lot of real-world physics into this HRW book, connecting it with the new edition of *The Flying Circus of Physics*. Much of the material comes from the HRW classes I teach, where I can judge from the faces and blunt comments what material and presentations work and what do not. The notes I make on my successes and failures there help form the basis of this book. My message here is the same as I had with every student I've met since Sharon so long ago: "Yes, you *can* reason from basic physics concepts all the way to valid conclusions about the real world, and that understanding of the real world is where the fun is."

I have many goals in writing this book but the overriding one is to provide instructors with tools by which they can teach students how to effectively read scientific material, identify fundamental concepts, reason through scientific questions, and solve quantitative problems. This process is not easy for either students or instructors. Indeed, the course associated with this book may be one of the most challenging of all the courses taken by a student. However, it can also be one of the most rewarding because it reveals the world's fundamental clockwork from which all scientific and engineering applications spring.

Many users of the eighth edition (both instructors and students) sent in comments and suggestions to improve the book. These improvements are now incorporated into the narrative and problems throughout the book. The publisher John Wiley & Sons and I regard the book as an ongoing project and encourage more input from users. You can send suggestions, corrections, and positive or negative comments to John Wiley & Sons or Jearl Walker (mail address: Physics Department, Cleveland State University, Cleveland, OH 44115 USA; or email address: physics@wiley.com; or the blog site at www.flyingcircusofphysics. com). We may not be able to respond to all suggestions, but we keep and study each of them.

LEARNINGS TOOLS

Animation

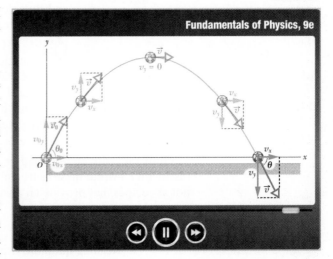

Fundamentals of Physics, 9e

Because today's students have a wide range of learning styles, I have produced a wide range of learning tools, both in this new edition and online in *WileyPLUS*:

ANIMATIONS of one of the key figures in each chapter. Here in the book, those figures are flagged with the swirling icon. In the online chapter in *WileyPLUS*, a mouse click begins the animation. I have chosen the figures that are rich in information so that a student can see the physics in action and played out over a minute or two instead of just being flat on a printed page. Not only does this give life to the physics, but the animation can be repeated as many times as a student wants.

WILEY PLUS **VIDEOS** I have made well over 1000 instructional videos, with more coming each semester. Students can watch me draw or type on the screen as they hear me talk about a solution, tutorial, sample problem, or review, very much as they would experience were they sitting next to me in my office while I worked out something on a notepad. An instructor's lectures and tutoring will always be the most valuable learning tools, but my videos are available 24 hours a day, 7 days a week, and can be repeated indefinitely.

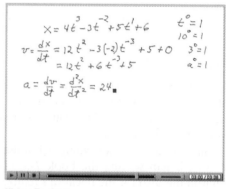

- **Video tutorials on subjects in the chapters**. I chose the subjects that challenge the students the most, the ones that my students scratch their heads about.

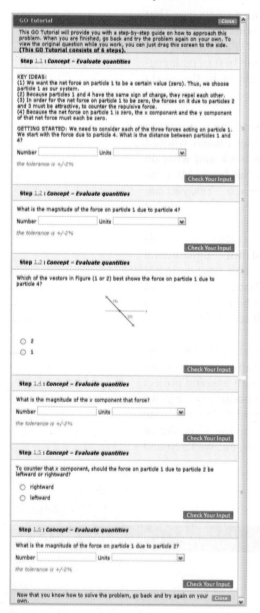

GO Tutorial

- **Video reviews of high school math,** such as basic algebraic manipulations, trig functions, and simultaneous equations.

Video Review

- **Video introductions to math,** such as vector multiplication, that will be new to the students.

- **Video presentations of every Sample Problem** in the textbook chapters (both 8e and 9e). My intent is to work out the physics, starting with the Key Ideas instead of just grabbing a formula. However, I also want to demonstrate how to read a sample problem, that is, how to read technical material to learn problem-solving procedures that can be transferred to other types of problems.

- **Video solutions to 20% of the end-of chapter problems.** The availability and timing of these solutions are controlled by the instructor. For example, they might be available after a homework deadline or a quiz. Each solution is not simply a plug-and-chug recipe. Rather I build a solution from the Key Ideas to the first step of reasoning and to a final solution. The student learns not just how to solve a particular problem but how to tackle any problem, even those that require *physics courage*.

- **Video examples of how to read data from graphs** (more than simply reading off a number with no comprehension of the physics).

READING MATERIAL I have written a large number of reading resources for *WileyPLUS*.

- **Every sample problem in the textbook** (both 8e and 9e) is available online in both reading and video formats.

- **Hundreds of additional sample problems.** These are available as standalone resources but (at the discretion of the instructor) they are also linked out of the homework problems. So, if a homework problem deals with, say, forces on a block on a ramp, a link to a related sample problem is provided. However, the sample problem is not just a replica of the homework problem and thus does not provide a solution that can be merely duplicated without comprehension.

- **GO Tutorials** for 10% of the end-of-chapter homework problems. In multiple steps, I lead a student through a homework problem, starting with the Key Ideas and giving hints when wrong answers are submitted. However, I purposely leave the last step (for the final answer) to the student so that they are responsible at the end. Some online tutorial systems trap a student when wrong answers are given, which can generate a lot of frustration. My GO Tutorials are not traps, because at any step along the way, a student can return to the main problem.

- **Hints on every end-of-chapter homework problem** are available online (at the discretion of the instructor). I wrote these as true hints about the main ideas and the general procedure for a solution, not as recipes that provide an answer without any comprehension.

EVALUATION MATERIALS Both self-evaluations and instructor evaluations are available.

- **Reading questions are available within each online section.** I wrote these so that they do not require analysis or any deep understanding; rather they simply test whether a student has read the section. When a student opens up a section, a randomly chosen reading question (from a bank of questions) appears at the end. The instructor can decide whether the question is part of the grading for that section or whether it is just for the benefit of the student.

- **Checkpoints are available within most sections.** I wrote these so that they require analysis and decisions about the physics in the section. *Answers to all checkpoints are in the back of the book.*

- **All end-of-chapter homework questions and problems** in the book (and many more problems) are available in *WileyPLUS*. The instructor can construct a homework assignment and control how it is graded when the answers are submitted online. For example, the instructor controls the deadline for submission and how many attempts a student is allowed on an answer. The instructor also controls which, if any, learning aids are available with each homework problem. Such links can include hints, sample problems, in-chapter reading materials, video tutorials, video math reviews, and even video solutions (which can be made available to the students after, say, a homework deadline).

- **Symbolic notation problems** are available in every chapter and require algebraic answers.

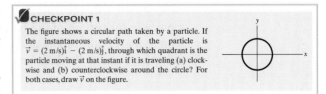

CHECKPOINT 1

The figure shows a circular path taken by a particle. If the instantaneous velocity of the particle is $\vec{v} = (2 \text{ m/s})\hat{i} - (2 \text{ m/s})\hat{j}$, through which quadrant is the particle moving at that instant if it is traveling (a) clockwise and (b) counterclockwise around the circle? For both cases, draw $\vec{v}$ on the figure.

Checkpoint

DEMONSTRATIONS AND INTERACTIVE SIMULATIONS These have been produced by a number of instructors, to provide the experience of a computerized lab and lecture-room demonstrations.

ART PROGRAM

- Many of the figures in the book have been modified to make the physics ideas more pronounced.

- At least one key figure per chapter has been greatly expanded so that its message is conveyed in steps.

Expanded Figure

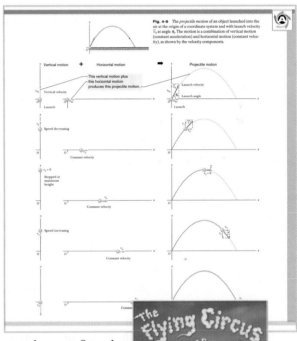

Fig. 4-9   The *projectile motion* of an object launched into the air at the origin of a coordinate system and with launch velocity $\vec{v}_0$ at angle θ_0. The motion is a combination of vertical motion (constant acceleration) and horizontal motion (constant velocity), as shown by the velocity components.

FLYING CIRCUS OF PHYSICS

- Flying Circus material has been incorporated into the text in several ways: Sample Problems, text examples, and end-of-chapter Problems. The purpose of this is two-fold: (1) make the subject more interesting and engaging, (2) show the student that the world around them can be examined and understood using the fundamental principles of physics.

- Links to *The Flying Circus of Physics* are shown throughout the text material and end-of-chapter problems with a biplane icon.   In the electronic version of this book, clicking on the icon takes you to the corresponding item in *Flying Circus*. The bibliography of *Flying Circus* (over 11 000 references to scientific and engineering journals) is located at www.flyingcircusofphysics.com.

SAMPLE PROBLEMS are chosen to demonstrate how problems can be solved with reasoned solutions rather than quick and simplistic plugging of numbers into an equation with no regard for what the equation means.

KEY IDEAS in the sample problems focus a student on the basic concepts at the root of the solution to a problem. In effect, these key ideas say, "We start our solution by using this basic concept, a procedure that prepares us for solving many other problems. We don't start by grabbing an equation for a quick plug-and-chug, a procedure that prepares us for nothing."

WHAT IS PHYSICS? The narrative of every chapter begins with this question, and with an answer that pertains to the subject of the chapter. (A plumber once asked me, "What do you do for a living?" I replied, "I teach physics." He thought for several minutes and then asked, "What is physics?" The plumber's career was entirely based on physics, yet he did not even know what physics is. Many students in introductory physics do not know what physics is but assume that it is irrelevant to their chosen career.)

ICONS FOR ADDITIONAL HELP. When worked-out solutions are provided either in print or electronically for certain of the odd-numbered problems, the statements for those problems include an icon to alert both student and instructor as to where the solutions are located. An icon guide is provided here and at the beginning of each set of problems

GO Tutoring problem available (at instructor's discretion) in *WileyPLUS* and WebAssign	
SSM Worked-out solution available in Student Solutions Manual	**WWW** Worked-out solution is at
• – ••• Number of dots indicates level of problem difficulty	**ILW** Interactive solution is at http://www.wiley.com/college/halliday
Additional information available in *The Flying Circus of Physics* and at flyingcircusofphysics.com	

Icon Guide

VERSIONS OF THE TEXT

To accommodate the individual needs of instructors and students, the ninth edition of *Fundamentals of Physics* is available in a number of different versions.

The **Regular Edition** consists of Chapters 1 through 37 (ISBN 978-0-470-04472-8).

The **Extended Edition** contains seven additional chapters on quantum physics and cosmology, Chapters 1–44 (ISBN 978-0-471-75801-3).

Both editions are available as single, hardcover books, or in the following alternative versions:
Volume 1 - Chapters 1–20 (Mechanics and Thermodynamics), hardcover, ISBN 978-0-47004473-5
Volume 2 - Chapters 21–44 (E&M, Optics, and Quantum Physics), hardcover, ISBN 978-0-470-04474-2

INSTRUCTOR SUPPLEMENTS

INSTRUCTOR'S SOLUTIONS MANUAL by Sen-Ben Liao, Lawrence Livermore National Laboratory. This manual provides worked-out solutions for all problems found at the end of each chapter.

INSTRUCTOR COMPANION SITE http://www.wiley.com/college/halliday

- **Instructor's Manual** This resource contains lecture notes outlining the most important topics of each chapter; demonstration experiments; laboratory and computer projects; film and video sources; answers to all Questions, Exercises, Problems, and Checkpoints; and a correlation guide to the Questions, Exercises, and Problems in the previous edition. It also contains a complete list of all problems for which solutions are available to students (SSM, WWW, and ILW).

- **Lecture PowerPoint Slides** by Sudipa Kirtley of The Rose Hulman Institute. These PowerPoint slides serve as a helpful starter pack for instructors, outlining key concepts and incorporating figures and equations from the text.

- **Classroom Response Systems ("Clicker") Questions** by David Marx, Illinois State University. There are two sets of questions available: Reading Quiz questions and Interactive Lecture questions. The Reading Quiz questions are intended to be relatively straightforward for any student who reads the assigned material. The Interactive Lecture questions are intended for use in an interactive lecture setting.

- **Wiley Physics Simulations** by Andrew Duffy, Boston University. This is a collection of 50 interactive simulations (Java applets) that can be used for classroom demonstrations.

- **Wiley Physics Demonstrations** by David Maiullo, Rutgers University. This is a collection of digital videos of 80 standard physics demonstrations. They can be shown in class or accessed from the Student Companion site. There is an accompanying Instructor's Guide that includes "clicker" questions.

- **Test Bank** The Test Bank includes more than 2200 multiple-choice questions. These items are also available in the Computerized Test Bank which provides full editing features to help you customize tests (available in both IBM and Macintosh versions). The Computerized Test Bank is offered in both Diploma and Respondus.

- *Instructor's Solutions Manual,* in both MSWord and PDF files.

- All text illustrations, suitable for both classroom projection and printing.

ONLINE HOMEWORK AND QUIZZING. In addition to *WileyPLUS*, *Fundamentals of Physics,* ninth edition, also supports WebAssignPLUS and LON-CAPA, which are other programs that give instructors the ability to deliver and grade homework and quizzes online. WebAssign PLUS also offers students an online version of the text.

STUDENT SUPPLEMENTS

STUDENT COMPANION SITE.   The web site http://www.wiley.com/college/halliday was developed specifically for *Fundamentals of Physics*, ninth edition, and is designed to further assist students in the study of physics. It includes solutions to selected end-of-chapter problems (which are identified with a www icon in the text); self-quizzes; simulation exercises; tips on how to make best use of a programmable calculator; and the Interactive LearningWare tutorials that are described below.

STUDENT STUDY GUIDE by Thomas Barrett of Ohio State University. The Student Study Guide consists of an overview of the chapter's important concepts, problem solving techniques and detailed examples.

STUDENT SOLUTIONS MANUAL by Sen-Ben Liao, Lawrence Livermore National Laboratory. This manual provides students with complete worked-out solutions to 15 percent of the problems found at the end of each chapter within the text. The Student Solutions Manual for the ninth edition is written using an innovative approach called TEAL which stands for Think, Express, Analyze, and Learn. This learning strategy was originally developed at the Massachusetts Institute of Technology and has proven to be an effective learning tool for students. These problems with TEAL solutions are indicated with an SSM icon in the text.

INTERACTIVE LEARNINGWARE. This software guides students through solutions to 200 of the end-of-chapter problems. These problems are indicated with an ILW icon in the text. The solutions process is developed interactively, with appropriate feedback and access to error-specific help for the most common mistakes.

INTRODUCTORY PHYSICS WITH CALCULUS AS A SECOND LANGUAGE: *Mastering Problem Solving* by Thomas Barrett of Ohio State University. This brief paperback teaches the student how to approach problems more efficiently and effectively. The student will learn how to recognize common patterns in physics problems, break problems down into manageable steps, and apply appropriate techniques. The book takes the student step by step through the solutions to numerous examples.

ACKNOWLEDGMENTS

A great many people have contributed to this book. J. Richard Christman, of the U.S. Coast Guard Academy, has once again created many fine supplements; his recommendations to this book have been invaluable. Sen-Ben Liao of Lawrence Livermore National Laboratory, James Whitenton of Southern Polytechnic State University, and Jerry Shi, of Pasadena City College, performed the Herculean task of working out solutions for every one of the homework problems in the book. At John Wiley publishers, the book received support from Stuart Johnson and Geraldine Osnato, the editors who oversaw the entire project from start to finish, and Tom Kulesa, who coordi-

nated the state-of-the-art media package. We thank Elizabeth Swain, the production editor, for pulling all the pieces together during the complex production process. We also thank Maddy Lesure for her design of the text and art direction of the cover; Lee Goldstein for her page make-up; and Lilian Brady for her proofreading. Hilary Newman was inspired in the search for unusual and interesting photographs. Both the publisher John Wiley & Sons, Inc. and Jearl Walker would like to thank the following for comments and ideas about the 8th edition: Jonathan Abramson, Portland State University; Omar Adawi, Parkland College; Edward Adelson, The Ohio State

University; Steven R. Baker, Naval Postgraduate School; George Caplan, Wellesley College; Richard Kass, The Ohio State University; M. R. Khoshbin-e- Khoshnazar, Research Institution for Curriculum Development & Educational Innovations (Tehran); Stuart Loucks, American River College; Laurence Lurio, Northern Illinois University; Ponn Maheswaranathan, Winthrop University; Joe McCullough, Cabrillo College; Don N. Page, University of Alberta; Elie Riachi, Fort Scott Community College; Andrew G. Rinzler, University of Florida; Dubravka Rupnik, Louisiana State University; Robert Schabinger, Rutgers University; Ruth

Schwartz, Milwaukee School of Engineering; Nora Thornber, Raritan Valley Community College; Frank Wang, LaGuardia Community College; Graham W. Wilson, University of Kansas; Roland Winkler, Northern Illinois University; Ulrich Zurcher, Cleveland State University. Finally, our external reviewers have been outstanding and we acknowledge here our debt to each member of that team.

Maris A. Abolins, *Michigan State University*

Edward Adelson, *Ohio State University*

Nural Akchurin, *Texas Tech*

Yildirim Aktas, *University of North Carolina- Charlotte*

Barbara Andereck, *Ohio Wesleyan University*

Tetyana Antimirova, *Ryerson University*

Mark Arnett, *Kirkwood Community College*

Arun Bansil, *Northeastern University*

Richard Barber, *Santa Clara University*

Neil Basecu, *Westchester Community College*

Anand Batra, *Howard University*

Richard Bone, *Florida International University*

Michael E. Browne, *University of Idaho*

Timothy J. Burns, *Leeward Community College*

Joseph Buschi, *Manhattan College*

Philip A. Casabella, *Rensselaer Polytechnic Institute*

Randall Caton, *Christopher Newport College*

Roger Clapp, *University of South Florida*

W. R. Conkie, *Queen's University*

Renate Crawford, *University of Massachusetts-Dartmouth*

Mike Crivello, *San Diego State University*

Robert N. Davie, Jr., *St. Petersburg Junior College*

Cheryl K. Dellai, *Glendale Community College*

Eric R. Dietz, *California State University at Chico*

N. John DiNardo, *Drexel University*

Eugene Dunnam, *University of Florida*

Robert Endorf, *University of Cincinnati*

F. Paul Esposito, *University of Cincinnati*

Jerry Finkelstein, *San Jose State University*

Robert H. Good, *California State University-Hayward*

Michael Gorman, *University of Houston*

Benjamin Grinstein, *University of California, San Diego*

John B. Gruber, *San Jose State University*

Ann Hanks, *American River College*

Randy Harris, *University of California-Davis*

Samuel Harris, *Purdue University*

Harold B. Hart, *Western Illinois University*

Rebecca Hartzler, *Seattle Central Community College*

John Hubisz, *North Carolina State University*

Joey Huston, *Michigan State University*

David Ingram, *Ohio University*

Shawn Jackson, *University of Tulsa*

Hector Jimenez, *University of Puerto Rico*

Sudhakar B. Joshi, *York University*

Leonard M. Kahn, *University of Rhode Island*

Sudipa Kirtley, *Rose-Hulman Institute*

Leonard Kleinman, *University of Texas at Austin*

Craig Kletzing, *University of Iowa*

Peter F. Koehler, *University of Pittsburgh*

Arthur Z. Kovacs, *Rochester Institute of Technology*

Kenneth Krane, *Oregon State University*

Priscilla Laws, *Dickinson College*

Edbertho Leal, *Polytechnic University of Puerto Rico*

Vern Lindberg, *Rochester Institute of Technology*

Peter Loly, *University of Manitoba*

James MacLaren, *Tulane University*

Andreas Mandelis, *University of Toronto*

Robert R. Marchini, *Memphis State University*

Andrea Markelz, *University at Buffalo, SUNY*

Paul Marquard, *Caspar College*

David Marx, *Illinois State University*

Dan Mazilu, *Washington and Lee University*

James H. McGuire, *Tulane University*

David M. McKinstry, *Eastern Washington University*

Jordon Morelli, *Queen's University*

Eugene Mosca, *United States Naval Academy*

Eric R. Murray, *Georgia Institute of Technology, School of Physics*

James Napolitano, *Rensselaer Polytechnic Institute*

Blaine Norum, *University of Virginia*

Michael O'Shea, *Kansas State University*

Patrick Papin, *San Diego State University*

Kiumars Parvin, *San Jose State University*

Robert Pelcovits, *Brown University*

Oren P. Quist, *South Dakota State University*

Joe Redish, *University of Maryland*

Timothy M. Ritter, *University of North Carolina at Pembroke*

Dan Styer, *Oberlin College*

Frank Wang, *LaGuardia Community College*

MEASUREMENT

1-1 WHAT IS PHYSICS?

Science and engineering are based on measurements and comparisons. Thus, we need rules about how things are measured and compared, and we need experiments to establish the units for those measurements and comparisons. One purpose of physics (and engineering) is to design and conduct those experiments.

For example, physicists strive to develop clocks of extreme accuracy so that any time or time interval can be precisely determined and compared. You may wonder whether such accuracy is actually needed or worth the effort. Here is one example of the worth: Without clocks of extreme accuracy, the Global Positioning System (GPS) that is now vital to worldwide navigation would be useless.

1-2 Measuring Things

We discover physics by learning how to measure the quantities involved in physics. Among these quantities are length, time, mass, temperature, pressure, and electric current.

We measure each physical quantity in its own units, by comparison with a **standard.** The **unit** is a unique name we assign to measures of that quantity—for example, meter (m) for the quantity length. The standard corresponds to exactly 1.0 unit of the quantity. As you will see, the standard for length, which corresponds to exactly 1.0 m, is the distance traveled by light in a vacuum during a certain fraction of a second. We can define a unit and its standard in any way we care to. However, the important thing is to do so in such a way that scientists around the world will agree that our definitions are both sensible and practical.

Once we have set up a standard—say, for length—we must work out procedures by which any length whatever, be it the radius of a hydrogen atom, the wheelbase of a skateboard, or the distance to a star, can be expressed in terms of the standard. Rulers, which approximate our length standard, give us one such procedure for measuring length. However, many of our comparisons must be indirect. You cannot use a ruler, for example, to measure the radius of an atom or the distance to a star.

There are so many physical quantities that it is a problem to organize them. Fortunately, they are not all independent; for example, speed is the ratio of a length to a time. Thus, what we do is pick out—by international agreement— a small number of physical quantities, such as length and time, and assign standards to them alone. We then define all other physical quantities in terms of these *base quantities* and their standards (called *base standards*). Speed, for example, is defined in terms of the base quantities length and time and their base standards.

Base standards must be both accessible and invariable. If we define the length standard as the distance between one's nose and the index finger on an outstretched arm, we certainly have an accessible standard—but it will, of course, vary from person to person. The demand for precision in science and engineering pushes us to aim first for invariability. We then exert great effort to make duplicates of the base standards that are accessible to those who need them.

Table 1-1

Units for Three SI Base Quantities

Quantity	Unit Name	Unit Symbol
Length	meter	m
Time	second	s
Mass	kilogram	kg

1-3 The International System of Units

In 1971, the 14th General Conference on Weights and Measures picked seven quantities as base quantities, thereby forming the basis of the International System of Units, abbreviated SI from its French name and popularly known as the *metric system*. Table 1-1 shows the units for the three base quantities— length, mass, and time—that we use in the early chapters of this book. These units were defined to be on a "human scale."

Many SI *derived units* are defined in terms of these base units. For example, the SI unit for power, called the **watt** (W), is defined in terms of the base units for mass, length, and time. Thus, as you will see in Chapter 7,

$$1 \text{ watt} = 1 \text{ W} = 1 \text{ kg} \cdot \text{m}^2/\text{s}^3, \tag{1-1}$$

where the last collection of unit symbols is read as kilogram-meter squared per second cubed.

To express the very large and very small quantities we often run into in physics, we use *scientific notation*, which employs powers of 10. In this notation,

$$3\,560\,000\,000 \text{ m} = 3.56 \times 10^9 \text{ m} \tag{1-2}$$

and

$$0.000\,000\,492 \text{ s} = 4.92 \times 10^{-7} \text{ s}. \tag{1-3}$$

Scientific notation on computers sometimes takes on an even briefer look, as in 3.56 E9 and 4.92 E−7, where E stands for "exponent of ten." It is briefer still on some calculators, where E is replaced with an empty space.

As a further convenience when dealing with very large or very small measurements, we use the prefixes listed in Table 1-2. As you can see, each prefix represents a certain power of 10, to be used as a multiplication factor. Attaching a prefix to an SI unit has the effect of multiplying by the associated factor. Thus, we can express a particular electric power as

$$1.27 \times 10^9 \text{ watts} = 1.27 \text{ gigawatts} = 1.27 \text{ GW} \tag{1-4}$$

or a particular time interval as

$$2.35 \times 10^{-9} \text{ s} = 2.35 \text{ nanoseconds} = 2.35 \text{ ns}. \tag{1-5}$$

Some prefixes, as used in milliliter, centimeter, kilogram, and megabyte, are probably familiar to you.

Table 1-2

Prefixes for SI Units

Factor	Prefix[a]	Symbol	Factor	Prefix[a]	Symbol
10^{24}	yotta-	Y	10^{-1}	deci-	d
10^{21}	zetta-	Z	10^{-2}	**centi-**	**c**
10^{18}	exa-	E	10^{-3}	**milli-**	**m**
10^{15}	peta-	P	10^{-6}	**micro-**	$\boldsymbol{\mu}$
10^{12}	tera-	T	10^{-9}	**nano-**	**n**
$\mathbf{10^9}$	**giga-**	**G**	10^{-12}	**pico-**	**p**
$\mathbf{10^6}$	**mega-**	**M**	10^{-15}	femto-	f
$\mathbf{10^3}$	**kilo-**	**k**	10^{-18}	atto-	a
10^2	hecto-	h	10^{-21}	zepto-	z
10^1	deka-	da	10^{-24}	yocto-	y

[a]The most frequently used prefixes are shown in bold type.

1-4 Changing Units

We often need to change the units in which a physical quantity is expressed. We do so by a method called *chain-link conversion*. In this method, we multiply the original measurement by a **conversion factor** (a ratio of units that is equal to unity). For example, because 1 min and 60 s are identical time intervals, we have

$$\frac{1 \text{ min}}{60 \text{ s}} = 1 \quad \text{and} \quad \frac{60 \text{ s}}{1 \text{ min}} = 1.$$

Thus, the ratios (1 min)/(60 s) and (60 s)/(1 min) can be used as conversion factors. This is *not* the same as writing $\frac{1}{60} = 1$ or $60 = 1$; each *number* and its *unit* must be treated together.

Because multiplying any quantity by unity leaves the quantity unchanged, we can introduce conversion factors wherever we find them useful. In chain-link conversion, we use the factors to cancel unwanted units. For example, to convert 2 min to seconds, we have

$$2 \text{ min} = (2 \text{ min})(1) = (2 \text{ min})\left(\frac{60 \text{ s}}{1 \text{ min}}\right) = 120 \text{ s}. \qquad (1\text{-}6)$$

If you introduce a conversion factor in such a way that unwanted units do *not* cancel, invert the factor and try again. In conversions, the units obey the same algebraic rules as variables and numbers.

Appendix D gives conversion factors between SI and other systems of units, including non-SI units still used in the United States. However, the conversion factors are written in the style of "1 min = 60 s" rather than as a ratio. So, you need to decide on the numerator and denominator in any needed ratio.

1-5 Length

In 1792, the newborn Republic of France established a new system of weights and measures. Its cornerstone was the meter, defined to be one ten-millionth of the distance from the north pole to the equator. Later, for practical reasons, this Earth standard was abandoned and the meter came to be defined as the distance between two fine lines engraved near the ends of a platinum–iridium bar, the **standard meter bar,** which was kept at the International Bureau of Weights and Measures near Paris. Accurate copies of the bar were sent to standardizing laboratories throughout the world. These **secondary standards** were used to produce other, still more accessible standards, so that ultimately every measuring device derived its authority from the standard meter bar through a complicated chain of comparisons.

Eventually, a standard more precise than the distance between two fine scratches on a metal bar was required. In 1960, a new standard for the meter, based on the wavelength of light, was adopted. Specifically, the standard for the meter was redefined to be 1 650 763.73 wavelengths of a particular orange-red light emitted by atoms of krypton-86 (a particular isotope, or type, of krypton) in a gas discharge tube that can be set up anywhere in the world. This awkward number of wavelengths was chosen so that the new standard would be close to the old meter-bar standard.

By 1983, however, the demand for higher precision had reached such a point that even the krypton-86 standard could not meet it, and in that year a bold step was taken. The meter was redefined as the distance traveled by light in a specified time interval. In the words of the 17th General Conference on Weights and Measures:

 The meter is the length of the path traveled by light in a vacuum during a time interval of 1/299 792 458 of a second.

This time interval was chosen so that the speed of light c is exactly

$$c = 299\ 792\ 458\ \text{m/s}.$$

Measurements of the speed of light had become extremely precise, so it made sense to adopt the speed of light as a defined quantity and to use it to redefine the meter.
Table 1-3 shows a wide range of lengths, from that of the universe (top line) to those of some very small objects.

Table 1-3	
Some Approximate Lengths	
Measurement	Length in Meters
Distance to the first galaxies formed	2×10^{26}
Distance to the Andromeda galaxy	2×10^{22}
Distance to the nearby star Proxima Centauri	4×10^{16}
Distance to Pluto	6×10^{12}
Radius of Earth	6×10^{6}
Height of Mt. Everest	9×10^{3}
Thickness of this page	1×10^{-4}
Length of a typical virus	1×10^{-8}
Radius of a hydrogen atom	5×10^{-11}
Radius of a proton	1×10^{-15}

Sample Problem

Estimating order of magnitude, ball of string

The world's largest ball of string is about 2 m in radius. To the nearest order of magnitude, what is the total length L of the string in the ball?

KEY IDEA

We could, of course, take the ball apart and measure the total length L, but that would take great effort and make the ball's builder most unhappy. Instead, because we want only the nearest order of magnitude, we can estimate any quantities required in the calculation.

Calculations: Let us assume the ball is spherical with radius $R = 2$ m. The string in the ball is not closely packed (there are uncountable gaps between adjacent sections of string). To allow for these gaps, let us somewhat overestimate the cross-sectional area of the string by assuming the cross section is square, with an edge length $d = 4$ mm.

Then, with a cross-sectional area of d^2 and a length L, the string occupies a total volume of

$$V = (\text{cross-sectional area})(\text{length}) = d^2 L.$$

This is approximately equal to the volume of the ball, given by $\frac{4}{3}\pi R^3$, which is about $4R^3$ because π is about 3. Thus, we have

$$d^2 L = 4R^3,$$

or $\qquad L = \dfrac{4R^3}{d^2} = \dfrac{4(2\ \text{m})^3}{(4 \times 10^{-3}\ \text{m})^2}$

$$= 2 \times 10^6\ \text{m} \approx 10^6\ \text{m} = 10^3\ \text{km}.$$

(Answer)

(Note that you do not need a calculator for such a simplified calculation.) To the nearest order of magnitude, the ball contains about 1000 km of string!

 Additional examples, video, and practice available at *WileyPLUS*

1-6 Time

Time has two aspects. For civil and some scientific purposes, we want to know the time of day so that we can order events in sequence. In much scientific work, we want to know how long an event lasts. Thus, any time standard must be able to answer two questions: "*When* did it happen?" and "What is its *duration*?" Table 1-4 shows some time intervals.

Table 1-4	
Some Approximate Time Intervals	
Measurement	Time Interval in Seconds
Lifetime of the proton (predicted)	3×10^{40}
Age of the universe	5×10^{17}
Age of the pyramid of Cheops	1×10^{11}
Human life expectancy	2×10^{9}
Length of a day	9×10^{4}
Time between human heartbeats	8×10^{-1}
Lifetime of the muon	2×10^{-6}
Shortest lab light pulse	1×10^{-16}
Lifetime of the most unstable particle	1×10^{-23}
The Planck time[a]	1×10^{-43}

[a]This is the earliest time after the big bang at which the laws of physics as we know them can be applied.

Any phenomenon that repeats itself is a possible time standard. Earth's rotation, which determines the length of the day, has been used in this way for centuries; Fig. 1-1 shows one novel example of a watch based on that rotation. A quartz clock, in which a quartz ring is made to vibrate continuously, can be calibrated against Earth's rotation via astronomical observations and used to measure time intervals in the laboratory. However, the calibration cannot be carried out with the accuracy called for by modern scientific and engineering technology.

To meet the need for a better time standard, atomic clocks have been developed. An atomic clock at the National Institute of Standards and Technology

Fig. 1-1 When the metric system was proposed in 1792, the hour was redefined to provide a 10-hour day. The idea did not catch on. The maker of this 10-hour watch wisely provided a small dial that kept conventional 12-hour time. Do the two dials indicate the same time? *(Steven Pitkin)*

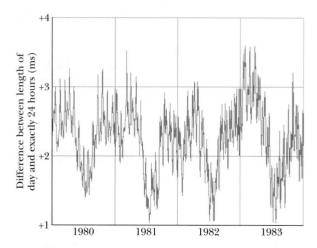

Fig. 1-2 Variations in the length of the day over a 4-year period. Note that the entire vertical scale amounts to only 3 ms (= 0.003 s).

(NIST) in Boulder, Colorado, is the standard for Coordinated Universal Time (UTC) in the United States. Its time signals are available by shortwave radio (stations WWV and WWVH) and by telephone (303-499-7111). Time signals (and related information) are also available from the United States Naval Observatory at website http://tycho.usno.navy.mil/time.html. (To set a clock extremely accurately at your particular location, you would have to account for the travel time required for these signals to reach you.)

Figure 1-2 shows variations in the length of one day on Earth over a 4-year period, as determined by comparison with a cesium (atomic) clock. Because the variation displayed by Fig. 1-2 is seasonal and repetitious, we suspect the rotating Earth when there is a difference between Earth and atom as timekeepers. The variation is due to tidal effects caused by the Moon and to large-scale winds.

The 13th General Conference on Weights and Measures in 1967 adopted a standard second based on the cesium clock:

 One second is the time taken by 9 192 631 770 oscillations of the light (of a specified wavelength) emitted by a cesium-133 atom.

Atomic clocks are so consistent that, in principle, two cesium clocks would have to run for 6000 years before their readings would differ by more than 1 s. Even such accuracy pales in comparison with that of clocks currently being developed; their precision may be 1 part in 10^{18} — that is, 1 s in 1×10^{18} s (which is about 3×10^{10} y).

1-7 **Mass**

The Standard Kilogram

The SI standard of mass is a platinum–iridium cylinder (Fig. 1-3) kept at the International Bureau of Weights and Measures near Paris and assigned, by international agreement, a mass of 1 kilogram. Accurate copies have been sent to standardizing laboratories in other countries, and the masses of other bodies can be determined by balancing them against a copy. Table 1-5 shows some masses expressed in kilograms, ranging over about 83 orders of magnitude.

The U.S. copy of the standard kilogram is housed in a vault at NIST. It is removed, no more than once a year, for the purpose of checking duplicate

Fig. 1-3 The international 1 kg standard of mass, a platinum–iridium cylinder 3.9 cm in height and in diameter. *(Courtesy Bureau International des Poids et Mesures, France)*

copies that are used elsewhere. Since 1889, it has been taken to France twice for recomparison with the primary standard.

A Second Mass Standard

The masses of atoms can be compared with one another more precisely than they can be compared with the standard kilogram. For this reason, we have a second mass standard. It is the carbon-12 atom, which, by international agreement, has been assigned a mass of 12 **atomic mass units** (u). The relation between the two units is

$$1 \text{ u} = 1.660\,538\,86 \times 10^{-27} \text{ kg}, \tag{1-7}$$

with an uncertainty of ± 10 in the last two decimal places. Scientists can, with reasonable precision, experimentally determine the masses of other atoms relative to the mass of carbon-12. What we presently lack is a reliable means of extending that precision to more common units of mass, such as a kilogram.

Density

As we shall discuss further in Chapter 14, **density** ρ (lowercase Greek letter rho) is the mass per unit volume:

$$\rho = \frac{m}{V}. \tag{1-8}$$

Densities are typically listed in kilograms per cubic meter or grams per cubic centimeter. The density of water (1.00 gram per cubic centimeter) is often used as a comparison. Fresh snow has about 10% of that density; platinum has a density that is about 21 times that of water.

Table 1-5

Some Approximate Masses

Object	Mass in Kilograms
Known universe	1×10^{53}
Our galaxy	2×10^{41}
Sun	2×10^{30}
Moon	7×10^{22}
Asteroid Eros	5×10^{15}
Small mountain	1×10^{12}
Ocean liner	7×10^{7}
Elephant	5×10^{3}
Grape	3×10^{-3}
Speck of dust	7×10^{-10}
Penicillin molecule	5×10^{-17}
Uranium atom	4×10^{-25}
Proton	2×10^{-27}
Electron	9×10^{-31}

Sample Problem

Density and liquefaction

A heavy object can sink into the ground during an earthquake if the shaking causes the ground to undergo *liquefaction*, in which the soil grains experience little friction as they slide over one another. The ground is then effectively quicksand. The possibility of liquefaction in sandy ground can be predicted in terms of the *void ratio e* for a sample of the ground:

$$e = \frac{V_{\text{voids}}}{V_{\text{grains}}}. \tag{1-9}$$

Here, V_{grains} is the total volume of the sand grains in the sample and V_{voids} is the total volume between the grains (in the *voids*). If e exceeds a critical value of 0.80, liquefaction can occur during an earthquake. What is the corresponding sand density ρ_{sand}? Solid silicon dioxide (the primary component of sand) has a density of $\rho_{\text{SiO}_2} = 2.600 \times 10^3 \text{ kg/m}^3$.

KEY IDEA

The density of the sand ρ_{sand} in a sample is the mass per unit volume—that is, the ratio of the total mass m_{sand} of the sand grains to the total volume V_{total} of the sample:

$$\rho_{\text{sand}} = \frac{m_{\text{sand}}}{V_{\text{total}}}. \tag{1-10}$$

Calculations: The total volume V_{total} of a sample is

$$V_{\text{total}} = V_{\text{grains}} + V_{\text{voids}}.$$

Substituting for V_{voids} from Eq. 1-9 and solving for V_{grains} lead to

$$V_{\text{grains}} = \frac{V_{\text{total}}}{1 + e}. \tag{1-11}$$

(continues on the next page)

From Eq. 1-8, the total mass m_{sand} of the sand grains is the product of the density of silicon dioxide and the total volume of the sand grains:

$$m_{sand} = \rho_{SiO_2} V_{grains}. \qquad (1\text{-}12)$$

Substituting this expression into Eq. 1-10 and then substituting for V_{grains} from Eq. 1-11 lead to

$$\rho_{sand} = \frac{\rho_{SiO_2}}{V_{total}} \frac{V_{total}}{1 + e} = \frac{\rho_{SiO_2}}{1 + e}. \qquad (1\text{-}13)$$

Substituting $\rho_{SiO_2} = 2.600 \times 10^3$ kg/m³ and the critical value of $e = 0.80$, we find that liquefaction occurs when the sand density is less than

$$\rho_{sand} = \frac{2.600 \times 10^3 \text{ kg/m}^3}{1.80} = 1.4 \times 10^3 \text{ kg/m}^3.$$

(Answer)

A building can sink several meters in such liquefaction.

 Additional examples, video, and practice available at *WileyPLUS*

REVIEW & SUMMARY

Measurement in Physics Physics is based on measurement of physical quantities. Certain physical quantities have been chosen as **base quantities** (such as length, time, and mass); each has been defined in terms of a **standard** and given a **unit** of measure (such as meter, second, and kilogram). Other physical quantities are defined in terms of the base quantities and their standards and units.

SI Units The unit system emphasized in this book is the International System of Units (SI). The three physical quantities displayed in Table 1-1 are used in the early chapters. Standards, which must be both accessible and invariable, have been established for these base quantities by international agreement. These standards are used in all physical measurement, for both the base quantities and the quantities derived from them. Scientific notation and the prefixes of Table 1-2 are used to simplify measurement notation.

Changing Units Conversion of units may be performed by using *chain-link conversions* in which the original data are multiplied successively by conversion factors written as unity and the units are manipulated like algebraic quantities until only the desired units remain.

Length The meter is defined as the distance traveled by light during a precisely specified time interval.

Time The second is defined in terms of the oscillations of light emitted by an atomic (cesium-133) source. Accurate time signals are sent worldwide by radio signals keyed to atomic clocks in standardizing laboratories.

Mass The kilogram is defined in terms of a platinum–iridium standard mass kept near Paris. For measurements on an atomic scale, the atomic mass unit, defined in terms of the atom carbon-12, is usually used.

Density The density ρ of a material is the mass per unit volume:

$$\rho = \frac{m}{V}. \qquad (1\text{-}8)$$

PROBLEMS

sec. 1-5 Length

•1 **SSM** Earth is approximately a sphere of radius 6.37×10^6 m. What are (a) its circumference in kilometers, (b) its surface area in square kilometers, and (c) its volume in cubic kilometers?

•2 A *gry* is an old English measure for length, defined as 1/10 of a line, where *line* is another old English measure for length, defined as 1/12 inch. A common measure for length in the publishing business is a *point,* defined as 1/72 inch. What is an area of 0.50 gry² in points squared (points²)?

•3 The micrometer (1 μm) is often called the *micron.* (a) How many microns make up 1.0 km? (b) What fraction of a centimeter equals 1.0 μm? (c) How many microns are in 1.0 yd?

•4 Spacing in this book was generally done in units of points and picas: 12 points = 1 pica, and 6 picas = 1 inch. If a figure was misplaced in the page proofs by 0.80 cm, what was the misplacement in (a) picas and (b) points?

•5 **SSM** **WWW** Horses are to race over a certain English meadow for a distance of 4.0 furlongs. What is the race distance in (a) rods and (b) chains? (1 furlong = 201.168 m, 1 rod = 5.0292 m, and 1 chain = 20.117 m.)

••6   You can easily convert common units and measures electronically, but you still should be able to use a conversion table, such as those in Appendix D. Table 1-6 is part of a conversion table for a system of volume measures once common in Spain; a volume of 1 fanega is equivalent to 55.501 dm³ (cubic decimeters). To complete the table, what numbers (to three significant figures) should be entered in (a) the cahiz column, (b) the fanega column, (c) the cuartilla column, and (d) the almude column, starting with the top blank? Express 7.00 almudes in (e) medios, (f) cahizes, and (g) cubic centimeters (cm³).

Table 1-6

Problem 6

	cahiz	fanega	cuartilla	almude	medio
1 cahiz =	1	12	48	144	288
1 fanega =		1	4	12	24
1 cuartilla =			1	3	6
1 almude =				1	2
1 medio =					1

••7 ILW Hydraulic engineers in the United States often use, as a unit of volume of water, the *acre-foot*, defined as the volume of water that will cover 1 acre of land to a depth of 1 ft. A severe thunderstorm dumped 2.0 in. of rain in 30 min on a town of area 26 km². What volume of water, in acre-feet, fell on the town?

••8 GO Harvard Bridge, which connects MIT with its fraternities across the Charles River, has a length of 364.4 Smoots plus one ear. The unit of one Smoot is based on the length of Oliver Reed Smoot, Jr., class of 1962, who was carried or dragged length by length across the bridge so that other pledge members of the Lambda Chi Alpha fraternity could mark off (with paint) 1-Smoot lengths along the bridge. The marks have been repainted biannually by fraternity pledges since the initial measurement, usually during times of traffic congestion so that the police cannot easily interfere. (Presumably, the police were originally upset because the Smoot is not an SI base unit, but these days they seem to have accepted the unit.) Figure 1-4 shows three parallel paths, measured in Smoots (S), Willies (W), and Zeldas (Z). What is the length of 50.0 Smoots in (a) Willies and (b) Zeldas?

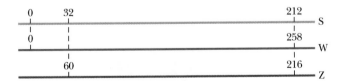

Fig. 1-4   Problem 8.

••9   Antarctica is roughly semicircular, with a radius of 2000 km (Fig. 1-5). The average thickness of its ice cover is 3000 m. How many cubic centimeters of ice does Antarctica contain? (Ignore the curvature of Earth.)

Fig. 1-5   Problem 9.

sec. 1-6   Time

•10   Until 1883, every city and town in the United States kept its own local time. Today, travelers reset their watches only when the time change equals 1.0 h. How far, on the average, must you travel in degrees of longitude between the time-zone boundaries at which your watch must be reset by 1.0 h? (*Hint:* Earth rotates 360° in about 24 h.)

•11   For about 10 years after the French Revolution, the French government attempted to base measures of time on multiples of ten: One week consisted of 10 days, one day consisted of 10 hours, one hour consisted of 100 minutes, and one minute consisted of 100 seconds. What are the ratios of (a) the French decimal week to the standard week and (b) the French decimal second to the standard second?

•12   The fastest growing plant on record is a *Hesperoyucca whipplei* that grew 3.7 m in 14 days. What was its growth rate in micrometers per second?

•13 GO Three digital clocks A, B, and C run at different rates and do not have simultaneous readings of zero. Figure 1-6 shows simultaneous readings on pairs of the clocks for four occasions. (At the earliest occasion, for example, B reads 25.0 s and C reads 92.0 s.) If two events are 600 s apart on clock A, how far apart are they on (a) clock B and (b) clock C? (c) When clock A reads 400 s, what does clock B read? (d) When clock C reads 15.0 s, what does clock B read? (Assume negative readings for prezero times.)

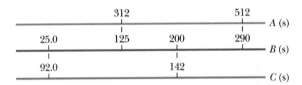

Fig. 1-6   Problem 13.

•14   A lecture period (50 min) is close to 1 microcentury. (a) How long is a microcentury in minutes? (b) Using

$$\text{percentage difference} = \left(\frac{\text{actual} - \text{approximation}}{\text{actual}} \right) 100,$$

find the percentage difference from the approximation.

•15   A fortnight is a charming English measure of time equal to 2.0 weeks (the word is a contraction of "fourteen nights"). That is a nice amount of time in pleasant company but perhaps a painful string of microseconds in unpleasant company. How many microseconds are in a fortnight?

•16   Time standards are now based on atomic clocks. A promising second standard is based on *pulsars*, which are rotating neutron stars (highly compact stars consisting only of neutrons). Some rotate at a rate that is highly stable, sending out a radio beacon that sweeps briefly across Earth once with each rotation, like a lighthouse beacon. Pulsar PSR 1937+21 is an example; it rotates once every 1.557 806 448 872 75 ± 3 ms, where the trailing ±3 indicates the uncertainty in the last decimal place (it does *not* mean ±3 ms). (a) How many rotations does PSR 1937+21 make in 7.00 days? (b) How much time does the pulsar take to rotate exactly one million times and (c) what is the associated uncertainty?

•17 SSM Five clocks are being tested in a laboratory. Exactly at noon, as determined by the WWV time signal, on successive days of a week the clocks read as in the following table. Rank the five clocks according to their relative value as good timekeepers, best to worst. Justify your choice.

Clock	Sun.	Mon.	Tues.	Wed.	Thurs.	Fri.	Sat.
A	12:36:40	12:36:56	12:37:12	12:37:27	12:37:44	12:37:59	12:38:14
B	11:59:59	12:00:02	11:59:57	12:00:07	12:00:02	11:59:56	12:00:03
C	15:50:45	15:51:43	15:52:41	15:53:39	15:54:37	15:55:35	15:56:33
D	12:03:59	12:02:52	12:01:45	12:00:38	11:59:31	11:58:24	11:57:17
E	12:03:59	12:02:49	12:01:54	12:01:52	12:01:32	12:01:22	12:01:12

••18 Because Earth's rotation is gradually slowing, the length of each day increases: The day at the end of 1.0 century is 1.0 ms longer than the day at the start of the century. In 20 centuries, what is the total of the daily increases in time?

•••19 Suppose that, while lying on a beach near the equator watching the Sun set over a calm ocean, you start a stopwatch just as the top of the Sun disappears. You then stand, elevating your eyes by a height $H = 1.70$ m, and stop the watch when the top of the Sun again disappears. If the elapsed time is $t = 11.1$ s, what is the radius r of Earth?

sec. 1-7 Mass

•20 GO The record for the largest glass bottle was set in 1992 by a team in Millville, New Jersey—they blew a bottle with a volume of 193 U.S. fluid gallons. (a) How much short of 1.0 million cubic centimeters is that? (b) If the bottle were filled with water at the leisurely rate of 1.8 g/min, how long would the filling take? Water has a density of 1000 kg/m³.

•21 Earth has a mass of 5.98×10^{24} kg. The average mass of the atoms that make up Earth is 40 u. How many atoms are there in Earth?

•22 Gold, which has a density of 19.32 g/cm³, is the most ductile metal and can be pressed into a thin leaf or drawn out into a long fiber. (a) If a sample of gold, with a mass of 27.63 g, is pressed into a leaf of 1.000 μm thickness, what is the area of the leaf? (b) If, instead, the gold is drawn out into a cylindrical fiber of radius 2.500 μm, what is the length of the fiber?

•23 SSM (a) Assuming that water has a density of exactly 1 g/cm³, find the mass of one cubic meter of water in kilograms. (b) Suppose that it takes 10.0 h to drain a container of 5700 m³ of water. What is the "mass flow rate," in kilograms per second, of water from the container?

••24 GO Grains of fine California beach sand are approximately spheres with an average radius of 50 μm and are made of silicon dioxide, which has a density of 2600 kg/m³. What mass of sand grains would have a total surface area (the total area of all the individual spheres) equal to the surface area of a cube 1.00 m on an edge?

••25 During heavy rain, a section of a mountainside measuring 2.5 km horizontally, 0.80 km up along the slope, and 2.0 m deep slips into a valley in a mud slide. Assume that the mud ends up uniformly distributed over a surface area of the valley measuring 0.40 km × 0.40 km and that mud has a density of 1900 kg/m³. What is the mass of the mud sitting above a 4.0 m² area of the valley floor?

••26 One cubic centimeter of a typical cumulus cloud contains 50 to 500 water drops, which have a typical radius of 10 μm. For that range, give the lower value and the higher value, respectively, for the following. (a) How many cubic meters of water are in a cylindrical cumulus cloud of height 3.0 km and radius 1.0 km? (b) How many 1-liter pop bottles would that water fill? (c) Water has a density of 1000 kg/m³. How much mass does the water in the cloud have?

••27 Iron has a density of 7.87 g/cm³, and the mass of an iron atom is 9.27×10^{-26} kg. If the atoms are spherical and tightly packed, (a) what is the volume of an iron atom and (b) what is the distance between the centers of adjacent atoms?

••28 A mole of atoms is 6.02×10^{23} atoms. To the nearest order of magnitude, how many moles of atoms are in a large domestic cat? The masses of a hydrogen atom, an oxygen atom, and a carbon atom are 1.0 u, 16 u, and 12 u, respectively. (*Hint:* Cats are sometimes known to kill a mole.)

••29 On a spending spree in Malaysia, you buy an ox with a weight of 28.9 piculs in the local unit of weights: 1 picul = 100 gins, 1 gin = 16 tahils, 1 tahil = 10 chees, and 1 chee = 10 hoons. The weight of 1 hoon corresponds to a mass of 0.3779 g. When you arrange to ship the ox home to your astonished family, how much mass in kilograms must you declare on the shipping manifest? (*Hint:* Set up multiple chain-link conversions.)

••30 GO Water is poured into a container that has a small leak. The mass m of the water is given as a function of time t by $m = 5.00t^{0.8} - 3.00t + 20.00$, with $t \geq 0$, m in grams, and t in seconds. (a) At what time is the water mass greatest, and (b) what is that greatest mass? In kilograms per minute, what is the rate of mass change at (c) $t = 2.00$ s and (d) $t = 5.00$ s?

•••31 A vertical container with base area measuring 14.0 cm by 17.0 cm is being filled with identical pieces of candy, each with a volume of 50.0 mm³ and a mass of 0.0200 g. Assume that the volume of the empty spaces between the candies is negligible. If the height of the candies in the container increases at the rate of 0.250 cm/s, at what rate (kilograms per minute) does the mass of the candies in the container increase?

Additional Problems

32 In the United States, a doll house has the scale of 1:12 of a real house (that is, each length of the doll house is $\frac{1}{12}$ that of the real house) and a miniature house (a doll house to fit within a doll house) has the scale of 1:144 of a real house. Suppose a real house (Fig. 1-7) has a front length of 20 m, a depth of 12 m, a height of 6.0 m, and a standard sloped roof (vertical triangular faces on the ends) of height 3.0 m. In cubic meters, what are the volumes of the corresponding (a) doll house and (b) miniature house?

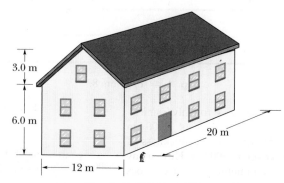

Fig. 1-7 Problem 32.

33 SSM A ton is a measure of volume frequently used in shipping, but that use requires some care because there are at least three types of tons: A *displacement ton* is equal to 7 barrels bulk, a *freight ton* is equal to 8 barrels bulk, and a *register ton* is equal to 20 barrels bulk. A *barrel bulk* is another measure of volume: 1 barrel bulk = 0.1415 m^3. Suppose you spot a shipping order for "73 tons" of M&M candies, and you are certain that the client who sent the order intended "ton" to refer to volume (instead of weight or mass, as discussed in Chapter 5). If the client actually meant displacement tons, how many extra U.S. bushels of the candies will you erroneously ship if you interpret the order as (a) 73 freight tons and (b) 73 register tons? (1 m^3 = 28.378 U.S. bushels.)

34 Two types of *barrel* units were in use in the 1920s in the United States. The apple barrel had a legally set volume of 7056 cubic inches; the cranberry barrel, 5826 cubic inches. If a merchant sells 20 cranberry barrels of goods to a customer who thinks he is receiving apple barrels, what is the discrepancy in the shipment volume in liters?

35 An old English children's rhyme states, "Little Miss Muffet sat on a tuffet, eating her curds and whey, when along came a spider who sat down beside her. . . ." The spider sat down not because of the curds and whey but because Miss Muffet had a stash of 11 tuffets of dried flies. The volume measure of a tuffet is given by 1 tuffet = 2 pecks = 0.50 Imperial bushel, where 1 Imperial bushel = 36.3687 liters (L). What was Miss Muffet's stash in (a) pecks, (b) Imperial bushels, and (c) liters?

36 Table 1-7 shows some old measures of liquid volume. To complete the table, what numbers (to three significant figures) should be entered in (a) the wey column, (b) the chaldron column, (c) the bag column, (d) the pottle column, and (e) the gill column, starting with the top blank? (f) The volume of 1 bag is equal to 0.1091 m^3. If an old story has a witch cooking up some vile liquid in a cauldron of volume 1.5 chaldrons, what is the volume in cubic meters?

Table 1-7

Problem 36

	wey	chaldron	bag	pottle	gill
1 wey =	1	10/9	40/3	640	120 240
1 chaldron =					
1 bag =					
1 pottle =					
1 gill =					

37 A typical sugar cube has an edge length of 1 cm. If you had a cubical box that contained a mole of sugar cubes, what would its edge length be? (One mole = 6.02×10^{23} units.)

38 An old manuscript reveals that a landowner in the time of King Arthur held 3.00 acres of plowed land plus a livestock area of 25.0 perches by 4.00 perches. What was the total area in (a) the old unit of roods and (b) the more modern unit of square meters? Here, 1 acre is an area of 40 perches by 4 perches, 1 rood is an area of 40 perches by 1 perch, and 1 perch is the length 16.5 ft.

39 SSM A tourist purchases a car in England and ships it home to the United States. The car sticker advertised that the car's fuel consumption was at the rate of 40 miles per gallon on the open road.

The tourist does not realize that the U.K. gallon differs from the U.S. gallon:

$$1 \text{ U.K. gallon} = 4.546\,090\,0 \text{ liters}$$
$$1 \text{ U.S. gallon} = 3.785\,411\,8 \text{ liters.}$$

For a trip of 750 miles (in the United States), how many gallons of fuel does (a) the mistaken tourist believe she needs and (b) the car actually require?

40 Using conversions and data in the chapter, determine the number of hydrogen atoms required to obtain 1.0 kg of hydrogen. A hydrogen atom has a mass of 1.0 u.

41 SSM A *cord* is a volume of cut wood equal to a stack 8 ft long, 4 ft wide, and 4 ft high. How many cords are in 1.0 m^3?

42 One molecule of water (H_2O) contains two atoms of hydrogen and one atom of oxygen. A hydrogen atom has a mass of 1.0 u and an atom of oxygen has a mass of 16 u, approximately. (a) What is the mass in kilograms of one molecule of water? (b) How many molecules of water are in the world's oceans, which have an estimated total mass of 1.4×10^{21} kg?

43 A person on a diet might lose 2.3 kg per week. Express the mass loss rate in milligrams per second, as if the dieter could sense the second-by-second loss.

44 What mass of water fell on the town in Problem 7? Water has a density of 1.0×10^3 kg/m^3.

45 (a) A unit of time sometimes used in microscopic physics is the *shake*. One shake equals 10^{-8} s. Are there more shakes in a second than there are seconds in a year? (b) Humans have existed for about 10^6 years, whereas the universe is about 10^{10} years old. If the age of the universe is defined as 1 "universe day," where a universe day consists of "universe seconds" as a normal day consists of normal seconds, how many universe seconds have humans existed?

46 A unit of area often used in measuring land areas is the *hectare*, defined as 10^4 m^2. An open-pit coal mine consumes 75 hectares of land, down to a depth of 26 m, each year. What volume of earth, in cubic kilometers, is removed in this time?

47 SSM An astronomical unit (AU) is the average distance between Earth and the Sun, approximately 1.50×10^8 km. The speed of light is about 3.0×10^8 m/s. Express the speed of light in astronomical units per minute.

48 The common Eastern mole, a mammal, typically has a mass of 75 g, which corresponds to about 7.5 moles of atoms. (A mole of atoms is 6.02×10^{23} atoms.) In atomic mass units (u), what is the average mass of the atoms in the common Eastern mole?

49 A traditional unit of length in Japan is the ken (1 ken = 1.97 m). What are the ratios of (a) square kens to square meters and (b) cubic kens to cubic meters? What is the volume of a cylindrical water tank of height 5.50 kens and radius 3.00 kens in (c) cubic kens and (d) cubic meters?

50 You receive orders to sail due east for 24.5 mi to put your salvage ship directly over a sunken pirate ship. However, when your divers probe the ocean floor at that location and find no evidence of a ship, you radio back to your source of information, only to discover that the sailing distance was supposed to be 24.5 *nautical miles*, not regular miles. Use the Length table in Appendix D to calculate how far horizontally you are from the pirate ship in kilometers.

51   The cubit is an ancient unit of length based on the distance between the elbow and the tip of the middle finger of the measurer. Assume that the distance ranged from 43 to 53 cm, and suppose that ancient drawings indicate that a cylindrical pillar was to have a length of 9 cubits and a diameter of 2 cubits. For the stated range, what are the lower value and the upper value, respectively, for (a) the cylinder's length in meters, (b) the cylinder's length in millimeters, and (c) the cylinder's volume in cubic meters?

52   As a contrast between the old and the modern and between the large and the small, consider the following: In old rural England 1 hide (between 100 and 120 acres) was the area of land needed to sustain one family with a single plough for one year. (An area of 1 acre is equal to 4047 m^2.) Also, 1 wapentake was the area of land needed by 100 such families. In quantum physics, the cross-sectional area of a nucleus (defined in terms of the chance of a particle hitting and being absorbed by it) is measured in units of barns, where 1 barn is 1×10^{-28} m^2. (In nuclear physics jargon, if a nucleus is "large," then shooting a particle at it is like shooting a bullet at a barn door, which can hardly be missed.) What is the ratio of 25 wapentakes to 11 barns?

53   **SSM**   An *astronomical unit* (AU) is equal to the average distance from Earth to the Sun, about 92.9×10^6 mi. A *parsec* (pc) is the distance at which a length of 1 AU would subtend an angle of exactly 1 second of arc (Fig. 1-8). A

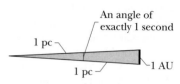

Fig. 1-8   Problem 53.

light-year (ly) is the distance that light, traveling through a vacuum with a speed of 186 000 mi/s, would cover in 1.0 year. Express the Earth–Sun distance in (a) parsecs and (b) light-years.

54   The description for a certain brand of house paint claims a coverage of 460 ft^2/gal. (a) Express this quantity in square meters per liter. (b) Express this quantity in an SI unit (see Appendices A and D). (c) What is the inverse of the original quantity, and (d) what is its physical significance?

MOTION ALONG A STRAIGHT LINE

2

WHAT IS PHYSICS?

One purpose of physics is to study the motion of objects—how fast they move, for example, and how far they move in a given amount of time. NASCAR engineers are fanatical about this aspect of physics as they determine the performance of their cars before and during a race. Geologists use this physics to measure tectonic-plate motion as they attempt to predict earthquakes. Medical researchers need this physics to map the blood flow through a patient when diagnosing a partially closed artery, and motorists use it to determine how they might slow sufficiently when their radar detector sounds a warning. There are countless other examples. In this chapter, we study the basic physics of motion where the object (race car, tectonic plate, blood cell, or any other object) moves along a single axis. Such motion is called *one-dimensional motion.*

2-2 Motion

The world, and everything in it, moves. Even seemingly stationary things, such as a roadway, move with Earth's rotation, Earth's orbit around the Sun, the Sun's orbit around the center of the Milky Way galaxy, and that galaxy's migration relative to other galaxies. The classification and comparison of motions (called **kinematics**) is often challenging. What exactly do you measure, and how do you compare?

Before we attempt an answer, we shall examine some general properties of motion that is restricted in three ways.

1. The motion is along a straight line only. The line may be vertical, horizontal, or slanted, but it must be straight.

2. Forces (pushes and pulls) cause motion but will not be discussed until Chapter 5. In this chapter we discuss only the motion itself and changes in the motion. Does the moving object speed up, slow down, stop, or reverse direction? If the motion does change, how is time involved in the change?

3. The moving object is either a **particle** (by which we mean a point-like object such as an electron) or an object that moves like a particle (such that every portion moves in the same direction and at the same rate). A stiff pig slipping down a straight playground slide might be considered to be moving like a particle; however, a tumbling tumbleweed would not.

2-3 Position and Displacement

To locate an object means to find its position relative to some reference point, often the **origin** (or zero point) of an axis such as the x axis in Fig. 2-1. The **positive direction** of the axis is in the direction of increasing numbers (coordinates), which is to the right in Fig. 2-1. The opposite is the **negative direction.**

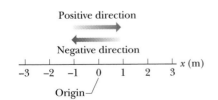

Fig. 2-1 Position is determined on an axis that is marked in units of length (here meters) and that extends indefinitely in opposite directions. The axis name, here x, is always on the positive side of the origin.

For example, a particle might be located at $x = 5$ m, which means it is 5 m in the positive direction from the origin. If it were at $x = -5$ m, it would be just as far from the origin but in the opposite direction. On the axis, a coordinate of -5 m is less than a coordinate of -1 m, and both coordinates are less than a coordinate of $+5$ m. A plus sign for a coordinate need not be shown, but a minus sign must always be shown.

A change from position x_1 to position x_2 is called a **displacement** Δx, where

$$\Delta x = x_2 - x_1. \tag{2-1}$$

(The symbol Δ, the Greek uppercase delta, represents a change in a quantity, and it means the final value of that quantity minus the initial value.) When numbers are inserted for the position values x_1 and x_2 in Eq. 2-1, a displacement in the positive direction (to the right in Fig. 2-1) always comes out positive, and a displacement in the opposite direction (left in the figure) always comes out negative. For example, if the particle moves from $x_1 = 5$ m to $x_2 = 12$ m, then the displacement is $\Delta x = (12 \text{ m}) - (5 \text{ m}) = +7$ m. The positive result indicates that the motion is in the positive direction. If, instead, the particle moves from $x_1 = 5$ m to $x_2 = 1$ m, then $\Delta x = (1 \text{ m}) - (5 \text{ m}) = -4$ m. The negative result indicates that the motion is in the negative direction.

The actual number of meters covered for a trip is irrelevant; displacement involves only the original and final positions. For example, if the particle moves from $x = 5$ m out to $x = 200$ m and then back to $x = 5$ m, the displacement from start to finish is $\Delta x = (5 \text{ m}) - (5 \text{ m}) = 0$.

A plus sign for a displacement need not be shown, but a minus sign must always be shown. If we ignore the sign (and thus the direction) of a displacement, we are left with the **magnitude** (or absolute value) of the displacement. For example, a displacement of $\Delta x = -4$ m has a magnitude of 4 m.

Displacement is an example of a **vector quantity,** which is a quantity that has both a direction and a magnitude. We explore vectors more fully in Chapter 3 (in fact, some of you may have already read that chapter), but here all we need is the idea that displacement has two features: (1) Its *magnitude* is the distance (such as the number of meters) between the original and final positions. (2) Its *direction,* from an original position to a final position, can be represented by a plus sign or a minus sign if the motion is along a single axis.

What follows is the first of many checkpoints you will see in this book. Each consists of one or more questions whose answers require some reasoning or a mental calculation, and each gives you a quick check of your understanding of a point just discussed. The answers are listed in the back of the book.

CHECKPOINT 1

Here are three pairs of initial and final positions, respectively, along an x axis. Which pairs give a negative displacement: (a) -3 m, $+5$ m; (b) -3 m, -7 m; (c) 7 m, -3 m?

2-4 Average Velocity and Average Speed

A compact way to describe position is with a graph of position x plotted as a function of time t—a graph of $x(t)$. (The notation $x(t)$ represents a function x of t, not the product x times t.) As a simple example, Fig. 2-2 shows the position function $x(t)$ for a stationary armadillo (which we treat as a particle) over a 7 s time interval. The animal's position stays at $x = -2$ m.

Figure 2-3 is more interesting, because it involves motion. The armadillo is apparently first noticed at $t = 0$ when it is at the position $x = -5$ m. It moves

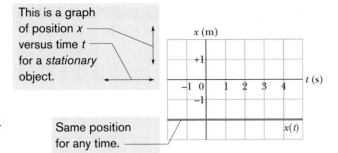

Fig. 2-2 The graph of $x(t)$ for an armadillo that is stationary at $x = -2$ m. The value of x is -2 m for all times t.

toward $x = 0$, passes through that point at $t = 3$ s, and then moves on to increasingly larger positive values of x. Figure 2-3 also depicts the straight-line motion of the armadillo (at three times) and is something like what you would see. The graph in Fig. 2-3 is more abstract and quite unlike what you would see, but it is richer in information. It also reveals how fast the armadillo moves.

Actually, several quantities are associated with the phrase "how fast." One of them is the **average velocity** v_{avg}, which is the ratio of the displacement Δx that occurs during a particular time interval Δt to that interval:

$$v_{avg} = \frac{\Delta x}{\Delta t} = \frac{x_2 - x_1}{t_2 - t_1}. \qquad (2\text{-}2)$$

The notation means that the position is x_1 at time t_1 and then x_2 at time t_2. A common unit for v_{avg} is the meter per second (m/s). You may see other units in the problems, but they are always in the form of length/time.

On a graph of x versus t, v_{avg} is the **slope** of the straight line that connects two particular points on the $x(t)$ curve: one is the point that corresponds to x_2 and t_2, and the other is the point that corresponds to x_1 and t_1. Like displacement, v_{avg} has both magnitude and direction (it is another vector quantity). Its magnitude is the magnitude of the line's slope. A positive v_{avg} (and slope) tells us that the line slants upward to the right; a negative v_{avg} (and slope) tells us that the line slants downward to the right. The average velocity v_{avg} always has the same sign as the displacement Δx because Δt in Eq. 2-2 is always positive.

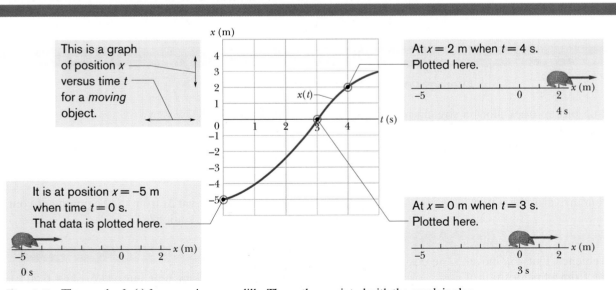

Fig. 2-3 The graph of $x(t)$ for a moving armadillo. The path associated with the graph is also shown, at three times.

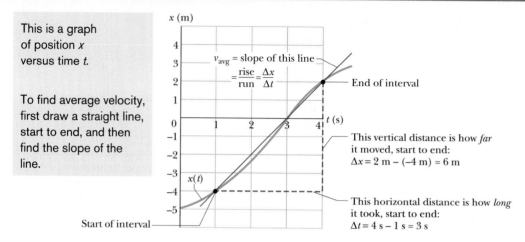

Fig. 2-4 Calculation of the average velocity between $t = 1$ s and $t = 4$ s as the slope of the line that connects the points on the $x(t)$ curve representing those times.

Figure 2-4 shows how to find v_{avg} in Fig. 2-3 for the time interval $t = 1$ s to $t = 4$ s. We draw the straight line that connects the point on the position curve at the beginning of the interval and the point on the curve at the end of the interval. Then we find the slope $\Delta x/\Delta t$ of the straight line. For the given time interval, the average velocity is

$$v_{avg} = \frac{6 \text{ m}}{3 \text{ s}} = 2 \text{ m/s}.$$

Average speed s_{avg} is a different way of describing "how fast" a particle moves. Whereas the average velocity involves the particle's displacement Δx, the average speed involves the total distance covered (for example, the number of meters moved), independent of direction; that is,

$$s_{avg} = \frac{\text{total distance}}{\Delta t}. \qquad (2\text{-}3)$$

Because average speed does *not* include direction, it lacks any algebraic sign. Sometimes s_{avg} is the same (except for the absence of a sign) as v_{avg}. However, the two can be quite different.

Sample Problem

Average velocity, beat-up pickup truck

You drive a beat-up pickup truck along a straight road for 8.4 km at 70 km/h, at which point the truck runs out of gasoline and stops. Over the next 30 min, you walk another 2.0 km farther along the road to a gasoline station.

(a) What is your overall displacement from the beginning of your drive to your arrival at the station?

KEY IDEA

Assume, for convenience, that you move in the positive direction of an x axis, from a first position of $x_1 = 0$ to a second position of x_2 at the station. That second position must be at $x_2 = 8.4$ km $+ 2.0$ km $= 10.4$ km. Then your displacement Δx along the x axis is the second position minus the first position.

Calculation: From Eq. 2-1, we have

$$\Delta x = x_2 - x_1 = 10.4 \text{ km} - 0 = 10.4 \text{ km.} \qquad \text{(Answer)}$$

Thus, your overall displacement is 10.4 km in the positive direction of the x axis.

(b) What is the time interval Δt from the beginning of your drive to your arrival at the station?

KEY IDEA

We already know the walking time interval Δt_{wlk} ($= 0.50$ h), but we lack the driving time interval Δt_{dr}. However, we know that for the drive the displacement Δx_{dr} is 8.4 km and the average velocity $v_{avg,dr}$ is 70 km/h. Thus, this average velocity is the ratio of the displacement for the drive to the time interval for the drive.

Calculations: We first write

$$v_{avg,dr} = \frac{\Delta x_{dr}}{\Delta t_{dr}}.$$

Rearranging and substituting data then give us

$$\Delta t_{dr} = \frac{\Delta x_{dr}}{v_{avg,dr}} = \frac{8.4 \text{ km}}{70 \text{ km/h}} = 0.12 \text{ h}.$$

So, $\Delta t = \Delta t_{dr} + \Delta t_{wlk}$

$$= 0.12 \text{ h} + 0.50 \text{ h} = 0.62 \text{ h}. \qquad \text{(Answer)}$$

(c) What is your average velocity v_{avg} from the beginning of your drive to your arrival at the station? Find it both numerically and graphically.

KEY IDEA

From Eq. 2-2 we know that v_{avg} *for the entire trip* is the ratio of the displacement of 10.4 km *for the entire trip* to the time interval of 0.62 h *for the entire trip*.

Calculation: Here we find

$$v_{avg} = \frac{\Delta x}{\Delta t} = \frac{10.4 \text{ km}}{0.62 \text{ h}}$$

$$= 16.8 \text{ km/h} \approx 17 \text{ km/h}. \qquad \text{(Answer)}$$

To find v_{avg} graphically, first we graph the function $x(t)$ as shown in Fig. 2-5, where the beginning and arrival points on the graph are the origin and the point labeled as "Station." Your average velocity is the slope of the straight line connecting those points; that is, v_{avg} is the ratio of the *rise* ($\Delta x = 10.4$ km) to the *run* ($\Delta t = 0.62$ h), which gives us $v_{avg} = 16.8$ km/h.

(d) Suppose that to pump the gasoline, pay for it, and walk back to the truck takes you another 45 min. What is your

average speed from the beginning of your drive to your return to the truck with the gasoline?

KEY IDEA

Your average speed is the ratio of the total distance you move to the total time interval you take to make that move.

Calculation: The total distance is 8.4 km + 2.0 km + 2.0 km = 12.4 km. The total time interval is 0.12 h + 0.50 h + 0.75 h = 1.37 h. Thus, Eq. 2-3 gives us

$$s_{avg} = \frac{12.4 \text{ km}}{1.37 \text{ h}} = 9.1 \text{ km/h}. \qquad \text{(Answer)}$$

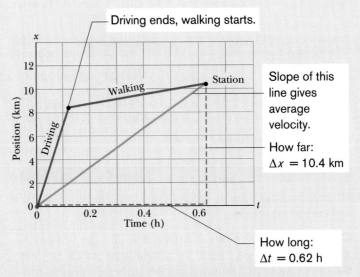

Fig. 2-5 The lines marked "Driving" and "Walking" are the position–time plots for the driving and walking stages. (The plot for the walking stage assumes a constant rate of walking.) The slope of the straight line joining the origin and the point labeled "Station" is the average velocity for the trip, from the beginning to the station.

 Additional examples, video, and practice available at *WileyPLUS*

2-5 Instantaneous Velocity and Speed

You have now seen two ways to describe how fast something moves: average velocity and average speed, both of which are measured over a time interval Δt. However, the phrase "how fast" more commonly refers to how fast a particle is moving at a given instant—its **instantaneous velocity** (or simply **velocity**) v.

The velocity at any instant is obtained from the average velocity by shrinking the time interval Δt closer and closer to 0. As Δt dwindles, the average velocity approaches a limiting value, which is the velocity at that instant:

$$v = \lim_{\Delta t \to 0} \frac{\Delta x}{\Delta t} = \frac{dx}{dt}. \qquad (2\text{-}4)$$

Note that v is the rate at which position x is changing with time at a given instant; that is, v is the derivative of x with respect to t. Also note that v at any instant is the slope of the position–time curve at the point representing that instant. Velocity is another vector quantity and thus has an associated direction.

Speed is the magnitude of velocity; that is, speed is velocity that has been stripped of any indication of direction, either in words or via an algebraic sign. (*Caution:* Speed and average speed can be quite different.) A velocity of $+5$ m/s and one of -5 m/s both have an associated speed of 5 m/s. The speedometer in a car measures speed, not velocity (it cannot determine the direction).

✔**CHECKPOINT 2**

The following equations give the position $x(t)$ of a particle in four situations (in each equation, x is in meters, t is in seconds, and $t > 0$): (1) $x = 3t - 2$; (2) $x = -4t^2 - 2$; (3) $x = 2/t^2$; and (4) $x = -2$. (a) In which situation is the velocity v of the particle constant? (b) In which is v in the negative x direction?

Sample Problem

Velocity and slope of x versus t, elevator cab

Figure 2-6a is an $x(t)$ plot for an elevator cab that is initially stationary, then moves upward (which we take to be the positive direction of x), and then stops. Plot $v(t)$.

KEY IDEA

We can find the velocity at any time from the slope of the $x(t)$ curve at that time.

Calculations: The slope of $x(t)$, and so also the velocity, is zero in the intervals from 0 to 1 s and from 9 s on, so then the cab is stationary. During the interval bc, the slope is constant and nonzero, so then the cab moves with constant velocity. We calculate the slope of $x(t)$ then as

$$\frac{\Delta x}{\Delta t} = v = \frac{24 \text{ m} - 4.0 \text{ m}}{8.0 \text{ s} - 3.0 \text{ s}} = +4.0 \text{ m/s}. \quad (2\text{-}5)$$

The plus sign indicates that the cab is moving in the positive x direction. These intervals (where $v = 0$ and $v = 4$ m/s) are plotted in Fig. 2-6b. In addition, as the cab initially begins to move and then later slows to a stop, v varies as indicated in the intervals 1 s to 3 s and 8 s to 9 s. Thus, Fig. 2-6b is the required plot. (Figure 2-6c is considered in Section 2-6.)

Given a $v(t)$ graph such as Fig. 2-6b, we could "work backward" to produce the shape of the associated $x(t)$ graph (Fig. 2-6a). However, we would not know the actual values for x at various times, because the $v(t)$ graph indicates only *changes* in x. To find such a change in x during any interval, we must, in the language of calculus, calculate the area "under the curve" on the $v(t)$ graph for that interval. For example, during the interval 3 s to 8 s in which the cab has a velocity of 4.0 m/s, the change in x is

$$\Delta x = (4.0 \text{ m/s})(8.0 \text{ s} - 3.0 \text{ s}) = +20 \text{ m}. \quad (2\text{-}6)$$

(This area is positive because the $v(t)$ curve is above the t axis.) Figure 2-6a shows that x does indeed increase by 20 m in that interval. However, Fig. 2-6b does not tell us the *values* of x at the beginning and end of the interval. For that, we need additional information, such as the value of x at some instant.

2-6 Acceleration

When a particle's velocity changes, the particle is said to undergo **acceleration** (or to accelerate). For motion along an axis, the **average acceleration** a_{avg} over a time interval Δt is

$$a_{avg} = \frac{v_2 - v_1}{t_2 - t_1} = \frac{\Delta v}{\Delta t}, \quad (2\text{-}7)$$

where the particle has velocity v_1 at time t_1 and then velocity v_2 at time t_2. The **instantaneous acceleration** (or simply **acceleration**) is

$$a = \frac{dv}{dt}. \quad (2\text{-}8)$$

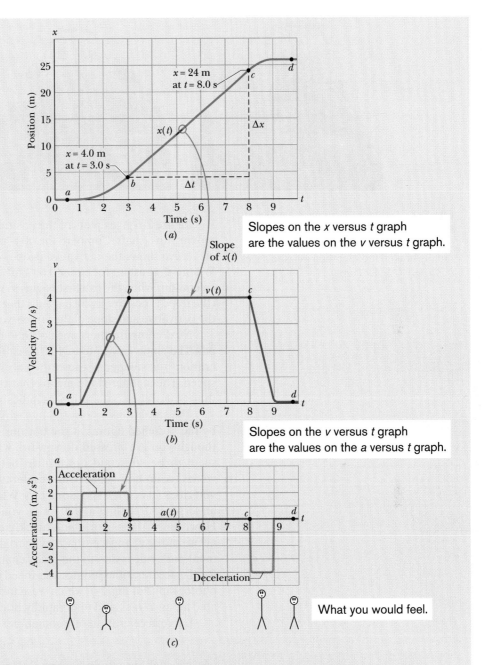

Fig. 2-6 (a) The x(t) curve for an elevator cab that moves upward along an x axis. (b) The v(t) curve for the cab. Note that it is the derivative of the x(t) curve (v = dx/dt). (c) The a(t) curve for the cab. It is the derivative of the v(t) curve (a = dv/dt). The stick figures along the bottom suggest how a passenger's body might feel during the accelerations.

 PLUS Additional examples, video, and practice available at *WileyPLUS*

In words, the acceleration of a particle at any instant is the rate at which its velocity is changing at that instant. Graphically, the acceleration at any point is the slope of the curve of v(t) at that point. We can combine Eq. 2-8 with Eq. 2-4 to write

$$ a = \frac{dv}{dt} = \frac{d}{dt}\left(\frac{dx}{dt}\right) = \frac{d^2x}{dt^2}. \qquad (2\text{-}9) $$

In words, the acceleration of a particle at any instant is the second derivative of its position x(t) with respect to time.

A common unit of acceleration is the meter per second per second: m/(s·s) or m/s². Other units are in the form of length/(time·time) or length/time². Acceleration has both magnitude and direction (it is yet another vector quantity). Its algebraic sign represents its direction on an axis just as for displacement and velocity; that is, acceleration with a positive value is in the positive direction of an axis, and acceleration with a negative value is in the negative direction.

Fig. 2-7 Colonel J. P. Stapp in a rocket sled as it is brought up to high speed (acceleration out of the page) and then very rapidly braked (acceleration into the page). *(Courtesy U.S. Air Force)*

Figure 2-6 gives plots of the position, velocity, and acceleration of an elevator moving up a shaft. Compare the $a(t)$ curve with the $v(t)$ curve—each point on the $a(t)$ curve shows the derivative (slope) of the $v(t)$ curve at the corresponding time. When v is constant (at either 0 or 4 m/s), the derivative is zero and so also is the acceleration. When the cab first begins to move, the $v(t)$ curve has a positive derivative (the slope is positive), which means that $a(t)$ is positive. When the cab slows to a stop, the derivative and slope of the $v(t)$ curve are negative; that is, $a(t)$ is negative.

Next compare the slopes of the $v(t)$ curve during the two acceleration periods. The slope associated with the cab's slowing down (commonly called "deceleration") is steeper because the cab stops in half the time it took to get up to speed. The steeper slope means that the magnitude of the deceleration is larger than that of the acceleration, as indicated in Fig. 2-6c.

The sensations you would feel while riding in the cab of Fig. 2-6 are indicated by the sketched figures at the bottom. When the cab first accelerates, you feel as though you are pressed downward; when later the cab is braked to a stop, you seem to be stretched upward. In between, you feel nothing special. In other words, your body reacts to accelerations (it is an accelerometer) but not to velocities (it is not a speedometer). When you are in a car traveling at 90 km/h or an airplane traveling at 900 km/h, you have no bodily awareness of the motion. However, if the car or plane quickly changes velocity, you may become keenly aware of the change, perhaps even frightened by it. Part of the thrill of an amusement park ride is due to the quick changes of velocity that you undergo (you pay for the accelerations, not for the speed). A more extreme example is shown in the photographs of Fig. 2-7, which were taken while a rocket sled was rapidly accelerated along a track and then rapidly braked to a stop.

Large accelerations are sometimes expressed in terms of g units, with

$$1g = 9.8 \text{ m/s}^2 \qquad (g \text{ unit}). \qquad (2\text{-}10)$$

(As we shall discuss in Section 2-9, g is the magnitude of the acceleration of a falling object near Earth's surface.) On a roller coaster, you may experience brief accelerations up to $3g$, which is $(3)(9.8 \text{ m/s}^2)$, or about 29 m/s², more than enough to justify the cost of the ride.

In common language, the sign of an acceleration has a nonscientific meaning: positive acceleration means that the speed of an object is increasing, and negative acceleration means that the speed is decreasing (the object is decelerating). In this book, however, the sign of an acceleration indicates a direction, not whether an object's speed is increasing or decreasing. For example, if a car with an initial velocity $v = -25$ m/s is braked to a stop in 5.0 s, then $a_{avg} = +5.0$ m/s². The acceleration is *positive*, but the car's speed has decreased. The reason is the difference in signs: the direction of the acceleration is opposite that of the velocity.

Here then is the proper way to interpret the signs:

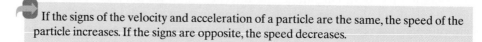

If the signs of the velocity and acceleration of a particle are the same, the speed of the particle increases. If the signs are opposite, the speed decreases.

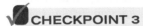

CHECKPOINT 3

A wombat moves along an x axis. What is the sign of its acceleration if it is moving (a) in the positive direction with increasing speed, (b) in the positive direction with decreasing speed, (c) in the negative direction with increasing speed, and (d) in the negative direction with decreasing speed?

Sample Problem

Acceleration and *dv/dt*

A particle's position on the x axis of Fig. 2-1 is given by

$$x = 4 - 27t + t^3,$$

with x in meters and t in seconds.

(a) Because position x depends on time t, the particle must be moving. Find the particle's velocity function $v(t)$ and acceleration function $a(t)$.

KEY IDEAS

(1) To get the velocity function $v(t)$, we differentiate the position function $x(t)$ with respect to time. (2) To get the acceleration function $a(t)$, we differentiate the velocity function $v(t)$ with respect to time.

Calculations: Differentiating the position function, we find

$$v = -27 + 3t^2, \text{(Answer)}$$

with v in meters per second. Differentiating the velocity function then gives us

$$a = +6t, \text{(Answer)}$$

with a in meters per second squared.

(b) Is there ever a time when $v = 0$?

Calculation: Setting $v(t) = 0$ yields

$$0 = -27 + 3t^2,$$

which has the solution

$$t = \pm 3 \text{ s}. \text{(Answer)}$$

Thus, the velocity is zero both 3 s before and 3 s after the clock reads 0.

(c) Describe the particle's motion for $t \geq 0$.

Reasoning: We need to examine the expressions for $x(t)$, $v(t)$, and $a(t)$.

At $t = 0$, the particle is at $x(0) = +4$ m and is moving with a velocity of $v(0) = -27$ m/s—that is, in the negative direction of the x axis. Its acceleration is $a(0) = 0$ because just then the particle's velocity is not changing.

For $0 < t < 3$ s, the particle still has a negative velocity, so it continues to move in the negative direction. However, its acceleration is no longer 0 but is increasing and positive. Because the signs of the velocity and the acceleration are opposite, the particle must be slowing.

Indeed, we already know that it stops momentarily at $t = 3$ s. Just then the particle is as far to the left of the origin in Fig. 2-1 as it will ever get. Substituting $t = 3$ s into the expression for $x(t)$, we find that the particle's position just then is $x = -50$ m. Its acceleration is still positive.

For $t > 3$ s, the particle moves to the right on the axis. Its acceleration remains positive and grows progressively larger in magnitude. The velocity is now positive, and it too grows progressively larger in magnitude.

 Additional examples, video, and practice available at *WileyPLUS*

2-7 Constant Acceleration: A Special Case

In many types of motion, the acceleration is either constant or approximately so. For example, you might accelerate a car at an approximately constant rate when a traffic light turns from red to green. Then graphs of your position, velocity, and acceleration would resemble those in Fig. 2-8. (Note that $a(t)$ in Fig. 2-8c is constant, which requires that $v(t)$ in Fig. 2-8b have a constant slope.) Later when you brake the car to a stop, the acceleration (or deceleration in common language) might also be approximately constant.

Such cases are so common that a special set of equations has been derived for dealing with them. One approach to the derivation of these equations is given in this section. A second approach is given in the next section. Throughout both sections and later when you work on the homework problems, keep in mind that *these equations are valid only for constant acceleration (or situations in which you can approximate the acceleration as being constant).*

When the acceleration is constant, the average acceleration and instantaneous acceleration are equal and we can write Eq. 2-7, with some changes in notation, as

$$a = a_{avg} = \frac{v - v_0}{t - 0}.$$

Here v_0 is the velocity at time $t = 0$ and v is the velocity at any later time t. We can recast this equation as

$$v = v_0 + at. \tag{2-11}$$

As a check, note that this equation reduces to $v = v_0$ for $t = 0$, as it must. As a further check, take the derivative of Eq. 2-11. Doing so yields $dv/dt = a$, which is the definition of a. Figure 2-8b shows a plot of Eq. 2-11, the $v(t)$ function; the function is linear and thus the plot is a straight line.

In a similar manner, we can rewrite Eq. 2-2 (with a few changes in notation) as

$$v_{avg} = \frac{x - x_0}{t - 0}$$

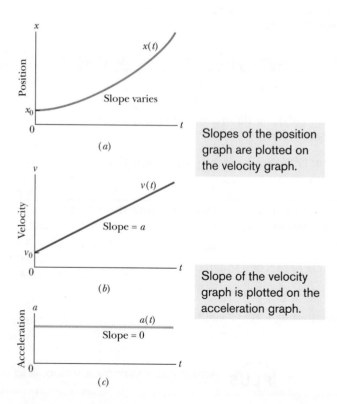

Fig. 2-8 (a) The position $x(t)$ of a particle moving with constant acceleration. (b) Its velocity $v(t)$, given at each point by the slope of the curve of $x(t)$. (c) Its (constant) acceleration, equal to the (constant) slope of the curve of $v(t)$.

Slopes of the position graph are plotted on the velocity graph.

Slope of the velocity graph is plotted on the acceleration graph.

and then as

$$x = x_0 + v_{avg}t,$$ (2-12)

in which x_0 is the position of the particle at $t = 0$ and v_{avg} is the average velocity between $t = 0$ and a later time t.

For the linear velocity function in Eq. 2-11, the *average* velocity over any time interval (say, from $t = 0$ to a later time t) is the average of the velocity at the beginning of the interval ($= v_0$) and the velocity at the end of the interval ($= v$). For the interval from $t = 0$ to the later time t then, the average velocity is

$$v_{avg} = \tfrac{1}{2}(v_0 + v).$$ (2-13)

Substituting the right side of Eq. 2-11 for v yields, after a little rearrangement,

$$v_{avg} = v_0 + \tfrac{1}{2}at.$$ (2-14)

Finally, substituting Eq. 2-14 into Eq. 2-12 yields

$$x - x_0 = v_0t + \tfrac{1}{2}at^2.$$ (2-15)

As a check, note that putting $t = 0$ yields $x = x_0$, as it must. As a further check, taking the derivative of Eq. 2-15 yields Eq. 2-11, again as it must. Figure 2-8a shows a plot of Eq. 2-15; the function is quadratic and thus the plot is curved.

Equations 2-11 and 2-15 are the *basic equations for constant acceleration;* they can be used to solve any constant acceleration problem in this book. However, we can derive other equations that might prove useful in certain specific situations. First, note that as many as five quantities can possibly be involved in any problem about constant acceleration—namely, $x - x_0$, v, t, a, and v_0. Usually, one of these quantities is *not* involved in the problem, *either as a given or as an unknown.* We are then presented with three of the remaining quantities and asked to find the fourth.

Equations 2-11 and 2-15 each contain four of these quantities, but not the same four. In Eq. 2-11, the "missing ingredient" is the displacement $x - x_0$. In Eq. 2-15, it is the velocity v. These two equations can also be combined in three ways to yield three additional equations, each of which involves a different "missing variable." First, we can eliminate t to obtain

$$v^2 = v_0^2 + 2a(x - x_0).$$ (2-16)

This equation is useful if we do not know t and are not required to find it. Second, we can eliminate the acceleration a between Eqs. 2-11 and 2-15 to produce an equation in which a does not appear:

$$x - x_0 = \tfrac{1}{2}(v_0 + v)t.$$ (2-17)

Finally, we can eliminate v_0, obtaining

$$x - x_0 = vt - \tfrac{1}{2}at^2.$$ (2-18)

Note the subtle difference between this equation and Eq. 2-15. One involves the initial velocity v_0; the other involves the velocity v at time t.

Table 2-1 lists the basic constant acceleration equations (Eqs. 2-11 and 2-15) as well as the specialized equations that we have derived. To solve a simple constant acceleration problem, you can usually use an equation from this list (*if* you have the list with you). Choose an equation for which the only unknown variable is the variable requested in the problem. A simpler plan is to remember only Eqs. 2-11 and 2-15, and then solve them as simultaneous equations whenever needed.

CHECKPOINT 4

The following equations give the position $x(t)$ of a particle in four situations: (1) $x = 3t - 4$; (2) $x = -5t^3 + 4t^2 + 6$; (3) $x = 2/t^2 - 4/t$; (4) $x = 5t^2 - 3$. To which of these situations do the equations of Table 2-1 apply?

Table 2-1

Equations for Motion with Constant Acceleration[a]

Equation Number	Equation	Missing Quantity
2-11	$v = v_0 + at$	$x - x_0$
2-15	$x - x_0 = v_0t + \tfrac{1}{2}at^2$	v
2-16	$v^2 = v_0^2 + 2a(x - x_0)$	t
2-17	$x - x_0 = \tfrac{1}{2}(v_0 + v)t$	a
2-18	$x - x_0 = vt - \tfrac{1}{2}at^2$	v_0

[a]Make sure that the acceleration is indeed constant before using the equations in this table.

Sample Problem

Constant acceleration, graph of v versus x

Figure 2-9 gives a particle's velocity v versus its position as it moves along an x axis with constant acceleration. What is its velocity at position $x = 0$?

KEY IDEA

We can use the constant-acceleration equations; in particular, we can use Eq. 2-16 ($v^2 = v_0^2 + 2a(x - x_0)$), which relates velocity and position.

First try: Normally we want to use an equation that includes the requested variable. In Eq. 2-16, we can identify x_0 as 0 and v_0 as being the requested variable. Then we can identify a second pair of values as being v and x. From the graph, we have

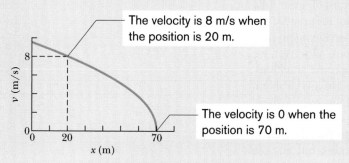

The velocity is 8 m/s when the position is 20 m.

The velocity is 0 when the position is 70 m.

Fig. 2-9 Velocity versus position.

two such pairs: (1) $v = 8$ m/s and $x = 20$ m, and (2) $v = 0$ and $x = 70$ m. For example, we can write Eq. 2-16 as

$$(8 \text{ m/s})^2 = v_0^2 + 2a(20 \text{ m} - 0). \tag{2-19}$$

However, we know neither v_0 nor a.

Second try: Instead of directly involving the requested variable, let's use Eq. 2-16 with the two pairs of known data, identifying $v_0 = 8$ m/s and $x_0 = 20$ m as the first pair and $v = 0$ m/s and $x = 70$ m as the second pair. Then we can write

$$(0 \text{ m/s})^2 = (8 \text{ m/s})^2 + 2a(70 \text{ m} - 20 \text{ m}),$$

which gives us $a = -0.64$ m/s². Substituting this value into Eq. 2-19 and solving for v_0 (the velocity associated with the position of $x = 0$), we find

$$v_0 = 9.5 \text{ m/s}. \tag{Answer}$$

Comment: Some problems involve an equation that includes the requested variable. A more challenging problem requires you to first use an equation that does *not* include the requested variable but that gives you a value needed to find it. Sometimes that procedure takes *physics courage* because it is so indirect. However, if you build your solving skills by solving lots of problems, the procedure gradually requires less courage and may even become obvious. Solving problems of any kind, whether physics or social, requires practice.

(WILEY PLUS) Additional examples, video, and practice available at *WileyPLUS*

2-8 Another Look at Constant Acceleration*

The first two equations in Table 2-1 are the basic equations from which the others are derived. Those two can be obtained by integration of the acceleration with the condition that a is constant. To find Eq. 2-11, we rewrite the definition of acceleration (Eq. 2-8) as

$$dv = a \, dt.$$

We next write the *indefinite integral* (or *antiderivative*) of both sides:

$$\int dv = \int a \, dt.$$

Since acceleration a is a constant, it can be taken outside the integration. We obtain

$$\int dv = a \int dt$$

or

$$v = at + C. \tag{2-20}$$

To evaluate the constant of integration C, we let $t = 0$, at which time $v = v_0$. Substituting these values into Eq. 2-20 (which must hold for all values of t,

---

*This section is intended for students who have had integral calculus.

including $t = 0$) yields

$$v_0 = (a)(0) + C = C.$$

Substituting this into Eq. 2-20 gives us Eq. 2-11.

To derive Eq. 2-15, we rewrite the definition of velocity (Eq. 2-4) as

$$dx = v \, dt$$

and then take the indefinite integral of both sides to obtain

$$\int dx = \int v \, dt.$$

Next, we substitute for v with Eq. 2-11:

$$\int dx = \int (v_0 + at) \, dt.$$

Since v_0 is a constant, as is the acceleration a, this can be rewritten as

$$\int dx = v_0 \int dt + a \int t \, dt.$$

Integration now yields

$$x = v_0 t + \tfrac{1}{2}at^2 + C', \tag{2-21}$$

where C' is another constant of integration. At time $t = 0$, we have $x = x_0$. Substituting these values in Eq. 2-21 yields $x_0 = C'$. Replacing C' with x_0 in Eq. 2-21 gives us Eq. 2-15.

2-9 Free-Fall Acceleration

If you tossed an object either up or down and could somehow eliminate the effects of air on its flight, you would find that the object accelerates downward at a certain constant rate. That rate is called the **free-fall acceleration,** and its magnitude is represented by g. The acceleration is independent of the object's characteristics, such as mass, density, or shape; it is the same for all objects.

Two examples of free-fall acceleration are shown in Fig. 2-10, which is a series of stroboscopic photos of a feather and an apple. As these objects fall, they accelerate downward—both at the same rate g. Thus, their speeds increase at the same rate, and they fall together.

The value of g varies slightly with latitude and with elevation. At sea level in Earth's midlatitudes the value is 9.8 m/s^2 (or 32 ft/s^2), which is what you should use as an exact number for the problems in this book unless otherwise noted.

The equations of motion in Table 2-1 for constant acceleration also apply to free fall near Earth's surface; that is, they apply to an object in vertical flight, either up or down, when the effects of the air can be neglected. However, note that for free fall: (1) The directions of motion are now along a vertical y axis instead of the x axis, with the positive direction of y upward. (This is important for later chapters when combined horizontal and vertical motions are examined.) (2) The free-fall acceleration is negative—that is, downward on the y axis, toward Earth's center—and so it has the value $-g$ in the equations.

The free-fall acceleration near Earth's surface is $a = -g = -9.8$ m/s^2, and the *magnitude* of the acceleration is $g = 9.8$ m/s^2. Do not substitute -9.8 m/s^2 for g.

Suppose you toss a tomato directly upward with an initial (positive) velocity v_0 and then catch it when it returns to the release level. During its *free-fall flight* (from just after its release to just before it is caught), the equations of Table 2-1 apply to its

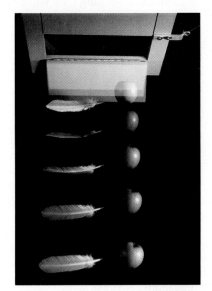

Fig. 2-10 A feather and an apple free fall in vacuum at the same magnitude of acceleration g. The acceleration increases the distance between successive images. In the absence of air, the feather and apple fall together. *(Jim Sugar/Corbis Images)*

motion. The acceleration is always $a = -g = -9.8$ m/s², negative and thus downward. The velocity, however, changes, as indicated by Eqs. 2-11 and 2-16: during the ascent, the magnitude of the positive velocity decreases, until it momentarily becomes zero. Because the tomato has then stopped, it is at its maximum height. During the descent, the magnitude of the (now negative) velocity increases.

✓ CHECKPOINT 5

(a) If you toss a ball straight up, what is the sign of the ball's displacement for the ascent, from the release point to the highest point? (b) What is it for the descent, from the highest point back to the release point? (c) What is the ball's acceleration at its highest point?

Sample Problem

Time for full up-down flight, baseball toss

In Fig. 2-11, a pitcher tosses a baseball up along a y axis, with an initial speed of 12 m/s.

(a) How long does the ball take to reach its maximum height?

KEY IDEAS

(1) Once the ball leaves the pitcher and before it returns to his hand, its acceleration is the free-fall acceleration $a = -g$. Because this is constant, Table 2-1 applies to the motion. (2) The velocity v at the maximum height must be 0.

Calculation: Knowing v, a, and the initial velocity $v_0 = 12$ m/s, and seeking t, we solve Eq. 2-11, which contains

those four variables. This yields

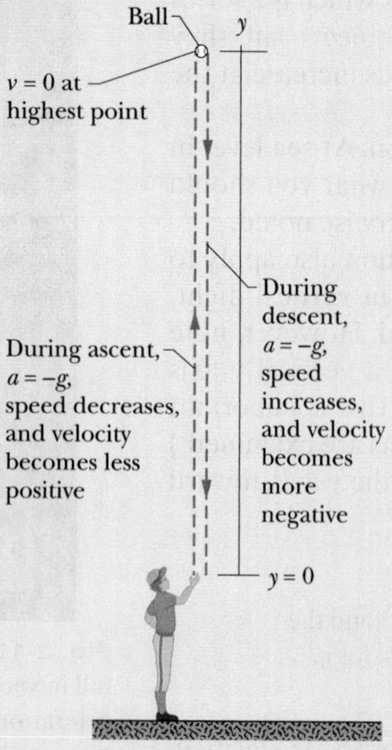

$$t = \frac{v - v_0}{a} = \frac{0 - 12 \text{ m/s}}{-9.8 \text{ m/s}^2} = 1.2 \text{ s.} \quad \text{(Answer)}$$

(b) What is the ball's maximum height above its release point?

Calculation: We can take the ball's release point to be $y_0 = 0$. We can then write Eq. 2-16 in y notation, set $y - y_0 = y$ and $v = 0$ (at the maximum height), and solve for y. We get

$$y = \frac{v^2 - v_0^2}{2a} = \frac{0 - (12 \text{ m/s})^2}{2(-9.8 \text{ m/s}^2)} = 7.3 \text{ m.} \quad \text{(Answer)}$$

(c) How long does the ball take to reach a point 5.0 m above its release point?

Calculations: We know v_0, $a = -g$, and displacement $y - y_0 = 5.0$ m, and we want t, so we choose Eq. 2-15. Rewriting it for y and setting $y_0 = 0$ give us

$$y = v_0 t - \tfrac{1}{2}gt^2,$$

or $\quad 5.0 \text{ m} = (12 \text{ m/s})t - (\tfrac{1}{2})(9.8 \text{ m/s}^2)t^2.$

If we temporarily omit the units (having noted that they are consistent), we can rewrite this as

$$4.9t^2 - 12t + 5.0 = 0.$$

Solving this quadratic equation for t yields

$$t = 0.53 \text{ s} \quad \text{and} \quad t = 1.9 \text{ s.} \quad \text{(Answer)}$$

There are two such times! This is not really surprising because the ball passes twice through $y = 5.0$ m, once on the way up and once on the way down.

Ball

y

$v = 0$ at highest point

During ascent, $a = -g$, speed decreases, and velocity becomes less positive

During descent, $a = -g$, speed increases, and velocity becomes more negative

$y = 0$

Fig. 2-11 A pitcher tosses a baseball straight up into the air. The equations of free fall apply for rising as well as for falling objects, provided any effects from the air can be neglected.

2-10 Graphical Integration in Motion Analysis

When we have a graph of an object's acceleration versus time, we can integrate on the graph to find the object's velocity at any given time. Because acceleration a is defined in terms of velocity as $a = dv/dt$, the Fundamental Theorem of Calculus tells us that

$$v_1 - v_0 = \int_{t_0}^{t_1} a\, dt. \qquad (2\text{-}22)$$

The right side of the equation is a definite integral (it gives a numerical result rather than a function), v_0 is the velocity at time t_0, and v_1 is the velocity at later time t_1. The definite integral can be evaluated from an $a(t)$ graph, such as in Fig. 2-12a. In particular,

$$\int_{t_0}^{t_1} a\, dt = \begin{pmatrix} \text{area between acceleration curve} \\ \text{and time axis, from } t_0 \text{ to } t_1 \end{pmatrix}. \qquad (2\text{-}23)$$

If a unit of acceleration is 1 m/s^2 and a unit of time is 1 s, then the corresponding unit of area on the graph is

$$(1 \text{ m/s}^2)(1 \text{ s}) = 1 \text{ m/s},$$

which is (properly) a unit of velocity. When the acceleration curve is above the time axis, the area is positive; when the curve is below the time axis, the area is negative.

Similarly, because velocity v is defined in terms of the position x as $v = dx/dt$, then

$$x_1 - x_0 = \int_{t_0}^{t_1} v\, dt, \qquad (2\text{-}24)$$

where x_0 is the position at time t_0 and x_1 is the position at time t_1. The definite integral on the right side of Eq. 2-24 can be evaluated from a $v(t)$ graph, like that shown in Fig. 2-12b. In particular,

$$\int_{t_0}^{t_1} v\, dt = \begin{pmatrix} \text{area between velocity curve} \\ \text{and time axis, from } t_0 \text{ to } t_1 \end{pmatrix}. \qquad (2\text{-}25)$$

If the unit of velocity is 1 m/s and the unit of time is 1 s, then the corresponding unit of area on the graph is

$$(1 \text{ m/s})(1 \text{ s}) = 1 \text{ m},$$

which is (properly) a unit of position and displacement. Whether this area is positive or negative is determined as described for the $a(t)$ curve of Fig. 2-12a.

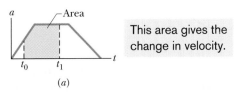

(a)

This area gives the change in velocity.

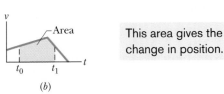

(b)

This area gives the change in position.

Fig. 2-12   The area between a plotted curve and the horizontal time axis, from time t_0 to time t_1, is indicated for (a) a graph of acceleration a versus t and (b) a graph of velocity v versus t.

Graphical integration *a* versus *t*, whiplash injury

"Whiplash injury" commonly occurs in a rear-end collision where a front car is hit from behind by a second car. In the 1970s, researchers concluded that the injury was due to the occupant's head being whipped back over the top of the seat as the car was slammed forward. As a result of this finding, head restraints were built into cars, yet neck injuries in rear-end collisions continued to occur.

In a recent test to study neck injury in rear-end collisions, a volunteer was strapped to a seat that was then moved abruptly to simulate a collision by a rear car moving at 10.5 km/h. Figure 2-13*a* gives the accelerations of the volunteer's torso and head during the collision, which began at time $t = 0$. The torso acceleration was delayed by 40 ms because during that time interval the seat back had to compress against the volunteer. The head acceleration was delayed by an additional 70 ms. What was the torso speed when the head began to accelerate?

KEY IDEA

We can calculate the torso speed at any time by finding an area on the torso $a(t)$ graph.

Calculations: We know that the initial torso speed is $v_0 = 0$ at time $t_0 = 0$, at the start of the "collision." We want the torso speed v_1 at time $t_1 = 110$ ms, which is when the head begins to accelerate.

Combining Eqs. 2-22 and 2-23, we can write

$$v_1 - v_0 = \left(\begin{array}{c} \text{area between acceleration curve} \\ \text{and time axis, from } t_0 \text{ to } t_1 \end{array} \right). \quad (2\text{-}26)$$

For convenience, let us separate the area into three regions (Fig. 2-13*b*). From 0 to 40 ms, region *A* has no area:

$$\text{area}_A = 0.$$

From 40 ms to 100 ms, region *B* has the shape of a triangle, with area

$$\text{area}_B = \tfrac{1}{2}(0.060 \text{ s})(50 \text{ m/s}^2) = 1.5 \text{ m/s}.$$

From 100 ms to 110 ms, region *C* has the shape of a rectangle, with area

$$\text{area}_C = (0.010 \text{ s})(50 \text{ m/s}^2) = 0.50 \text{ m/s}.$$

Substituting these values and $v_0 = 0$ into Eq. 2-26 gives us

$$v_1 - 0 = 0 + 1.5 \text{ m/s} + 0.50 \text{ m/s},$$

or $\qquad\qquad v_1 = 2.0 \text{ m/s} = 7.2 \text{ km/h}.$ (Answer)

Comments: When the head is just starting to move forward, the torso already has a speed of 7.2 km/h. Researchers argue that it is this difference in speeds during the early stage of a rear-end collision that injures the neck. The backward whipping of the head happens later and could, especially if there is no head restraint, increase the injury.

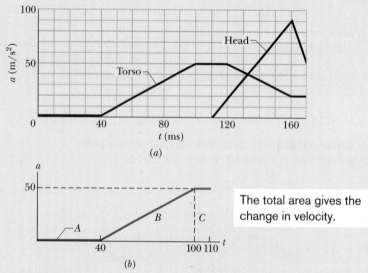

Fig. 2-13 (*a*) The $a(t)$ curve of the torso and head of a volunteer in a simulation of a rear-end collision. (*b*) Breaking up the region between the plotted curve and the time axis to calculate the area.

 Additional examples, video, and practice available at *WileyPLUS*

REVIEW & SUMMARY

Position The *position x* of a particle on an *x* axis locates the particle with respect to the **origin,** or zero point, of the axis. The position is either positive or negative, according to which side of the origin the particle is on, or zero if the particle is at the origin. The **positive direction** on an axis is the direction of increasing positive numbers; the opposite direction is the **negative direction** on the axis.

Displacement The *displacement* Δx of a particle is the change in its position:

$$\Delta x = x_2 - x_1. \tag{2-1}$$

Displacement is a vector quantity. It is positive if the particle has moved in the positive direction of the *x* axis and negative if the particle has moved in the negative direction.

Average Velocity When a particle has moved from position x_1 to position x_2 during a time interval $\Delta t = t_2 - t_1$, its *average velocity* during that interval is

$$v_{\text{avg}} = \frac{\Delta x}{\Delta t} = \frac{x_2 - x_1}{t_2 - t_1}. \tag{2-2}$$

The algebraic sign of v_{avg} indicates the direction of motion (v_{avg} is a vector quantity). Average velocity does not depend on the actual distance a particle moves, but instead depends on its original and final positions.

On a graph of *x* versus *t*, the average velocity for a time interval Δt is the slope of the straight line connecting the points on the curve that represent the two ends of the interval.

Average Speed The *average speed* s_{avg} of a particle during a time interval Δt depends on the total distance the particle moves in that time interval:

$$s_{\text{avg}} = \frac{\text{total distance}}{\Delta t}. \tag{2-3}$$

Instantaneous Velocity The *instantaneous velocity* (or simply **velocity**) *v* of a moving particle is

$$v = \lim_{\Delta t \to 0} \frac{\Delta x}{\Delta t} = \frac{dx}{dt}, \tag{2-4}$$

where Δx and Δt are defined by Eq. 2-2. The instantaneous velocity (at a particular time) may be found as the slope (at that particular time) of the graph of *x* versus *t*. **Speed** is the magnitude of instantaneous velocity.

Average Acceleration *Average acceleration* is the ratio of a change in velocity Δv to the time interval Δt in which the change occurs:

$$a_{\text{avg}} = \frac{\Delta v}{\Delta t}. \tag{2-7}$$

The algebraic sign indicates the direction of a_{avg}.

Instantaneous Acceleration *Instantaneous acceleration* (or simply **acceleration**) *a* is the first time derivative of velocity $v(t)$ and the second time derivative of position $x(t)$:

$$a = \frac{dv}{dt} = \frac{d^2x}{dt^2}. \tag{2-8, 2-9}$$

On a graph of *v* versus *t*, the acceleration *a* at any time *t* is the slope of the curve at the point that represents *t*.

Constant Acceleration The five equations in Table 2-1 describe the motion of a particle with constant acceleration:

$$v = v_0 + at, \tag{2-11}$$

$$x - x_0 = v_0 t + \tfrac{1}{2}at^2, \tag{2-15}$$

$$v^2 = v_0^2 + 2a(x - x_0), \tag{2-16}$$

$$x - x_0 = \tfrac{1}{2}(v_0 + v)t, \tag{2-17}$$

$$x - x_0 = vt - \tfrac{1}{2}at^2. \tag{2-18}$$

These are *not* valid when the acceleration is not constant.

Free-Fall Acceleration An important example of straight-line motion with constant acceleration is that of an object rising or falling freely near Earth's surface. The constant acceleration equations describe this motion, but we make two changes in notation: (1) we refer the motion to the vertical *y* axis with $+y$ vertically *up*; (2) we replace *a* with $-g$, where *g* is the magnitude of the free-fall acceleration. Near Earth's surface, $g = 9.8 \text{ m/s}^2 (= 32 \text{ ft/s}^2)$.

QUESTIONS

1 Figure 2-14 gives the velocity of a particle moving on an *x* axis. What are (a) the initial and (b) the final directions of travel? (c) Does the particle stop momentarily? (d) Is the acceleration positive or negative? (e) Is it constant or varying?

2 Figure 2-15 gives the acceleration *a(t)* of a Chihuahua as it chases

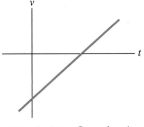

Fig. 2-14 Question 1.

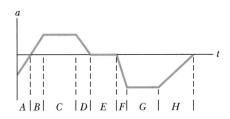

Fig. 2-15 Question 2.

a German shepherd along an axis. In which of the time periods indicated does the Chihuahua move at constant speed?

3 Figure 2-16 shows four paths along which objects move from a starting point to a final point, all in the same time interval. The paths pass over a grid of equally spaced straight lines. Rank the paths according to (a) the average velocity of the objects and (b) the average speed of the objects, greatest first.

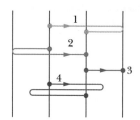

Fig. 2-16 Question 3.

4 Figure 2-17 is a graph of a particle's position along an x axis versus time. (a) At time $t = 0$, what is the sign of the particle's position? Is the particle's velocity positive, negative, or 0 at (b) $t = 1$ s, (c) $t = 2$ s, and (d) $t = 3$ s? (e) How many times does the particle go through the point $x = 0$?

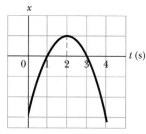

Fig. 2-17 Question 4.

5 Figure 2-18 gives the velocity of a particle moving along an axis. Point 1 is at the highest point on the curve; point 4 is at the lowest point; and points 2 and 6 are at the same height. What is the direction of travel at (a) time $t = 0$ and (b) point 4? (c) At which of the six numbered points does the particle reverse its direction of travel? (d) Rank the six points according to the magnitude of the acceleration, greatest first.

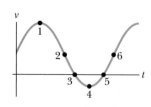

Fig. 2-18 Question 5.

6 At $t = 0$, a particle moving along an x axis is at position $x_0 = -20$ m. The signs of the particle's initial velocity v_0 (at time t_0) and constant acceleration a are, respectively, for four situations: (1) +, +; (2) +, −; (3) −, +; (4) −, −. In which situations will the particle (a) stop momentarily, (b) pass through the origin, and (c) never pass through the origin?

7 Hanging over the railing of a bridge, you drop an egg (no initial velocity) as you throw a second egg downward. Which curves in Fig. 2-19 give the velocity $v(t)$ for (a) the dropped egg and (b) the thrown egg? (Curves A and B are parallel; so are C, D, and E; so are F and G.)

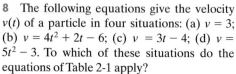

8 The following equations give the velocity $v(t)$ of a particle in four situations: (a) $v = 3$; (b) $v = 4t^2 + 2t - 6$; (c) $v = 3t - 4$; (d) $v = 5t^2 - 3$. To which of these situations do the equations of Table 2-1 apply?

9 In Fig. 2-20, a cream tangerine is thrown directly upward past three evenly spaced windows of equal heights. Rank the windows according to (a) the average speed of the cream tangerine while passing them, (b) the time the cream tangerine takes to pass them, (c) the magnitude of the acceleration of the cream tangerine while passing them, and (d) the change Δv in the speed of the cream tangerine during the passage, greatest first.

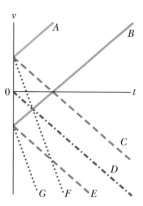

Fig. 2-19 Question 7.

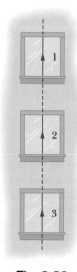

Fig. 2-20 Question 9.

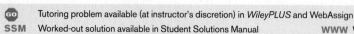

PROBLEMS

WILEY PLUS

sec. 2-4 Average Velocity and Average Speed

•1 During a hard sneeze, your eyes might shut for 0.50 s. If you are driving a car at 90 km/h during such a sneeze, how far does the car move during that time?

•2 Compute your average velocity in the following two cases: (a) You walk 73.2 m at a speed of 1.22 m/s and then run 73.2 m at a speed of 3.05 m/s along a straight track. (b) You walk for 1.00 min at a speed of 1.22 m/s and then run for 1.00 min at 3.05 m/s along a straight track. (c) Graph x versus t for both cases and indicate how the average velocity is found on the graph.

•3 **SSM** **WWW** An automobile travels on a straight road for 40 km at 30 km/h. It then continues in the same direction for another 40 km at 60 km/h. (a) What is the average velocity of the car during the full 80 km trip? (Assume that it moves in the positive x direc-

tion.) (b) What is the average speed? (c) Graph x versus t and indicate how the average velocity is found on the graph.

•4 A car travels up a hill at a constant speed of 40 km/h and returns down the hill at a constant speed of 60 km/h. Calculate the average speed for the round trip.

•5 **SSM** The position of an object moving along an x axis is given by $x = 3t - 4t^2 + t^3$, where x is in meters and t in seconds. Find the position of the object at the following values of t: (a) 1 s, (b) 2 s, (c) 3 s, and (d) 4 s. (e) What is the object's displacement between $t = 0$ and $t = 4$ s? (f) What is its average velocity for the time interval from $t = 2$ s to $t = 4$ s? (g) Graph x versus t for $0 \le t \le 4$ s and indicate how the answer for (f) can be found on the graph.

•6 The 1992 world speed record for a bicycle (human-powered vehicle) was set by Chris Huber. His time through the measured

200 m stretch was a sizzling 6.509 s, at which he commented, "Cogito ergo zoom!" (I think, therefore I go fast!). In 2001, Sam Whittingham beat Huber's record by 19.0 km/h. What was Whittingham's time through the 200 m?

••7 Two trains, each having a speed of 30 km/h, are headed at each other on the same straight track. A bird that can fly 60 km/h flies off the front of one train when they are 60 km apart and heads directly for the other train. On reaching the other train, the bird flies directly back to the first train, and so forth. (We have no idea *why* a bird would behave in this way.) What is the total distance the bird travels before the trains collide?

••8 ✈ GO *Panic escape.* Figure 2-21 shows a general situation in which a stream of people attempt to escape through an exit door that turns out to be locked. The people move toward the door at speed $v_s = 3.50$ m/s, are each $d = 0.25$ m in depth, and are separated by $L = 1.75$ m. The arrangement in Fig. 2-21 occurs at time $t = 0$. (a) At what average rate does the layer of people at the door increase? (b) At what time does the layer's depth reach 5.0 m? (The answers reveal how quickly such a situation becomes dangerous.)

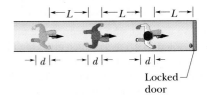

Fig. 2-21   Problem 8.

••9 ILW In 1 km races, runner 1 on track 1 (with time 2 min, 27.95 s) appears to be faster than runner 2 on track 2 (2 min, 28.15 s). However, length L_2 of track 2 might be slightly greater than length L_1 of track 1. How large can $L_2 - L_1$ be for us still to conclude that runner 1 is faster?

••10 ✈ To set a speed record in a measured (straight-line) distance d, a race car must be driven first in one direction (in time t_1) and then in the opposite direction (in time t_2). (a) To eliminate the effects of the wind and obtain the car's speed v_c in a windless situation, should we find the average of d/t_1 and d/t_2 (method 1) or should we divide d by the average of t_1 and t_2? (b) What is the fractional difference in the two methods when a steady wind blows along the car's route and the ratio of the wind speed v_w to the car's speed v_c is 0.0240?

••11 You are to drive to an interview in another town, at a distance of 300 km on an expressway. The interview is at 11:15 A.M. You plan to drive at 100 km/h, so you leave at 8:00 A.M. to allow some extra time. You drive at that speed for the first 100 km, but then construction work forces you to slow to 40 km/h for 40 km. What would be the least speed needed for the rest of the trip to arrive in time for the interview?

•••12 ✈ *Traffic shock wave.* An abrupt slowdown in concentrated traffic can travel as a pulse, termed a *shock wave,* along the line of cars, either downstream (in the traffic direction) or upstream, or it can be stationary. Figure 2-22 shows a uniformly spaced line of cars moving at speed $v = 25.0$ m/s toward a uniformly spaced line of slow cars moving at speed $v_s = 5.00$ m/s. Assume that each faster car adds length $L = 12.0$ m (car length plus buffer zone) to the line of slow cars when it joins the line, and assume it slows abruptly at the last instant. (a) For what separation

distance d between the faster cars does the shock wave remain stationary? If the separation is twice that amount, what are the (b) speed and (c) direction (upstream or downstream) of the shock wave?

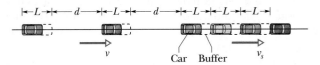

Fig. 2-22   Problem 12.

•••13 ILW You drive on Interstate 10 from San Antonio to Houston, half the *time* at 55 km/h and the other half at 90 km/h. On the way back you travel half the *distance* at 55 km/h and the other half at 90 km/h. What is your average speed (a) from San Antonio to Houston, (b) from Houston back to San Antonio, and (c) for the entire trip? (d) What is your average velocity for the entire trip? (e) Sketch x versus t for (a), assuming the motion is all in the positive x direction. Indicate how the average velocity can be found on the sketch.

sec. 2-5 Instantaneous Velocity and Speed

•14 GO An electron moving along the x axis has a position given by $x = 16te^{-t}$ m, where t is in seconds. How far is the electron from the origin when it momentarily stops?

•15 GO (a) If a particle's position is given by $x = 4 - 12t + 3t^2$ (where t is in seconds and x is in meters), what is its velocity at $t = 1$ s? (b) Is it moving in the positive or negative direction of x just then? (c) What is its speed just then? (d) Is the speed increasing or decreasing just then? (Try answering the next two questions without further calculation.) (e) Is there ever an instant when the velocity is zero? If so, give the time t; if not, answer no. (f) Is there a time after $t = 3$ s when the particle is moving in the negative direction of x? If so, give the time t; if not, answer no.

•16 The position function $x(t)$ of a particle moving along an x axis is $x = 4.0 - 6.0t^2$, with x in meters and t in seconds. (a) At what time and (b) where does the particle (momentarily) stop? At what (c) negative time and (d) positive time does the particle pass through the origin? (e) Graph x versus t for the range -5 s to $+5$ s. (f) To shift the curve rightward on the graph, should we include the term $+20t$ or the term $-20t$ in $x(t)$? (g) Does that inclusion increase or decrease the value of x at which the particle momentarily stops?

••17 The position of a particle moving along the x axis is given in centimeters by $x = 9.75 + 1.50t^3$, where t is in seconds. Calculate (a) the average velocity during the time interval $t = 2.00$ s to $t = 3.00$ s; (b) the instantaneous velocity at $t = 2.00$ s; (c) the instantaneous velocity at $t = 3.00$ s; (d) the instantaneous velocity at $t = 2.50$ s; and (e) the instantaneous velocity when the particle is midway between its positions at $t = 2.00$ s and $t = 3.00$ s. (f) Graph x versus t and indicate your answers graphically.

sec. 2-6 Acceleration

•18 The position of a particle moving along an x axis is given by $x = 12t^2 - 2t^3$, where x is in meters and t is in seconds. Determine (a) the position, (b) the velocity, and (c) the acceleration of the particle at $t = 3.0$ s. (d) What is the maximum positive coordinate reached by the particle and (e) at what time is it reached? (f) What is the maximum positive velocity reached by the particle and (g) at what time is it reached? (h) What is the acceleration of the particle at the instant the particle is not moving (other than at $t = 0$)? (i) Determine the average velocity of the particle between $t = 0$ and $t = 3$ s.

•19 SSM At a certain time a particle had a speed of 18 m/s in the positive x direction, and 2.4 s later its speed was 30 m/s in the opposite direction. What is the average acceleration of the particle during this 2.4 s interval?

•20 (a) If the position of a particle is given by $x = 20t - 5t^3$, where x is in meters and t is in seconds, when, if ever, is the particle's velocity zero? (b) When is its acceleration a zero? (c) For what time range (positive or negative) is a negative? (d) Positive? (e) Graph $x(t)$, $v(t)$, and $a(t)$.

••21 From $t = 0$ to $t = 5.00$ min, a man stands still, and from $t = 5.00$ min to $t = 10.0$ min, he walks briskly in a straight line at a constant speed of 2.20 m/s. What are (a) his average velocity v_{avg} and (b) his average acceleration a_{avg} in the time interval 2.00 min to 8.00 min? What are (c) v_{avg} and (d) a_{avg} in the time interval 3.00 min to 9.00 min? (e) Sketch x versus t and v versus t, and indicate how the answers to (a) through (d) can be obtained from the graphs.

••22 The position of a particle moving along the x axis depends on the time according to the equation $x = ct^2 - bt^3$, where x is in meters and t in seconds. What are the units of (a) constant c and (b) constant b? Let their numerical values be 3.0 and 2.0, respectively. (c) At what time does the particle reach its maximum positive x position? From $t = 0.0$ s to $t = 4.0$ s, (d) what distance does the particle move and (e) what is its displacement? Find its velocity at times (f) 1.0 s, (g) 2.0 s, (h) 3.0 s, and (i) 4.0 s. Find its acceleration at times (j) 1.0 s, (k) 2.0 s, (l) 3.0 s, and (m) 4.0 s.

sec. 2-7   Constant Acceleration: A Special Case

•23 SSM An electron with an initial velocity $v_0 = 1.50 \times 10^5$ m/s enters a region of length $L = 1.00$ cm where it is electrically accelerated (Fig. 2-23). It emerges with $v = 5.70 \times 10^6$ m/s. What is its acceleration, assumed constant?

•24 🚀 *Catapulting mush-rooms.* Certain mushrooms launch their spores by a catapult mechanism. As water condenses from the air onto a spore that is attached to the mushroom, a drop grows on one side of the spore and a film grows on the other side. The spore is bent over by the drop's weight, but when the film reaches the drop, the drop's water suddenly spreads into the film and the spore springs upward so rapidly that it is slung off into the air. Typically, the spore reaches a speed of 1.6 m/s in a 5.0 μm launch; its speed is then reduced to zero in 1.0 mm by the air. Using that data and assuming constant accelerations, find the acceleration in terms of g during (a) the launch and (b) the speed reduction.

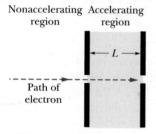

Nonaccelerating region   Accelerating region

Path of electron

Fig. 2-23   Problem 23.

•25 An electric vehicle starts from rest and accelerates at a rate of 2.0 m/s² in a straight line until it reaches a speed of 20 m/s. The vehicle then slows at a constant rate of 1.0 m/s² until it stops. (a) How much time elapses from start to stop? (b) How far does the vehicle travel from start to stop?

•26 A muon (an elementary particle) enters a region with a speed of 5.00×10^6 m/s and then is slowed at the rate of $1.25 \times$ 10^{14} m/s². (a) How far does the muon take to stop? (b) Graph x versus t and v versus t for the muon.

•27 An electron has a constant acceleration of +3.2 m/s². At a certain instant its velocity is +9.6 m/s. What is its velocity (a) 2.5 s earlier and (b) 2.5 s later?

•28 On a dry road, a car with good tires may be able to brake with a constant deceleration of 4.92 m/s². (a) How long does such a car, initially traveling at 24.6 m/s, take to stop? (b) How far does it travel in this time? (c) Graph x versus t and v versus t for the deceleration.

•29 ILW A certain elevator cab has a total run of 190 m and a maximum speed of 305 m/min, and it accelerates from rest and then back to rest at 1.22 m/s². (a) How far does the cab move while accelerating to full speed from rest? (b) How long does it take to make the nonstop 190 m run, starting and ending at rest?

•30 The brakes on your car can slow you at a rate of 5.2 m/s². (a) If you are going 137 km/h and suddenly see a state trooper, what is the minimum time in which you can get your car under the 90 km/h speed limit? (The answer reveals the futility of braking to keep your high speed from being detected with a radar or laser gun.) (b) Graph x versus t and v versus t for such a slowing.

•31 SSM Suppose a rocket ship in deep space moves with constant acceleration equal to 9.8 m/s², which gives the illusion of normal gravity during the flight. (a) If it starts from rest, how long will it take to acquire a speed one-tenth that of light, which travels at 3.0×10^8 m/s? (b) How far will it travel in so doing?

•32 🚀 A world's land speed record was set by Colonel John P. Stapp when in March 1954 he rode a rocket-propelled sled that moved along a track at 1020 km/h. He and the sled were brought to a stop in 1.4 s. (See Fig. 2-7.) In terms of g, what acceleration did he experience while stopping?

•33 SSM ILW A car traveling 56.0 km/h is 24.0 m from a barrier when the driver slams on the brakes. The car hits the barrier 2.00 s later. (a) What is the magnitude of the car's constant acceleration before impact? (b) How fast is the car traveling at impact?

••34 GO In Fig. 2-24, a red car and a green car, identical except for the color, move toward each other in adjacent lanes and parallel to an x axis. At time $t = 0$, the red car is at $x_r = 0$ and the green car is at $x_g = 220$ m. If the red car has a constant velocity of 20 km/h, the cars pass each other at $x = 44.5$ m, and if it has a constant velocity of 40 km/h, they pass each other at $x = 76.6$ m. What are (a) the initial velocity and (b) the constant acceleration of the green car?

Red car x_r Green car x_g x

Fig. 2-24   Problems 34 and 35.

••35 Figure 2-24 shows a red car and a green car that move toward each other. Figure 2-25 is a graph of their motion, showing the positions $x_{g0} = 270$ m and $x_{r0} = -35.0$ m at time $t = 0$. The green car has a constant speed of 20.0 m/s and the red car begins from rest. What is the acceleration magnitude of the red car?

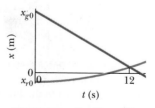

Fig. 2-25   Problem 35.

••**36** A car moves along an x axis through a distance of 900 m, starting at rest (at $x = 0$) and ending at rest (at $x = 900$ m). Through the first $\frac{1}{4}$ of that distance, its acceleration is $+2.25$ m/s². Through the rest of that distance, its acceleration is -0.750 m/s². What are (a) its travel time through the 900 m and (b) its maximum speed? (c) Graph position x, velocity v, and acceleration a versus time t for the trip.

••**37** Figure 2-26 depicts the motion of a particle moving along an x axis with a constant acceleration. The figure's vertical scaling is set by $x_s = 6.0$ m.What are the (a) magnitude and (b) direction of the particle's acceleration?

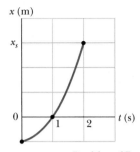

Fig. 2-26 Problem 37.

••**38** (a) If the maximum acceleration that is tolerable for passengers in a subway train is 1.34 m/s² and subway stations are located 806 m apart, what is the maximum speed a subway train can attain between stations? (b) What is the travel time between stations? (c) If a subway train stops for 20 s at each station, what is the maximum average speed of the train, from one start-up to the next? (d) Graph x, v, and a versus t for the interval from one start-up to the next.

••**39** Cars A and B move in the same direction in adjacent lanes. The position x of car A is given in Fig. 2-27, from time $t = 0$ to $t = 7.0$ s. The figure's vertical scaling is set by $x_s = 32.0$ m. At $t = 0$, car B is at $x = 0$, with a velocity of 12 m/s and a negative constant acceleration a_B. (a) What must a_B be such that the cars are (momentarily) side by side (momentarily at the same value of x) at $t = 4.0$ s? (b) For that value of a_B, how many times are the cars side by side? (c) Sketch the position x of car B versus time t on Fig. 2-27. How many times will the cars be side by side if the magnitude of acceleration a_B is (d) more than and (e) less than the answer to part (a)?

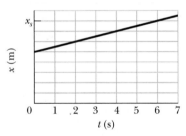

Fig. 2-27 Problem 39.

••**40** You are driving toward a traffic signal when it turns yellow. Your speed is the legal speed limit of $v_0 = 55$ km/h; your best deceleration rate has the magnitude $a = 5.18$ m/s². Your best reaction time to begin braking is $T = 0.75$ s. To avoid having the front of your car enter the intersection after the light turns red, should you brake to a stop or continue to move at 55 km/h if the distance to the intersection and the duration of the yellow light are (a) 40 m and 2.8 s, and (b) 32 m and 1.8 s? Give an answer of brake, continue, either (if either strategy works), or neither (if neither strategy works and the yellow duration is inappropriate).

•••**41** As two trains move along a track, their conductors suddenly notice that they are headed toward each other. Figure 2-28 gives their velocities v as functions of time t as the conductors slow the trains. The figure's vertical scaling is set by $v_s = 40.0$ m/s. The slowing

processes begin when the trains are 200 m apart. What is their separation when both trains have stopped?

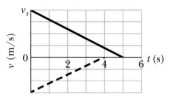

Fig. 2-28 Problem 41.

•••**42** You are arguing over a cell phone while trailing an unmarked police car by 25 m; both your car and the police car are traveling at 110 km/h. Your argument diverts your attention from the police car for 2.0 s (long enough for you to look at the phone and yell, "I won't do that!"). At the beginning of that 2.0 s, the police officer begins braking suddenly at 5.0 m/s². (a) What is the separation between the two cars when your attention finally returns? Suppose that you take another 0.40 s to realize your danger and begin braking. (b) If you too brake at 5.0 m/s², what is your speed when you hit the police car?

•••**43** GO When a high-speed passenger train traveling at 161 km/h rounds a bend, the engineer is shocked to see that a locomotive has improperly entered onto the track from a siding and is a distance $D = 676$ m ahead (Fig. 2-29). The locomotive is moving at 29.0 km/h. The engineer of the high-speed train immediately applies the brakes. (a) What must be the magnitude of the resulting constant deceleration if a collision is to be just avoided? (b) Assume that the engineer is at $x = 0$ when, at $t = 0$, he first spots the locomotive. Sketch $x(t)$ curves for the locomotive and high-speed train for the cases in which a collision is just avoided and is not quite avoided.

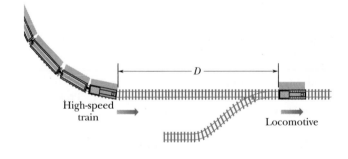

Fig. 2-29 Problem 43.

sec. 2-9 Free-Fall Acceleration

•**44** When startled, an armadillo will leap upward. Suppose it rises 0.544 m in the first 0.200 s. (a) What is its initial speed as it leaves the ground? (b) What is its speed at the height of 0.544 m? (c) How much higher does it go?

•**45** SSM WWW (a) With what speed must a ball be thrown vertically from ground level to rise to a maximum height of 50 m? (b) How long will it be in the air? (c) Sketch graphs of y, v, and a versus t for the ball. On the first two graphs, indicate the time at which 50 m is reached.

•**46** Raindrops fall 1700 m from a cloud to the ground. (a) If they were not slowed by air resistance, how fast would the drops be moving when they struck the ground? (b) Would it be safe to walk outside during a rainstorm?

•**47** SSM At a construction site a pipe wrench struck the ground with a speed of 24 m/s. (a) From what height was it inadvertently dropped? (b) How long was it falling? (c) Sketch graphs of y, v, and a versus t for the wrench.

•48 A hoodlum throws a stone vertically downward with an initial speed of 12.0 m/s from the roof of a building, 30.0 m above the ground. (a) How long does it take the stone to reach the ground? (b) What is the speed of the stone at impact?

•49 SSM A hot-air balloon is ascending at the rate of 12 m/s and is 80 m above the ground when a package is dropped over the side. (a) How long does the package take to reach the ground? (b) With what speed does it hit the ground?

••50 At time $t = 0$, apple 1 is dropped from a bridge onto a roadway beneath the bridge; somewhat later, apple 2 is thrown down from the same height. Figure 2-30 gives the vertical positions y of the apples versus t during the falling, until both apples have hit the roadway. The scaling is set by $t_s = 2.0$ s. With approximately what speed is apple 2 thrown down?

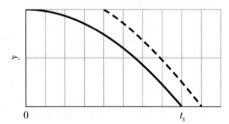

Fig. 2-30 Problem 50.

••51 As a runaway scientific balloon ascends at 19.6 m/s, one of its instrument packages breaks free of a harness and free-falls. Figure 2-31 gives the vertical velocity of the package versus time, from before it breaks free to when it reaches the ground. (a) What maximum height above the break-free point does it rise? (b) How high is the break-free point above the ground?

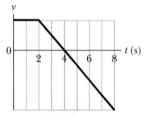

Fig. 2-31 Problem 51.

••52 GO A bolt is dropped from a bridge under construction, falling 90 m to the valley below the bridge. (a) In how much time does it pass through the last 20% of its fall? What is its speed (b) when it begins that last 20% of its fall and (c) when it reaches the valley beneath the bridge?

••53 SSM ILW A key falls from a bridge that is 45 m above the water. It falls directly into a model boat, moving with constant velocity, that is 12 m from the point of impact when the key is released. What is the speed of the boat?

••54 A stone is dropped into a river from a bridge 43.9 m above the water. Another stone is thrown vertically down 1.00 s after the first is dropped. The stones strike the water at the same time. (a) What is the initial speed of the second stone? (b) Plot velocity versus time on a graph for each stone, taking zero time as the instant the first stone is released.

••55 SSM A ball of moist clay falls 15.0 m to the ground. It is in contact with the ground for 20.0 ms before stopping. (a) What is the magnitude of the average acceleration of the ball during the time it is in contact with the ground? (Treat the ball as a particle.) (b) Is the average acceleration up or down?

••56 GO Figure 2-32 shows the speed v versus height y of a ball tossed directly upward, along a y axis. Distance d is 0.40 m. The speed at height y_A is v_A. The speed at height y_B is $\frac{1}{3}v_A$. What is speed v_A?

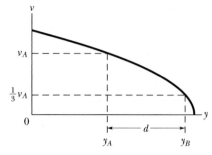

Fig. 2-32 Problem 56.

••57 To test the quality of a tennis ball, you drop it onto the floor from a height of 4.00 m. It rebounds to a height of 2.00 m. If the ball is in contact with the floor for 12.0 ms, (a) what is the magnitude of its average acceleration during that contact and (b) is the average acceleration up or down?

••58 An object falls a distance h from rest. If it travels $0.50h$ in the last 1.00 s, find (a) the time and (b) the height of its fall. (c) Explain the physically unacceptable solution of the quadratic equation in t that you obtain.

••59 Water drips from the nozzle of a shower onto the floor 200 cm below. The drops fall at regular (equal) intervals of time, the first drop striking the floor at the instant the fourth drop begins to fall. When the first drop strikes the floor, how far below the nozzle are the (a) second and (b) third drops?

••60 A rock is thrown vertically upward from ground level at time $t = 0$. At $t = 1.5$ s it passes the top of a tall tower, and 1.0 s later it reaches its maximum height. What is the height of the tower?

•••61 GO A steel ball is dropped from a building's roof and passes a window, taking 0.125 s to fall from the top to the bottom of the window, a distance of 1.20 m. It then falls to a sidewalk and bounces back past the window, moving from bottom to top in 0.125 s. Assume that the upward flight is an exact reverse of the fall. The time the ball spends below the bottom of the window is 2.00 s. How tall is the building?

•••62 ~~~ A basketball player grabbing a rebound jumps 76.0 cm vertically. How much total time (ascent and descent) does the player spend (a) in the top 15.0 cm of this jump and (b) in the bottom 15.0 cm? Do your results explain why such players seem to hang in the air at the top of a jump?

•••63 GO A drowsy cat spots a flowerpot that sails first up and then down past an open window. The pot is in view for a total of 0.50 s, and the top-to-bottom height of the window is 2.00 m. How high above the window top does the flowerpot go?

•••64 A ball is shot vertically upward from the surface of another planet. A plot of y versus t for the ball is shown in Fig. 2-33, where y is the

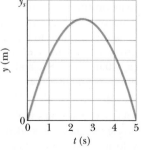

Fig. 2-33 Problem 64.

height of the ball above its starting point and $t = 0$ at the instant the ball is shot. The figure's vertical scaling is set by $y_s = 30.0$ m. What are the magnitudes of (a) the free-fall acceleration on the planet and (b) the initial velocity of the ball?

sec. 2-10   Graphical Integration in Motion Analysis

•65   Figure 2-13a gives the acceleration of a volunteer's head and torso during a rear-end collision. At maximum head acceleration, what is the speed of (a) the head and (b) the torso?

••66   In a forward punch in karate, the fist begins at rest at the waist and is brought rapidly forward until the arm is fully extended. The speed $v(t)$ of the fist is given in Fig. 2-34 for someone skilled in karate. The vertical scaling is set by $v_s = 8.0$ m/s. How far has the fist moved at (a) time $t = 50$ ms and (b) when the speed of the fist is maximum?

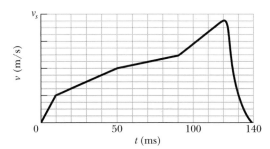

Fig. 2-34   Problem 66.

••67   When a soccer ball is kicked toward a player and the player deflects the ball by "heading" it, the acceleration of the head during the collision can be significant. Figure 2-35 gives the measured acceleration $a(t)$ of a soccer player's head for a bare head and a helmeted head, starting from rest. The scaling on the vertical axis is set by $a_s = 200$ m/s². At time $t = 7.0$ ms, what is the difference in the speed acquired by the bare head and the speed acquired by the helmeted head?

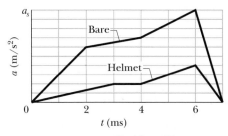

Fig. 2-35   Problem 67.

••68   A salamander of the genus *Hydromantes* captures prey by launching its tongue as a projectile: The skeletal part of the tongue is shot forward, unfolding the rest of the tongue, until the outer portion lands on the prey, sticking to it. Figure 2-36 shows the acceleration magnitude a versus time t for the acceleration phase of the

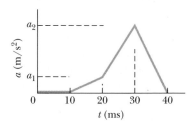

Fig. 2-36   Problem 68.

launch in a typical situation. The indicated accelerations are $a_2 = 400$ m/s² and $a_1 = 100$ m/s². What is the outward speed of the tongue at the end of the acceleration phase?

••69   ILW   How far does the runner whose velocity–time graph is shown in Fig. 2-37 travel in 16 s? The figure's vertical scaling is set by $v_s = 8.0$ m/s.

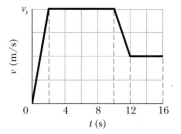

Fig. 2-37   Problem 69.

•••70   Two particles move along an x axis. The position of particle 1 is given by $x = 6.00t^2 + 3.00t + 2.00$ (in meters and seconds); the acceleration of particle 2 is given by $a = -8.00t$ (in meters per second squared and seconds) and, at $t = 0$, its velocity is 20 m/s. When the velocities of the particles match, what is their velocity?

Additional Problems

71   In an arcade video game, a spot is programmed to move across the screen according to $x = 9.00t - 0.750t^3$, where x is distance in centimeters measured from the left edge of the screen and t is time in seconds. When the spot reaches a screen edge, at either $x = 0$ or $x = 15.0$ cm, t is reset to 0 and the spot starts moving again according to $x(t)$. (a) At what time after starting is the spot instantaneously at rest? (b) At what value of x does this occur? (c) What is the spot's acceleration (including sign) when this occurs? (d) Is it moving right or left just prior to coming to rest? (e) Just after? (f) At what time $t > 0$ does it first reach an edge of the screen?

72   A rock is shot vertically upward from the edge of the top of a tall building. The rock reaches its maximum height above the top of the building 1.60 s after being shot. Then, after barely missing the edge of the building as it falls downward, the rock strikes the ground 6.00 s after it is launched. In SI units: (a) with what upward velocity is the rock shot, (b) what maximum height above the top of the building is reached by the rock, and (c) how tall is the building?

73   GO   At the instant the traffic light turns green, an automobile starts with a constant acceleration a of 2.2 m/s². At the same instant a truck, traveling with a constant speed of 9.5 m/s, overtakes and passes the automobile. (a) How far beyond the traffic signal will the automobile overtake the truck? (b) How fast will the automobile be traveling at that instant?

74   A pilot flies horizontally at 1300 km/h, at height $h = 35$ m above initially level ground. However, at time $t = 0$, the pilot begins to fly over ground sloping upward at angle $\theta = 4.3°$ (Fig. 2-38). If the pilot does not change the airplane's heading, at what time t does the plane strike the ground?

Fig. 2-38   Problem 74.

75   To stop a car, first you require a certain reaction time to begin braking; then the car slows at a constant rate. Suppose that the total distance moved by your car during these two phases is 56.7 m when its initial speed is 80.5 km/h, and 24.4 m when its initial speed

is 48.3 km/h. What are (a) your reaction time and (b) the magnitude of the acceleration?

76 Figure 2-39 shows part of a street where traffic flow is to be controlled to allow a *platoon* of cars to move smoothly along the street. Suppose that the platoon leaders have just reached intersection 2, where the green appeared when they were distance d from the intersection. They continue to travel at a certain speed v_p (the speed limit) to reach intersection 3, where the green appears when they are distance d from it. The intersections are separated by distances D_{23} and D_{12}. (a) What should be the time delay of the onset of green at intersection 3 relative to that at intersection 2 to keep the platoon moving smoothly?

Suppose, instead, that the platoon had been stopped by a red light at intersection 1. When the green comes on there, the leaders require a certain time t_r to respond to the change and an additional time to accelerate at some rate a to the cruising speed v_p. (b) If the green at intersection 2 is to appear when the leaders are distance d from that intersection, how long after the light at intersection 1 turns green should the light at intersection 2 turn green?

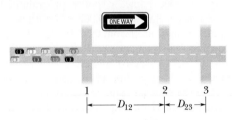

Fig. 2-39 Problem 76.

77 SSM A hot rod can accelerate from 0 to 60 km/h in 5.4 s. (a) What is its average acceleration, in m/s², during this time? (b) How far will it travel during the 5.4 s, assuming its acceleration is constant? (c) From rest, how much time would it require to go a distance of 0.25 km if its acceleration could be maintained at the value in (a)?

78 A red train traveling at 72 km/h and a green train traveling at 144 km/h are headed toward each other along a straight, level track. When they are 950 m apart, each engineer sees the other's train and applies the brakes. The brakes slow each train at the rate of 1.0 m/s². Is there a collision? If so, answer yes and give the speed of the red train and the speed of the green train at impact, respectively. If not, answer no and give the separation between the trains when they stop.

79 At time $t = 0$, a rock climber accidentally allows a piton to fall freely from a high point on the rock wall to the valley below him. Then, after a short delay, his climbing partner, who is 10 m higher on the wall, throws a piton downward. The positions y of the pitons versus t during the falling are given in Fig. 2-40. With what speed is the second piton thrown?

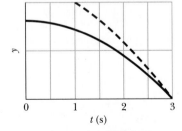

Fig. 2-40 Problem 79.

80 A train started from rest and moved with constant acceleration. At one time it was traveling 30 m/s, and 160 m farther on it was traveling 50 m/s. Calculate (a) the acceleration, (b) the time required to travel the 160 m mentioned, (c) the time required to attain the speed of 30 m/s, and (d) the distance moved from rest to the time the train had a speed of 30 m/s. (e) Graph x versus t and v versus t for the train, from rest.

81 SSM A particle's acceleration along an x axis is $a = 5.0t$, with t in seconds and a in meters per second squared. At $t = 2.0$ s, its velocity is $+17$ m/s. What is its velocity at $t = 4.0$ s?

82 Figure 2-41 gives the acceleration a versus time t for a particle moving along an x axis. The a-axis scale is set by $a_s = 12.0$ m/s². At $t = -2.0$ s, the particle's velocity is 7.0 m/s. What is its velocity at $t = 6.0$ s?

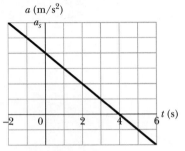

Fig. 2-41 Problem 82.

83 Figure 2-42 shows a simple device for measuring your reaction time. It consists of a cardboard strip marked with a scale and two large dots. A friend holds the strip *vertically,* with thumb and forefinger at the dot on the right in Fig. 2-42. You then position your thumb and forefinger at the other dot (on the left in Fig. 2-42), being careful not to touch the strip. Your friend releases the strip, and you try to pinch it as soon as possible after you see it begin to fall. The mark at the place where you pinch the strip gives your reaction time. (a) How far from the lower dot should you place the 50.0 ms mark? How much higher should you place the marks for (b) 100, (c) 150, (d) 200, and (e) 250 ms? (For example, should the 100 ms marker be 2 times as far from the dot as the 50 ms marker? If so, give an answer of 2 times. Can you find any pattern in the answers?)

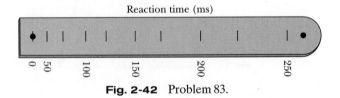

Fig. 2-42 Problem 83.

84 A rocket-driven sled running on a straight, level track is used to investigate the effects of large accelerations on humans. One such sled can attain a speed of 1600 km/h in 1.8 s, starting from rest. Find (a) the acceleration (assumed constant) in terms of g and (b) the distance traveled.

85 A mining cart is pulled up a hill at 20 km/h and then pulled back down the hill at 35 km/h through its original level. (The time required for the cart's reversal at the top of its climb is negligible.) What is the average speed of the cart for its round trip, from its original level back to its original level?

86 A motorcyclist who is moving along an x axis directed toward the east has an acceleration given by $a = (6.1 - 1.2t)$ m/s² for $0 \leq t \leq 6.0$ s. At $t = 0$, the velocity and position of the cyclist are 2.7 m/s and 7.3 m. (a) What is the maximum speed achieved by the cyclist? (b) What total distance does the cyclist travel between $t = 0$ and 6.0 s?

87 SSM When the legal speed limit for the New York Thruway was increased from 55 mi/h to 65 mi/h, how much time was saved by a motorist who drove the 700 km between the Buffalo entrance and the New York City exit at the legal speed limit?

88 A car moving with constant acceleration covered the distance between two points 60.0 m apart in 6.00 s. Its speed as it passed the second point was 15.0 m/s. (a) What was the speed at the first point? (b) What was the magnitude of the acceleration? (c) At what prior distance from the first point was the car at rest? (d) Graph x versus t and v versus t for the car, from rest ($t = 0$).

89 SSM [icon] A certain juggler usually tosses balls vertically to a height H. To what height must they be tossed if they are to spend twice as much time in the air?

90 A particle starts from the origin at $t = 0$ and moves along the positive x axis. A graph of the velocity of the particle as a function of the time is shown in Fig. 2-43; the v-axis scale is set by $v_s = 4.0$ m/s. (a) What is the coordinate of the particle at $t = 5.0$ s? (b) What is the velocity of the particle at $t = 5.0$ s? (c) What is the acceleration of the particle at $t = 5.0$ s? (d) What is the average velocity of the particle between $t = 1.0$ s and $t = 5.0$ s? (e) What is the average acceleration of the particle between $t = 1.0$ s and $t = 5.0$ s?

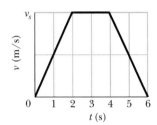

Fig. 2-43   Problem 90.

91 A rock is dropped from a 100-m-high cliff. How long does it take to fall (a) the first 50 m and (b) the second 50 m?

92 Two subway stops are separated by 1100 m. If a subway train accelerates at $+1.2$ m/s² from rest through the first half of the distance and decelerates at -1.2 m/s² through the second half, what are (a) its travel time and (b) its maximum speed? (c) Graph x, v, and a versus t for the trip.

93 A stone is thrown vertically upward. On its way up it passes point A with speed v, and point B, 3.00 m higher than A, with speed $\frac{1}{2}v$. Calculate (a) the speed v and (b) the maximum height reached by the stone above point B.

94 A rock is dropped (from rest) from the top of a 60-m-tall building. How far above the ground is the rock 1.2 s before it reaches the ground?

95 SSM An iceboat has a constant velocity toward the east when a sudden gust of wind causes the iceboat to have a constant acceleration toward the east for a period of 3.0 s. A plot of x versus t is shown in Fig. 2-44, where $t = 0$ is taken to be the instant the wind starts to blow and the positive x axis is toward the east. (a) What is the acceleration of the iceboat during the 3.0 s interval? (b) What is the velocity of the iceboat at the end of the 3.0 s interval? (c) If

the acceleration remains constant for an additional 3.0 s, how far does the iceboat travel during this second 3.0 s interval?

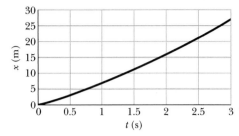

Fig. 2-44   Problem 95.

96 A lead ball is dropped in a lake from a diving board 5.20 m above the water. It hits the water with a certain velocity and then sinks to the bottom with this same constant velocity. It reaches the bottom 4.80 s after it is dropped. (a) How deep is the lake? What are the (b) magnitude and (c) direction (up or down) of the average velocity of the ball for the entire fall? Suppose that all the water is drained from the lake. The ball is now thrown from the diving board so that it again reaches the bottom in 4.80 s. What are the (d) magnitude and (e) direction of the initial velocity of the ball?

97 The single cable supporting an unoccupied construction elevator breaks when the elevator is at rest at the top of a 120-m-high building. (a) With what speed does the elevator strike the ground? (b) How long is it falling? (c) What is its speed when it passes the halfway point on the way down? (d) How long has it been falling when it passes the halfway point?

98 Two diamonds begin a free fall from rest from the same height, 1.0 s apart. How long after the first diamond begins to fall will the two diamonds be 10 m apart?

99 A ball is thrown vertically downward from the top of a 36.6-m-tall building. The ball passes the top of a window that is 12.2 m above the ground 2.00 s after being thrown. What is the speed of the ball as it passes the top of the window?

100 A parachutist bails out and freely falls 50 m. Then the parachute opens, and thereafter she decelerates at 2.0 m/s². She reaches the ground with a speed of 3.0 m/s. (a) How long is the parachutist in the air? (b) At what height does the fall begin?

101 A ball is thrown *down* vertically with an initial *speed* of v_0 from a height of h. (a) What is its speed just before it strikes the ground? (b) How long does the ball take to reach the ground? What would be the answers to (c) part a and (d) part b if the ball were thrown *upward* from the same height and with the same initial speed? Before solving any equations, decide whether the answers to (c) and (d) should be greater than, less than, or the same as in (a) and (b).

102 The sport with the fastest moving ball is jai alai, where measured speeds have reached 303 km/h. If a professional jai alai player faces a ball at that speed and involuntarily blinks, he blacks out the scene for 100 ms. How far does the ball move during the blackout?

3

VECTORS

3-1 WHAT IS PHYSICS?

Physics deals with a great many quantities that have both size and direction, and it needs a special mathematical language—the language of vectors—to describe those quantities. This language is also used in engineering, the other sciences, and even in common speech. If you have ever given directions such as "Go five blocks down this street and then hang a left," you have used the language of vectors. In fact, navigation of any sort is based on vectors, but physics and engineering also need vectors in special ways to explain phenomena involving rotation and magnetic forces, which we get to in later chapters. In this chapter, we focus on the basic language of vectors.

3-2 Vectors and Scalars

A particle moving along a straight line can move in only two directions. We can take its motion to be positive in one of these directions and negative in the other. For a particle moving in three dimensions, however, a plus sign or minus sign is no longer enough to indicate a direction. Instead, we must use a *vector*.

A **vector** has magnitude as well as direction, and vectors follow certain (vector) rules of combination, which we examine in this chapter. A **vector quantity** is a quantity that has both a magnitude and a direction and thus can be represented with a vector. Some physical quantities that are vector quantities are displacement, velocity, and acceleration. You will see many more throughout this book, so learning the rules of vector combination now will help you greatly in later chapters.

Not all physical quantities involve a direction. Temperature, pressure, energy, mass, and time, for example, do not "point" in the spatial sense. We call such quantities **scalars,** and we deal with them by the rules of ordinary algebra. A single value, with a sign (as in a temperature of $-40°F$), specifies a scalar.

The simplest vector quantity is displacement, or change of position. A vector that represents a displacement is called, reasonably, a **displacement vector.** (Similarly, we have velocity vectors and acceleration vectors.) If a particle changes its position by moving from A to B in Fig. 3-1a, we say that it undergoes a displacement from A to B, which we represent with an arrow pointing from A to B. The arrow specifies the vector graphically. To distinguish vector symbols from other kinds of arrows in this book, we use the outline of a triangle as the arrowhead.

In Fig. 3-1a, the arrows from A to B, from A' to B', and from A'' to B'' have the same magnitude and direction. Thus, they specify identical displacement vectors and represent the same *change of position* for the particle. A vector can be shifted without changing its value *if* its length and direction are not changed.

The displacement vector tells us nothing about the actual path that the particle takes. In Fig. 3-1b, for example, all three paths connecting points A and B correspond to the same displacement vector, that of Fig. 3-1a. Displacement vectors represent only the overall effect of the motion, not the motion itself.

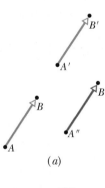

(a)

(b)

Fig. 3-1 (a) All three arrows have the same magnitude and direction and thus represent the same displacement. (b) All three paths connecting the two points correspond to the same displacement vector.

3-3 Adding Vectors Geometrically

Suppose that, as in the vector diagram of Fig. 3-2a, a particle moves from A to B and then later from B to C. We can represent its overall displacement (no matter what its actual path) with two successive displacement vectors, AB and BC. The *net* displacement of these two displacements is a single displacement from A to C. We call AC the **vector sum** (or **resultant**) of the vectors AB and BC. This sum is not the usual algebraic sum.

In Fig. 3-2b, we redraw the vectors of Fig. 3-2a and relabel them in the way that we shall use from now on, namely, with an arrow over an italic symbol, as in $\vec{a}$. If we want to indicate only the magnitude of the vector (a quantity that lacks a sign or direction), we shall use the italic symbol, as in a, b, and s. (You can use just a handwritten symbol.) A symbol with an overhead arrow always implies both properties of a vector, magnitude and direction.

We can represent the relation among the three vectors in Fig. 3-2b with the *vector equation*

$$\vec{s} = \vec{a} + \vec{b}, \tag{3-1}$$

which says that the vector $\vec{s}$ is the vector sum of vectors $\vec{a}$ and $\vec{b}$. The symbol + in Eq. 3-1 and the words "sum" and "add" have different meanings for vectors than they do in the usual algebra because they involve both magnitude *and* direction.

Figure 3-2 suggests a procedure for adding two-dimensional vectors $\vec{a}$ and $\vec{b}$ geometrically. (1) On paper, sketch vector $\vec{a}$ to some convenient scale and at the proper angle. (2) Sketch vector $\vec{b}$ to the same scale, with its tail at the head of vector $\vec{a}$, again at the proper angle. (3) The vector sum $\vec{s}$ is the vector that extends from the tail of $\vec{a}$ to the head of $\vec{b}$.

Vector addition, defined in this way, has two important properties. First, the order of addition does not matter. Adding $\vec{a}$ to $\vec{b}$ gives the same result as adding $\vec{b}$ to $\vec{a}$ (Fig. 3-3); that is,

$$\vec{a} + \vec{b} = \vec{b} + \vec{a} \qquad \text{(commutative law).} \tag{3-2}$$

Second, when there are more than two vectors, we can group them in any order as we add them. Thus, if we want to add vectors $\vec{a}$, $\vec{b}$, and $\vec{c}$, we can add $\vec{a}$ and $\vec{b}$ first and then add their vector sum to $\vec{c}$. We can also add $\vec{b}$ and $\vec{c}$ first and then add *that* sum to $\vec{a}$. We get the same result either way, as shown in Fig. 3-4. That is,

$$(\vec{a} + \vec{b}) + \vec{c} = \vec{a} + (\vec{b} + \vec{c}) \qquad \text{(associative law).} \tag{3-3}$$

The vector $-\vec{b}$ is a vector with the same magnitude as $\vec{b}$ but the opposite direction (see Fig. 3-5). Adding the two vectors in Fig. 3-5 would yield

$$\vec{b} + (-\vec{b}) = 0.$$

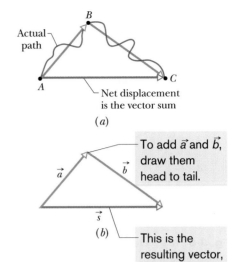

(a)

(b)

Fig. 3-2 (*a*) AC is the vector sum of the vectors AB and BC. (*b*) The same vectors relabeled.

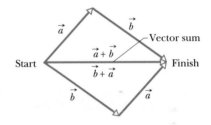

You get the same vector result for either order of adding vectors.

Fig. 3-3 The two vectors $\vec{a}$ and $\vec{b}$ can be added in either order; see Eq. 3-2.

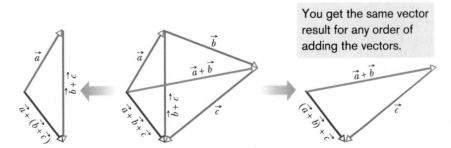

You get the same vector result for any order of adding the vectors.

Fig. 3-4 The three vectors $\vec{a}$, $\vec{b}$, and $\vec{c}$ can be grouped in any way as they are added; see Eq. 3-3.

Fig. 3-5 The vectors $\vec{b}$ and $-\vec{b}$ have the same magnitude and opposite directions.

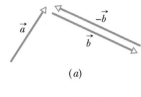

(a)

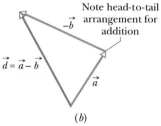

Note head-to-tail arrangement for addition

$\vec{d} = \vec{a} - \vec{b}$

(b)

Fig. 3-6 (a) Vectors $\vec{a}$, $\vec{b}$, and $-\vec{b}$. (b) To subtract vector $\vec{b}$ from vector $\vec{a}$, add vector $-\vec{b}$ to vector $\vec{a}$.

Thus, adding $-\vec{b}$ has the effect of subtracting $\vec{b}$. We use this property to define the difference between two vectors: let $\vec{d} = \vec{a} - \vec{b}$. Then

$$\vec{d} = \vec{a} - \vec{b} = \vec{a} + (-\vec{b}) \quad \text{(vector subtraction);} \quad (3\text{-}4)$$

that is, we find the difference vector $\vec{d}$ by adding the vector $-\vec{b}$ to the vector $\vec{a}$. Figure 3-6 shows how this is done geometrically.

As in the usual algebra, we can move a term that includes a vector symbol from one side of a vector equation to the other, but we must change its sign. For example, if we are given Eq. 3-4 and need to solve for $\vec{a}$, we can rearrange the equation as

$$\vec{d} + \vec{b} = \vec{a} \quad \text{or} \quad \vec{a} = \vec{d} + \vec{b}.$$

Remember that, although we have used displacement vectors here, the rules for addition and subtraction hold for vectors of all kinds, whether they represent velocities, accelerations, or any other vector quantity. However, we can add only vectors of the same kind. For example, we can add two displacements, or two velocities, but adding a displacement and a velocity makes no sense. In the arithmetic of scalars, that would be like trying to add 21 s and 12 m.

> ☑ **CHECKPOINT 1**
>
> The magnitudes of displacements $\vec{a}$ and $\vec{b}$ are 3 m and 4 m, respectively, and $\vec{c} = \vec{a} + \vec{b}$. Considering various orientations of $\vec{a}$ and $\vec{b}$, what is (a) the maximum possible magnitude for $\vec{c}$ and (b) the minimum possible magnitude?

Sample Problem

Adding vectors in a drawing, orienteering

In an orienteering class, you have the goal of moving as far (straight-line distance) from base camp as possible by making three straight-line moves. You may use the following displacements in any order: (a) $\vec{a}$, 2.0 km due east (directly toward the east); (b) $\vec{b}$, 2.0 km 30° north of east (at an angle of 30° toward the north from due east); (c) $\vec{c}$, 1.0 km due west. Alternatively, you may substitute either $-\vec{b}$ for $\vec{b}$ or $-\vec{c}$ for $\vec{c}$. What is the greatest distance you can be from base camp at the end of the third displacement?

Reasoning: Using a convenient scale, we draw vectors $\vec{a}$, $\vec{b}$, $\vec{c}$, $-\vec{b}$, and $-\vec{c}$ as in Fig. 3-7a. We then mentally slide the vectors over the page, connecting three of them at a time in head-to-tail arrangements to find their vector sum $\vec{d}$. The tail of the first vector represents base camp. The head of the third vector represents the point at which you stop. The vector sum $\vec{d}$ extends from the tail of the first vector to the head of the third vector. Its magnitude d is your distance from base camp.

We find that distance d is greatest for a head-to-tail arrangement of vectors $\vec{a}$, $\vec{b}$, and $-\vec{c}$. They can be in any order, because their vector sum is the same for any order.

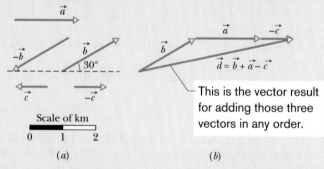

Scale of km

This is the vector result for adding those three vectors in any order.

(a) (b)

Fig. 3-7 (a) Displacement vectors; three are to be used. (b) Your distance from base camp is greatest if you undergo displacements $\vec{a}$, $\vec{b}$, and $-\vec{c}$, in any order.

The order shown in Fig. 3-7b is for the vector sum

$$\vec{d} = \vec{b} + \vec{a} + (-\vec{c}).$$

Using the scale given in Fig. 3-7a, we measure the length d of this vector sum, finding

$$d = 4.8 \text{ m.} \quad \text{(Answer)}$$

 Additional examples, video, and practice available at *WileyPLUS*

3-4 Components of Vectors

Adding vectors geometrically can be tedious. A neater and easier technique involves algebra but requires that the vectors be placed on a rectangular coordinate system. The x and y axes are usually drawn in the plane of the page, as shown in Fig. 3-8a. The z axis comes directly out of the page at the origin; we ignore it for now and deal only with two-dimensional vectors.

A **component** of a vector is the projection of the vector on an axis. In Fig. 3-8a, for example, a_x is the component of vector $\vec{a}$ on (or along) the x axis and a_y is the component along the y axis. To find the projection of a vector along an axis, we draw perpendicular lines from the two ends of the vector to the axis, as shown. The projection of a vector on an x axis is its x *component*, and similarly the projection on the y axis is the y *component*. The process of finding the components of a vector is called **resolving the vector.**

A component of a vector has the same direction (along an axis) as the vector. In Fig. 3-8, a_x and a_y are both positive because $\vec{a}$ extends in the positive direction of both axes. (Note the small arrowheads on the components, to indicate their direction.) If we were to reverse vector $\vec{a}$, then both components would be negative and their arrowheads would point toward negative x and y. Resolving vector $\vec{b}$ in Fig. 3-9 yields a positive component b_x and a negative component b_y.

In general, a vector has three components, although for the case of Fig. 3-8a the component along the z axis is zero. As Figs. 3-8a and b show, if you shift a vector without changing its direction, its components do not change.

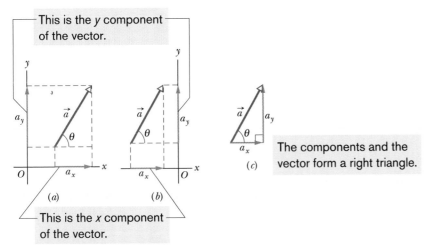

This is the y component of the vector.

The components and the vector form a right triangle.

(a)   (b)   (c)

This is the x component of the vector.

Fig. 3-8   (a) The components a_x and a_y of vector $\vec{a}$. (b) The components are unchanged if the vector is shifted, as long as the magnitude and orientation are maintained. (c) The components form the legs of a right triangle whose hypotenuse is the magnitude of the vector.

We can find the components of $\vec{a}$ in Fig. 3-8a geometrically from the right triangle there:

$$a_x = a\cos\theta \quad \text{and} \quad a_y = a\sin\theta, \tag{3-5}$$

where θ is the angle that the vector $\vec{a}$ makes with the positive direction of the x axis, and a is the magnitude of $\vec{a}$. Figure 3-8c shows that $\vec{a}$ and its x and y components form a right triangle. It also shows how we can reconstruct a vector from its components: we arrange those components *head to tail.* Then we complete a right triangle with the vector forming the hypotenuse, from the tail of one component to the head of the other component.

Once a vector has been resolved into its components along a set of axes, the components themselves can be used in place of the vector. For example, $\vec{a}$ in

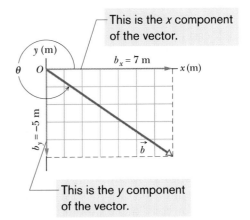

This is the x component of the vector.

$b_x = 7$ m

This is the y component of the vector.

Fig. 3-9   The component of $\vec{b}$ on the x axis is positive, and that on the y axis is negative.

Fig. 3-8a is given (completely determined) by a and θ. It can also be given by its components a_x and a_y. Both pairs of values contain the same information. If we know a vector in *component notation* (a_x and a_y) and want it in *magnitude-angle notation* (a and θ), we can use the equations

$$a = \sqrt{a_x^2 + a_y^2} \quad \text{and} \quad \tan \theta = \frac{a_y}{a_x} \tag{3-6}$$

to transform it.

In the more general three-dimensional case, we need a magnitude and two angles (say, a, θ, and ϕ) or three components (a_x, a_y, and a_z) to specify a vector.

✔ CHECKPOINT 2

In the figure, which of the indicated methods for combining the x and y components of vector $\vec{a}$ are proper to determine that vector?

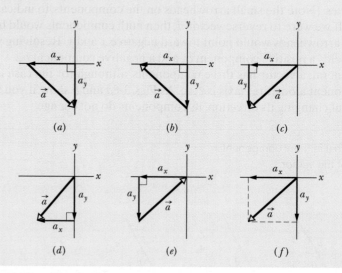

Sample Problem

Finding components, airplane flight

A small airplane leaves an airport on an overcast day and is later sighted 215 km away, in a direction making an angle of 22° east of due north. How far east and north is the airplane from the airport when sighted?

KEY IDEA

We are given the magnitude (215 km) and the angle (22° east of due north) of a vector and need to find the components of the vector.

Calculations: We draw an xy coordinate system with the positive direction of x due east and that of y due north (Fig. 3-10). For convenience, the origin is placed at the airport. The airplane's displacement $\vec{d}$ points from the origin to where the airplane is sighted.

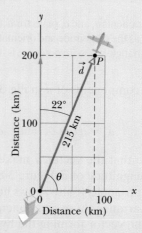

Fig. 3-10 A plane takes off from an airport at the origin and is later sighted at P.

To find the components of $\vec{d}$, we use Eq. 3-5 with $\theta = 68° (= 90° - 22°)$:

$$d_x = d \cos \theta = (215 \text{ km})(\cos 68°)$$
$$= 81 \text{ km} \qquad \text{(Answer)}$$

$$d_y = d \sin \theta = (215 \text{ km})(\sin 68°)$$
$$= 199 \text{ km} \approx 2.0 \times 10^2 \text{ km.} \qquad \text{(Answer)}$$

Thus, the airplane is 81 km east and 2.0×10^2 km north of the airport.

Problem-Solving Tactics

Angles, trig functions, and inverse trig functions

Tactic 1: Angles—Degrees and Radians Angles that are measured relative to the positive direction of the x axis are positive if they are measured in the counterclockwise direction and negative if measured clockwise. For example, 210° and −150° are the same angle.

Angles may be measured in degrees or radians (rad). To relate the two measures, recall that a full circle is 360° and 2π rad. To convert, say, 40° to radians, write

$$40° \frac{2\pi \text{ rad}}{360°} = 0.70 \text{ rad.}$$

Tactic 2: Trig Functions You need to know the definitions of the common trigonometric functions—sine, cosine, and tangent—because they are part of the language of science and engineering. They are given in Fig. 3-11 in a form that does not depend on how the triangle is labeled.

You should also be able to sketch how the trig functions vary with angle, as in Fig. 3-12, in order to be able to judge whether a calculator result is reasonable. Even knowing the signs of the functions in the various quadrants can be of help.

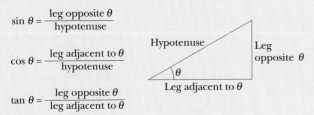

Fig. 3-11 A triangle used to define the trigonometric functions. See also Appendix E.

Tactic 3: Inverse Trig Functions When the inverse trig functions $\sin^{-1}$, $\cos^{-1}$, and $\tan^{-1}$ are taken on a calculator, you must consider the reasonableness of the answer you get, because there is usually another possible answer that the calculator does not give. The range of operation for a calculator in taking each inverse trig function is indicated in Fig. 3-12. As an example, $\sin^{-1} 0.5$ has associated angles of 30° (which is displayed by the calculator, since 30° falls within its range of operation) and 150°. To see both values, draw a horizontal line through 0.5 in Fig. 3-12a and note where it cuts the sine curve. How do you distinguish a correct answer? It is the one that seems more reasonable for the given situation.

Tactic 4: Measuring Vector Angles The equations for $\cos \theta$ and $\sin \theta$ in Eq. 3-5 and for $\tan \theta$ in Eq. 3-6 are valid only if

the angle is measured from the positive direction of the x axis. If it is measured relative to some other direction, then the trig functions in Eq. 3-5 may have to be interchanged and the ratio in Eq. 3-6 may have to be inverted. A safer method is to convert the angle to one measured from the positive direction of the x axis.

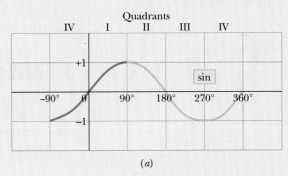

(a)

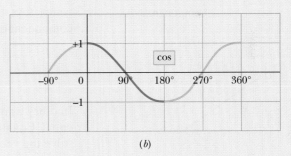

(b)

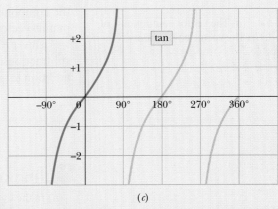

(c)

Fig. 3-12 Three useful curves to remember. A calculator's range of operation for taking *inverse* trig functions is indicated by the darker portions of the colored curves.

 Additional examples, video, and practice available at *WileyPLUS*

The unit vectors point along axes.

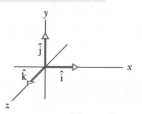

Fig. 3-13 Unit vectors $\hat{i}, \hat{j}$, and $\hat{k}$ define the directions of a right-handed coordinate system.

This is the y vector component.

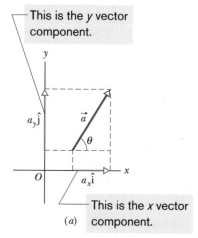

This is the x vector component.

(a)

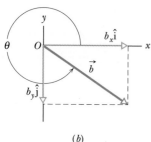

(b)

Fig. 3-14 (a) The vector components of vector $\vec{a}$. (b) The vector components of vector $\vec{b}$.

3-5 Unit Vectors

A **unit vector** is a vector that has a magnitude of exactly 1 and points in a particular direction. It lacks both dimension and unit. Its sole purpose is to point—that is, to specify a direction. The unit vectors in the positive directions of the x, y, and z axes are labeled $\hat{i}, \hat{j}$, and $\hat{k}$, where the hat ^ is used instead of an overhead arrow as for other vectors (Fig. 3-13). The arrangement of axes in Fig. 3-13 is said to be a **right-handed coordinate system.** The system remains right-handed if it is rotated rigidly. We use such coordinate systems exclusively in this book.

Unit vectors are very useful for expressing other vectors; for example, we can express $\vec{a}$ and $\vec{b}$ of Figs. 3-8 and 3-9 as

$$\vec{a} = a_x\hat{i} + a_y\hat{j} \qquad (3\text{-}7)$$

and

$$\vec{b} = b_x\hat{i} + b_y\hat{j}. \qquad (3\text{-}8)$$

These two equations are illustrated in Fig. 3-14. The quantities $a_x\hat{i}$ and $a_y\hat{j}$ are vectors, called the **vector components** of $\vec{a}$. The quantities a_x and a_y are scalars, called the **scalar components** of $\vec{a}$ (or, as before, simply its **components**).

3-6 Adding Vectors by Components

Using a sketch, we can add vectors geometrically. On a vector-capable calculator, we can add them directly on the screen. A third way to add vectors is to combine their components axis by axis, which is the way we examine here.

To start, consider the statement

$$\vec{r} = \vec{a} + \vec{b}, \qquad (3\text{-}9)$$

which says that the vector $\vec{r}$ is the same as the vector $(\vec{a} + \vec{b})$. Thus, each component of $\vec{r}$ must be the same as the corresponding component of $(\vec{a} + \vec{b})$:

$$r_x = a_x + b_x \qquad (3\text{-}10)$$
$$r_y = a_y + b_y \qquad (3\text{-}11)$$
$$r_z = a_z + b_z. \qquad (3\text{-}12)$$

In other words, two vectors must be equal if their corresponding components are equal. Equations 3-9 to 3-12 tell us that to add vectors $\vec{a}$ and $\vec{b}$, we must (1) resolve the vectors into their scalar components; (2) combine these scalar components, axis by axis, to get the components of the sum $\vec{r}$; and (3) combine the components of $\vec{r}$ to get $\vec{r}$ itself. We have a choice in step 3. We can express $\vec{r}$ in unit-vector notation or in magnitude-angle notation.

This procedure for adding vectors by components also applies to vector subtractions. Recall that a subtraction such as $\vec{d} = \vec{a} - \vec{b}$ can be rewritten as an addition $\vec{d} = \vec{a} + (-\vec{b})$. To subtract, we add $\vec{a}$ and $-\vec{b}$ by components, to get

$$d_x = a_x - b_x, \quad d_y = a_y - b_y, \quad \text{and} \quad d_z = a_z - b_z,$$

where

$$\vec{d} = d_x\hat{i} + d_y\hat{j} + d_z\hat{k}. \qquad (3\text{-}13)$$

✔CHECKPOINT 3

(a) In the figure here, what are the signs of the x components of $\vec{d}_1$ and $\vec{d}_2$? (b) What are the signs of the y components of $\vec{d}_1$ and $\vec{d}_2$? (c) What are the signs of the x and y components of $\vec{d}_1 + \vec{d}_2$?

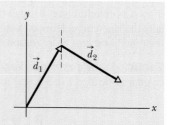

Adding vectors, unit-vector components

Figure 3-15a shows the following three vectors:

$$\vec{a} = (4.2 \text{ m})\hat{i} - (1.5 \text{ m})\hat{j},$$
$$\vec{b} = (-1.6 \text{ m})\hat{i} + (2.9 \text{ m})\hat{j},$$

and $\qquad \vec{c} = (-3.7 \text{ m})\hat{j}.$

What is their vector sum $\vec{r}$ which is also shown?

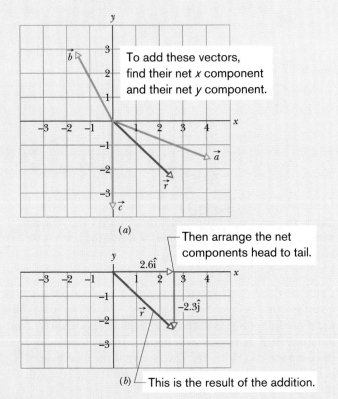

To add these vectors, find their net x component and their net y component.

(a)

Then arrange the net components head to tail.

(b) This is the result of the addition.

Fig. 3-15 Vector $\vec{r}$ is the vector sum of the other three vectors.

We can add the three vectors by components, axis by axis, and then combine the components to write the vector sum $\vec{r}$.

Calculations: For the x axis, we add the x components of $\vec{a}$, $\vec{b}$, and $\vec{c}$, to get the x component of the vector sum $\vec{r}$:

$$r_x = a_x + b_x + c_x$$
$$= 4.2 \text{ m} - 1.6 \text{ m} + 0 = 2.6 \text{ m}.$$

Similarly, for the y axis,

$$r_y = a_y + b_y + c_y$$
$$= -1.5 \text{ m} + 2.9 \text{ m} - 3.7 \text{ m} = -2.3 \text{ m}.$$

We then combine these components of $\vec{r}$ to write the vector in unit-vector notation:

$$\vec{r} = (2.6 \text{ m})\hat{i} - (2.3 \text{ m})\hat{j}, \qquad \text{(Answer)}$$

where $(2.6 \text{ m})\hat{i}$ is the vector component of $\vec{r}$ along the x axis and $-(2.3 \text{ m})\hat{j}$ is that along the y axis. Figure 3-15b shows one way to arrange these vector components to form $\vec{r}$. (Can you sketch the other way?)

We can also answer the question by giving the magnitude and an angle for $\vec{r}$. From Eq. 3-6, the magnitude is

$$r = \sqrt{(2.6 \text{ m})^2 + (-2.3 \text{ m})^2} \approx 3.5 \text{ m} \qquad \text{(Answer)}$$

and the angle (measured from the +x direction) is

$$\theta = \tan^{-1}\left(\frac{-2.3 \text{ m}}{2.6 \text{ m}}\right) = -41°, \qquad \text{(Answer)}$$

where the minus sign means clockwise.

Adding vectors by components, desert ant

The desert ant *Cataglyphis fortis* lives in the plains of the Sahara desert. When one of the ants forages for food, it travels from its home nest along a haphazard search path, over flat, featureless sand that contains no landmarks. Yet, when the ant decides to return home, it turns and then runs directly home. According to experiments, the ant keeps track of its movements along a mental coordinate system. When it wants to return to its home nest, it effectively sums its displacements along the axes of the system to calculate a vector that points directly home. As an example of the calculation, let's consider an ant making five runs of 6.0 cm each on an *xy* coordinate system, in the directions shown in Fig. 3-16a, starting from home. At the end of the fifth run, what are the magnitude and angle of the ant's net displacement vector $\vec{d}_{net}$, and what are those of the homeward vector $\vec{d}_{home}$ that extends from the ant's final position back to home? In a real situation, such vector calculations might involve thousands of such runs.

(1) To find the net displacement $\vec{d}_{net}$, we need to sum the five individual displacement vectors:

$$\vec{d}_{net} = \vec{d}_1 + \vec{d}_2 + \vec{d}_3 + \vec{d}_4 + \vec{d}_5.$$

(2) We evaluate this sum for the x components alone,

$$d_{net,x} = d_{1x} + d_{2x} + d_{3x} + d_{4x} + d_{5x}, \quad (3\text{-}14)$$

and for the y components alone,

$$d_{net,y} = d_{1y} + d_{2y} + d_{3y} + d_{4y} + d_{5y}. \quad (3\text{-}15)$$

(3) We construct $\vec{d}_{net}$ from its x and y components.

Calculations: To evaluate Eq. 3-14, we apply the x part of Eq. 3-5 to each run:

$$d_{1x} = (6.0 \text{ cm}) \cos 0° = +6.0 \text{ cm}$$
$$d_{2x} = (6.0 \text{ cm}) \cos 150° = -5.2 \text{ cm}$$
$$d_{3x} = (6.0 \text{ cm}) \cos 180° = -6.0 \text{ cm}$$
$$d_{4x} = (6.0 \text{ cm}) \cos(-120°) = -3.0 \text{ cm}$$
$$d_{5x} = (6.0 \text{ cm}) \cos 90° = 0.$$

Equation 3-14 then gives us

$$\begin{aligned} d_{net,x} &= +6.0 \text{ cm} + (-5.2 \text{ cm}) + (-6.0 \text{ cm}) \\ &\quad + (-3.0 \text{ cm}) + 0 \\ &= -8.2 \text{ cm}. \end{aligned}$$

Similarly, we evaluate the individual y components of the five runs using the y part of Eq. 3-5. The results are shown in Table 3-1. Substituting the results into Eq. 3-15 then gives us

$$d_{net,y} = +3.8 \text{ cm}.$$

Table 3-1

Run	d_x (cm)	d_y (cm)
1	+6.0	0
2	-5.2	+3.0
3	-6.0	0
4	-3.0	-5.2
5	0	+6.0
net	-8.2	+3.8

Vector $\vec{d}_{net}$ and its x and y components are shown in Fig. 3-16b. To find the magnitude and angle of $\vec{d}_{net}$ from its components, we use Eq. 3-6. The magnitude is

$$\begin{aligned} d_{net} &= \sqrt{d_{net,x}^2 + d_{net,y}^2} \\ &= \sqrt{(-8.2 \text{ cm})^2 + (3.8 \text{ cm})^2} = 9.0 \text{ cm}. \end{aligned}$$

To find the angle (measured from the positive direction of x), we take an inverse tangent:

$$\begin{aligned} \theta &= \tan^{-1}\left(\frac{d_{net,y}}{d_{net,x}}\right) \quad (3\text{-}16) \\ &= \tan^{-1}\left(\frac{3.8 \text{ cm}}{-8.2 \text{ cm}}\right) = -24.86°. \end{aligned}$$

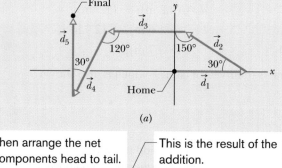

To add these vectors, find their net x component and their net y component.

Final

$\vec{d}_5$ $\vec{d}_3$ 120° 150° $\vec{d}_2$

30° 30°

$\vec{d}_4$ Home $\vec{d}_1$

x

(a)

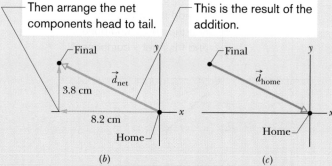

Then arrange the net components head to tail.

This is the result of the addition.

Final $\vec{d}_{net}$

3.8 cm

8.2 cm

Home

(b)

Final $\vec{d}_{home}$

Home

(c)

Fig. 3-16 (a) A search path of five runs. (b) The x and y components of $\vec{d}_{net}$. (c) Vector $\vec{d}_{home}$ points the way to the home nest.

Caution: Taking an inverse tangent on a calculator may not give the correct answer. The answer $-24.86°$ indicates that the direction of $\vec{d}_{net}$ is in the fourth quadrant of our xy coordinate system. However, when we construct the vector from its components (Fig. 3-16b), we see that the direction of $\vec{d}_{net}$ is in the second quadrant. Thus, we must "fix" the calculator's answer by adding 180°:

$$\theta = -24.86° + 180° = 155.14° \approx 155°. \quad (3\text{-}17)$$

Thus, the ant's displacement $\vec{d}_{net}$ has magnitude and angle

$$d_{net} = 9.0 \text{ cm at } 155°. \quad \text{(Answer)}$$

Vector $\vec{d}_{home}$ directed from the ant to its home has the same magnitude as $\vec{d}_{net}$ but the opposite direction (Fig. 3-16c). We already have the angle ($-24.86° \approx -25°$) for the direction opposite $\vec{d}_{net}$. Thus, $\vec{d}_{home}$ has magnitude and angle

$$d_{home} = 9.0 \text{ cm at } -25°. \quad \text{(Answer)}$$

A desert ant traveling more than 500 m from its home will actually make thousands of individual runs. Yet, it somehow knows how to calculate $\vec{d}_{home}$ (without studying this chapter).

 Additional examples, video, and practice available at *WileyPLUS*

3-7 Vectors and the Laws of Physics

So far, in every figure that includes a coordinate system, the x and y axes are parallel to the edges of the book page. Thus, when a vector $\vec{a}$ is included, its components a_x and a_y are also parallel to the edges (as in Fig. 3-17a). The only reason for that orientation of the axes is that it looks "proper"; there is no deeper reason. We could, instead, rotate the axes (but not the vector $\vec{a}$) through an angle ϕ as in Fig. 3-17b, in which case the components would have new values, call them a'_x and a'_y. Since there are an infinite number of choices of ϕ, there are an infinite number of different pairs of components for $\vec{a}$.

Which then is the "right" pair of components? The answer is that they are all equally valid because each pair (with its axes) just gives us a different way of describing the same vector $\vec{a}$; all produce the same magnitude and direction for the vector. In Fig. 3-17 we have

$$a = \sqrt{a_x^2 + a_y^2} = \sqrt{a'^2_x + a'^2_y} \tag{3-18}$$

and

$$\theta = \theta' + \phi. \tag{3-19}$$

The point is that we have great freedom in choosing a coordinate system, because the relations among vectors do not depend on the location of the origin or on the orientation of the axes. This is also true of the relations of physics; they are all independent of the choice of coordinate system. Add to that the simplicity and richness of the language of vectors and you can see why the laws of physics are almost always presented in that language: one equation, like Eq. 3-9, can represent three (or even more) relations, like Eqs. 3-10, 3-11, and 3-12.

Rotating the axes
changes the components
but not the vector.

Fig. 3-17 (a) The vector $\vec{a}$ and its components. (b) The same vector, with the axes of the coordinate system rotated through an angle ϕ.

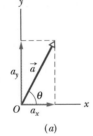

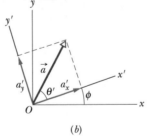

(a)

(b)

3-8 Multiplying Vectors*

There are three ways in which vectors can be multiplied, but none is exactly like the usual algebraic multiplication. As you read this section, keep in mind that a vector-capable calculator will help you multiply vectors only if you understand the basic rules of that multiplication.

Multiplying a Vector by a Scalar

If we multiply a vector $\vec{a}$ by a scalar s, we get a new vector. Its magnitude is the product of the magnitude of $\vec{a}$ and the absolute value of s. Its direction is the

*This material will not be employed until later (Chapter 7 for scalar products and Chapter 11 for vector products), and so your instructor may wish to postpone assignment of this section.

direction of $\vec{a}$ if s is positive but the opposite direction if s is negative. To divide $\vec{a}$ by s, we multiply $\vec{a}$ by $1/s$.

Multiplying a Vector by a Vector

There are two ways to multiply a vector by a vector: one way produces a scalar (called the *scalar product*), and the other produces a new vector (called the *vector product*). (Students commonly confuse the two ways.)

The Scalar Product

The **scalar product** of the vectors $\vec{a}$ and $\vec{b}$ in Fig. 3-18a is written as $\vec{a} \cdot \vec{b}$ and defined to be

$$\vec{a} \cdot \vec{b} = ab \cos \phi, \tag{3-20}$$

where a is the magnitude of $\vec{a}$, b is the magnitude of $\vec{b}$, and ϕ is the angle between $\vec{a}$ and $\vec{b}$ (or, more properly, between the directions of $\vec{a}$ and $\vec{b}$). There are actually two such angles: ϕ and $360° - \phi$. Either can be used in Eq. 3-20, because their cosines are the same.

Note that there are only scalars on the right side of Eq. 3-20 (including the value of $\cos \phi$). Thus $\vec{a} \cdot \vec{b}$ on the left side represents a *scalar* quantity. Because of the notation, $\vec{a} \cdot \vec{b}$ is also known as the **dot product** and is spoken as "a dot b."

A dot product can be regarded as the product of two quantities: (1) the magnitude of one of the vectors and (2) the scalar component of the second vector along the direction of the first vector. For example, in Fig. 3-18b, $\vec{a}$ has a scalar component $a \cos \phi$ along the direction of $\vec{b}$; note that a perpendicular dropped from the head of $\vec{a}$ onto $\vec{b}$ determines that component. Similarly, $\vec{b}$ has a scalar component $b \cos \phi$ along the direction of $\vec{a}$.

If the angle ϕ between two vectors is $0°$, the component of one vector along the other is maximum, and so also is the dot product of the vectors. If, instead, ϕ is $90°$, the component of one vector along the other is zero, and so is the dot product.

Equation 3-20 can be rewritten as follows to emphasize the components:

$$\vec{a} \cdot \vec{b} = (a \cos \phi)(b) = (a)(b \cos \phi). \tag{3-21}$$

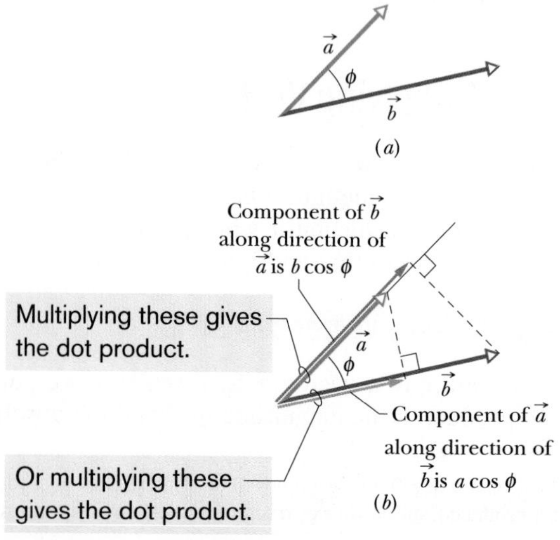

Component of $\vec{b}$ along direction of $\vec{a}$ is $b \cos \phi$

Multiplying these gives the dot product.

Component of $\vec{a}$ along direction of $\vec{b}$ is $a \cos \phi$

Or multiplying these gives the dot product.

Fig. 3-18 (a) Two vectors $\vec{a}$ and $\vec{b}$, with an angle ϕ between them. (b) Each vector has a component along the direction of the other vector.

The commutative law applies to a scalar product, so we can write

$$\vec{a} \cdot \vec{b} = \vec{b} \cdot \vec{a}.$$

When two vectors are in unit-vector notation, we write their dot product as

$$\vec{a} \cdot \vec{b} = (a_x\hat{i} + a_y\hat{j} + a_z\hat{k}) \cdot (b_x\hat{i} + b_y\hat{j} + b_z\hat{k}), \qquad (3\text{-}22)$$

which we can expand according to the distributive law: Each vector component of the first vector is to be dotted with each vector component of the second vector. By doing so, we can show that

$$\vec{a} \cdot \vec{b} = a_xb_x + a_yb_y + a_zb_z. \qquad (3\text{-}23)$$

 CHECKPOINT 4

Vectors $\vec{C}$ and $\vec{D}$ have magnitudes of 3 units and 4 units, respectively. What is the angle between the directions of $\vec{C}$ and $\vec{D}$ if $\vec{C} \cdot \vec{D}$ equals (a) zero, (b) 12 units, and (c) −12 units?

Sample Problem

Angle between two vectors using dot products

What is the angle ϕ between $\vec{a} = 3.0\hat{i} - 4.0\hat{j}$ and $\vec{b} = -2.0\hat{i} + 3.0\hat{k}$? (*Caution:* Although many of the following steps can be bypassed with a vector-capable calculator, you will learn more about scalar products if, at least here, you use these steps.)

KEY IDEA

The angle between the directions of two vectors is included in the definition of their scalar product (Eq. 3-20):

$$\vec{a} \cdot \vec{b} = ab \cos \phi. \qquad (3\text{-}24)$$

Calculations: In Eq. 3-24, a is the magnitude of $\vec{a}$, or

$$a = \sqrt{3.0^2 + (-4.0)^2} = 5.00, \qquad (3\text{-}25)$$

and b is the magnitude of $\vec{b}$, or

$$b = \sqrt{(-2.0)^2 + 3.0^2} = 3.61. \qquad (3\text{-}26)$$

We can separately evaluate the left side of Eq. 3-24 by writ-

ing the vectors in unit-vector notation and using the distributive law:

$$\begin{aligned} \vec{a} \cdot \vec{b} &= (3.0\hat{i} - 4.0\hat{j}) \cdot (-2.0\hat{i} + 3.0\hat{k}) \\ &= (3.0\hat{i}) \cdot (-2.0\hat{i}) + (3.0\hat{i}) \cdot (3.0\hat{k}) \\ &\quad + (-4.0\hat{j}) \cdot (-2.0\hat{i}) + (-4.0\hat{j}) \cdot (3.0\hat{k}). \end{aligned}$$

We next apply Eq. 3-20 to each term in this last expression. The angle between the unit vectors in the first term ($\hat{i}$ and $\hat{i}$) is 0°, and in the other terms it is 90°. We then have

$$\begin{aligned} \vec{a} \cdot \vec{b} &= -(6.0)(1) + (9.0)(0) + (8.0)(0) - (12)(0) \\ &= -6.0. \end{aligned}$$

Substituting this result and the results of Eqs. 3-25 and 3-26 into Eq. 3-24 yields

$$-6.0 = (5.00)(3.61) \cos \phi,$$

so $$\phi = \cos^{-1} \frac{-6.0}{(5.00)(3.61)} = 109° \approx 110°. \qquad \text{(Answer)}$$

 Additional examples, video, and practice available at *WileyPLUS*

The Vector Product

The **vector product** of $\vec{a}$ and $\vec{b}$, written $\vec{a} \times \vec{b}$, produces a third vector $\vec{c}$ whose magnitude is

$$c = ab \sin \phi, \qquad (3\text{-}27)$$

where ϕ is the *smaller* of the two angles between $\vec{a}$ and $\vec{b}$. (You must use the smaller of the two angles between the vectors because $\sin \phi$ and $\sin(360° - \phi)$ differ in algebraic sign.) Because of the notation, $\vec{a} \times \vec{b}$ is also known as the **cross product,** and in speech it is "a cross b."

If $\vec{a}$ and $\vec{b}$ are parallel or antiparallel, $\vec{a} \times \vec{b} = 0$. The magnitude of $\vec{a} \times \vec{b}$, which can be written as $|\vec{a} \times \vec{b}|$, is maximum when $\vec{a}$ and $\vec{b}$ are perpendicular to each other.

The direction of $\vec{c}$ is perpendicular to the plane that contains $\vec{a}$ and $\vec{b}$. Figure 3-19a shows how to determine the direction of $\vec{c} = \vec{a} \times \vec{b}$ with what is known as a **right-hand rule.** Place the vectors $\vec{a}$ and $\vec{b}$ tail to tail without altering their orientations, and imagine a line that is perpendicular to their plane where they meet. Pretend to place your *right* hand around that line in such a way that your fingers would sweep $\vec{a}$ into $\vec{b}$ through the smaller angle between them. Your outstretched thumb points in the direction of $\vec{c}$.

The order of the vector multiplication is important. In Fig. 3-19b, we are determining the direction of $\vec{c}' = \vec{b} \times \vec{a}$, so the fingers are placed to sweep $\vec{b}$ into $\vec{a}$ through the smaller angle. The thumb ends up in the opposite direction from

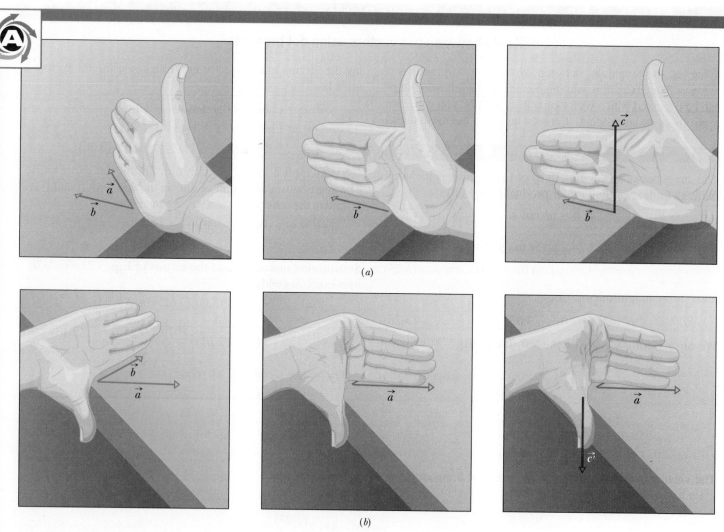

(a)

(b)

Fig. 3-19 Illustration of the right-hand rule for vector products. (a) Sweep vector $\vec{a}$ into vector $\vec{b}$ with the fingers of your right hand. Your outstretched thumb shows the direction of vector $\vec{c} = \vec{a} \times \vec{b}$. (b) Showing that $\vec{b} \times \vec{a}$ is the reverse of $\vec{a} \times \vec{b}$.

previously, and so it must be that $\vec{c}' = -\vec{c}$; that is,

$$\vec{b} \times \vec{a} = -(\vec{a} \times \vec{b}). \qquad (3\text{-}28)$$

In other words, the commutative law does not apply to a vector product.

In unit-vector notation, we write

$$\vec{a} \times \vec{b} = (a_x\hat{i} + a_y\hat{j} + a_z\hat{k}) \times (b_x\hat{i} + b_y\hat{j} + b_z\hat{k}), \qquad (3\text{-}29)$$

which can be expanded according to the distributive law; that is, each component of the first vector is to be crossed with each component of the second vector. The cross products of unit vectors are given in Appendix E (see "Products of Vectors"). For example, in the expansion of Eq. 3-29, we have

$$a_x\hat{i} \times b_x\hat{i} = a_xb_x(\hat{i} \times \hat{i}) = 0,$$

because the two unit vectors $\hat{i}$ and $\hat{i}$ are parallel and thus have a zero cross product. Similarly, we have

$$a_x\hat{i} \times b_y\hat{j} = a_xb_y(\hat{i} \times \hat{j}) = a_xb_y\hat{k}.$$

In the last step we used Eq. 3-27 to evaluate the magnitude of $\hat{i} \times \hat{j}$ as unity. (These vectors $\hat{i}$ and $\hat{j}$ each have a magnitude of unity, and the angle between them is 90°.) Also, we used the right-hand rule to get the direction of $\hat{i} \times \hat{j}$ as being in the positive direction of the z axis (thus in the direction of $\hat{k}$).

Continuing to expand Eq. 3-29, you can show that

$$\vec{a} \times \vec{b} = (a_yb_z - b_ya_z)\hat{i} + (a_zb_x - b_za_x)\hat{j} + (a_xb_y - b_xa_y)\hat{k}. \qquad (3\text{-}30)$$

A determinant (Appendix E) or a vector-capable calculator can also be used.

To check whether any xyz coordinate system is a right-handed coordinate system, use the right-hand rule for the cross product $\hat{i} \times \hat{j} = \hat{k}$ with that system. If your fingers sweep $\hat{i}$ (positive direction of x) into $\hat{j}$ (positive direction of y) with the outstretched thumb pointing in the positive direction of z (not the negative direction), then the system is right-handed.

 **CHECKPOINT 5**

Vectors $\vec{C}$ and $\vec{D}$ have magnitudes of 3 units and 4 units, respectively. What is the angle between the directions of $\vec{C}$ and $\vec{D}$ if the magnitude of the vector product $\vec{C} \times \vec{D}$ is (a) zero and (b) 12 units?

Sample Problem

Cross product, right-hand rule

In Fig. 3-20, vector $\vec{a}$ lies in the xy plane, has a magnitude of 18 units and points in a direction 250° from the positive direction of the x axis. Also, vector $\vec{b}$ has a magnitude of 12 units and points in the positive direction of the z axis. What is the vector product $\vec{c} = \vec{a} \times \vec{b}$?

KEY IDEA

When we have two vectors in magnitude-angle notation, we find the magnitude of their cross product with Eq. 3-27 and the direction of their cross product with the right-hand rule of Fig. 3-19.

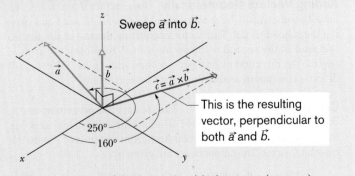

Fig. 3-20 Vector $\vec{c}$ (in the xy plane) is the vector (or cross) product of vectors $\vec{a}$ and $\vec{b}$.

Calculations: For the magnitude we write

$$c = ab \sin \phi = (18)(12)(\sin 90°) = 216. \qquad \text{(Answer)}$$

To determine the direction in Fig. 3-20, imagine placing the fingers of your right hand around a line perpendicular to the plane of $\vec{a}$ and $\vec{b}$ (the line on which $\vec{c}$ is shown) such that your fingers sweep $\vec{a}$ into $\vec{b}$. Your outstretched thumb then gives the direction of $\vec{c}$. Thus, as shown in the figure, $\vec{c}$ lies in the xy plane. Because its direction is perpendicular to the direction of $\vec{a}$ (a cross product always gives a perpendicular vector), it is at an angle of

$$250° - 90° = 160° \qquad \text{(Answer)}$$

from the positive direction of the x axis.

Sample Problem

Cross product, unit-vector notation

If $\vec{a} = 3\hat{i} - 4\hat{j}$ and $\vec{b} = -2\hat{i} + 3\hat{k}$, what is $\vec{c} = \vec{a} \times \vec{b}$?

KEY IDEA

When two vectors are in unit-vector notation, we can find their cross product by using the distributive law.

Calculations: Here we write

$$\vec{c} = (3\hat{i} - 4\hat{j}) \times (-2\hat{i} + 3\hat{k})$$
$$= 3\hat{i} \times (-2\hat{i}) + 3\hat{i} \times 3\hat{k} + (-4\hat{j}) \times (-2\hat{i})$$
$$+ (-4\hat{j}) \times 3\hat{k}.$$

We next evaluate each term with Eq. 3-27, finding the direction with the right-hand rule. For the first term here, the angle ϕ between the two vectors being crossed is 0. For the other terms, ϕ is 90°. We find

$$\vec{c} = -6(0) + 9(-\hat{j}) + 8(-\hat{k}) - 12\hat{i}$$
$$= -12\hat{i} - 9\hat{j} - 8\hat{k}. \qquad \text{(Answer)}$$

This vector $\vec{c}$ is perpendicular to both $\vec{a}$ and $\vec{b}$, a fact you can check by showing that $\vec{c} \cdot \vec{a} = 0$ and $\vec{c} \cdot \vec{b} = 0$; that is, there is no component of $\vec{c}$ along the direction of either $\vec{a}$ or $\vec{b}$.

 Additional examples, video, and practice available at *WileyPLUS*

REVIEW & SUMMARY

Scalars and Vectors *Scalars,* such as temperature, have magnitude only. They are specified by a number with a unit (10°C) and obey the rules of arithmetic and ordinary algebra. *Vectors,* such as displacement, have both magnitude and direction (5 m, north) and obey the rules of vector algebra.

Adding Vectors Geometrically Two vectors $\vec{a}$ and $\vec{b}$ may be added geometrically by drawing them to a common scale and placing them head to tail. The vector connecting the tail of the first to the head of the second is the vector sum $\vec{s}$. To subtract $\vec{b}$ from $\vec{a}$, reverse the direction of $\vec{b}$ to get $-\vec{b}$; then add $-\vec{b}$ to $\vec{a}$. Vector addition is commutative and obeys the associative law.

Components of a Vector The (scalar) *components* a_x and a_y of any two-dimensional vector $\vec{a}$ along the coordinate axes are found by dropping perpendicular lines from the ends of $\vec{a}$ onto the coordinate axes. The components are given by

$$a_x = a \cos \theta \quad \text{and} \quad a_y = a \sin \theta, \qquad (3\text{-}5)$$

where θ is the angle between the positive direction of the x axis and the direction of $\vec{a}$. The algebraic sign of a component indicates its direction along the associated axis. Given its components, we can find the magnitude and orientation of the vector $\vec{a}$ with

$$a = \sqrt{a_x^2 + a_y^2} \quad \text{and} \quad \tan \theta = \frac{a_y}{a_x}. \qquad (3\text{-}6)$$

Unit-Vector Notation *Unit vectors* $\hat{i}$, $\hat{j}$, and $\hat{k}$ have magnitudes of unity and are directed in the positive directions of the x, y, and z axes, respectively, in a right-handed coordinate system. We can write a vector $\vec{a}$ in terms of unit vectors as

$$\vec{a} = a_x\hat{i} + a_y\hat{j} + a_z\hat{k}, \qquad (3\text{-}7)$$

in which $a_x\hat{i}$, $a_y\hat{j}$, and $a_z\hat{k}$ are the **vector components** of $\vec{a}$ and $a_x, a_y,$ and a_z are its **scalar components.**

Adding Vectors in Component Form To add vectors in component form, we use the rules

$$r_x = a_x + b_x \quad r_y = a_y + b_y \quad r_z = a_z + b_z. \qquad (3\text{-}10 \text{ to } 3\text{-}12)$$

Here $\vec{a}$ and $\vec{b}$ are the vectors to be added, and $\vec{r}$ is the vector sum. Note that we add components axis by axis.

Product of a Scalar and a Vector   The product of a scalar s and a vector $\vec{v}$ is a new vector whose magnitude is sv and whose direction is the same as that of $\vec{v}$ if s is positive, and opposite that of $\vec{v}$ if s is negative. To divide $\vec{v}$ by s, multiply $\vec{v}$ by $1/s$.

The Scalar Product   The **scalar** (or **dot**) **product** of two vectors $\vec{a}$ and $\vec{b}$ is written $\vec{a} \cdot \vec{b}$ and is the *scalar* quantity given by

$$\vec{a} \cdot \vec{b} = ab \cos \phi, \tag{3-20}$$

in which ϕ is the angle between the directions of $\vec{a}$ and $\vec{b}$. A scalar product is the product of the magnitude of one vector and the scalar component of the second vector along the direction of the first vector. In unit-vector notation,

$$\vec{a} \cdot \vec{b} = (a_x\hat{i} + a_y\hat{j} + a_z\hat{k}) \cdot (b_x\hat{i} + b_y\hat{j} + b_z\hat{k}), \tag{3-22}$$

which may be expanded according to the distributive law. Note that $\vec{a} \cdot \vec{b} = \vec{b} \cdot \vec{a}$.

The Vector Product   The **vector** (or **cross**) **product** of two vectors $\vec{a}$ and $\vec{b}$ is written $\vec{a} \times \vec{b}$ and is a *vector* $\vec{c}$ whose magnitude c is given by

$$c = ab \sin \phi, \tag{3-27}$$

in which ϕ is the smaller of the angles between the directions of $\vec{a}$ and $\vec{b}$. The direction of $\vec{c}$ is perpendicular to the plane defined by $\vec{a}$ and $\vec{b}$ and is given by a right-hand rule, as shown in Fig. 3-19. Note that $\vec{a} \times \vec{b} = -(\vec{b} \times \vec{a})$. In unit-vector notation,

$$\vec{a} \times \vec{b} = (a_x\hat{i} + a_y\hat{j} + a_z\hat{k}) \times (b_x\hat{i} + b_y\hat{j} + b_z\hat{k}), \tag{3-29}$$

which we may expand with the distributive law.

QUESTIONS

1   Can the sum of the magnitudes of two vectors ever be equal to the magnitude of the sum of the same two vectors? If no, why not? If yes, when?

2   The two vectors shown in Fig. 3-21 lie in an xy plane. What are the signs of the x and y components, respectively, of (a) $\vec{d}_1 + \vec{d}_2$, (b) $\vec{d}_1 - \vec{d}_2$, and (c) $\vec{d}_2 - \vec{d}_1$?

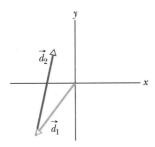

Fig. 3-21   Question 2.

3   Being part of the "Gators," the University of Florida golfing team must play on a putting green with an alligator pit. Figure 3-22 shows an overhead view of one putting challenge of the team; an xy coordinate system is superimposed. Team members must putt from the origin to the hole, which is at xy coordinates (8 m, 12 m), but they can putt the golf ball using only one or more of the following displacements, one or more times:

$$\vec{d}_1 = (8 \text{ m})\hat{i} + (6 \text{ m})\hat{j}, \quad \vec{d}_2 = (6 \text{ m})\hat{j}, \quad \vec{d}_3 = (8 \text{ m})\hat{i}.$$

The pit is at coordinates (8 m, 6 m). If a team member putts the ball into or through the pit, the member is automatically transferred to Florida State University, the arch rival. What sequence of displacements should a team member use to avoid the pit?

4   Equation 3-2 shows that the addition of two vectors $\vec{a}$ and $\vec{b}$ is commutative. Does that mean subtraction is commutative, so that $\vec{a} - \vec{b} = \vec{b} - \vec{a}$?

5   Which of the arrangements of axes in Fig. 3-23 can be labeled "right-handed coordinate system"? As usual, each axis label indicates the positive side of the axis.

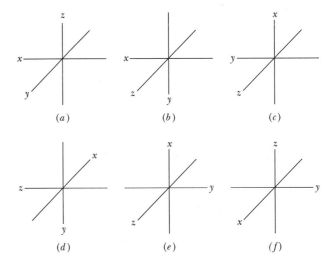

Fig. 3-23   Question 5.

6   Describe two vectors $\vec{a}$ and $\vec{b}$ such that

(a) $\vec{a} + \vec{b} = \vec{c}$ and $a + b = c$;

(b) $\vec{a} + \vec{b} = \vec{a} - \vec{b}$;

(c) $\vec{a} + \vec{b} = \vec{c}$ and $a^2 + b^2 = c^2$.

7   If $\vec{d} = \vec{a} + \vec{b} + (-\vec{c})$, does (a) $\vec{a} + (-\vec{d}) = \vec{c} + (-\vec{b})$, (b) $\vec{a} = (-\vec{b}) + \vec{d} + \vec{c}$, and (c) $\vec{c} + (-\vec{d}) = \vec{a} + \vec{b}$?

8   If $\vec{a} \cdot \vec{b} = \vec{a} \cdot \vec{c}$, must $\vec{b}$ equal $\vec{c}$?

9   If $\vec{F} = q(\vec{v} \times \vec{B})$ and $\vec{v}$ is perpendicular to $\vec{B}$, then what is the

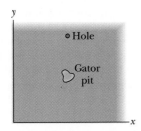

Fig. 3-22   Question 3.

direction of $\vec{B}$ in the three situations shown in Fig. 3-24 when constant q is (a) positive and (b) negative?

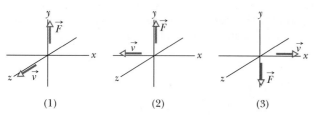

(1)   (2)   (3)

Fig. 3-24   Question 9.

10   Figure 3-25 shows vector $\vec{A}$ and four other vectors that have the same magnitude but differ in orientation. (a) Which of those other four vectors have the same dot product with $\vec{A}$? (b) Which have a negative dot product with $\vec{A}$?

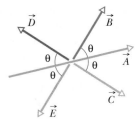

Fig. 3-25   Question 10.

PROBLEMS

sec. 3-4   Components of Vectors

•**1** **SSM**   What are (a) the x component and (b) the y component of a vector $\vec{a}$ in the xy plane if its direction is 250° counterclockwise from the positive direction of the x axis and its magnitude is 7.3 m?

•**2**   A displacement vector $\vec{r}$ in the xy plane is 15 m long and directed at angle $\theta = 30°$ in Fig. 3-26. Determine (a) the x component and (b) the y component of the vector.

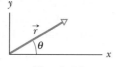

Fig. 3-26
Problem 2.

•**3** **SSM**   The x component of vector $\vec{A}$ is -25.0 m and the y component is $+40.0$ m. (a) What is the magnitude of $\vec{A}$? (b) What is the angle between the direction of $\vec{A}$ and the positive direction of x?

•**4**   Express the following angles in radians: (a) 20.0°, (b) 50.0°, (c) 100°. Convert the following angles to degrees: (d) 0.330 rad, (e) 2.10 rad, (f) 7.70 rad.

•**5**   A ship sets out to sail to a point 120 km due north. An unexpected storm blows the ship to a point 100 km due east of its starting point. (a) How far and (b) in what direction must it now sail to reach its original destination?

•**6**   In Fig. 3-27, a heavy piece of machinery is raised by sliding it a distance $d = 12.5$ m along a plank oriented at angle $\theta = 20.0°$ to the horizontal. How far is it moved (a) vertically and (b) horizontally?

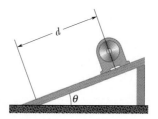

Fig. 3-27   Problem 6.

••**7** **SSM** **WWW**   A room has dimensions 3.00 m (height) $\times$ 3.70 m $\times$ 4.30 m. A fly starting at one corner flies around, ending up at the diagonally opposite corner. (a) What is the magnitude of its displacement? (b) Could the length of its path be less than this magnitude? (c) Greater? (d) Equal? (e) Choose a suitable coordinate system and express the components of the displacement vector in that system in unit-vector notation. (f) If the fly walks, what is the length of the shortest path? (*Hint:* This can be answered without calculus. The room is like a box. Unfold its walls to flatten them into a plane.)

sec. 3-6   Adding Vectors by Components

•**8**   A person walks in the following pattern: 3.1 km north, then 2.4 km west, and finally 5.2 km south. (a) Sketch the vector diagram that represents this motion. (b) How far and (c) in what direction would a bird fly in a straight line from the same starting point to the same final point?

•**9**   Two vectors are given by

$$\vec{a} = (4.0 \text{ m})\hat{i} - (3.0 \text{ m})\hat{j} + (1.0 \text{ m})\hat{k}$$

and

$$\vec{b} = (-1.0 \text{ m})\hat{i} + (1.0 \text{ m})\hat{j} + (4.0 \text{ m})\hat{k}.$$

In unit-vector notation, find (a) $\vec{a} + \vec{b}$, (b) $\vec{a} - \vec{b}$, and (c) a third vector $\vec{c}$ such that $\vec{a} - \vec{b} + \vec{c} = 0$.

•**10**   Find the (a) x, (b) y, and (c) z components of the sum $\vec{r}$ of the displacements $\vec{c}$ and $\vec{d}$ whose components in meters are $c_x = 7.4, c_y = -3.8, c_z = -6.1; d_x = 4.4, d_y = -2.0, d_z = 3.3$.

•**11** **SSM**   (a) In unit-vector notation, what is the sum $\vec{a} + \vec{b}$ if $\vec{a} = (4.0 \text{ m})\hat{i} + (3.0 \text{ m})\hat{j}$ and $\vec{b} = (-13.0 \text{ m})\hat{i} + (7.0 \text{ m})\hat{j}$? What are the (b) magnitude and (c) direction of $\vec{a} + \vec{b}$?

•**12**   A car is driven east for a distance of 50 km, then north for 30 km, and then in a direction 30° east of north for 25 km. Sketch the vector diagram and determine (a) the magnitude and (b) the angle of the car's total displacement from its starting point.

•**13**   A person desires to reach a point that is 3.40 km from her present location and in a direction that is 35.0° north of east. However, she must travel along streets that are oriented either north–south or east–west. What is the minimum distance she could travel to reach her destination?

•**14**   You are to make four straight-line moves over a flat desert floor, starting at the origin of an xy coordinate system and ending at the xy coordinates $(-140 \text{ m}, 30 \text{ m})$. The x component and y component of your moves are the following, respectively, in meters: (20 and 60), then (b_x and -70), then (-20 and c_y), then (-60 and -70). What are (a) component b_x and (b) component c_y? What are (c) the magnitude and (d) the angle (relative to the positive direction of the x axis) of the overall displacement?

•**15** **SSM** **ILW** **WWW**   The two vectors $\vec{a}$ and $\vec{b}$ in Fig. 3-28 have equal magnitudes of 10.0 m and the angles are $\theta_1 = 30°$ and $\theta_2 = $

105°. Find the (a) x and (b) y components of their vector sum $\vec{r}$, (c) the magnitude of $\vec{r}$, and (d) the angle $\vec{r}$ makes with the positive direction of the x axis.

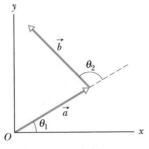

Fig. 3-28   Problem 15.

•16   For the displacement vectors $\vec{a} = (3.0\text{ m})\hat{i} + (4.0\text{ m})\hat{j}$ and $\vec{b} = (5.0\text{ m})\hat{i} + (-2.0\text{ m})\hat{j}$, give $\vec{a} + \vec{b}$ in (a) unit-vector notation, and as (b) a magnitude and (c) an angle (relative to $\hat{i}$). Now give $\vec{b} - \vec{a}$ in (d) unit-vector notation, and as (e) a magnitude and (f) an angle.

•17 GO ILW   Three vectors $\vec{a}, \vec{b}$, and $\vec{c}$ each have a magnitude of 50 m and lie in an xy plane. Their directions relative to the positive direction of the x axis are 30°, 195°, and 315°, respectively. What are (a) the magnitude and (b) the angle of the vector $\vec{a} + \vec{b} + \vec{c}$, and (c) the magnitude and (d) the angle of $\vec{a} - \vec{b} + \vec{c}$? What are the (e) magnitude and (f) angle of a fourth vector $\vec{d}$ such that $(\vec{a} + \vec{b}) - (\vec{c} + \vec{d}) = 0$?

•18   In the sum $\vec{A} + \vec{B} = \vec{C}$, vector $\vec{A}$ has a magnitude of 12.0 m and is angled 40.0° counterclockwise from the +x direction, and vector $\vec{C}$ has a magnitude of 15.0 m and is angled 20.0° counterclockwise from the −x direction. What are (a) the magnitude and (b) the angle (relative to +x) of $\vec{B}$?

•19   In a game of lawn chess, where pieces are moved between the centers of squares that are each 1.00 m on edge, a knight is moved in the following way: (1) two squares forward, one square rightward; (2) two squares leftward, one square forward; (3) two squares forward, one square leftward. What are (a) the magnitude and (b) the angle (relative to "forward") of the knight's overall displacement for the series of three moves?

••20 [airplane] An explorer is caught in a whiteout (in which the snowfall is so thick that the ground cannot be distinguished from the sky) while returning to base camp. He was supposed to travel due north for 5.6 km, but when the snow clears, he discovers that he actually traveled 7.8 km at 50° north of due east. (a) How far and (b) in what direction must he now travel to reach base camp?

••21 GO   An ant, crazed by the Sun on a hot Texas afternoon, darts over an xy plane scratched in the dirt. The x and y components of four consecutive darts are the following, all in centimeters: (30.0, 40.0), $(b_x, -70.0)$, $(-20.0, c_y)$, $(-80.0, -70.0)$. The overall displacement of the four darts has the xy components $(-140, -20.0)$. What are (a) b_x and (b) c_y? What are the (c) magnitude and (d) angle (relative to the positive direction of the x axis) of the overall displacement?

••22   (a) What is the sum of the following four vectors in unit-vector notation? For that sum, what are (b) the magnitude, (c) the angle in degrees, and (d) the angle in radians?

$\vec{E}$: 6.00 m at +0.900 rad $\vec{F}$: 5.00 m at −75.0°

$\vec{G}$: 4.00 m at +1.20 rad $\vec{H}$: 6.00 m at −210°

••23   If $\vec{B}$ is added to $\vec{C} = 3.0\hat{i} + 4.0\hat{j}$, the result is a vector in the positive direction of the y axis, with a magnitude equal to that of $\vec{C}$. What is the magnitude of $\vec{B}$?

••24   Vector $\vec{A}$, which is directed along an x axis, is to be added to vector $\vec{B}$, which has a magnitude of 7.0 m. The sum is a third vector that is directed along the y axis, with a magnitude that is 3.0 times that of $\vec{A}$. What is that magnitude of $\vec{A}$?

••25 GO   Oasis B is 25 km due east of oasis A. Starting from oasis A, a camel walks 24 km in a direction 15° south of east and then walks 8.0 km due north. How far is the camel then from oasis B?

••26   What is the sum of the following four vectors in (a) unit-vector notation, and as (b) a magnitude and (c) an angle?

$\vec{A} = (2.00\text{ m})\hat{i} + (3.00\text{ m})\hat{j}$ $\vec{B}$: 4.00 m, at +65.0°

$\vec{C} = (-4.00\text{ m})\hat{i} + (-6.00\text{ m})\hat{j}$ $\vec{D}$: 5.00 m, at −235°

••27 GO   If $\vec{d}_1 + \vec{d}_2 = 5\vec{d}_3, \vec{d}_1 - \vec{d}_2 = 3\vec{d}_3$, and $\vec{d}_3 = 2\hat{i} + 4\hat{j}$, then what are, in unit-vector notation, (a) $\vec{d}_1$ and (b) $\vec{d}_2$?

••28   Two beetles run across flat sand, starting at the same point. Beetle 1 runs 0.50 m due east, then 0.80 m at 30° north of due east. Beetle 2 also makes two runs; the first is 1.6 m at 40° east of due north. What must be (a) the magnitude and (b) the direction of its second run if it is to end up at the new location of beetle 1?

••29 [airplane] GO   Typical backyard ants often create a network of chemical trails for guidance. Extending outward from the nest, a trail branches (bifurcates) repeatedly, with 60° between the branches. If a roaming ant chances upon a trail, it can tell the way to the nest at any branch point: If it is moving away from the nest, it has two choices of path requiring a small turn in its travel direction, either 30° leftward or 30° rightward. If it is moving toward the nest, it has only one such choice. Figure 3-29 shows a typical ant trail, with lettered straight sections of 2.0 cm length and symmetric bifurcation of 60°. Path v is parallel to the y axis. What are the (a) magnitude and (b) angle (relative to the positive direction of the superimposed x axis) of an ant's displacement from the nest (find it in the figure) if the ant enters the trail at point A? What are the (c) magnitude and (d) angle if it enters at point B?

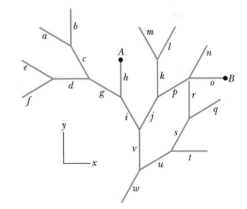

Fig. 3-29
Problem 29.

••30   Here are two vectors:

$\vec{a} = (4.0\text{ m})\hat{i} - (3.0\text{ m})\hat{j}$ and $\vec{b} = (6.0\text{ m})\hat{i} + (8.0\text{ m})\hat{j}$.

What are (a) the magnitude and (b) the angle (relative to $\hat{i}$) of $\vec{a}$? What are (c) the magnitude and (d) the angle of $\vec{b}$? What are (e) the magnitude and (f) the angle of $\vec{a} + \vec{b}$; (g) the magnitude and (h) the angle of $\vec{b} - \vec{a}$; and (i) the magnitude and (j) the angle of $\vec{a} - \vec{b}$? (k) What is the angle between the directions of $\vec{b} - \vec{a}$ and $\vec{a} - \vec{b}$?

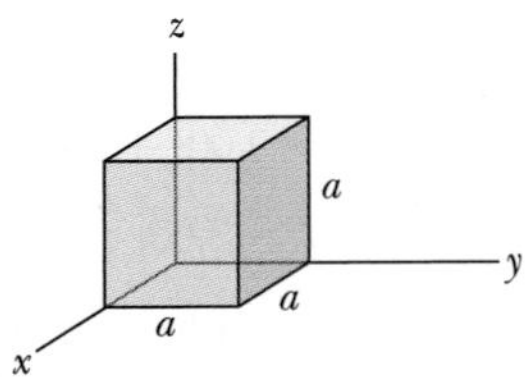

•••31   In Fig. 3-30, a cube of edge length a sits with one corner at the ori-

Fig. 3-30 Problem 31.

gin of an *xyz* coordinate system. A *body diagonal* is a line that extends from one corner to another through the center. In unit-vector notation, what is the body diagonal that extends from the corner at (a) coordinates $(0, 0, 0)$, (b) coordinates $(a, 0, 0)$, (c) coordinates $(0, a, 0)$, and (d) coordinates $(a, a, 0)$? (e) Determine the angles that the body diagonals make with the adjacent edges. (f) Determine the length of the body diagonals in terms of a.

sec. 3-7　Vectors and the Laws of Physics

•**32**　In Fig. 3-31, a vector $\vec{a}$ with a magnitude of 17.0 m is directed at angle $\theta = 56.0°$ counterclockwise from the $+x$ axis. What are the components (a) a_x and (b) a_y of the vector? A second coordinate system is inclined by angle $\theta' = 18.0°$ with respect to the first. What are the components (c) a'_x and (d) a'_y in this primed coordinate system?

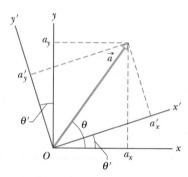

Fig. 3-31　Problem 32.

sec. 3-8　Multiplying Vectors

•**33**　For the vectors in Fig. 3-32, with $a = 4$, $b = 3$, and $c = 5$, what are (a) the magnitude and (b) the direction of $\vec{a} \times \vec{b}$, (c) the magnitude and (d) the direction of $\vec{a} \times \vec{c}$, and (e) the magnitude and (f) the direction of $\vec{b} \times \vec{c}$? (The z axis is not shown.)

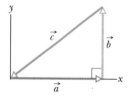

Fig. 3-32
Problems 33 and 54.

•**34**　Two vectors are presented as $\vec{a} = 3.0\hat{i} + 5.0\hat{j}$ and $\vec{b} = 2.0\hat{i} + 4.0\hat{j}$. Find (a) $\vec{a} \times \vec{b}$, (b) $\vec{a} \cdot \vec{b}$, (c) $(\vec{a} + \vec{b}) \cdot \vec{b}$, and (d) the component of $\vec{a}$ along the direction of $\vec{b}$. (*Hint:* For (d), consider Eq. 3-20 and Fig. 3-18.)

•**35**　Two vectors, $\vec{r}$ and $\vec{s}$, lie in the xy plane. Their magnitudes are 4.50 and 7.30 units, respectively, and their directions are 320° and 85.0°, respectively, as measured counterclockwise from the positive x axis. What are the values of (a) $\vec{r} \cdot \vec{s}$ and (b) $\vec{r} \times \vec{s}$?

•**36**　If $\vec{d}_1 = 3\hat{i} - 2\hat{j} + 4\hat{k}$ and $\vec{d}_2 = -5\hat{i} + 2\hat{j} - \hat{k}$, then what is $(\vec{d}_1 + \vec{d}_2) \cdot (\vec{d}_1 \times 4\vec{d}_2)$?

•**37**　Three vectors are given by $\vec{a} = 3.0\hat{i} + 3.0\hat{j} - 2.0\hat{k}$, $\vec{b} = -1.0\hat{i} - 4.0\hat{j} + 2.0\hat{k}$, and $\vec{c} = 2.0\hat{i} + 2.0\hat{j} + 1.0\hat{k}$. Find (a) $\vec{a} \cdot (\vec{b} \times \vec{c})$, (b) $\vec{a} \cdot (\vec{b} + \vec{c})$, and (c) $\vec{a} \times (\vec{b} + \vec{c})$.

••**38**　**GO**　For the following three vectors, what is $3\vec{C} \cdot (2\vec{A} \times \vec{B})$?

$$\vec{A} = 2.00\hat{i} + 3.00\hat{j} - 4.00\hat{k}$$

$$\vec{B} = -3.00\hat{i} + 4.00\hat{j} + 2.00\hat{k} \quad \vec{C} = 7.00\hat{i} - 8.00\hat{j}$$

••**39**　Vector $\vec{A}$ has a magnitude of 6.00 units, vector $\vec{B}$ has a magnitude of 7.00 units, and $\vec{A} \cdot \vec{B}$ has a value of 14.0. What is the angle between the directions of $\vec{A}$ and $\vec{B}$?

••**40**　Displacement $\vec{d}_1$ is in the yz plane 63.0° from the positive direction of the y axis, has a positive z component, and has a magnitude of 4.50 m. Displacement $\vec{d}_2$ is in the xz plane 30.0° from the positive direction of the x axis, has a positive z component, and has magnitude 1.40 m. What are (a) $\vec{d}_1 \cdot \vec{d}_2$, (b) $\vec{d}_1 \times \vec{d}_2$, and (c) the angle between $\vec{d}_1$ and $\vec{d}_2$?

••**41**　**SSM**　**ILW**　**WWW**　Use the definition of scalar product, $\vec{a} \cdot \vec{b} = ab \cos \theta$, and the fact that $\vec{a} \cdot \vec{b} = a_x b_x + a_y b_y + a_z b_z$ to calculate the angle between the two vectors given by $\vec{a} = 3.0\hat{i} + 3.0\hat{j} + 3.0\hat{k}$ and $\vec{b} = 2.0\hat{i} + 1.0\hat{j} + 3.0\hat{k}$.

••**42**　In a meeting of mimes, mime 1 goes through a displacement $\vec{d}_1 = (4.0 \text{ m})\hat{i} + (5.0 \text{ m})\hat{j}$ and mime 2 goes through a displacement $\vec{d}_2 = (-3.0 \text{ m})\hat{i} + (4.0 \text{ m})\hat{j}$. What are (a) $\vec{d}_1 \times \vec{d}_2$, (b) $\vec{d}_1 \cdot \vec{d}_2$, (c) $(\vec{d}_1 + \vec{d}_2) \cdot \vec{d}_2$, and (d) the component of $\vec{d}_1$ along the direction of $\vec{d}_2$? (*Hint:* For (d), see Eq. 3-20 and Fig. 3-18.)

••**43**　**SSM**　**ILW**　The three vectors in Fig. 3-33 have magnitudes $a = 3.00$ m, $b = 4.00$ m, and $c = 10.0$ m and angle $\theta = 30.0°$. What are (a) the x component and (b) the y component of $\vec{a}$; (c) the x component and (d) the y component of $\vec{b}$; and (e) the x component and (f) the y component of $\vec{c}$? If $\vec{c} = p\vec{a} + q\vec{b}$, what are the values of (g) p and (h) q?

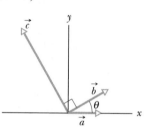

Fig. 3-33　Problem 43.

••**44**　**GO**　In the product $\vec{F} = q\vec{v} \times \vec{B}$, take $q = 2$,

$$\vec{v} = 2.0\hat{i} + 4.0\hat{j} + 6.0\hat{k} \quad \text{and} \quad \vec{F} = 4.0\hat{i} - 20\hat{j} + 12\hat{k}.$$

What then is $\vec{B}$ in unit-vector notation if $B_x = B_y$?

Additional Problems

45　Vectors $\vec{A}$ and $\vec{B}$ lie in an xy plane. $\vec{A}$ has magnitude 8.00 and angle 130°; $\vec{B}$ has components $B_x = -7.72$ and $B_y = -9.20$. (a) What is $5\vec{A} \cdot \vec{B}$? What is $4\vec{A} \times 3\vec{B}$ in (b) unit-vector notation and (c) magnitude-angle notation with spherical coordinates (see Fig. 3-34)? (d) What is the angle between the directions of $\vec{A}$ and $4\vec{A} \times 3\vec{B}$? (*Hint:* Think a bit before you resort to a calculation.) What is $\vec{A} + 3.00\hat{k}$ in (e) unit-vector notation and (f) magnitude-angle notation with spherical coordinates?

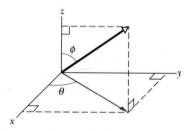

Fig. 3-34　Problem 45.

46　Vector $\vec{a}$ has a magnitude of 5.0 m and is directed east. Vector $\vec{b}$ has a magnitude of 4.0 m and is directed 35° west of due north. What are (a) the magnitude and (b) the direction of $\vec{a} + \vec{b}$? What are (c) the magnitude and (d) the direction of $\vec{b} - \vec{a}$? (e) Draw a vector diagram for each combination.

47　Vectors $\vec{A}$ and $\vec{B}$ lie in an xy plane. $\vec{A}$ has magnitude 8.00 and angle 130°; $\vec{B}$ has components $B_x = -7.72$ and $B_y = -9.20$. What are the angles between the negative direction of the y axis and (a) the direction of $\vec{A}$, (b) the direction of the product $\vec{A} \times \vec{B}$, and (c) the direction of $\vec{A} \times (\vec{B} + 3.00\hat{k})$?

48　Two vectors $\vec{a}$ and $\vec{b}$ have the components, in meters, $a_x = 3.2$, $a_y = 1.6$, $b_x = 0.50$, $b_y = 4.5$. (a) Find the angle between the directions of $\vec{a}$ and $\vec{b}$. There are two vectors in the xy plane that are perpendicular to $\vec{a}$ and have a magnitude of 5.0 m. One, vector $\vec{c}$, has a positive x component and the other, vector $\vec{d}$, a negative x component. What are (b) the x component and (c) the y component of vector $\vec{c}$, and (d) the x component and (e) the y component of vector $\vec{d}$?

49 SSM A sailboat sets out from the U.S. side of Lake Erie for a point on the Canadian side, 90.0 km due north. The sailor, however, ends up 50.0 km due east of the starting point. (a) How far and (b) in what direction must the sailor now sail to reach the original destination?

50 Vector $\vec{d}_1$ is in the negative direction of a y axis, and vector $\vec{d}_2$ is in the positive direction of an x axis. What are the directions of (a) $\vec{d}_2/4$ and (b) $\vec{d}_1/(-4)$? What are the magnitudes of products (c) $\vec{d}_1 \cdot \vec{d}_2$ and (d) $\vec{d}_1 \cdot (\vec{d}_2/4)$? What is the direction of the vector resulting from (e) $\vec{d}_1 \times \vec{d}_2$ and (f) $\vec{d}_2 \times \vec{d}_1$? What is the magnitude of the vector product in (g) part (e) and (h) part (f)? What are the (i) magnitude and (j) direction of $\vec{d}_1 \times (\vec{d}_2/4)$?

51 Rock *faults* are ruptures along which opposite faces of rock have slid past each other. In Fig. 3-35, points A and B coincided before the rock in the foreground slid down to the right. The net displacement $\overrightarrow{AB}$ is along the plane of the fault. The horizontal component of $\overrightarrow{AB}$ is the *strike-slip AC*. The component of $\overrightarrow{AB}$ that is directed down the plane of the fault is the *dip-slip AD*. (a) What is the magnitude of the net displacement $\overrightarrow{AB}$ if the strike-slip is 22.0 m and the dip-slip is 17.0 m? (b) If the plane of the fault is inclined at angle $\phi = 52.0°$ to the horizontal, what is the vertical component of $\overrightarrow{AB}$?

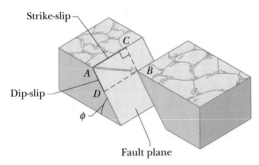

Fig. 3-35 Problem 51.

52 Here are three displacements, each measured in meters: $\vec{d}_1 = 4.0\hat{i} + 5.0\hat{j} - 6.0\hat{k}$, $\vec{d}_2 = -1.0\hat{i} + 2.0\hat{j} + 3.0\hat{k}$, and $\vec{d}_3 = 4.0\hat{i} + 3.0\hat{j} + 2.0\hat{k}$. (a) What is $\vec{r} = \vec{d}_1 - \vec{d}_2 + \vec{d}_3$? (b) What is the angle between $\vec{r}$ and the positive z axis? (c) What is the component of $\vec{d}_1$ along the direction of $\vec{d}_2$? (d) What is the component of $\vec{d}_1$ that is perpendicular to the direction of $\vec{d}_2$ and in the plane of $\vec{d}_1$ and $\vec{d}_2$? (*Hint:* For (c), consider Eq. 3-20 and Fig. 3-18; for (d), consider Eq. 3-27.)

53 SSM A vector $\vec{a}$ of magnitude 10 units and another vector $\vec{b}$ of magnitude 6.0 units differ in directions by 60°. Find (a) the scalar product of the two vectors and (b) the magnitude of the vector product $\vec{a} \times \vec{b}$.

54 For the vectors in Fig. 3-32, with $a = 4$, $b = 3$, and $c = 5$, calculate (a) $\vec{a} \cdot \vec{b}$, (b) $\vec{a} \cdot \vec{c}$, and (c) $\vec{b} \cdot \vec{c}$.

55 A particle undergoes three successive displacements in a plane, as follows: $\vec{d}_1$, 4.00 m southwest; then $\vec{d}_2$, 5.00 m east; and finally $\vec{d}_3$, 6.00 m in a direction 60.0° north of east. Choose a coordinate system with the y axis pointing north and the x axis pointing east. What are (a) the x component and (b) the y component of $\vec{d}_1$? What are (c) the x component and (d) the y component of $\vec{d}_2$? What are (e) the x component and (f) the y component of $\vec{d}_3$? Next, consider the *net* displacement of the particle for the three successive displacements. What are (g) the x component, (h) the y component, (i) the magnitude, and (j) the direction of the net displacement? If the particle is to return directly to the starting point, (k) how far and (l) in what direction should it move?

56 Find the sum of the following four vectors in (a) unit-vector notation, and as (b) a magnitude and (c) an angle relative to $+x$.

$\vec{P}$: 10.0 m, at 25.0° counterclockwise from $+x$

$\vec{Q}$: 12.0 m, at 10.0° counterclockwise from $+y$

$\vec{R}$: 8.00 m, at 20.0° clockwise from $-y$

$\vec{S}$: 9.00 m, at 40.0° counterclockwise from $-y$

57 SSM If $\vec{B}$ is added to $\vec{A}$, the result is $6.0\hat{i} + 1.0\hat{j}$. If $\vec{B}$ is subtracted from $\vec{A}$, the result is $-4.0\hat{i} + 7.0\hat{j}$. What is the magnitude of $\vec{A}$?

58 A vector $\vec{d}$ has a magnitude of 2.5 m and points north. What are (a) the magnitude and (b) the direction of $4.0\vec{d}$? What are (c) the magnitude and (d) the direction of $-3.0\vec{d}$?

59 $\vec{A}$ has the magnitude 12.0 m and is angled 60.0° counterclockwise from the positive direction of the x axis of an xy coordinate system. Also, $\vec{B} = (12.0 \text{ m})\hat{i} + (8.00 \text{ m})\hat{j}$ on that same coordinate system. We now rotate the system counterclockwise about the origin by 20.0° to form an $x'y'$ system. On this new system, what are (a) $\vec{A}$ and (b) $\vec{B}$, both in unit-vector notation?

60 If $\vec{a} - \vec{b} = 2\vec{c}$, $\vec{a} + \vec{b} = 4\vec{c}$, and $\vec{c} = 3\hat{i} + 4\hat{j}$, then what are (a) $\vec{a}$ and (b) $\vec{b}$?

61 (a) In unit-vector notation, what is $\vec{r} = \vec{a} - \vec{b} + \vec{c}$ if $\vec{a} = 5.0\hat{i} + 4.0\hat{j} - 6.0\hat{k}$, $\vec{b} = -2.0\hat{i} + 2.0\hat{j} + 3.0\hat{k}$, and $\vec{c} = 4.0\hat{i} + 3.0\hat{j} + 2.0\hat{k}$? (b) Calculate the angle between $\vec{r}$ and the positive z axis. (c) What is the component of $\vec{a}$ along the direction of $\vec{b}$? (d) What is the component of $\vec{a}$ perpendicular to the direction of $\vec{b}$ but in the plane of $\vec{a}$ and $\vec{b}$? (*Hint:* For (c), see Eq. 3-20 and Fig. 3-18; for (d), see Eq. 3-27.)

62 A golfer takes three putts to get the ball into the hole. The first putt displaces the ball 3.66 m north, the second 1.83 m southeast, and the third 0.91 m southwest. What are (a) the magnitude and (b) the direction of the displacement needed to get the ball into the hole on the first putt?

63 Here are three vectors in meters:

$$\vec{d}_1 = -3.0\hat{i} + 3.0\hat{j} + 2.0\hat{k}$$
$$\vec{d}_2 = -2.0\hat{i} - 4.0\hat{j} + 2.0\hat{k}$$
$$\vec{d}_3 = 2.0\hat{i} + 3.0\hat{j} + 1.0\hat{k}.$$

What results from (a) $\vec{d}_1 \cdot (\vec{d}_2 + \vec{d}_3)$, (b) $\vec{d}_1 \cdot (\vec{d}_2 \times \vec{d}_3)$, and (c) $\vec{d}_1 \times (\vec{d}_2 + \vec{d}_3)$?

64 Consider two displacements, one of magnitude 3 m and another of magnitude 4 m. Show how the displacement vectors may be combined to get a resultant displacement of magnitude (a) 7 m, (b) 1 m, and (c) 5 m.

65 A protester carries his sign of protest, starting from the origin of an xyz coordinate system, with the xy plane horizontal. He moves 40 m in the negative direction of the x axis, then 20 m along a perpendicular path to his left, and then 25 m up a water tower. (a) In unit-vector notation, what is the displacement of the sign from start to end? (b) The sign then falls to the foot of the tower. What is the magnitude of the displacement of the sign from start to this new end?

4

MOTION IN TWO AND THREE DIMENSIONS

4-1 WHAT IS PHYSICS?

In this chapter we continue looking at the aspect of physics that analyzes motion, but now the motion can be in two or three dimensions. For example, medical researchers and aeronautical engineers might concentrate on the physics of the two- and three-dimensional turns taken by fighter pilots in dogfights because a modern high-performance jet can take a tight turn so quickly that the pilot immediately loses consciousness. A sports engineer might focus on the physics of basketball. For example, in a *free throw* (where a player gets an uncontested shot at the basket from about 4.3 m), a player might employ the *overhand push shot,* in which the ball is pushed away from about shoulder height and then released. Or the player might use an *underhand loop shot,* in which the ball is brought upward from about the belt-line level and released. The first technique is the overwhelming choice among professional players, but the legendary Rick Barry set the record for free-throw shooting with the underhand technique.

Motion in three dimensions is not easy to understand. For example, you are probably good at driving a car along a freeway (one-dimensional motion) but would probably have a difficult time in landing an airplane on a runway (three-dimensional motion) without a lot of training.

In our study of two- and three-dimensional motion, we start with position and displacement.

4-2 Position and Displacement

One general way of locating a particle (or particle-like object) is with a **position vector** $\vec{r}$, which is a vector that extends from a reference point (usually the origin) to the particle. In the unit-vector notation of Section 3-5, $\vec{r}$ can be written

$$\vec{r} = x\hat{i} + y\hat{j} + z\hat{k}, \tag{4-1}$$

where $x\hat{i}, y\hat{j},$ and $z\hat{k}$ are the vector components of $\vec{r}$ and the coefficients $x, y,$ and z are its scalar components.

The coefficients $x, y,$ and z give the particle's location along the coordinate axes and relative to the origin; that is, the particle has the rectangular coordinates (x, y, z). For instance, Fig. 4-1 shows a particle with position vector

$$\vec{r} = (-3 \text{ m})\hat{i} + (2 \text{ m})\hat{j} + (5 \text{ m})\hat{k}$$

and rectangular coordinates $(-3 \text{ m}, 2 \text{ m}, 5 \text{ m})$. Along the x axis the particle is 3 m from the origin, in the $-\hat{i}$ direction. Along the y axis it is 2 m from the origin, in the $+\hat{j}$ direction. Along the z axis it is 5 m from the origin, in the $+\hat{k}$ direction.

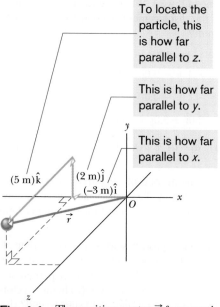

To locate the particle, this is how far parallel to z.

This is how far parallel to y.

This is how far parallel to x.

Fig. 4-1 The position vector $\vec{r}$ for a particle is the vector sum of its vector components.

As a particle moves, its position vector changes in such a way that the vector always extends to the particle from the reference point (the origin). If the position vector changes — say, from $\vec{r}_1$ to $\vec{r}_2$ during a certain time interval — then the particle's **displacement** $\Delta \vec{r}$ during that time interval is

$$\Delta \vec{r} = \vec{r}_2 - \vec{r}_1. \tag{4-2}$$

Using the unit-vector notation of Eq. 4-1, we can rewrite this displacement as

$$\Delta \vec{r} = (x_2 \hat{i} + y_2 \hat{j} + z_2 \hat{k}) - (x_1 \hat{i} + y_1 \hat{j} + z_1 \hat{k})$$

or as $\qquad \Delta \vec{r} = (x_2 - x_1)\hat{i} + (y_2 - y_1)\hat{j} + (z_2 - z_1)\hat{k}, \tag{4-3}$

where coordinates (x_1, y_1, z_1) correspond to position vector $\vec{r}_1$ and coordinates (x_2, y_2, z_2) correspond to position vector $\vec{r}_2$. We can also rewrite the displacement by substituting Δx for $(x_2 - x_1)$, Δy for $(y_2 - y_1)$, and Δz for $(z_2 - z_1)$:

$$\Delta \vec{r} = \Delta x \hat{i} + \Delta y \hat{j} + \Delta z \hat{k}. \tag{4-4}$$

Sample Problem

Two-dimensional position vector, rabbit run

A rabbit runs across a parking lot on which a set of coordinate axes has, strangely enough, been drawn. The coordinates (meters) of the rabbit's position as functions of time t (seconds) are given by

$$x = -0.31t^2 + 7.2t + 28 \tag{4-5}$$

and $\qquad y = 0.22t^2 - 9.1t + 30. \tag{4-6}$

(a) At $t = 15$ s, what is the rabbit's position vector $\vec{r}$ in unit-vector notation and in magnitude-angle notation?

KEY IDEA

The x and y coordinates of the rabbit's position, as given by Eqs. 4-5 and 4-6, are the scalar components of the rabbit's position vector $\vec{r}$.

Calculations: We can write

$$\vec{r}(t) = x(t)\hat{i} + y(t)\hat{j}. \tag{4-7}$$

(We write $\vec{r}(t)$ rather than $\vec{r}$ because the components are functions of t, and thus $\vec{r}$ is also.)

At $t = 15$ s, the scalar components are

$$x = (-0.31)(15)^2 + (7.2)(15) + 28 = 66 \text{ m}$$

and $\quad y = (0.22)(15)^2 - (9.1)(15) + 30 = -57$ m,

so $\qquad\qquad \vec{r} = (66 \text{ m})\hat{i} - (57 \text{ m})\hat{j},$ (Answer)

which is drawn in Fig. 4-2a. To get the magnitude and angle of $\vec{r}$, we use Eq. 3-6:

$$r = \sqrt{x^2 + y^2} = \sqrt{(66 \text{ m})^2 + (-57 \text{ m})^2}$$

$$= 87 \text{ m}, \qquad\qquad \text{(Answer)}$$

and $\quad \theta = \tan^{-1} \dfrac{y}{x} = \tan^{-1}\left(\dfrac{-57 \text{ m}}{66 \text{ m}}\right) = -41°.$ (Answer)

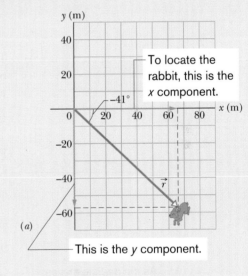

To locate the rabbit, this is the x component.

This is the y component.

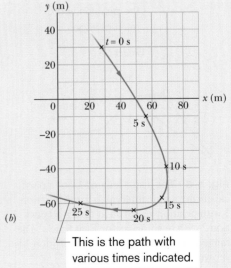

Fig. 4-2 (a) A rabbit's position vector $\vec{r}$ at time $t = 15$ s. The scalar components of $\vec{r}$ are shown along the axes. (b) The rabbit's path and its position at six values of t.

This is the path with various times indicated.

Check: Although $\theta = 139°$ has the same tangent as $-41°$, the components of position vector $\vec{r}$ indicate that the desired angle is $139° - 180° = -41°$.

(b) Graph the rabbit's path for $t = 0$ to $t = 25$ s.

Graphing: We have located the rabbit at one instant, but to see its path we need a graph. So we repeat part (a) for several values of t and then plot the results. Figure 4-2b shows the plots for six values of t and the path connecting them. We can also plot Eqs. 4-5 and 4-6 on a calculator.

 Additional examples, video, and practice available at *WileyPLUS*

4-3 Average Velocity and Instantaneous Velocity

If a particle moves from one point to another, we might need to know how fast it moves. Just as in Chapter 2, we can define two quantities that deal with "how fast": *average velocity* and *instantaneous velocity*. However, here we must consider these quantities as vectors and use vector notation.

If a particle moves through a displacement $\Delta\vec{r}$ in a time interval Δt, then its **average velocity** $\vec{v}_{\text{avg}}$ is

$$\text{average velocity} = \frac{\text{displacement}}{\text{time interval}},$$

or

$$\vec{v}_{\text{avg}} = \frac{\Delta\vec{r}}{\Delta t}. \qquad (4\text{-}8)$$

This tells us that the direction of $\vec{v}_{\text{avg}}$ (the vector on the left side of Eq. 4-8) must be the same as that of the displacement $\Delta\vec{r}$ (the vector on the right side). Using Eq. 4-4, we can write Eq. 4-8 in vector components as

$$\vec{v}_{\text{avg}} = \frac{\Delta x\hat{i} + \Delta y\hat{j} + \Delta z\hat{k}}{\Delta t} = \frac{\Delta x}{\Delta t}\hat{i} + \frac{\Delta y}{\Delta t}\hat{j} + \frac{\Delta z}{\Delta t}\hat{k}. \qquad (4\text{-}9)$$

For example, if a particle moves through displacement $(12 \text{ m})\hat{i} + (3.0 \text{ m})\hat{k}$ in 2.0 s, then its average velocity during that move is

$$\vec{v}_{\text{avg}} = \frac{\Delta\vec{r}}{\Delta t} = \frac{(12 \text{ m})\hat{i} + (3.0 \text{ m})\hat{k}}{2.0 \text{ s}} = (6.0 \text{ m/s})\hat{i} + (1.5 \text{ m/s})\hat{k}.$$

That is, the average velocity (a vector quantity) has a component of 6.0 m/s along the x axis and a component of 1.5 m/s along the z axis.

When we speak of the **velocity** of a particle, we usually mean the particle's **instantaneous velocity** $\vec{v}$ at some instant. This $\vec{v}$ is the value that $\vec{v}_{\text{avg}}$ approaches in the limit as we shrink the time interval Δt to 0 about that instant. Using the language of calculus, we may write $\vec{v}$ as the derivative

$$\vec{v} = \frac{d\vec{r}}{dt}. \qquad (4\text{-}10)$$

Figure 4-3 shows the path of a particle that is restricted to the xy plane. As the particle travels to the right along the curve, its position vector sweeps to the right. During time interval Δt, the position vector changes from $\vec{r}_1$ to $\vec{r}_2$ and the particle's displacement is $\Delta\vec{r}$.

To find the instantaneous velocity of the particle at, say, instant t_1 (when the particle is at position 1), we shrink interval Δt to 0 about t_1. Three things happen as we do so. (1) Position vector $\vec{r}_2$ in Fig. 4-3 moves toward $\vec{r}_1$ so that $\Delta\vec{r}$ shrinks toward zero. (2) The direction of $\Delta\vec{r}/\Delta t$ (and thus of $\vec{v}_{\text{avg}}$) approaches the direction of the line tangent to the particle's path at position 1. (3) The average velocity $\vec{v}_{\text{avg}}$ approaches the instantaneous velocity $\vec{v}$ at t_1.

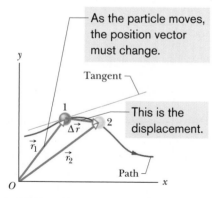

Fig. 4-3 The displacement $\Delta\vec{r}$ of a particle during a time interval Δt, from position 1 with position vector $\vec{r}_1$ at time t_1 to position 2 with position vector $\vec{r}_2$ at time t_2. The tangent to the particle's path at position 1 is shown.

In the limit as $\Delta t \to 0$, we have $\vec{v}_{avg} \to \vec{v}$ and, most important here, $\vec{v}_{avg}$ takes on the direction of the tangent line. Thus, $\vec{v}$ has that direction as well:

 The direction of the instantaneous velocity $\vec{v}$ of a particle is always tangent to the particle's path at the particle's position.

The result is the same in three dimensions: $\vec{v}$ is always tangent to the particle's path.

To write Eq. 4-10 in unit-vector form, we substitute for $\vec{r}$ from Eq. 4-1:

$$\vec{v} = \frac{d}{dt}(x\hat{i} + y\hat{j} + z\hat{k}) = \frac{dx}{dt}\hat{i} + \frac{dy}{dt}\hat{j} + \frac{dz}{dt}\hat{k}.$$

This equation can be simplified somewhat by writing it as

$$\vec{v} = v_x\hat{i} + v_y\hat{j} + v_z\hat{k}, \tag{4-11}$$

where the scalar components of $\vec{v}$ are

$$v_x = \frac{dx}{dt}, \quad v_y = \frac{dy}{dt}, \quad \text{and} \quad v_z = \frac{dz}{dt}. \tag{4-12}$$

For example, dx/dt is the scalar component of $\vec{v}$ along the x axis. Thus, we can find the scalar components of $\vec{v}$ by differentiating the scalar components of $\vec{r}$.

Figure 4-4 shows a velocity vector $\vec{v}$ and its scalar x and y components. Note that $\vec{v}$ is tangent to the particle's path at the particle's position. *Caution:* When a position vector is drawn, as in Figs. 4-1 through 4-3, it is an arrow that extends from one point (a "here") to another point (a "there"). However, when a velocity vector is drawn, as in Fig. 4-4, it does *not* extend from one point to another. Rather, it shows the instantaneous direction of travel of a particle at the tail, and its length (representing the velocity magnitude) can be drawn to any scale.

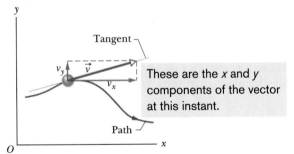

Fig. 4-4 The velocity $\vec{v}$ of a particle, along with the scalar components of $\vec{v}$.

**CHECKPOINT 1**

The figure shows a circular path taken by a particle. If the instantaneous velocity of the particle is $\vec{v} = (2\ \text{m/s})\hat{i} - (2\ \text{m/s})\hat{j}$, through which quadrant is the particle moving at that instant if it is traveling (a) clockwise and (b) counterclockwise around the circle? For both cases, draw $\vec{v}$ on the figure.

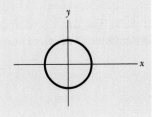

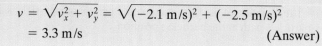

Sample Problem

Two-dimensional velocity, rabbit run

For the rabbit in the preceding Sample Problem, find the velocity $\vec{v}$ at time $t = 15$ s.

We can find $\vec{v}$ by taking derivatives of the components of the rabbit's position vector.

Calculations: Applying the v_x part of Eq. 4-12 to Eq. 4-5, we find the x component of $\vec{v}$ to be

$$v_x = \frac{dx}{dt} = \frac{d}{dt}(-0.31t^2 + 7.2t + 28)$$

$$= -0.62t + 7.2. \qquad (4-13)$$

At $t = 15$ s, this gives $v_x = -2.1$ m/s. Similarly, applying the v_y part of Eq. 4-12 to Eq. 4-6, we find

$$v_y = \frac{dy}{dt} = \frac{d}{dt}(0.22t^2 - 9.1t + 30)$$

$$= 0.44t - 9.1. \qquad (4-14)$$

At $t = 15$ s, this gives $v_y = -2.5$ m/s. Equation 4-11 then yields

$$\vec{v} = (-2.1 \text{ m/s})\hat{i} + (-2.5 \text{ m/s})\hat{j}, \qquad \text{(Answer)}$$

which is shown in Fig. 4-5, tangent to the rabbit's path and in the direction the rabbit is running at $t = 15$ s.

To get the magnitude and angle of $\vec{v}$, either we use a vector-capable calculator or we follow Eq. 3-6 to write

$$v = \sqrt{v_x^2 + v_y^2} = \sqrt{(-2.1 \text{ m/s})^2 + (-2.5 \text{ m/s})^2}$$

$$= 3.3 \text{ m/s} \qquad \text{(Answer)}$$

and

$$\theta = \tan^{-1}\frac{v_y}{v_x} = \tan^{-1}\left(\frac{-2.5 \text{ m/s}}{-2.1 \text{ m/s}}\right)$$

$$= \tan^{-1}1.19 = -130°. \qquad \text{(Answer)}$$

Check: Is the angle $-130°$ or $-130° + 180° = 50°$?

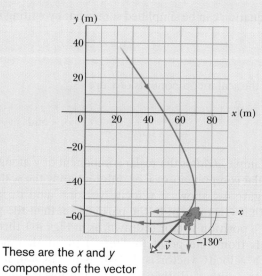

These are the x and y components of the vector at this instant.

Fig. 4-5 The rabbit's velocity $\vec{v}$ at $t = 15$ s.

 Additional examples, video, and practice available at *WileyPLUS*

4-4 Average Acceleration and Instantaneous Acceleration

When a particle's velocity changes from $\vec{v}_1$ to $\vec{v}_2$ in a time interval Δt, its **average acceleration** $\vec{a}_{\text{avg}}$ during Δt is

$$\frac{\text{average}}{\text{acceleration}} = \frac{\text{change in velocity}}{\text{time interval}},$$

or

$$\vec{a}_{\text{avg}} = \frac{\vec{v}_2 - \vec{v}_1}{\Delta t} = \frac{\Delta \vec{v}}{\Delta t}. \qquad (4-15)$$

If we shrink Δt to zero about some instant, then in the limit $\vec{a}_{\text{avg}}$ approaches the **instantaneous acceleration** (or **acceleration**) $\vec{a}$ at that instant; that is,

$$\vec{a} = \frac{d\vec{v}}{dt}. \qquad (4-16)$$

If the velocity changes in *either* magnitude *or* direction (or both), the particle must have an acceleration.

We can write Eq. 4-16 in unit-vector form by substituting Eq. 4-11 for $\vec{v}$ to obtain

$$\vec{a} = \frac{d}{dt}(v_x\hat{i} + v_y\hat{j} + v_z\hat{k})$$

$$= \frac{dv_x}{dt}\hat{i} + \frac{dv_y}{dt}\hat{j} + \frac{dv_z}{dt}\hat{k}.$$

We can rewrite this as

$$\vec{a} = a_x\hat{i} + a_y\hat{j} + a_z\hat{k}, \qquad (4\text{-}17)$$

where the scalar components of $\vec{a}$ are

$$a_x = \frac{dv_x}{dt}, \quad a_y = \frac{dv_y}{dt}, \quad \text{and} \quad a_z = \frac{dv_z}{dt}. \qquad (4\text{-}18)$$

To find the scalar components of $\vec{a}$, we differentiate the scalar components of $\vec{v}$.

Figure 4-6 shows an acceleration vector $\vec{a}$ and its scalar components for a particle moving in two dimensions. *Caution:* When an acceleration vector is drawn, as in Fig. 4-6, it does *not* extend from one position to another. Rather, it shows the direction of acceleration for a particle located at its tail, and its length (representing the acceleration magnitude) can be drawn to any scale.

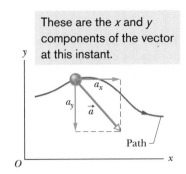

These are the x and y components of the vector at this instant.

Fig. 4-6 The acceleration $\vec{a}$ of a particle and the scalar components of $\vec{a}$.

✔ CHECKPOINT 2

Here are four descriptions of the position (in meters) of a puck as it moves in an xy plane:

(1) $x = -3t^2 + 4t - 2$ and $y = 6t^2 - 4t$ (3) $\vec{r} = 2t^2\hat{i} - (4t + 3)\hat{j}$

(2) $x = -3t^3 - 4t$ and $y = -5t^2 + 6$ (4) $\vec{r} = (4t^3 - 2t)\hat{i} + 3\hat{j}$

Are the x and y acceleration components constant? Is acceleration $\vec{a}$ constant?

Sample Problem

Two-dimensional acceleration, rabbit run

For the rabbit in the preceding two Sample Problems, find the acceleration $\vec{a}$ at time $t = 15$ s.

KEY IDEA

We can find $\vec{a}$ by taking derivatives of the rabbit's velocity components.

Calculations: Applying the a_x part of Eq. 4-18 to Eq. 4-13, we find the x component of $\vec{a}$ to be

$$a_x = \frac{dv_x}{dt} = \frac{d}{dt}(-0.62t + 7.2) = -0.62 \text{ m/s}^2.$$

Similarly, applying the a_y part of Eq. 4-18 to Eq. 4-14 yields the y component as

$$a_y = \frac{dv_y}{dt} = \frac{d}{dt}(0.44t - 9.1) = 0.44 \text{ m/s}^2.$$

We see that the acceleration does not vary with time (it is a constant) because the time variable t does not appear in the expression for either acceleration component. Equation 4-17 then yields

$$\vec{a} = (-0.62 \text{ m/s}^2)\hat{i} + (0.44 \text{ m/s}^2)\hat{j}, \quad \text{(Answer)}$$

which is superimposed on the rabbit's path in Fig. 4-7.

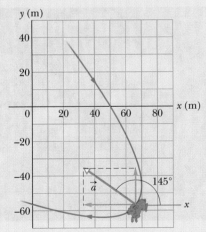

Fig. 4-7 The acceleration $\vec{a}$ of the rabbit at $t = 15$ s. The rabbit happens to have this same acceleration at all points on its path.

These are the x and y components of the vector at this instant.

To get the magnitude and angle of $\vec{a}$, either we use a vector-capable calculator or we follow Eq. 3-6. For the magnitude we have

$$a = \sqrt{a_x^2 + a_y^2} = \sqrt{(-0.62 \text{ m/s}^2)^2 + (0.44 \text{ m/s}^2)^2}$$
$$= 0.76 \text{ m/s}^2. \quad \text{(Answer)}$$

For the angle we have

$$\theta = \tan^{-1}\frac{a_y}{a_x} = \tan^{-1}\left(\frac{0.44 \text{ m/s}^2}{-0.62 \text{ m/s}^2}\right) = -35°.$$

However, this angle, which is the one displayed on a calcula-

tor, indicates that $\vec{a}$ is directed to the right and downward in Fig. 4-7. Yet, we know from the components that $\vec{a}$ must be directed to the left and upward. To find the other angle that has the same tangent as $-35°$ but is not displayed on a calculator, we add 180°:

$$-35° + 180° = 145°. \quad \text{(Answer)}$$

This *is* consistent with the components of $\vec{a}$ because it gives a vector that is to the left and upward. Note that $\vec{a}$ has the same magnitude and direction throughout the rabbit's run because the acceleration is constant.

 Additional examples, video, and practice available at *WileyPLUS*

4-5 Projectile Motion

We next consider a special case of two-dimensional motion: A particle moves in a vertical plane with some initial velocity $\vec{v}_0$ but its acceleration is always the free-fall acceleration $\vec{g}$, which is downward. Such a particle is called a **projectile** (meaning that it is projected or launched), and its motion is called **projectile motion.** A projectile might be a tennis ball (Fig. 4-8) or baseball in flight, but it is not an airplane or a duck in flight. Many sports (from golf and football to lacrosse and racquetball) involve the projectile motion of a ball, and much effort is spent in trying to control that motion for an advantage. For example, the racquetball player who discovered the Z-shot in the 1970s easily won his games because the ball's peculiar flight to the rear of the court always perplexed his opponents.

Our goal here is to analyze projectile motion using the tools for two-dimensional motion described in Sections 4-2 through 4-4 and making the assumption that air has no effect on the projectile. Figure 4-9, which is analyzed in the next section, shows the path followed by a projectile when the air has no effect. The projectile is launched with an initial velocity $\vec{v}_0$ that can be written as

$$\vec{v}_0 = v_{0x}\hat{i} + v_{0y}\hat{j}. \quad (4\text{-}19)$$

The components v_{0x} and v_{0y} can then be found if we know the angle θ_0 between $\vec{v}_0$ and the positive x direction:

$$v_{0x} = v_0 \cos \theta_0 \quad \text{and} \quad v_{0y} = v_0 \sin \theta_0. \quad (4\text{-}20)$$

During its two-dimensional motion, the projectile's position vector $\vec{r}$ and velocity vector $\vec{v}$ change continuously, but its acceleration vector $\vec{a}$ is constant and *always* directed vertically downward. The projectile has *no* horizontal acceleration.

Projectile motion, like that in Figs. 4-8 and 4-9, looks complicated, but we have the following simplifying feature (known from experiment):

> In projectile motion, the horizontal motion and the vertical motion are independent of each other; that is, neither motion affects the other.

Fig. 4-8 A stroboscopic photograph of a yellow tennis ball bouncing off a hard surface. Between impacts, the ball has projectile motion. *Source:* Richard Megna/Fundamental Photographs.

This feature allows us to break up a problem involving two-dimensional motion into two separate and easier one-dimensional problems, one for the horizontal motion (with *zero acceleration*) and one for the vertical motion (with *constant downward acceleration*). Here are two experiments that show that the horizontal motion and the vertical motion are independent.

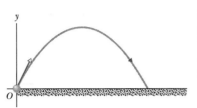

Fig. 4-9 The *projectile motion* of an object launched into the air at the origin of a coordinate system and with launch velocity $\vec{v}_0$ at angle θ_0. The motion is a combination of vertical motion (constant acceleration) and horizontal motion (constant velocity), as shown by the velocity components.

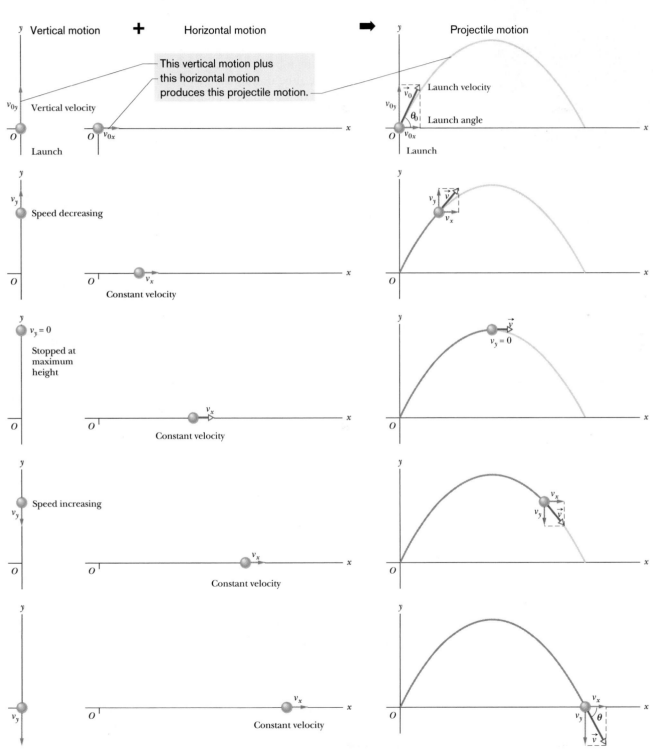

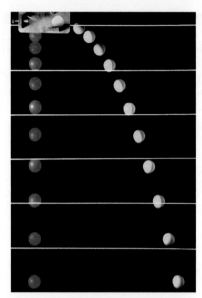

Fig. 4-10 One ball is released from rest at the same instant that another ball is shot horizontally to the right. Their vertical motions are identical. *Source:* Richard Megna/ Fundamental Photographs.

Two Golf Balls

Figure 4-10 is a stroboscopic photograph of two golf balls, one simply released and the other shot horizontally by a spring. The golf balls have the same vertical motion, both falling through the same vertical distance in the same interval of time. *The fact that one ball is moving horizontally while it is falling has no effect on its vertical motion;* that is, the horizontal and vertical motions are independent of each other.

A Great Student Rouser

Figure 4-11 shows a demonstration that has enlivened many a physics lecture. It involves a blowgun G, using a ball as a projectile. The target is a can suspended from a magnet M, and the tube of the blowgun is aimed directly at the can. The experiment is arranged so that the magnet releases the can just as the ball leaves the blowgun.

If *g* (the magnitude of the free-fall acceleration) were zero, the ball would follow the straight-line path shown in Fig. 4-11 and the can would float in place after the magnet released it. The ball would certainly hit the can.

However, *g* is *not* zero, but the ball *still* hits the can! As Fig. 4-11 shows, during the time of flight of the ball, both ball and can fall the same distance *h* from their zero-*g* locations. The harder the demonstrator blows, the greater is the ball's initial speed, the shorter the flight time, and the smaller the value of *h*.

**CHECKPOINT 3**

At a certain instant, a fly ball has velocity $\vec{v} = 25\hat{i} - 4.9\hat{j}$ (the *x* axis is horizontal, the *y* axis is upward, and $\vec{v}$ is in meters per second). Has the ball passed its highest point?

The ball and the can fall the same distance *h*.

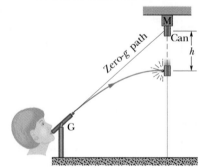

Fig. 4-11 The projectile ball always hits the falling can. Each falls a distance *h* from where it would be were there no free-fall acceleration.

4-6 Projectile Motion Analyzed

Now we are ready to analyze projectile motion, horizontally and vertically.

The Horizontal Motion

Because there is *no acceleration* in the horizontal direction, the horizontal component v_x of the projectile's velocity remains unchanged from its initial value v_{0x} throughout the motion, as demonstrated in Fig. 4-12. At any time *t*, the projec-

Fig. 4-12 The vertical component of this skateboarder's velocity is changing but not the horizontal component, which matches the skateboard's velocity. As a result, the skateboard stays underneath him, allowing him to land on it. *Source:* Jamie Budge/ Liaison/Getty Images, Inc.

tile's horizontal displacement $x - x_0$ from an initial position x_0 is given by Eq. 2-15 with $a = 0$, which we write as

$$x - x_0 = v_{0x}t.$$

Because $v_{0x} = v_0 \cos \theta_0$, this becomes

$$x - x_0 = (v_0 \cos \theta_0)t. \tag{4-21}$$

The Vertical Motion

The vertical motion is the motion we discussed in Section 2-9 for a particle in free fall. Most important is that the acceleration is constant. Thus, the equations of Table 2-1 apply, provided we substitute $-g$ for a and switch to y notation. Then, for example, Eq. 2-15 becomes

$$y - y_0 = v_{0y}t - \tfrac{1}{2}gt^2$$
$$= (v_0 \sin \theta_0)t - \tfrac{1}{2}gt^2, \tag{4-22}$$

where the initial vertical velocity component v_{0y} is replaced with the equivalent $v_0 \sin \theta_0$. Similarly, Eqs. 2-11 and 2-16 become

$$v_y = v_0 \sin \theta_0 - gt \tag{4-23}$$

and

$$v_y^2 = (v_0 \sin \theta_0)^2 - 2g(y - y_0). \tag{4-24}$$

As is illustrated in Fig. 4-9 and Eq. 4-23, the vertical velocity component behaves just as for a ball thrown vertically upward. It is directed upward initially, and its magnitude steadily decreases to zero, *which marks the maximum height of the path*. The vertical velocity component then reverses direction, and its magnitude becomes larger with time.

The Equation of the Path

We can find the equation of the projectile's path (its **trajectory**) by eliminating time t between Eqs. 4-21 and 4-22. Solving Eq. 4-21 for t and substituting into Eq. 4-22, we obtain, after a little rearrangement,

$$y = (\tan \theta_0)x - \frac{gx^2}{2(v_0 \cos \theta_0)^2} \qquad \text{(trajectory).} \tag{4-25}$$

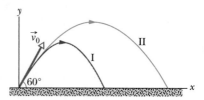

Fig. 4-13   (I) The path of a fly ball calculated by taking air resistance into account. (II) The path the ball would follow in a vacuum, calculated by the methods of this chapter. See Table 4-1 for corresponding data. (Adapted from "The Trajectory of a Fly Ball," by Peter J. Brancazio, *The Physics Teacher,* January 1985.)

Table 4-1		
Two Fly Balls[a]		
	Path I (Air)	Path II (Vacuum)
Range	98.5 m	177 m
Maximum height	53.0 m	76.8 m
Time of flight	6.6 s	7.9 s

[a]See Fig. 4-13. The launch angle is 60° and the launch speed is 44.7 m/s.

This is the equation of the path shown in Fig. 4-9. In deriving it, for simplicity we let $x_0 = 0$ and $y_0 = 0$ in Eqs. 4-21 and 4-22, respectively. Because g, θ_0, and v_0 are constants, Eq. 4-25 is of the form $y = ax + bx^2$, in which a and b are constants. This is the equation of a parabola, so the path is *parabolic.*

The Horizontal Range

The *horizontal range R* of the projectile is the *horizontal* distance the projectile has traveled when it returns to its initial height (the height at which it is launched). To find range R, let us put $x - x_0 = R$ in Eq. 4-21 and $y - y_0 = 0$ in Eq. 4-22, obtaining

$$R = (v_0 \cos \theta_0)t$$

and
$$0 = (v_0 \sin \theta_0)t - \tfrac{1}{2}gt^2.$$

Eliminating t between these two equations yields

$$R = \frac{2v_0^2}{g} \sin \theta_0 \cos \theta_0.$$

Using the identity $\sin 2\theta_0 = 2 \sin \theta_0 \cos \theta_0$ (see Appendix E), we obtain

$$R = \frac{v_0^2}{g} \sin 2\theta_0. \tag{4-26}$$

Caution: This equation does *not* give the horizontal distance traveled by a projectile when the final height is not the launch height.

Note that R in Eq. 4-26 has its maximum value when $\sin 2\theta_0 = 1$, which corresponds to $2\theta_0 = 90°$ or $\theta_0 = 45°$.

 The horizontal range R is maximum for a launch angle of 45°.

However, when the launch and landing heights differ, as in shot put, hammer throw, and basketball, a launch angle of 45° does not yield the maximum horizontal distance.

The Effects of the Air

We have assumed that the air through which the projectile moves has no effect on its motion. However, in many situations, the disagreement between our calculations and the actual motion of the projectile can be large because the air resists (opposes) the motion. Figure 4-13, for example, shows two paths for a fly ball that leaves the bat at an angle of 60° with the horizontal and an initial speed of 44.7 m/s. Path I (the baseball player's fly ball) is a calculated path that approximates normal conditions of play, in air. Path II (the physics professor's fly ball) is the path the ball would follow in a vacuum.

✓CHECKPOINT 4

A fly ball is hit to the outfield. During its flight (ignore the effects of the air), what happens to its (a) horizontal and (b) vertical components of velocity? What are the (c) horizontal and (d) vertical components of its acceleration during ascent, during descent, and at the topmost point of its flight?

Sample Problem

Projectile dropped from airplane

In Fig. 4-14, a rescue plane flies at 198 km/h ($= 55.0$ m/s) and constant height $h = 500$ m toward a point directly over a victim, where a rescue capsule is to land.

(a) What should be the angle ϕ of the pilot's line of sight to the victim when the capsule release is made?

KEY IDEAS

Once released, the capsule is a projectile, so its horizontal and vertical motions can be considered separately (we need not consider the actual curved path of the capsule).

Calculations: In Fig. 4-14, we see that ϕ is given by

$$\phi = \tan^{-1}\frac{x}{h}, \qquad (4\text{-}27)$$

where x is the horizontal coordinate of the victim (and of the capsule when it hits the water) and $h = 500$ m. We should be able to find x with Eq. 4-21:

$$x - x_0 = (v_0 \cos \theta_0)t. \qquad (4\text{-}28)$$

Here we know that $x_0 = 0$ because the origin is placed at the point of release. Because the capsule is *released* and not shot from the plane, its initial velocity $\vec{v}_0$ is equal to the plane's velocity. Thus, we know also that the initial velocity has magnitude $v_0 = 55.0$ m/s and angle $\theta_0 = 0°$ (measured relative to the positive direction of the x axis). However, we do not know the time t the capsule takes to move from the plane to the victim.

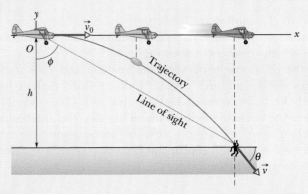

Fig. 4-14 A plane drops a rescue capsule while moving at constant velocity in level flight. While falling, the capsule remains under the plane.

To find t, we next consider the *vertical* motion and specifically Eq. 4-22:

$$y - y_0 = (v_0 \sin \theta_0)t - \tfrac{1}{2}gt^2. \qquad (4\text{-}29)$$

Here the vertical displacement $y - y_0$ of the capsule is -500 m (the negative value indicates that the capsule moves *downward*). So,

$$-500 \text{ m} = (55.0 \text{ m/s})(\sin 0°)t - \tfrac{1}{2}(9.8 \text{ m/s}^2)t^2. \quad (4\text{-}30)$$

Solving for t, we find $t = 10.1$ s. Using that value in Eq. 4-28 yields

$$x - 0 = (55.0 \text{ m/s})(\cos 0°)(10.1 \text{ s}), \qquad (4\text{-}31)$$

or $x = 555.5$ m.

Then Eq. 4-27 gives us

$$\phi = \tan^{-1}\frac{555.5 \text{ m}}{500 \text{ m}} = 48.0°. \qquad \text{(Answer)}$$

(b) As the capsule reaches the water, what is its velocity $\vec{v}$ in unit-vector notation and in magnitude-angle notation?

KEY IDEAS

(1) The horizontal and vertical components of the capsule's velocity are independent. (2) Component v_x does not change from its initial value $v_{0x} = v_0 \cos \theta_0$ because there is no horizontal acceleration. (3) Component v_y changes from its initial value $v_{0y} = v_0 \sin \theta_0$ because there is a vertical acceleration.

Calculations: When the capsule reaches the water,

$$v_x = v_0 \cos \theta_0 = (55.0 \text{ m/s})(\cos 0°) = 55.0 \text{ m/s}.$$

Using Eq. 4-23 and the capsule's time of fall $t = 10.1$ s, we also find that when the capsule reaches the water,

$$v_y = v_0 \sin \theta_0 - gt \qquad (4\text{-}32)$$
$$= (55.0 \text{ m/s})(\sin 0°) - (9.8 \text{ m/s}^2)(10.1 \text{ s})$$
$$= -99.0 \text{ m/s}.$$

Thus, at the water

$$\vec{v} = (55.0 \text{ m/s})\hat{i} - (99.0 \text{ m/s})\hat{j}. \qquad \text{(Answer)}$$

Using Eq. 3-6 as a guide, we find that the magnitude and the angle of $\vec{v}$ are

$$v = 113 \text{ m/s} \quad \text{and} \quad \theta = -60.9°. \qquad \text{(Answer)}$$

 Additional examples, video, and practice available at *WileyPLUS*

Sample Problem

Cannonball to pirate ship

Figure 4-15 shows a pirate ship 560 m from a fort defending a harbor entrance. A defense cannon, located at sea level, fires balls at initial speed $v_0 = 82$ m/s.

(a) At what angle θ_0 from the horizontal must a ball be fired to hit the ship?

KEY IDEAS

(1) A fired cannonball is a projectile. We want an equation that relates the launch angle θ_0 to the ball's horizontal displacement as it moves from cannon to ship. (2) Because the cannon and the ship are at the same height, the horizontal displacement is the range.

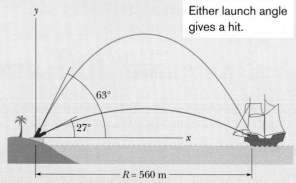

Either launch angle gives a hit.

Fig. 4-15 A pirate ship under fire.

Calculations: We can relate the launch angle θ_0 to the range R with Eq. 4-26 which, after rearrangement, gives

$$\theta_0 = \frac{1}{2}\sin^{-1}\frac{gR}{v_0^2} = \frac{1}{2}\sin^{-1}\frac{(9.8 \text{ m/s}^2)(560 \text{ m})}{(82 \text{ m/s})^2}$$

$$= \frac{1}{2}\sin^{-1} 0.816. \tag{4-33}$$

One solution of $\sin^{-1}$ (54.7°) is displayed by a calculator; we subtract it from 180° to get the other solution (125.3°). Thus, Eq. 4-33 gives us

$$\theta_0 = 27° \quad \text{and} \quad \theta_0 = 63°. \quad \text{(Answer)}$$

(b) What is the maximum range of the cannonballs?

Calculations: We have seen that maximum range corresponds to an elevation angle θ_0 of 45°. Thus,

$$R = \frac{v_0^2}{g}\sin 2\theta_0 = \frac{(82 \text{ m/s})^2}{9.8 \text{ m/s}^2}\sin (2 \times 45°)$$

$$= 686 \text{ m} \approx 690 \text{ m}. \quad \text{(Answer)}$$

As the pirate ship sails away, the two elevation angles at which the ship can be hit draw together, eventually merging at $\theta_0 = 45°$ when the ship is 690 m away. Beyond that distance the ship is safe. However, the cannonballs could go farther if the cannon were higher.

PLUS Additional examples, video, and practice available at *WileyPLUS*

4-7 Uniform Circular Motion

A particle is in **uniform circular motion** if it travels around a circle or a circular arc at constant (*uniform*) speed. Although the speed does not vary, *the particle is accelerating* because the velocity changes in direction.

Figure 4-16 shows the relationship between the velocity and acceleration vectors at various stages during uniform circular motion. Both vectors have constant magnitude, but their directions change continuously. The velocity is always directed tangent to the circle in the direction of motion. The acceleration is always directed *radially inward*. Because of this, the acceleration associated with uniform circular motion is called a **centripetal** (meaning "center seeking") **acceleration.** As we prove next, the magnitude of this acceleration $\vec{a}$ is

$$a = \frac{v^2}{r} \quad \text{(centripetal acceleration)}, \tag{4-34}$$

where r is the radius of the circle and v is the speed of the particle.

In addition, during this acceleration at constant speed, the particle travels the circumference of the circle (a distance of $2\pi r$) in time

$$T = \frac{2\pi r}{v} \quad \text{(period)}. \tag{4-35}$$

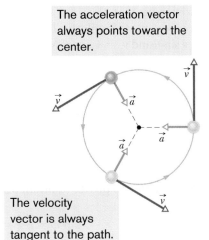

The acceleration vector always points toward the center.

The velocity vector is always tangent to the path.

Fig. 4-16 Velocity and acceleration vectors for uniform circular motion.

T is called the *period of revolution,* or simply the *period,* of the motion. It is, in general, the time for a particle to go around a closed path exactly once.

Proof of Eq. 4-34

To find the magnitude and direction of the acceleration for uniform circular motion, we consider Fig. 4-17. In Fig. 4-17a, particle p moves at constant speed v around a circle of radius r. At the instant shown, p has coordinates x_p and y_p.

Recall from Section 4-3 that the velocity $\vec{v}$ of a moving particle is always tangent to the particle's path at the particle's position. In Fig. 4-17a, that means $\vec{v}$ is perpendicular to a radius r drawn to the particle's position. Then the angle θ that $\vec{v}$ makes with a vertical at p equals the angle θ that radius r makes with the x axis.

The scalar components of $\vec{v}$ are shown in Fig. 4-17b. With them, we can write the velocity $\vec{v}$ as

$$\vec{v} = v_x\hat{i} + v_y\hat{j} = (-v\sin\theta)\hat{i} + (v\cos\theta)\hat{j}. \qquad (4\text{-}36)$$

Now, using the right triangle in Fig. 4-17a, we can replace $\sin\theta$ with y_p/r and $\cos\theta$ with x_p/r to write

$$\vec{v} = \left(-\frac{vy_p}{r}\right)\hat{i} + \left(\frac{vx_p}{r}\right)\hat{j}. \qquad (4\text{-}37)$$

To find the acceleration $\vec{a}$ of particle p, we must take the time derivative of this equation. Noting that speed v and radius r do not change with time, we obtain

$$\vec{a} = \frac{d\vec{v}}{dt} = \left(-\frac{v}{r}\frac{dy_p}{dt}\right)\hat{i} + \left(\frac{v}{r}\frac{dx_p}{dt}\right)\hat{j}. \qquad (4\text{-}38)$$

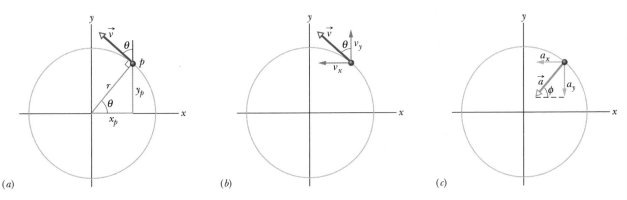

(a)   (b)   (c)

Fig. 4-17 Particle p moves in counterclockwise uniform circular motion. (a) Its position and velocity $\vec{v}$ at a certain instant. (b) Velocity $\vec{v}$. (c) Acceleration $\vec{a}$.

Now note that the rate dy_p/dt at which y_p changes is equal to the velocity component v_y. Similarly, $dx_p/dt = v_x$, and, again from Fig. 4-17b, we see that $v_x = -v \sin \theta$ and $v_y = v \cos \theta$. Making these substitutions in Eq. 4-38, we find

$$\vec{a} = \left(-\frac{v^2}{r} \cos \theta\right)\hat{i} + \left(-\frac{v^2}{r} \sin \theta\right)\hat{j}. \tag{4-39}$$

This vector and its components are shown in Fig. 4-17c. Following Eq. 3-6, we find

$$a = \sqrt{a_x^2 + a_y^2} = \frac{v^2}{r} \sqrt{(\cos \theta)^2 + (\sin \theta)^2} = \frac{v^2}{r} \sqrt{1} = \frac{v^2}{r},$$

as we wanted to prove. To orient $\vec{a}$, we find the angle ϕ shown in Fig. 4-17c:

$$\tan \phi = \frac{a_y}{a_x} = \frac{-(v^2/r) \sin \theta}{-(v^2/r) \cos \theta} = \tan \theta.$$

Thus, $\phi = \theta$, which means that $\vec{a}$ is directed along the radius r of Fig. 4-17a, toward the circle's center, as we wanted to prove.

✓ CHECKPOINT 5

An object moves at constant speed along a circular path in a horizontal xy plane, with the center at the origin. When the object is at $x = -2$ m, its velocity is $-(4$ m/s$)\hat{j}$. Give the object's (a) velocity and (b) acceleration at $y = 2$ m.

Sample Problem

Top gun pilots in turns

"Top gun" pilots have long worried about taking a turn too tightly. As a pilot's body undergoes centripetal acceleration, with the head toward the center of curvature, the blood pressure in the brain decreases, leading to loss of brain function.

There are several warning signs. When the centripetal acceleration is 2g or 3g, the pilot feels heavy. At about 4g, the pilot's vision switches to black and white and narrows to "tunnel vision." If that acceleration is sustained or increased, vision ceases and, soon after, the pilot is unconscious—a condition known as g-LOC for "g-induced loss of consciousness."

What is the magnitude of the acceleration, in g units, of a pilot whose aircraft enters a horizontal circular turn with a velocity of $\vec{v}_i = (400\hat{i} + 500\hat{j})$ m/s and 24.0 s later leaves the turn with a velocity of $\vec{v}_f = (-400\hat{i} - 500\hat{j})$ m/s?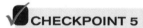

KEY IDEAS

We assume the turn is made with uniform circular motion. Then the pilot's acceleration is centripetal and has magnitude a given by Eq. 4-34 ($a = v^2/R$), where R is the cir-

cle's radius. Also, the time required to complete a full circle is the period given by Eq. 4-35 ($T = 2\pi R/v$).

Calculations: Because we do not know radius R, let's solve Eq. 4-35 for R and substitute into Eq. 4-34. We find

$$a = \frac{2\pi v}{T}.$$

Speed v here is the (constant) magnitude of the velocity during the turning. Let's substitute the components of the initial velocity into Eq. 3-6:

$$v = \sqrt{(400 \text{ m/s})^2 + (500 \text{ m/s})^2} = 640.31 \text{ m/s}.$$

To find the period T of the motion, first note that the final velocity is the reverse of the initial velocity. This means the aircraft leaves on the opposite side of the circle from the initial point and must have completed half a circle in the given 24.0 s. Thus a full circle would have taken $T = 48.0$ s. Substituting these values into our equation for a, we find

$$a = \frac{2\pi(640.31 \text{ m/s})}{48.0 \text{ s}} = 83.81 \text{ m/s}^2 \approx 8.6g. \quad \text{(Answer)}$$

 Additional examples, video, and practice available at *WileyPLUS*

4-8 Relative Motion in One Dimension

Suppose you see a duck flying north at 30 km/h. To another duck flying alongside, the first duck seems to be stationary. In other words, the velocity of a particle depends on the **reference frame** of whoever is observing or measuring the velocity. For our purposes, a reference frame is the physical object to which we attach our coordinate system. In everyday life, that object is the ground. For example, the speed listed on a speeding ticket is always measured relative to the ground. The speed relative to the police officer would be different if the officer were moving while making the speed measurement.

Suppose that Alex (at the origin of frame A in Fig. 4-18) is parked by the side of a highway, watching car P (the "particle") speed past. Barbara (at the origin of frame B) is driving along the highway at constant speed and is also watching car P. Suppose that they both measure the position of the car at a given moment. From Fig. 4-18 we see that

$$x_{PA} = x_{PB} + x_{BA}. \tag{4-40}$$

The equation is read: "The coordinate x_{PA} of P as measured by A *is equal to* the coordinate x_{PB} of P as measured by B *plus* the coordinate x_{BA} of B as measured by A." Note how this reading is supported by the sequence of the subscripts.

Taking the time derivative of Eq. 4-40, we obtain

$$\frac{d}{dt}(x_{PA}) = \frac{d}{dt}(x_{PB}) + \frac{d}{dt}(x_{BA}).$$

Thus, the velocity components are related by

$$v_{PA} = v_{PB} + v_{BA}. \tag{4-41}$$

This equation is read: "The velocity v_{PA} of P as measured by A *is equal to* the velocity v_{PB} of P as measured by B *plus* the velocity v_{BA} of B as measured by A." The term v_{BA} is the velocity of frame B relative to frame A.

Here we consider only frames that move at constant velocity relative to each other. In our example, this means that Barbara (frame B) drives always at constant velocity v_{BA} relative to Alex (frame A). Car P (the moving particle), however, can change speed and direction (that is, it can accelerate).

To relate an acceleration of P as measured by Barbara and by Alex, we take the time derivative of Eq. 4-41:

$$\frac{d}{dt}(v_{PA}) = \frac{d}{dt}(v_{PB}) + \frac{d}{dt}(v_{BA}).$$

Because v_{BA} is constant, the last term is zero and we have

$$a_{PA} = a_{PB}. \tag{4-42}$$

In other words,

> Observers on different frames of reference that move at constant velocity relative to each other will measure the same acceleration for a moving particle.

Frame B moves past frame A while both observe P.

Fig. 4-18 Alex (frame A) and Barbara (frame B) watch car P, as both B and P move at different velocities along the common x axis of the two frames. At the instant shown, x_{BA} is the coordinate of B in the A frame. Also, P is at coordinate x_{PB} in the B frame and coordinate $x_{PA} = x_{PB} + x_{BA}$ in the A frame.

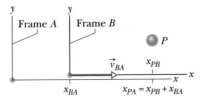

Sample Problem

Relative motion, one dimensional, Alex and Barbara

In Fig. 4-18, suppose that Barbara's velocity relative to Alex is a constant $v_{BA} = 52$ km/h and car P is moving in the negative direction of the x axis.

(a) If Alex measures a constant $v_{PA} = -78$ km/h for car P, what velocity v_{PB} will Barbara measure?

KEY IDEAS

We can attach a frame of reference A to Alex and a frame of reference B to Barbara. Because the frames move at constant velocity relative to each other along one axis, we can use Eq. 4-41 ($v_{PA} = v_{PB} + v_{BA}$) to relate v_{PB} to v_{PA} and v_{BA}.

Calculation: We find

$$-78 \text{ km/h} = v_{PB} + 52 \text{ km/h}.$$

Thus, $$v_{PB} = -130 \text{ km/h}. \qquad \text{(Answer)}$$

Comment: If car P were connected to Barbara's car by a cord wound on a spool, the cord would be unwinding at a speed of 130 km/h as the two cars separated.

(b) If car P brakes to a stop relative to Alex (and thus relative to the ground) in time $t = 10$ s at constant acceleration, what is its acceleration a_{PA} relative to Alex?

KEY IDEAS

To calculate the acceleration of car P *relative to Alex,* we must use the car's velocities *relative to Alex.* Because the

acceleration is constant, we can use Eq. 2-11 ($v = v_0 + at$) to relate the acceleration to the initial and final velocities of P.

Calculation: The initial velocity of P relative to Alex is $v_{PA} = -78$ km/h and the final velocity is 0. Thus, the acceleration relative to Alex is

$$a_{PA} = \frac{v - v_0}{t} = \frac{0 - (-78 \text{ km/h})}{10 \text{ s}} \frac{1 \text{ m/s}}{3.6 \text{ km/h}}$$

$$= 2.2 \text{ m/s}^2. \qquad \text{(Answer)}$$

(c) What is the acceleration a_{PB} of car P relative to Barbara during the braking?

KEY IDEA

To calculate the acceleration of car P *relative to Barbara,* we must use the car's velocities *relative to Barbara.*

Calculation: We know the initial velocity of P relative to Barbara from part (a) ($v_{PB} = -130$ km/h). The final velocity of P relative to Barbara is -52 km/h (this is the velocity of the stopped car relative to the moving Barbara). Thus,

$$a_{PB} = \frac{v - v_0}{t} = \frac{-52 \text{ km/h} - (-130 \text{ km/h})}{10 \text{ s}} \frac{1 \text{ m/s}}{3.6 \text{ km/h}}$$

$$= 2.2 \text{ m/s}^2. \qquad \text{(Answer)}$$

Comment: We should have foreseen this result: Because Alex and Barbara have a constant relative velocity, they must measure the same acceleration for the car.

 Additional examples, video, and practice available at *WileyPLUS*

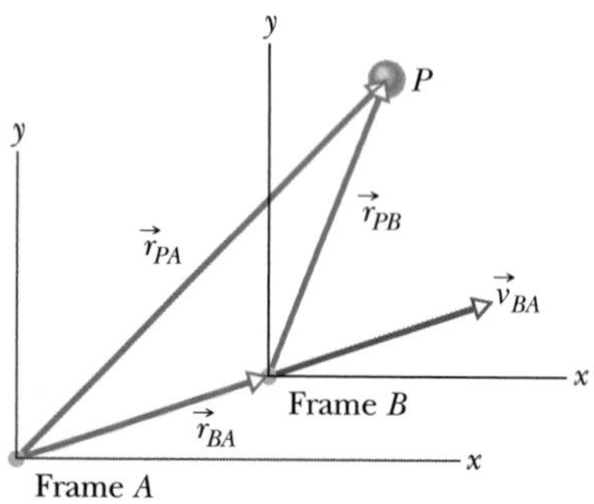

Fig. 4-19   Frame B has the constant two-dimensional velocity $\vec{v}_{BA}$ relative to frame A. The position vector of B relative to A is $\vec{r}_{BA}$. The position vectors of particle P are $\vec{r}_{PA}$ relative to A and $\vec{r}_{PB}$ relative to B.

4-9 Relative Motion in Two Dimensions

Our two observers are again watching a moving particle P from the origins of reference frames A and B, while B moves at a constant velocity $\vec{v}_{BA}$ relative to A. (The corresponding axes of these two frames remain parallel.) Figure 4-19 shows a certain instant during the motion. At that instant, the position vector of the origin of B relative to the origin of A is $\vec{r}_{BA}$. Also, the position vectors of particle P are $\vec{r}_{PA}$ relative to the origin of A and $\vec{r}_{PB}$ relative to the origin of B. From the arrangement of heads and tails of those three position vectors, we can relate the vectors with

$$\vec{r}_{PA} = \vec{r}_{PB} + \vec{r}_{BA}. \qquad (4\text{-}43)$$

By taking the time derivative of this equation, we can relate the velocities $\vec{v}_{PA}$ and $\vec{v}_{PB}$ of particle P relative to our observers:

$$\vec{v}_{PA} = \vec{v}_{PB} + \vec{v}_{BA}. \qquad (4\text{-}44)$$

By taking the time derivative of this relation, we can relate the accelerations $\vec{a}_{PA}$ and $\vec{a}_{PB}$ of the particle P relative to our observers. However, note that because $\vec{v}_{BA}$ is constant, its time derivative is zero. Thus, we get

$$\vec{a}_{PA} = \vec{a}_{PB}. \qquad (4\text{-}45)$$

As for one-dimensional motion, we have the following rule: Observers on different frames of reference that move at constant velocity relative to each other will measure the *same* acceleration for a moving particle.

Sample Problem

Relative motion, two dimensional, airplanes

In Fig. 4-20a, a plane moves due east while the pilot points the plane somewhat south of east, toward a steady wind that blows to the northeast. The plane has velocity $\vec{v}_{PW}$ relative to the wind, with an airspeed (speed relative to the wind) of 215 km/h, directed at angle θ south of east. The wind has velocity $\vec{v}_{WG}$ relative to the ground with speed 65.0 km/h, directed 20.0° east of north. What is the magnitude of the velocity $\vec{v}_{PG}$ of the plane relative to the ground, and what is θ?

Similarly, for the x components we find

$$v_{PG,x} = v_{PW,x} + v_{WG,x}.$$

Here, because $\vec{v}_{PG}$ is parallel to the x axis, the component $v_{PG,x}$ is equal to the magnitude v_{PG}. Substituting this notation and the value $\theta = 16.5°$, we find

$$v_{PG} = (215 \text{ km/h})(\cos 16.5°) + (65.0 \text{ km/h})(\sin 20.0°)$$
$$= 228 \text{ km/h}. \qquad \text{(Answer)}$$

KEY IDEAS

The situation is like the one in Fig. 4-19. Here the moving particle P is the plane, frame A is attached to the ground (call it G), and frame B is "attached" to the wind (call it W). We need a vector diagram like Fig. 4-19 but with three velocity vectors.

Calculations: First we construct a sentence that relates the three vectors shown in Fig. 4-20b:

$$\begin{matrix} \text{velocity of plane} \\ \text{relative to ground} \\ (PG) \end{matrix} = \begin{matrix} \text{velocity of plane} \\ \text{relative to wind} \\ (PW) \end{matrix} + \begin{matrix} \text{velocity of wind} \\ \text{relative to ground.} \\ (WG) \end{matrix}$$

This relation is written in vector notation as

$$\vec{v}_{PG} = \vec{v}_{PW} + \vec{v}_{WG}. \qquad (4\text{-}46)$$

We need to resolve the vectors into components on the coordinate system of Fig. 4-20b and then solve Eq. 4-46 axis by axis. For the y components, we find

$$v_{PG,y} = v_{PW,y} + v_{WG,y}$$

or $0 = -(215 \text{ km/h}) \sin \theta + (65.0 \text{ km/h})(\cos 20.0°).$

Solving for θ gives us

$$\theta = \sin^{-1} \frac{(65.0 \text{ km/h})(\cos 20.0°)}{215 \text{ km/h}} = 16.5°. \qquad \text{(Answer)}$$

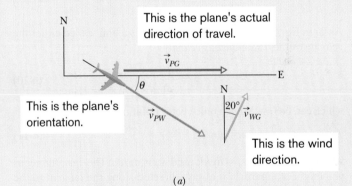

This is the plane's actual direction of travel.

This is the plane's orientation.

This is the wind direction.

(a)

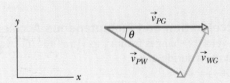

The actual direction is the vector sum of the other two vectors (head-to-tail arrangement).

(b)

Fig. 4-20 A plane flying in a wind.

 Additional examples, video, and practice available at *WileyPLUS*

Position Vector The location of a particle relative to the origin of a coordinate system is given by a *position vector* $\vec{r}$, which in unit-vector notation is

$$\vec{r} = x\hat{i} + y\hat{j} + z\hat{k}. \tag{4-1}$$

Here $x\hat{i}$, $y\hat{j}$, and $z\hat{k}$ are the vector components of position vector $\vec{r}$, and x, y, and z are its scalar components (as well as the coordinates of the particle). A position vector is described either by a magnitude and one or two angles for orientation, or by its vector or scalar components.

Displacement If a particle moves so that its position vector changes from $\vec{r}_1$ to $\vec{r}_2$, the particle's *displacement* $\Delta\vec{r}$ is

$$\Delta\vec{r} = \vec{r}_2 - \vec{r}_1. \tag{4-2}$$

The displacement can also be written as

$$\Delta\vec{r} = (x_2 - x_1)\hat{i} + (y_2 - y_1)\hat{j} + (z_2 - z_1)\hat{k} \tag{4-3}$$
$$= \Delta x\hat{i} + \Delta y\hat{j} + \Delta z\hat{k}. \tag{4-4}$$

Average Velocity and Instantaneous Velocity If a particle undergoes a displacement $\Delta\vec{r}$ in time interval Δt, its *average velocity* $\vec{v}_{avg}$ for that time interval is

$$\vec{v}_{avg} = \frac{\Delta\vec{r}}{\Delta t}. \tag{4-8}$$

As Δt in Eq. 4-8 is shrunk to 0, $\vec{v}_{avg}$ reaches a limit called either the *velocity* or the *instantaneous velocity* $\vec{v}$:

$$\vec{v} = \frac{d\vec{r}}{dt}, \tag{4-10}$$

which can be rewritten in unit-vector notation as

$$\vec{v} = v_x\hat{i} + v_y\hat{j} + v_z\hat{k}, \tag{4-11}$$

where $v_x = dx/dt$, $v_y = dy/dt$, and $v_z = dz/dt$. The instantaneous velocity $\vec{v}$ of a particle is always directed along the tangent to the particle's path at the particle's position.

Average Acceleration and Instantaneous Acceleration If a particle's velocity changes from $\vec{v}_1$ to $\vec{v}_2$ in time interval Δt, its *average acceleration* during Δt is

$$\vec{a}_{avg} = \frac{\vec{v}_2 - \vec{v}_1}{\Delta t} = \frac{\Delta\vec{v}}{\Delta t}. \tag{4-15}$$

As Δt in Eq. 4-15 is shrunk to 0, $\vec{a}_{avg}$ reaches a limiting value called either the *acceleration* or the *instantaneous acceleration* $\vec{a}$:

$$\vec{a} = \frac{d\vec{v}}{dt}. \tag{4-16}$$

In unit-vector notation,

$$\vec{a} = a_x\hat{i} + a_y\hat{j} + a_z\hat{k}, \tag{4-17}$$

where $a_x = dv_x/dt$, $a_y = dv_y/dt$, and $a_z = dv_z/dt$.

Projectile Motion *Projectile motion* is the motion of a particle that is launched with an initial velocity $\vec{v}_0$. During its flight, the particle's horizontal acceleration is zero and its vertical acceleration is the free-fall acceleration $-g$. (Upward is taken to be a positive direction.) If $\vec{v}_0$ is expressed as a magnitude (the speed v_0) and an angle θ_0 (measured from the horizontal), the particle's equations of motion along the horizontal x axis and vertical y axis are

$$x - x_0 = (v_0 \cos\theta_0)t, \tag{4-21}$$
$$y - y_0 = (v_0 \sin\theta_0)t - \tfrac{1}{2}gt^2, \tag{4-22}$$
$$v_y = v_0 \sin\theta_0 - gt, \tag{4-23}$$
$$v_y^2 = (v_0 \sin\theta_0)^2 - 2g(y - y_0). \tag{4-24}$$

The **trajectory** (path) of a particle in projectile motion is parabolic and is given by

$$y = (\tan\theta_0)x - \frac{gx^2}{2(v_0 \cos\theta_0)^2}, \tag{4-25}$$

if x_0 and y_0 of Eqs. 4-21 to 4-24 are zero. The particle's **horizontal range** R, which is the horizontal distance from the launch point to the point at which the particle returns to the launch height, is

$$R = \frac{v_0^2}{g} \sin 2\theta_0. \tag{4-26}$$

Uniform Circular Motion If a particle travels along a circle or circular arc of radius r at constant speed v, it is said to be in *uniform circular motion* and has an acceleration $\vec{a}$ of constant magnitude

$$a = \frac{v^2}{r}. \tag{4-34}$$

The direction of $\vec{a}$ is toward the center of the circle or circular arc, and $\vec{a}$ is said to be *centripetal*. The time for the particle to complete a circle is

$$T = \frac{2\pi r}{v}. \tag{4-35}$$

T is called the *period of revolution*, or simply the *period*, of the motion.

Relative Motion When two frames of reference A and B are moving relative to each other at constant velocity, the velocity of a particle P as measured by an observer in frame A usually differs from that measured from frame B. The two measured velocities are related by

$$\vec{v}_{PA} = \vec{v}_{PB} + \vec{v}_{BA}, \tag{4-44}$$

where $\vec{v}_{BA}$ is the velocity of B with respect to A. Both observers measure the same acceleration for the particle:

$$\vec{a}_{PA} = \vec{a}_{PB}. \tag{4-45}$$

1 Figure 4-21 shows the path taken by a skunk foraging for trash food, from initial point i. The skunk took the same time T to go from each labeled point to the next along its path. Rank points a, b, and c according to the magnitude of the average velocity of the skunk to reach them from initial point i, greatest first.

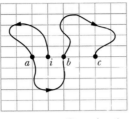

Fig. 4-21   Question 1.

2 Figure 4-22 shows the initial position i and the final position f of a particle. What are the (a) initial position vector $\vec{r}_i$ and (b) final position vector $\vec{r}_f$, both in unit-vector notation? (c) What is the x component of displacement $\Delta\vec{r}$?

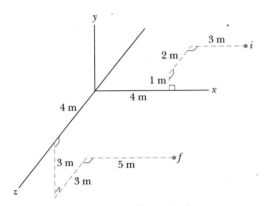

Fig. 4-22   Question 2.

3 When Paris was shelled from 100 km away with the WWI long-range artillery piece "Big Bertha," the shells were fired at an angle greater than 45° to give them a greater range, possibly even twice as long as at 45°. Does that result mean that the air density at high altitudes increases with altitude or decreases?

4 You are to launch a rocket, from just above the ground, with one of the following initial velocity vectors: (1) $\vec{v}_0 = 20\hat{i} + 70\hat{j}$, (2) $\vec{v}_0 = -20\hat{i} + 70\hat{j}$, (3) $\vec{v}_0 = 20\hat{i} - 70\hat{j}$, (4) $\vec{v}_0 = -20\hat{i} - 70\hat{j}$. In your coordinate system, x runs along level ground and y increases upward. (a) Rank the vectors according to the launch speed of the projectile, greatest first. (b) Rank the vectors according to the time of flight of the projectile, greatest first.

5 Figure 4-23 shows three situations in which identical projectiles are launched (at the same level) at identical initial speeds and angles. The projectiles do not land on the same terrain, however. Rank the situations according to the final speeds of the projectiles just before they land, greatest first.

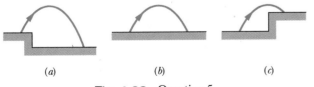

Fig. 4-23   Question 5.

6 The only good use of a fruitcake is in catapult practice. Curve 1 in Fig. 4-24 gives the height y of a catapulted fruitcake versus the

angle θ between its velocity vector and its acceleration vector during flight. (a) Which of the lettered points on that curve corresponds to the landing of the fruitcake on the ground? (b) Curve 2 is a similar plot for the same launch speed but for a different launch angle. Does the fruitcake now land farther away or closer to the launch point?

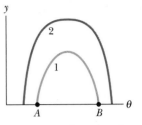

Fig. 4-24   Question 6.

7 An airplane flying horizontally at a constant speed of 350 km/h over level ground releases a bundle of food supplies. Ignore the effect of the air on the bundle. What are the bundle's initial (a) vertical and (b) horizontal components of velocity? (c) What is its horizontal component of velocity just before hitting the ground? (d) If the airplane's speed were, instead, 450 km/h, would the time of fall be longer, shorter, or the same?

8 In Fig. 4-25, a cream tangerine is thrown up past windows 1, 2, and 3, which are identical in size and regularly spaced vertically. Rank those three windows according to (a) the time the cream tangerine takes to pass them and (b) the average speed of the cream tangerine during the passage, greatest first.

The cream tangerine then moves down past windows 4, 5, and 6, which are identical in size and irregularly spaced horizontally. Rank those three windows according to (c) the time the cream tangerine takes to pass them and (d) the average speed of the cream tangerine during the passage, greatest first.

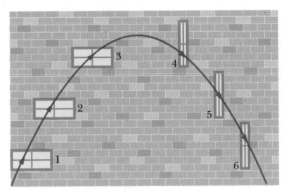

Fig. 4-25   Question 8.

9 Figure 4-26 shows three paths for a football kicked from ground level. Ignoring the effects of air, rank the paths according to (a) time of flight, (b) initial vertical velocity component, (c) initial horizontal velocity component, and (d) initial speed, greatest first.

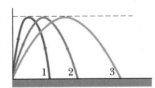

Fig. 4-26   Question 9.

10 A ball is shot from ground level over level ground at a certain initial speed. Figure 4-27 gives the range R of the ball versus its launch angle θ_0. Rank the three lettered points on the plot according to (a) the total flight time of the ball and (b) the

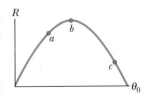

Fig. 4-27   Question 10.

ball's speed at maximum height, greatest first.

11 Figure 4-28 shows four tracks (either half- or quarter-circles) that can be taken by a train, which moves at a constant speed. Rank the tracks according to the magnitude of a train's acceleration on the curved portion, greatest first.

12 In Fig. 4-29, particle *P* is in uniform circular motion, centered on the origin of an *xy* coordinate system. (a)

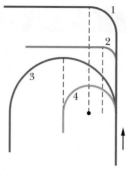

Fig. 4-28 Question 11.

At what values of θ is the vertical component r_y of the position vector greatest in magnitude? (b) At what values of θ is the vertical component v_y of the particle's velocity greatest in magnitude? (c) At what values of θ is the vertical component a_y of the particle's acceleration greatest in magnitude?

13 (a) Is it possible to be accelerating while traveling at constant speed? Is it possible to round a curve with (b) zero acceleration and (c) a constant magnitude of acceleration?

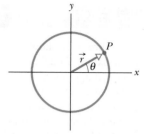

Fig. 4-29 Question 12.

PROBLEMS

sec. 4-2 Position and Displacement

•1 The position vector for an electron is $\vec{r} = (5.0 \text{ m})\hat{i} - (3.0 \text{ m})\hat{j} + (2.0 \text{ m})\hat{k}$. (a) Find the magnitude of $\vec{r}$. (b) Sketch the vector on a right-handed coordinate system.

•2 A watermelon seed has the following coordinates: $x = -5.0$ m, $y = 8.0$ m, and $z = 0$ m. Find its position vector (a) in unit-vector notation and as (b) a magnitude and (c) an angle relative to the positive direction of the *x* axis. (d) Sketch the vector on a right-handed coordinate system. If the seed is moved to the *xyz* coordinates (3.00 m, 0 m, 0 m), what is its displacement (e) in unit-vector notation and as (f) a magnitude and (g) an angle relative to the positive *x* direction?

•3 A positron undergoes a displacement $\Delta\vec{r} = 2.0\hat{i} - 3.0\hat{j} + 6.0\hat{k}$, ending with the position vector $\vec{r} = 3.0\hat{j} - 4.0\hat{k}$, in meters. What was the positron's initial position vector?

••4 The minute hand of a wall clock measures 10 cm from its tip to the axis about which it rotates. The magnitude and angle of the displacement vector of the tip are to be determined for three time intervals. What are the (a) magnitude and (b) angle from a quarter after the hour to half past, the (c) magnitude and (d) angle for the next half hour, and the (e) magnitude and (f) angle for the hour after that?

sec. 4-3 Average Velocity and Instantaneous Velocity

•5 **SSM** A train at a constant 60.0 km/h moves east for 40.0 min, then in a direction 50.0° east of due north for 20.0 min, and then west for 50.0 min. What are the (a) magnitude and (b) angle of its average velocity during this trip?

•6 An electron's position is given by $\vec{r} = 3.00t\hat{i} - 4.00t^2\hat{j} + 2.00\hat{k}$, with *t* in seconds and $\vec{r}$ in meters. (a) In unit-vector notation, what is the electron's velocity $\vec{v}(t)$? At *t* = 2.00 s, what is $\vec{v}$ (b) in unit-vector notation and as (c) a magnitude and (d) an angle relative to the positive direction of the *x* axis?

•7 An ion's position vector is initially $\vec{r} = 5.0\hat{i} - 6.0\hat{j} + 2.0\hat{k}$, and 10 s later it is $\vec{r} = -2.0\hat{i} + 8.0\hat{j} - 2.0\hat{k}$, all in meters. In unit-vector notation, what is its $\vec{v}_{avg}$ during the 10 s?

••8 A plane flies 483 km east from city *A* to city *B* in 45.0 min and then 966 km south from city *B* to city *C* in 1.50 h. For the total trip,

what are the (a) magnitude and (b) direction of the plane's displacement, the (c) magnitude and (d) direction of its average velocity, and (e) its average speed?

••9 Figure 4-30 gives the path of a squirrel moving about on level ground, from point *A* (at time *t* = 0), to points *B* (at *t* = 5.00 min), *C* (at *t* = 10.0 min), and finally *D* (at *t* = 15.0 min). Consider the average velocities of the squirrel from point *A* to each of the other three points. Of them, what are the (a) magnitude and (b) angle of the one with the least magnitude and the (c) magnitude and (d) angle of the one with the greatest magnitude?

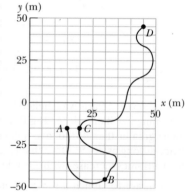

Fig. 4-30 Problem 9.

•••10 The position vector $\vec{r} = 5.00t\hat{i} + (et + ft^2)\hat{j}$ locates a particle as a function of time *t*. Vector $\vec{r}$ is in meters, *t* is in seconds, and factors *e* and *f* are constants. Figure 4-31 gives the angle θ of the particle's direction of travel as a function of *t* (θ is measured from the positive *x* direction). What are (a) *e* and (b) *f*, including units?

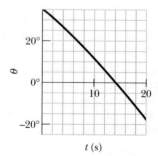

Fig. 4-31 Problem 10.

sec. 4-4 Average Acceleration and Instantaneous Acceleration

•11 **GO** The position $\vec{r}$ of a particle moving in an *xy* plane is given by $\vec{r} = (2.00t^3 - 5.00t)\hat{i} + (6.00 - 7.00t^4)\hat{j}$, with $\vec{r}$ in meters and *t* in seconds. In unit-vector notation, calculate (a) $\vec{r}$, (b) $\vec{v}$, and (c) $\vec{a}$

for $t = 2.00$ s. (d) What is the angle between the positive direction of the x axis and a line tangent to the particle's path at $t = 2.00$ s?

•12 At one instant a bicyclist is 40.0 m due east of a park's flagpole, going due south with a speed of 10.0 m/s. Then 30.0 s later, the cyclist is 40.0 m due north of the flagpole, going due east with a speed of 10.0 m/s. For the cyclist in this 30.0 s interval, what are the (a) magnitude and (b) direction of the displacement, the (c) magnitude and (d) direction of the average velocity, and the (e) magnitude and (f) direction of the average acceleration?

•13 SSM A particle moves so that its position (in meters) as a function of time (in seconds) is $\vec{r} = \hat{i} + 4t^2\hat{j} + t\hat{k}$. Write expressions for (a) its velocity and (b) its acceleration as functions of time.

•14 A proton initially has $\vec{v} = 4.0\hat{i} - 2.0\hat{j} + 3.0\hat{k}$ and then 4.0 s later has $\vec{v} = -2.0\hat{i} - 2.0\hat{j} + 5.0\hat{k}$ (in meters per second). For that 4.0 s, what are (a) the proton's average acceleration $\vec{a}_{avg}$ in unit-vector notation, (b) the magnitude of $\vec{a}_{avg}$, and (c) the angle between $\vec{a}_{avg}$ and the positive direction of the x axis?

••15 SSM ILW A particle leaves the origin with an initial velocity $\vec{v} = (3.00\hat{i})$ m/s and a constant acceleration $\vec{a} = (-1.00\hat{i} - 0.500\hat{j})$ m/s². When it reaches its maximum x coordinate, what are its (a) velocity and (b) position vector?

••16 GO The velocity $\vec{v}$ of a particle moving in the xy plane is given by $\vec{v} = (6.0t - 4.0t^2)\hat{i} + 8.0\hat{j}$, with $\vec{v}$ in meters per second and t (> 0) in seconds. (a) What is the acceleration when $t = 3.0$ s? (b) When (if ever) is the acceleration zero? (c) When (if ever) is the velocity zero? (d) When (if ever) does the speed equal 10 m/s?

••17 A cart is propelled over an xy plane with acceleration components $a_x = 4.0$ m/s² and $a_y = -2.0$ m/s². Its initial velocity has components $v_{0x} = 8.0$ m/s and $v_{0y} = 12$ m/s. In unit-vector notation, what is the velocity of the cart when it reaches its greatest y coordinate?

••18 A moderate wind accelerates a pebble over a horizontal xy plane with a constant acceleration $\vec{a} = (5.00 \text{ m/s}^2)\hat{i} + (7.00 \text{ m/s}^2)\hat{j}$. At time $t = 0$, the velocity is $(4.00 \text{ m/s})\hat{i}$. What are the (a) magnitude and (b) angle of its velocity when it has been displaced by 12.0 m parallel to the x axis?

•••19 The acceleration of a particle moving only on a horizontal xy plane is given by $\vec{a} = 3t\hat{i} + 4t\hat{j}$, where $\vec{a}$ is in meters per second-squared and t is in seconds. At $t = 0$, the position vector $\vec{r} = (20.0 \text{ m})\hat{i} + (40.0 \text{ m})\hat{j}$ locates the particle, which then has the velocity vector $\vec{v} = (5.00 \text{ m/s})\hat{i} + (2.00 \text{ m/s})\hat{j}$. At $t = 4.00$ s, what are (a) its position vector in unit-vector notation and (b) the angle between its direction of travel and the positive direction of the x axis?

•••20 In Fig. 4-32, particle A moves along the line $y = 30$ m with a constant velocity $\vec{v}$ of magnitude 3.0 m/s and parallel to the x axis. At the instant particle A passes the y axis, particle B leaves the origin with a zero initial speed and a constant acceleration $\vec{a}$ of magnitude 0.40 m/s². What angle θ between $\vec{a}$ and the positive direction of the y axis would result in a collision?

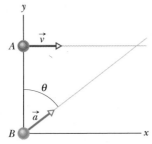

Fig. 4-32 Problem 20.

sec. 4-6 Projectile Motion Analyzed

•21 A dart is thrown horizontally with an initial speed of 10 m/s toward point P, the bull's-eye on a dart board. It hits at point Q on the rim, vertically below P, 0.19 s later. (a) What is the distance PQ? (b) How far away from the dart board is the dart released?

•22 A small ball rolls horizontally off the edge of a tabletop that is 1.20 m high. It strikes the floor at a point 1.52 m horizontally from the table edge. (a) How long is the ball in the air? (b) What is its speed at the instant it leaves the table?

•23 A projectile is fired horizontally from a gun that is 45.0 m above flat ground, emerging from the gun with a speed of 250 m/s. (a) How long does the projectile remain in the air? (b) At what horizontal distance from the firing point does it strike the ground? (c) What is the magnitude of the vertical component of its velocity as it strikes the ground?

•24 ✈ In the 1991 World Track and Field Championships in Tokyo, Mike Powell jumped 8.95 m, breaking by a full 5 cm the 23-year long-jump record set by Bob Beamon. Assume that Powell's speed on takeoff was 9.5 m/s (about equal to that of a sprinter) and that $g = 9.80$ m/s² in Tokyo. How much less was Powell's range than the maximum possible range for a particle launched at the same speed?

•25 ✈ The current world-record motorcycle jump is 77.0 m, set by Jason Renie. Assume that he left the take-off ramp at 12.0° to the horizontal and that the take-off and landing heights are the same. Neglecting air drag, determine his take-off speed.

•26 A stone is catapulted at time $t = 0$, with an initial velocity of magnitude 20.0 m/s and at an angle of 40.0° above the horizontal. What are the magnitudes of the (a) horizontal and (b) vertical components of its displacement from the catapult site at $t = 1.10$ s? Repeat for the (c) horizontal and (d) vertical components at $t = 1.80$ s, and for the (e) horizontal and (f) vertical components at $t = 5.00$ s.

••27 ILW A certain airplane has a speed of 290.0 km/h and is diving at an angle of $\theta = 30.0°$ below the horizontal when the pilot releases a radar decoy (Fig. 4-33). The horizontal distance between the release point and the point where the decoy strikes the ground is $d = 700$ m. (a) How long is the decoy in the air? (b) How high was the release point?

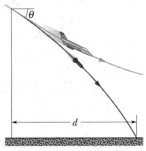

Fig. 4-33 Problem 27.

••28 In Fig. 4-34, a stone is projected at a cliff of height h with an initial speed of 42.0 m/s directed at angle $\theta_0 = 60.0°$ above the horizontal. The stone strikes at A, 5.50 s after launching. Find (a) the height h of the cliff, (b) the speed of the stone just before impact at A, and (c) the maximum height H reached above the ground.

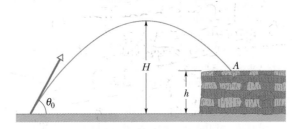

Fig. 4-34 Problem 28.

••29 A projectile's launch speed is five times its speed at maximum height. Find launch angle θ_0.

••30 ⬡ A soccer ball is kicked from the ground with an initial speed of 19.5 m/s at an upward angle of 45°. A player 55 m away in the direction of the kick starts running to meet the ball at that instant. What must be his average speed if he is to meet the ball just before it hits the ground?

••31 ✈ In a jump spike, a volleyball player slams the ball from overhead and toward the opposite floor. Controlling the angle of the spike is difficult. Suppose a ball is spiked from a height of 2.30 m with an initial speed of 20.0 m/s at a downward angle of 18.00°. How much farther on the opposite floor would it have landed if the downward angle were, instead, 8.00°?

••32 ⬡ You throw a ball toward a wall at speed 25.0 m/s and at angle $\theta_0 = 40.0°$ above the horizontal (Fig. 4-35). The wall is distance $d = 22.0$ m from the release point of the ball. (a) How far above the release point does the ball hit the wall? What are the (b) horizontal and (c) vertical

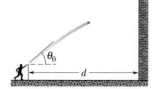

Fig. 4-35 Problem 32.

components of its velocity as it hits the wall? (d) When it hits, has it passed the highest point on its trajectory?

••33 SSM A plane, diving with constant speed at an angle of 53.0° with the vertical, releases a projectile at an altitude of 730 m. The projectile hits the ground 5.00 s after release. (a) What is the speed of the plane? (b) How far does the projectile travel horizontally during its flight? What are the (c) horizontal and (d) vertical components of its velocity just before striking the ground?

••34 ✈ A trebuchet was a hurling machine built to attack the walls of a castle under siege. A large stone could be hurled against a wall to break apart the wall. The machine was not placed near the wall because then arrows could reach it from the castle wall. Instead, it was positioned so that the stone hit the wall during the second half of its flight. Suppose a stone is launched with a speed of $v_0 = 28.0$ m/s and at an angle of $\theta_0 = 40.0°$. What is the speed of the stone if it hits the wall (a) just as it reaches the top of its parabolic path and (b) when it has descended to half that height? (c) As a percentage, how much faster is it moving in part (b) than in part (a)?

••35 SSM A rifle that shoots bullets at 460 m/s is to be aimed at a target 45.7 m away. If the center of the target is level with the rifle, how high above the target must the rifle barrel be pointed so that the bullet hits dead center?

••36 During a tennis match, a player serves the ball at 23.6 m/s, with the center of the ball leaving the racquet horizontally 2.37 m above the court surface. The net is 12 m away and 0.90 m high. When the ball reaches the net, (a) does the ball clear it and (b) what is the distance between the center of the ball and the top of the net? Suppose that, instead, the ball is served as before but now it leaves the racquet at 5.00° below the horizontal. When the ball reaches the net, (c) does the ball clear it and (d) what now is the distance between the center of the ball and the top of the net?

••37 SSM WWW A lowly high diver pushes off horizontally with a speed of 2.00 m/s from the platform edge 10.0 m above the surface of the water. (a) At what horizontal distance from the edge is the diver 0.800 s after pushing off? (b) At what vertical distance above the surface of the water is the diver just then? (c) At what horizontal distance from the edge does the diver strike the water?

••38 A golf ball is struck at ground level. The speed of the golf ball as a function of the time is shown in Fig. 4-36, where $t = 0$ at the instant the ball is struck. The scaling on the vertical axis is set by $v_a = 19$ m/s and $v_b = 31$ m/s. (a) How far does the golf ball travel horizontally before returning to ground

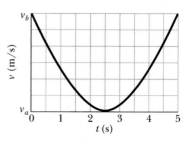

Fig. 4-36 Problem 38.

level? (b) What is the maximum height above ground level attained by the ball?

••39 In Fig. 4-37, a ball is thrown leftward from the left edge of the roof, at height h above the ground. The ball hits the ground 1.50 s later, at distance $d = 25.0$ m from the building and at angle $\theta = 60.0°$ with the horizontal. (a) Find h. (Hint: One way is to reverse the motion, as if on video.) What are the (b) magnitude and (c) angle relative to the horizontal of the velocity at which the ball is thrown? (d) Is the angle above or below the horizontal?

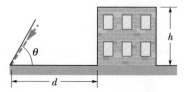

Fig. 4-37 Problem 39.

••40 ✈ Suppose that a shot putter can put a shot at the world-class speed $v_0 = 15.00$ m/s and at a height of 2.160 m. What horizontal distance would the shot travel if the launch angle θ_0 is (a) 45.00° and (b) 42.00°? The answers indicate that the angle of 45°, which maximizes the range of projectile motion, does not maximize the horizontal distance when the launch and landing are at different heights.

••41 ⬡ ✈ Upon spotting an insect on a twig overhanging water, an archer fish squirts water drops at the insect to knock it into the water (Fig. 4-38). Although the fish sees the insect along a straight-line path at angle ϕ and distance d, a drop must be launched at a different angle θ_0 if its parabolic path is to intersect the insect. If $\phi = 36.0°$

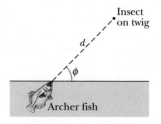

Fig. 4-38 Problem 41.

and $d = 0.900$ m, what launch angle θ_0 is required for the drop to be at the top of the parabolic path when it reaches the insect?

••42 ✈ In 1939 or 1940, Emanuel Zacchini took his human-cannonball act to an extreme: After being shot from a cannon, he soared over three Ferris wheels and into a net (Fig. 4-39). Assume that

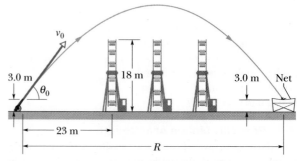

Fig. 4-39 Problem 42.

he is launched with a speed of 26.5 m/s and at an angle of 53.0°. (a) Treating him as a particle, calculate his clearance over the first wheel. (b) If he reached maximum height over the middle wheel, by how much did he clear it? (c) How far from the cannon should the net's center have been positioned (neglect air drag)?

••43 ILW A ball is shot from the ground into the air. At a height of 9.1 m, its velocity is $\vec{v} = (7.6\hat{i} + 6.1\hat{j})$ m/s, with $\hat{i}$ horizontal and $\hat{j}$ upward. (a) To what maximum height does the ball rise? (b) What total horizontal distance does the ball travel? What are the (c) magnitude and (d) angle (below the horizontal) of the ball's velocity just before it hits the ground?

••44 A baseball leaves a pitcher's hand horizontally at a speed of 161 km/h. The distance to the batter is 18.3 m. (a) How long does the ball take to travel the first half of that distance? (b) The second half? (c) How far does the ball fall freely during the first half? (d) During the second half? (e) Why aren't the quantities in (c) and (d) equal?

••45 In Fig. 4-40, a ball is launched with a velocity of magnitude 10.0 m/s, at an angle of 50.0° to the horizontal. The launch point is at the base of a ramp of horizontal length $d_1 = 6.00$ m and height $d_2 = 3.60$ m.

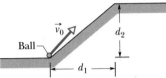

Fig. 4-40 Problem 45.

A plateau is located at the top of the ramp. (a) Does the ball land on the ramp or the plateau? When it lands, what are the (b) magnitude and (c) angle of its displacement from the launch point?

••46 GO In basketball, *hang* is an illusion in which a player seems to weaken the gravitational acceleration while in midair. The illusion depends much on a skilled player's ability to rapidly shift the ball between hands during the flight, but it might also be supported by the longer horizontal distance the player travels in the upper part of the jump than in the lower part. If a player jumps with an initial speed of $v_0 = 7.00$ m/s at an angle of $\theta_0 = 35.0°$, what percent of the jump's range does the player spend in the upper half of the jump (between maximum height and half maximum height)?

••47 SSM WWW A batter hits a pitched ball when the center of the ball is 1.22 m above the ground. The ball leaves the bat at an angle of 45° with the ground. With that launch, the ball should have a horizontal range (returning to the *launch* level) of 107 m. (a) Does the ball clear a 7.32-m-high fence that is 97.5 m horizontally from the launch point? (b) At the fence, what is the distance between the fence top and the ball center?

••48 GO In Fig. 4-41, a ball is thrown up onto a roof, landing 4.00 s later at height $h = 20.0$ m above the release level. The ball's path just before landing is angled at $\theta = 60.0°$ with the roof. (a) Find the horizontal distance d it travels. (See the hint to Problem 39.) What are the (b) magnitude and (c) angle (relative to the horizontal) of the ball's initial velocity?

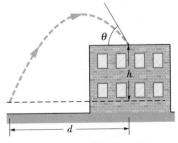

Fig. 4-41 Problem 48.

•••49 SSM A football kicker can give the ball an initial speed of 25 m/s. What are the (a) least and (b) greatest elevation angles at which he can kick the ball to score a field goal from a point 50 m in

front of goalposts whose horizontal bar is 3.44 m above the ground?

•••50 GO Two seconds after being projected from ground level, a projectile is displaced 40 m horizontally and 53 m vertically above its launch point. What are the (a) horizontal and (b) vertical components of the initial velocity of the projectile? (c) At the instant the projectile achieves its maximum height above ground level, how far is it displaced horizontally from the launch point?

•••51 A skilled skier knows to jump upward before reaching a downward slope. Consider a jump in which the launch speed is $v_0 = 10$ m/s, the launch angle is $\theta_0 = 9.0°$, the initial course is approximately flat, and the steeper track has a slope of 11.3°. Figure 4-42a shows a *prejump* that allows the skier to land on the top portion of the steeper track. Figure 4-42b shows a jump at the edge of the steeper track. In Fig. 4-42a, the skier lands at approximately the launch level. (a) In the landing, what is the angle ϕ between the skier's path and the slope? In Fig. 4-42b, (b) how far below the launch level does the skier land and (c) what is ϕ? (The greater fall and greater ϕ can result in loss of control in the landing.)

(a)   (b)

Fig. 4-42 Problem 51.

•••52 A ball is to be shot from level ground toward a wall at distance x (Fig. 4-43a). Figure 4-43b shows the y component v_y of the ball's velocity just as it would reach the wall, as a function of that distance x. The scaling is set by $v_{ys} = 5.0$ m/s and $x_s = 20$ m. What is the launch angle?

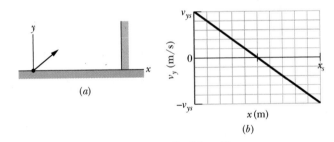

(a)   (b)

Fig. 4-43 Problem 52.

•••53 In Fig. 4-44, a baseball is hit at a height $h = 1.00$ m and then caught at the same height. It travels alongside a wall, moving up past the top of the wall 1.00 s after it is hit and then down past the top of the wall 4.00 s later, at distance $D = 50.0$ m farther along the wall. (a) What horizontal distance is traveled by the ball from hit to catch? What are the (b) magnitude and (c) angle (relative to the horizontal) of the ball's velocity just after being hit? (d) How high is the wall?

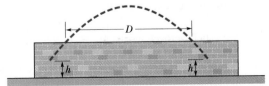

Fig. 4-44 Problem 53.

•••54 A ball is to be shot from level ground with a certain speed. Figure 4-45 shows the range R it will have versus the launch angle θ_0. The value of θ_0 determines the flight time; let t_{max} represent the maximum flight time. What is the least speed the ball will have during its flight if θ_0 is chosen such that the flight time is $0.500t_{max}$?

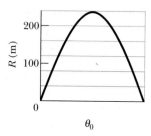

Fig. 4-45 Problem 54.

•••55 SSM A ball rolls horizontally off the top of a stairway with a speed of 1.52 m/s. The steps are 20.3 cm high and 20.3 cm wide. Which step does the ball hit first?

sec. 4-7 Uniform Circular Motion

•56 An Earth satellite moves in a circular orbit 640 km above Earth's surface with a period of 98.0 min. What are the (a) speed and (b) magnitude of the centripetal acceleration of the satellite?

•57 A carnival merry-go-round rotates about a vertical axis at a constant rate. A man standing on the edge has a constant speed of 3.66 m/s and a centripetal acceleration $\vec{a}$ of magnitude 1.83 m/s². Position vector $\vec{r}$ locates him relative to the rotation axis. (a) What is the magnitude of $\vec{r}$? What is the direction of $\vec{r}$ when $\vec{a}$ is directed (b) due east and (c) due south?

•58 A rotating fan completes 1200 revolutions every minute. Consider the tip of a blade, at a radius of 0.15 m. (a) Through what distance does the tip move in one revolution? What are (b) the tip's speed and (c) the magnitude of its acceleration? (d) What is the period of the motion?

•59 ILW A woman rides a carnival Ferris wheel at radius 15 m, completing five turns about its horizontal axis every minute. What are (a) the period of the motion, the (b) magnitude and (c) direction of her centripetal acceleration at the highest point, and the (d) magnitude and (e) direction of her centripetal acceleration at the lowest point?

•60 A centripetal-acceleration addict rides in uniform circular motion with period $T = 2.0$ s and radius $r = 3.00$ m. At t_1 his acceleration is $\vec{a} = (6.00 \text{ m/s}^2)\hat{i} + (-4.00 \text{ m/s}^2)\hat{j}$. At that instant, what are the values of (a) $\vec{v} \cdot \vec{a}$ and (b) $\vec{r} \times \vec{a}$?

•61 When a large star becomes a *supernova*, its core may be compressed so tightly that it becomes a *neutron star*, with a radius of about 20 km (about the size of the San Francisco area). If a neutron star rotates once every second, (a) what is the speed of a particle on the star's equator and (b) what is the magnitude of the particle's centripetal acceleration? (c) If the neutron star rotates faster, do the answers to (a) and (b) increase, decrease, or remain the same?

•62 What is the magnitude of the acceleration of a sprinter running at 10 m/s when rounding a turn of radius 25 m?

••63 GO At $t_1 = 2.00$ s, the acceleration of a particle in counterclockwise circular motion is $(6.00 \text{ m/s}^2)\hat{i} + (4.00 \text{ m/s}^2)\hat{j}$. It moves at constant speed. At time $t_2 = 5.00$ s, the particle's acceleration is $(4.00 \text{ m/s}^2)\hat{i} + (-6.00 \text{ m/s}^2)\hat{j}$. What is the radius of the path taken by the particle if $t_2 - t_1$ is less than one period?

••64 GO A particle moves horizontally in uniform circular motion, over a horizontal xy plane. At one instant, it moves through the point at coordinates (4.00 m, 4.00 m) with a velocity of $-5.00\hat{i}$ m/s and an acceleration of $+12.5\hat{j}$ m/s². What are the (a) x and (b) y coordinates of the center of the circular path?

••65 A purse at radius 2.00 m and a wallet at radius 3.00 m travel in uniform circular motion on the floor of a merry-go-round as the ride turns. They are on the same radial line. At one instant, the acceleration of the purse is $(2.00 \text{ m/s}^2)\hat{i} + (4.00 \text{ m/s}^2)\hat{j}$. At that instant and in unit-vector notation, what is the acceleration of the wallet?

••66 A particle moves along a circular path over a horizontal xy coordinate system, at constant speed. At time $t_1 = 4.00$ s, it is at point (5.00 m, 6.00 m) with velocity $(3.00 \text{ m/s})\hat{j}$ and acceleration in the positive x direction. At time $t_2 = 10.0$ s, it has velocity $(-3.00 \text{ m/s})\hat{i}$ and acceleration in the positive y direction. What are the (a) x and (b) y coordinates of the center of the circular path if $t_2 - t_1$ is less than one period?

•••67 SSM WWW A boy whirls a stone in a horizontal circle of radius 1.5 m and at height 2.0 m above level ground. The string breaks, and the stone flies off horizontally and strikes the ground after traveling a horizontal distance of 10 m. What is the magnitude of the centripetal acceleration of the stone during the circular motion?

•••68 GO A cat rides a merry-go-round turning with uniform circular motion. At time $t_1 = 2.00$ s, the cat's velocity is $\vec{v}_1 = (3.00 \text{ m/s})\hat{i} + (4.00 \text{ m/s})\hat{j}$, measured on a horizontal xy coordinate system. At $t_2 = 5.00$ s, the cat's velocity is $\vec{v}_2 = (-3.00 \text{ m/s})\hat{i} + (-4.00 \text{ m/s})\hat{j}$. What are (a) the magnitude of the cat's centripetal acceleration and (b) the cat's average acceleration during the time interval $t_2 - t_1$, which is less than one period?

sec. 4-8 Relative Motion in One Dimension

•69 A cameraman on a pickup truck is traveling westward at 20 km/h while he records a cheetah that is moving westward 30 km/h faster than the truck. Suddenly, the cheetah stops, turns, and then runs at 45 km/h eastward, as measured by a suddenly nervous crew member who stands alongside the cheetah's path. The change in the animal's velocity takes 2.0 s. What are the (a) magnitude and (b) direction of the animal's acceleration according to the cameraman and the (c) magnitude and (d) direction according to the nervous crew member?

•70 A boat is traveling upstream in the positive direction of an x axis at 14 km/h with respect to the water of a river. The water is flowing at 9.0 km/h with respect to the ground. What are the (a) magnitude and (b) direction of the boat's velocity with respect to the ground? A child on the boat walks from front to rear at 6.0 km/h with respect to the boat. What are the (c) magnitude and (d) direction of the child's velocity with respect to the ground?

••71 A suspicious-looking man runs as fast as he can along a moving sidewalk from one end to the other, taking 2.50 s. Then security agents appear, and the man runs as fast as he can back along the sidewalk to his starting point, taking 10.0 s. What is the ratio of the man's running speed to the sidewalk's speed?

sec. 4-9 Relative Motion in Two Dimensions

•72 A rugby player runs with the ball directly toward his opponent's goal, along the positive direction of an x axis. He can legally pass the ball to a teammate as long as the ball's velocity relative to the field does not have a positive x component. Suppose the player runs at speed 4.0 m/s relative to the field while he passes the ball with velocity $\vec{v}_{BP}$ relative to himself. If $\vec{v}_{BP}$ has magnitude 6.0 m/s, what is the smallest angle it can have for the pass to be legal?

••73 Two highways intersect as shown in Fig. 4-46. At the instant shown, a police car P is distance $d_P = 800$ m from the intersection and moving at speed $v_P = 80$ km/h. Motorist M is distance $d_M = 600$ m from the intersection and moving at speed $v_M = 60$ km/h. (a) In unit-vector notation, what is the velocity of the motorist with respect to the police car? (b) For the instant shown in Fig. 4-46, what is the angle between the velocity found in (a) and the line of sight between the two cars? (c) If the cars maintain their velocities, do the answers to (a) and (b) change as the cars move nearer the intersection?

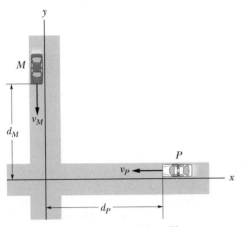

Fig. 4-46   Problem 73.

••74   After flying for 15 min in a wind blowing 42 km/h at an angle of 20° south of east, an airplane pilot is over a town that is 55 km due north of the starting point. What is the speed of the airplane relative to the air?

••75 SSM A train travels due south at 30 m/s (relative to the ground) in a rain that is blown toward the south by the wind. The path of each raindrop makes an angle of 70° with the vertical, as measured by an observer stationary on the ground. An observer on the train, however, sees the drops fall perfectly vertically. Determine the speed of the raindrops relative to the ground.

••76   A light plane attains an airspeed of 500 km/h. The pilot sets out for a destination 800 km due north but discovers that the plane must be headed 20.0° east of due north to fly there directly. The plane arrives in 2.00 h. What were the (a) magnitude and (b) direction of the wind velocity?

••77 SSM Snow is falling vertically at a constant speed of 8.0 m/s. At what angle from the vertical do the snowflakes appear to be falling as viewed by the driver of a car traveling on a straight, level road with a speed of 50 km/h?

••78   In the overhead view of Fig. 4-47, Jeeps P and B race along straight lines, across flat terrain, and past stationary border guard A. Relative to the guard, B travels at a constant speed of 20.0 m/s, at the angle $\theta_2 = 30.0°$. Relative to the guard, P has accelerated from rest at a constant rate of 0.400 m/s^2 at the angle $\theta_1 = 60.0°$. At a certain time during the acceleration, P has a speed of 40.0 m/s. At that time, what are the (a) magnitude and (b) direction of the velocity of P relative to B and the (c) magnitude and (d) direction of the acceleration of P relative to B?

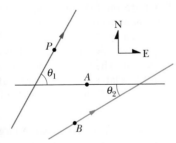

Fig. 4-47   Problem 78.

••79 SSM ILW Two ships, A and B, leave port at the same time. Ship A travels northwest at 24 knots, and ship B travels at 28 knots in a direction 40° west of south. (1 knot = 1 nautical mile per hour; see Appendix D.) What are the (a) magnitude and (b) direction of the velocity of ship A relative to B? (c) After what time will the ships be 160 nautical miles apart? (d) What will be the bearing of B (the direction of B's position) relative to A at that time?

••80 GO A 200-m-wide river flows due east at a uniform speed of 2.0 m/s. A boat with a speed of 8.0 m/s relative to the water leaves the south bank pointed in a direction 30° west of north. What are the (a) magnitude and (b) direction of the boat's velocity relative to the ground? (c) How long does the boat take to cross the river?

•••81   Ship A is located 4.0 km north and 2.5 km east of ship B. Ship A has a velocity of 22 km/h toward the south, and ship B has a velocity of 40 km/h in a direction 37° north of east. (a) What is the velocity of A relative to B in unit-vector notation with $\hat{i}$ toward the east? (b) Write an expression (in terms of $\hat{i}$ and $\hat{j}$) for the position of A relative to B as a function of t, where $t = 0$ when the ships are in the positions described above. (c) At what time is the separation between the ships least? (d) What is that least separation?

•••82   A 200-m-wide river has a uniform flow speed of 1.1 m/s through a jungle and toward the east. An explorer wishes to leave a small clearing on the south bank and cross the river in a powerboat that moves at a constant speed of 4.0 m/s with respect to the water. There is a clearing on the north bank 82 m upstream from a point directly opposite the clearing on the south bank. (a) In what direction must the boat be pointed in order to travel in a straight line and land in the clearing on the north bank? (b) How long will the boat take to cross the river and land in the clearing?

Additional Problems

83   A woman who can row a boat at 6.4 km/h in still water faces a long, straight river with a width of 6.4 km and a current of 3.2 km/h. Let $\hat{i}$ point directly across the river and $\hat{j}$ point directly downstream. If she rows in a straight line to a point directly opposite her starting position, (a) at what angle to $\hat{i}$ must she point the boat and (b) how long will she take? (c) How long will she take if, instead, she rows 3.2 km *down* the river and then back to her starting point? (d) How long if she rows 3.2 km *up* the river and then back to her starting point? (e) At what angle to $\hat{i}$ should she point the boat if she wants to cross the river in the shortest possible time? (f) How long is that shortest time?

84   In Fig. 4-48a, a sled moves in the negative x direction at constant speed v_s while a ball of ice is shot from the sled with a velocity $\vec{v}_0 = v_{0x}\hat{i} + v_{0y}\hat{j}$ relative to the sled. When the ball lands, its

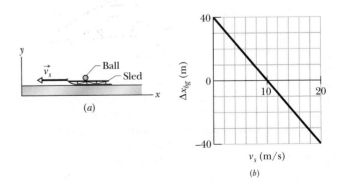

Fig. 4-48   Problem 84.

horizontal displacement Δx_{bg} relative to the ground (from its launch position to its landing position) is measured. Figure 4-48*b* gives Δx_{bg} as a function of v_s. Assume the ball lands at approximately its launch height. What are the values of (a) v_{0x} and (b) v_{0y}? The ball's displacement Δx_{bs} relative to the sled can also be measured. Assume that the sled's velocity is not changed when the ball is shot. What is Δx_{bs} when v_s is (c) 5.0 m/s and (d) 15 m/s?

85 You are kidnapped by political-science majors (who are upset because you told them political science is not a real science). Although blindfolded, you can tell the speed of their car (by the whine of the engine), the time of travel (by mentally counting off seconds), and the direction of travel (by turns along the rectangular street system). From these clues, you know that you are taken along the following course: 50 km/h for 2.0 min, turn 90° to the right, 20 km/h for 4.0 min, turn 90° to the right, 20 km/h for 60 s, turn 90° to the left, 50 km/h for 60 s, turn 90° to the right, 20 km/h for 2.0 min, turn 90° to the left, 50 km/h for 30 s. At that point, (a) how far are you from your starting point, and (b) in what direction relative to your initial direction of travel are you?

86 In Fig. 4-49, a radar station detects an airplane approaching directly from the east. At first observation, the airplane is at distance $d_1 = 360$ m from the station and at angle $\theta_1 = 40°$ above the horizon. The airplane is tracked through an angular change $\Delta\theta = 123°$ in the vertical east–west plane; its distance is then $d_2 = 790$ m. Find the (a) magnitude and (b) direction of the airplane's displacement during this period.

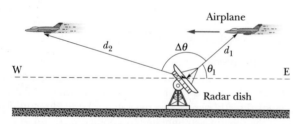

Fig. 4-49 Problem 86.

87 **SSM** A baseball is hit at ground level. The ball reaches its maximum height above ground level 3.0 s after being hit. Then 2.5 s after reaching its maximum height, the ball barely clears a fence that is 97.5 m from where it was hit. Assume the ground is level. (a) What maximum height above ground level is reached by the ball? (b) How high is the fence? (c) How far beyond the fence does the ball strike the ground?

88 Long flights at midlatitudes in the Northern Hemisphere encounter the jet stream, an eastward airflow that can affect a plane's speed relative to Earth's surface. If a pilot maintains a certain speed relative to the air (the plane's *airspeed*), the speed relative to the surface (the plane's *ground speed*) is more when the flight is in the direction of the jet stream and less when the flight is opposite the jet stream. Suppose a round-trip flight is scheduled between two cities separated by 4000 km, with the outgoing flight in the direction of the jet stream and the return flight opposite it. The airline computer advises an airspeed of 1000 km/h, for which the difference in flight times for the outgoing and return flights is 70.0 min. What jet-stream speed is the computer using?

89 **SSM** A particle starts from the origin at $t = 0$ with a velocity of $8.0\hat{j}$ m/s and moves in the xy plane with constant acceleration

$(4.0\hat{i} + 2.0\hat{j})$ m/s². When the particle's x coordinate is 29 m, what are its (a) y coordinate and (b) speed?

90 At what initial speed must the basketball player in Fig. 4-50 throw the ball, at angle $\theta_0 = 55°$ above the horizontal, to make the foul shot? The horizontal distances are $d_1 = 1.0$ ft and $d_2 = 14$ ft, and the heights are $h_1 = 7.0$ ft and $h_2 = 10$ ft.

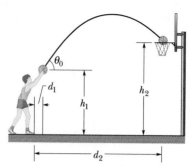

Fig. 4-50 Problem 90.

91 During volcanic eruptions, chunks of solid rock can be blasted out of the volcano; these projectiles are called *volcanic bombs*. Figure 4-51 shows a cross section of Mt. Fuji, in Japan. (a) At what initial speed would a bomb have to be ejected, at angle $\theta_0 = 35°$ to the horizontal, from the vent at A in order to fall at the foot of the volcano at B, at vertical distance $h = 3.30$ km and horizontal distance $d = 9.40$ km? Ignore, for the moment, the effects of air on the bomb's travel. (b) What would be the time of flight? (c) Would the effect of the air increase or decrease your answer in (a)?

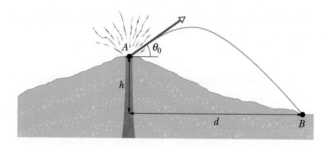

Fig. 4-51 Problem 91.

92 An astronaut is rotated in a horizontal centrifuge at a radius of 5.0 m. (a) What is the astronaut's speed if the centripetal acceleration has a magnitude of $7.0g$? (b) How many revolutions per minute are required to produce this acceleration? (c) What is the period of the motion?

93 **SSM** Oasis A is 90 km due west of oasis B. A desert camel leaves A and takes 50 h to walk 75 km at 37° north of due east. Next it takes 35 h to walk 65 km due south. Then it rests for 5.0 h. What are the (a) magnitude and (b) direction of the camel's displacement relative to A at the resting point? From the time the camel leaves A until the end of the rest period, what are the (c) magnitude and (d) direction of its average velocity and (e) its average speed? The camel's last drink was at A; it must be at B no more than 120 h later for its next drink. If it is to reach B just in time, what must be the (f) magnitude and (g) direction of its average velocity after the rest period?

94 ✈ *Curtain of death.* A large metallic asteroid strikes Earth and quickly digs a crater into the rocky material below ground level by launching rocks upward and outward. The following table gives five pairs of launch speeds and angles (from the horizontal) for such rocks, based on a model of crater formation. (Other rocks, with intermediate speeds and angles, are also launched.) Suppose that you are at $x = 20$ km when the asteroid strikes the ground at time $t = 0$ and position $x = 0$ (Fig. 4-52). (a) At $t = 20$ s, what are the x and y coordinates of the rocks headed in your direction from launches A through E? (b) Plot these coordinates and then sketch a curve through the points to include rocks with intermediate launch speeds and angles. The curve should indicate what you would see as you look up into the approaching rocks and what dinosaurs must have seen during asteroid strikes long ago.

Launch	Speed (m/s)	Angle (degrees)
A	520	14.0
B	630	16.0
C	750	18.0
D	870	20.0
E	1000	22.0

Fig. 4-52 Problem 94.

95 Figure 4-53 shows the straight path of a particle across an xy coordinate system as the particle is accelerated from rest during time interval Δt_1. The acceleration is constant. The xy coordinates for point A are (4.00 m, 6.00 m); those for point B are (12.0 m, 18.0 m). (a) What is the ratio a_y/a_x of the acceleration components? (b) What are the coordinates of the particle if the motion is continued for another interval equal to Δt_1?

Fig. 4-53 Problem 95.

96 For women's volleyball the top of the net is 2.24 m above the floor and the court measures 9.0 m by 9.0 m on each side of the net. Using a jump serve, a player strikes the ball at a point that is 3.0 m above the floor and a horizontal distance of 8.0 m from the net. If the initial velocity of the ball is horizontal, (a) what minimum magnitude must it have if the ball is to clear the net and (b) what maximum magnitude can it have if the ball is to strike the floor inside the back line on the other side of the net?

97 SSM A rifle is aimed horizontally at a target 30 m away. The bullet hits the target 1.9 cm below the aiming point. What are (a) the bullet's time of flight and (b) its speed as it emerges from the rifle?

98 A particle is in uniform circular motion about the origin of an xy coordinate system, moving clockwise with a period of 7.00 s. At one instant, its position vector (measured from the origin) is $\vec{r} = (2.00 \text{ m})\hat{i} - (3.00 \text{ m})\hat{j}$. At that instant, what is its velocity in unit-vector notation?

99 In Fig. 4-54, a lump of wet putty moves in uniform circular motion as it rides at a radius of 20.0 cm on the rim of a wheel rotating counterclockwise with a period of 5.00 ms. The lump then happens to fly off the rim at the 5 o'clock position (as if on a clock face). It leaves the rim at a

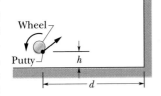

Fig. 4-54 Problem 99.

height of $h = 1.20$ m from the floor and at a distance $d = 2.50$ m from a wall. At what height on the wall does the lump hit?

100 An iceboat sails across the surface of a frozen lake with constant acceleration produced by the wind. At a certain instant the boat's velocity is $(6.30\hat{i} - 8.42\hat{j})$ m/s. Three seconds later, because of a wind shift, the boat is instantaneously at rest. What is its average acceleration for this 3.00 s interval?

101 In Fig. 4-55, a ball is shot directly upward from the ground with an initial speed of $v_0 = 7.00$ m/s. Simultaneously, a construction elevator cab begins to move upward from the ground with a constant speed of $v_c = 3.00$ m/s. What maximum height does the ball reach relative to (a) the ground and (b) the cab floor? At what rate does the speed of the ball change relative to (c) the ground and (d) the cab floor?

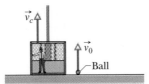

Fig. 4-55 Problem 101.

102 A magnetic field can force a charged particle to move in a circular path. Suppose that an electron moving in a circle experiences a radial acceleration of magnitude 3.0×10^{14} m/s² in a particular magnetic field. (a) What is the speed of the electron if the radius of its circular path is 15 cm? (b) What is the period of the motion?

103 In 3.50 h, a balloon drifts 21.5 km north, 9.70 km east, and 2.88 km upward from its release point on the ground. Find (a) the magnitude of its average velocity and (b) the angle its average velocity makes with the horizontal.

104 A ball is thrown horizontally from a height of 20 m and hits the ground with a speed that is three times its initial speed. What is the initial speed?

105 A projectile is launched with an initial speed of 30 m/s at an angle of 60° above the horizontal. What are the (a) magnitude and (b) angle of its velocity 2.0 s after launch, and (c) is the angle above or below the horizontal? What are the (d) magnitude and (e) angle of its velocity 5.0 s after launch, and (f) is the angle above or below the horizontal?

106 The position vector for a proton is initially $\vec{r} = 5.0\hat{i} - 6.0\hat{j} + 2.0\hat{k}$ and then later is $\vec{r} = -2.0\hat{i} + 6.0\hat{j} + 2.0\hat{k}$, all in meters. (a) What is the proton's displacement vector, and (b) to what plane is that vector parallel?

107 A particle P travels with constant speed on a circle of radius $r = 3.00$ m (Fig. 4-56) and completes one revolution in 20.0 s. The particle passes through O at time $t = 0$. State the following vectors in magnitude-angle notation (angle relative to the positive direction of x). With respect to O, find the particle's position vector at the times t of (a) 5.00 s, (b) 7.50 s, and (c) 10.0 s.

(d) For the 5.00 s interval from the end of the fifth second to the end of the tenth second, find the particle's displacement. For that interval, find (e) its average velocity and its velocity at the (f) beginning and (g) end. Next, find the acceleration at the (h) beginning and (i) end of that interval.

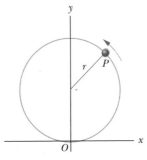

Fig. 4-56 Problem 107.

108 The fast French train known as the TGV (Train à Grande Vitesse) has a scheduled average speed of 216 km/h. (a) If the train goes around a curve at that speed and the magnitude of the acceleration experienced by the passengers is to be limited to 0.050g, what is the smallest radius of curvature for the track that can be tolerated? (b) At what speed must the train go around a curve with a 1.00 km radius to be at the acceleration limit?

109 (a) If an electron is projected horizontally with a speed of 3.0×10^6 m/s, how far will it fall in traversing 1.0 m of horizontal distance? (b) Does the answer increase or decrease if the initial speed is increased?

110 A person walks up a stalled 15-m-long escalator in 90 s. When standing on the same escalator, now moving, the person is carried up in 60 s. How much time would it take that person to walk up the moving escalator? Does the answer depend on the length of the escalator?

111 (a) What is the magnitude of the centripetal acceleration of an object on Earth's equator due to the rotation of Earth? (b) What would Earth's rotation period have to be for objects on the equator to have a centripetal acceleration of magnitude 9.8 m/s^2?

112 ✈ The range of a projectile depends not only on v_0 and θ_0 but also on the value g of the free-fall acceleration, which varies from place to place. In 1936, Jesse Owens established a world's running broad jump record of 8.09 m at the Olympic Games at Berlin (where $g = 9.8128$ m/s^2). Assuming the same values of v_0 and θ_0, by how much would his record have differed if he had competed instead in 1956 at Melbourne (where $g = 9.7999$ m/s^2)?

113 Figure 4-57 shows the path taken by a drunk skunk over level ground, from initial point i to final point f. The angles are $\theta_1 = 30.0°$, $\theta_2 = 50.0°$, and $\theta_3 = 80.0°$, and the distances are $d_1 = 5.00$ m, $d_2 = 8.00$ m, and $d_3 = 12.0$ m. What are the (a) magnitude and (b) angle of the skunk's displacement from i to f?

114 The position vector $\vec{r}$ of a particle moving in the xy plane is $\vec{r} = 2t\hat{i} + 2\sin[(\pi/4 \text{ rad/s})t]\hat{j}$, with $\vec{r}$ in meters and t in seconds. (a) Calculate the x and y components of the particle's position at $t = 0, 1.0, 2.0, 3.0,$ and 4.0 s and sketch the particle's path in the xy plane for the interval $0 \le t \le 4.0$ s.

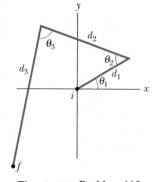

Fig. 4-57 Problem 113.

(b) Calculate the components of the particle's velocity at $t = 1.0$, 2.0, and 3.0 s. Show that the velocity is tangent to the path of the particle and in the direction the particle is moving at each time by drawing the velocity vectors on the plot of the particle's path in part (a). (c) Calculate the components of the particle's acceleration at $t = 1.0, 2.0,$ and 3.0 s.

115 An electron having an initial horizontal velocity of magnitude 1.00×10^9 cm/s travels into the region between two horizontal metal plates that are electrically charged. In that region, the electron travels a horizontal distance of 2.00 cm and has a constant downward acceleration of magnitude 1.00×10^{17} cm/s^2 due to the charged plates. Find (a) the time the electron takes to travel the 2.00 cm, (b) the vertical distance it travels during that time, and the magnitudes of its (c) horizontal and (d) vertical velocity components as it emerges from the region.

116 An elevator without a ceiling is ascending with a constant speed of 10 m/s. A boy on the elevator shoots a ball directly upward, from a height of 2.0 m above the elevator floor, just as the elevator floor is 28 m above the ground. The initial speed of the ball with respect to the elevator is 20 m/s. (a) What maximum height above the ground does the ball reach? (b) How long does the ball take to return to the elevator floor?

117 A football player punts the football so that it will have a "hang time" (time of flight) of 4.5 s and land 46 m away. If the ball leaves the player's foot 150 cm above the ground, what must be the (a) magnitude and (b) angle (relative to the horizontal) of the ball's initial velocity?

118 An airport terminal has a moving sidewalk to speed passengers through a long corridor. Larry does not use the moving sidewalk; he takes 150 s to walk through the corridor. Curly, who simply stands on the moving sidewalk, covers the same distance in 70 s. Moe boards the sidewalk and walks along it. How long does Moe take to move through the corridor? Assume that Larry and Moe walk at the same speed.

119 A wooden boxcar is moving along a straight railroad track at speed v_1. A sniper fires a bullet (initial speed v_2) at it from a high-powered rifle. The bullet passes through both lengthwise walls of the car, its entrance and exit holes being exactly opposite each other as viewed from within the car. From what direction, relative to the track, is the bullet fired? Assume that the bullet is not deflected upon entering the car, but that its speed decreases by 20%. Take $v_1 = 85$ km/h and $v_2 = 650$ m/s. (Why don't you need to know the width of the boxcar?)

FORCE AND MOTION–I

5-1 WHAT IS PHYSICS?

We have seen that part of physics is a study of motion, including accelerations, which are changes in velocities. Physics is also a study of what can *cause* an object to accelerate. That cause is a **force,** which is, loosely speaking, a push or pull on the object. The force is said to *act* on the object to change its velocity. For example, when a dragster accelerates, a force from the track acts on the rear tires to cause the dragster's acceleration. When a defensive guard knocks down a quarterback, a force from the guard acts on the quarterback to cause the quarterback's backward acceleration. When a car slams into a telephone pole, a force on the car from the pole causes the car to stop. Science, engineering, legal, and medical journals are filled with articles about forces on objects, including people.

5-2 Newtonian Mechanics

The relation between a force and the acceleration it causes was first understood by Isaac Newton (1642–1727) and is the subject of this chapter. The study of that relation, as Newton presented it, is called *Newtonian mechanics.* We shall focus on its three primary laws of motion.

Newtonian mechanics does not apply to all situations. If the speeds of the interacting bodies are very large—an appreciable fraction of the speed of light—we must replace Newtonian mechanics with Einstein's special theory of relativity, which holds at any speed, including those near the speed of light. If the interacting bodies are on the scale of atomic structure (for example, they might be electrons in an atom), we must replace Newtonian mechanics with quantum mechanics. Physicists now view Newtonian mechanics as a special case of these two more comprehensive theories. Still, it is a very important special case because it applies to the motion of objects ranging in size from the very small (almost on the scale of atomic structure) to astronomical (galaxies and clusters of galaxies).

5-3 Newton's First Law

Before Newton formulated his mechanics, it was thought that some influence, a "force," was needed to keep a body moving at constant velocity. Similarly, a body was thought to be in its "natural state" when it was at rest. For a body to move with constant velocity, it seemingly had to be propelled in some way, by a push or a pull. Otherwise, it would "naturally" stop moving.

These ideas were reasonable. If you send a puck sliding across a wooden floor, it does indeed slow and then stop. If you want to make it move across the floor with constant velocity, you have to continuously pull or push it.

Send a puck sliding over the ice of a skating rink, however, and it goes a lot farther. You can imagine longer and more slippery surfaces, over which the puck would slide farther and farther. In the limit you can think of a long, extremely slippery surface (said to be a **frictionless surface**), over which the puck would hardly slow. (We can in fact come close to this situation by sending a puck sliding over a horizontal air table, across which it moves on a film of air.)

From these observations, we can conclude that a body will keep moving with constant velocity if no force acts on it. That leads us to the first of Newton's three laws of motion:

> **Newton's First Law:** If no force acts on a body, the body's velocity cannot change; that is, the body cannot accelerate.

In other words, if the body is at rest, it stays at rest. If it is moving, it continues to move with the same velocity (same magnitude *and* same direction).

5-4 **Force**

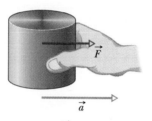

Fig. 5-1 A force $\vec{F}$ on the standard kilogram gives that body an acceleration $\vec{a}$.

We now wish to define the unit of force. We know that a force can cause the acceleration of a body. Thus, we shall define the unit of force in terms of the acceleration that a force gives to a standard reference body, which we take to be the standard kilogram of Fig. 1-3. This body has been assigned, exactly and by definition, a mass of 1 kg.

We put the standard body on a horizontal frictionless table and pull the body to the right (Fig. 5-1) so that, by trial and error, it eventually experiences a measured acceleration of 1 m/s². We then declare, as a matter of definition, that the force we are exerting on the standard body has a magnitude of 1 newton (abbreviated N).

We can exert a 2 N force on our standard body by pulling it so that its measured acceleration is 2 m/s², and so on. Thus in general, if our standard body of 1 kg mass has an acceleration of magnitude a, we know that a force F must be acting on it and that the magnitude of the force (in newtons) is equal to the magnitude of the acceleration (in meters per second per second).

Thus, a force is measured by the acceleration it produces. However, acceleration is a vector quantity, with both magnitude and direction. Is force also a vector quantity? We can easily assign a direction to a force (just assign the direction of the acceleration), but that is not sufficient. We must prove by experiment that forces are vector quantities. Actually, that has been done: forces are indeed vector quantities; they have magnitudes and directions, and they combine according to the vector rules of Chapter 3.

This means that when two or more forces act on a body, we can find their **net force,** or **resultant force,** by adding the individual forces vectorially. A single force that has the magnitude and direction of the net force has the same effect on the body as all the individual forces together. This fact is called the **principle of superposition for forces.** The world would be quite strange if, for example, you and a friend were to pull on the standard body in the same direction, each with a force of 1 N, and yet somehow the net pull was 14 N.

In this book, forces are most often represented with a vector symbol such as $\vec{F}$, and a net force is represented with the vector symbol $\vec{F}_{net}$. As with other vectors, a force or a net force can have components along coordinate axes. When forces act only along a single axis, they are single-component forces. Then we can drop the overhead arrows on the force symbols and just use signs to indicate the directions of the forces along that axis.

Instead of the wording used in Section 5-3, the more proper statement of Newton's First Law is in terms of a *net* force:

 Newton's First Law: If no *net* force acts on a body ($\vec{F}_{net} = 0$), the body's velocity cannot change; that is, the body cannot accelerate.

There may be multiple forces acting on a body, but if their net force is zero, the body cannot accelerate.

Inertial Reference Frames

Newton's first law is not true in all reference frames, but we can always find reference frames in which it (as well as the rest of Newtonian mechanics) is true. Such special frames are referred to as **inertial reference frames,** or simply **inertial frames.**

 An inertial reference frame is one in which Newton's laws hold.

For example, we can assume that the ground is an inertial frame provided we can neglect Earth's astronomical motions (such as its rotation).

That assumption works well if, say, a puck is sent sliding along a *short* strip of frictionless ice—we would find that the puck's motion obeys Newton's laws. However, suppose the puck is sent sliding along a *long* ice strip extending from the north pole (Fig. 5-2a). If we view the puck from a stationary frame in space, the puck moves south along a simple straight line because Earth's rotation around the north pole merely slides the ice beneath the puck. However, if we view the puck from a point on the ground so that we rotate with Earth, the puck's path is not a simple straight line. Because the eastward speed of the ground beneath the puck is greater the farther south the puck slides, from our ground-based view the puck appears to be deflected westward (Fig. 5-2b). However, this apparent deflection is caused not by a force as required by Newton's laws but by the fact that we see the puck from a rotating frame. In this situation, the ground is a **noninertial frame.**

In this book we usually assume that the ground is an inertial frame and that measured forces and accelerations are from this frame. If measurements

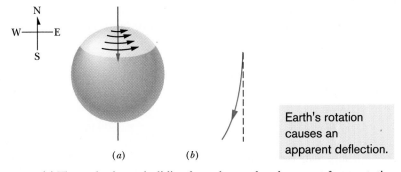

Earth's rotation causes an apparent deflection.

Fig. 5-2 (a) The path of a puck sliding from the north pole as seen from a stationary point in space. Earth rotates to the east. (b) The path of the puck as seen from the ground.

are made in, say, an elevator that is accelerating relative to the ground, then the measurements are being made in a noninertial frame and the results can be surprising.

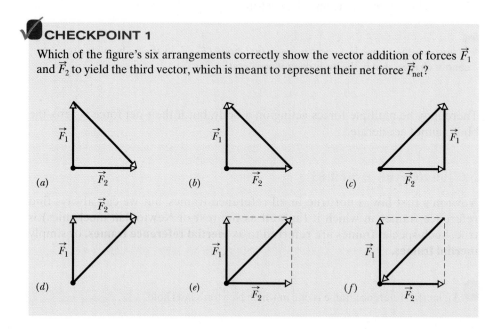

✔ **CHECKPOINT 1**

Which of the figure's six arrangements correctly show the vector addition of forces $\vec{F}_1$ and $\vec{F}_2$ to yield the third vector, which is meant to represent their net force $\vec{F}_{net}$?

5-5 **Mass**

Everyday experience tells us that a given force produces different magnitudes of acceleration for different bodies. Put a baseball and a bowling ball on the floor and give both the same sharp kick. Even if you don't actually do this, you know the result: The baseball receives a noticeably larger acceleration than the bowling ball. The two accelerations differ because the mass of the baseball differs from the mass of the bowling ball—but what, exactly, is mass?

We can explain how to measure mass by imagining a series of experiments in an inertial frame. In the first experiment we exert a force on a standard body, whose mass m_0 is defined to be 1.0 kg. Suppose that the standard body accelerates at 1.0 m/s². We can then say the force on that body is 1.0 N.

We next apply that same force (we would need some way of being certain it is the same force) to a second body, body X, whose mass is not known. Suppose we find that this body X accelerates at 0.25 m/s². We know that a *less massive* baseball receives a *greater acceleration* than a more massive bowling ball when the same force (kick) is applied to both. Let us then make the following conjecture: The ratio of the masses of two bodies is equal to the inverse of the ratio of their accelerations when the same force is applied to both. For body X and the standard body, this tells us that

$$\frac{m_X}{m_0} = \frac{a_0}{a_X}.$$

Solving for m_X yields

$$m_X = m_0 \frac{a_0}{a_X} = (1.0 \text{ kg}) \frac{1.0 \text{ m/s}^2}{0.25 \text{ m/s}^2} = 4.0 \text{ kg}.$$

Our conjecture will be useful, of course, only if it continues to hold when we change the applied force to other values. For example, if we apply an 8.0 N force to the standard body, we obtain an acceleration of 8.0 m/s². When the 8.0 N force is applied to body X, we obtain an acceleration of 2.0 m/s^2. Our conjecture then gives us

$$m_X = m_0 \frac{a_0}{a_X} = (1.0 \text{ kg}) \frac{8.0 \text{ m/s}^2}{2.0 \text{ m/s}^2} = 4.0 \text{ kg},$$

consistent with our first experiment. Many experiments yielding similar results indicate that our conjecture provides a consistent and reliable means of assigning a mass to any given body.

Our measurement experiments indicate that mass is an *intrinsic* characteristic of a body—that is, a characteristic that automatically comes with the existence of the body. They also indicate that mass is a scalar quantity. However, the nagging question remains: What, exactly, is mass?

Since the word *mass* is used in everyday English, we should have some intuitive understanding of it, maybe something that we can physically sense. Is it a body's size, weight, or density? The answer is no, although those characteristics are sometimes confused with mass. We can say only that *the mass of a body is the characteristic that relates a force on the body to the resulting acceleration.* Mass has no more familiar definition; you can have a physical sensation of mass only when you try to accelerate a body, as in the kicking of a baseball or a bowling ball.

5-6 Newton's Second Law

All the definitions, experiments, and observations we have discussed so far can be summarized in one neat statement:

 Newton's Second Law: The net force on a body is equal to the product of the body's mass and its acceleration.

In equation form,

$$\vec{F}_{\text{net}} = m\vec{a} \qquad \text{(Newton's second law)}. \qquad (5\text{-}1)$$

This equation is simple, but we must use it cautiously. First, we must be certain about which body we are applying it to. Then $\vec{F}_{\text{net}}$ must be the vector sum of *all* the forces that act on *that* body. Only forces that act on *that* body are to be included in the vector sum, not forces acting on other bodies that might be involved in the given situation. For example, if you are in a rugby scrum, the net force on *you* is the vector sum of all the pushes and pulls on *your* body. It does not include any push or pull on another player from you or from anyone else. Every time you work a force problem, your first step is to clearly state the body to which you are applying Newton's law.

Like other vector equations, Eq. 5-1 is equivalent to three component equations, one for each axis of an *xyz* coordinate system:

$$F_{\text{net},x} = ma_x, \quad F_{\text{net},y} = ma_y, \quad \text{and} \quad F_{\text{net},z} = ma_z. \qquad (5\text{-}2)$$

Each of these equations relates the net force component along an axis to the acceleration along that same axis. For example, the first equation tells us that the sum of all the force components along the x axis causes the x component a_x of the body's acceleration, but causes no acceleration in the y and z directions. Turned around, the acceleration component a_x is caused only by the sum of the force components along the x axis. In general,

> The acceleration component along a given axis is caused *only by* the sum of the force components along that *same* axis, and not by force components along any other axis.

Equation 5-1 tells us that if the net force on a body is zero, the body's acceleration $\vec{a} = 0$. If the body is at rest, it stays at rest; if it is moving, it continues to move at constant velocity. In such cases, any forces on the body *balance* one another, and both the forces and the body are said to be in *equilibrium*. Commonly, the forces are also said to *cancel* one another, but the term "cancel" is tricky. It does *not* mean that the forces cease to exist (canceling forces is not like canceling dinner reservations). The forces still act on the body.

For SI units, Eq. 5-1 tells us that

$$1 \text{ N} = (1 \text{ kg})(1 \text{ m/s}^2) = 1 \text{ kg} \cdot \text{m/s}^2. \tag{5-3}$$

Some force units in other systems of units are given in Table 5-1 and Appendix D.

Table 5-1

Units in Newton's Second Law (Eqs. 5-1 and 5-2)

System	Force	Mass	Acceleration
SI	newton (N)	kilogram (kg)	m/s^2
CGS[a]	dyne	gram (g)	cm/s^2
British[b]	pound (lb)	slug	ft/s^2

[a] 1 dyne = $1 \text{ g} \cdot \text{cm/s}^2$.
[b] 1 lb = $1 \text{ slug} \cdot \text{ft/s}^2$.

To solve problems with Newton's second law, we often draw a **free-body diagram** in which the only body shown is the one for which we are summing forces. A sketch of the body itself is preferred by some teachers but, to save space in these chapters, we shall usually represent the body with a dot. Also, each force on the body is drawn as a vector arrow with its tail on the body. A coordinate system is usually included, and the acceleration of the body is sometimes shown with a vector arrow (labeled as an acceleration).

A **system** consists of one or more bodies, and any force on the bodies inside the system from bodies outside the system is called an **external force.** If the bodies making up a system are rigidly connected to one another, we can treat the system as one composite body, and the net force $\vec{F}_{net}$ on it is the vector sum of all external forces. (We do not include **internal forces**—that is, forces between two bodies inside the system.) For example, a connected railroad engine and car form a system. If, say, a tow line pulls on the front of the engine, the force due to the tow line acts on the whole engine–car system. Just as for a single body, we can relate the net external force on a system to its acceleration with Newton's second law, $\vec{F}_{net} = m\vec{a}$, where m is the total mass of the system.

✔ CHECKPOINT 2

The figure here shows two horizontal forces acting on a block on a frictionless floor. If a third horizontal force $\vec{F}_3$ also acts on the block, what are the magnitude and direction of $\vec{F}_3$ when the block is (a) stationary and (b) moving to the left with a constant speed of 5 m/s?

3 N ← ☐ → 5 N

Sample Problem

One- and two-dimensional forces, puck

Parts A, B, and C of Fig. 5-3 show three situations in which one or two forces act on a puck that moves over frictionless ice along an x axis, in one-dimensional motion. The puck's mass is $m = 0.20$ kg. Forces $\vec{F}_1$ and $\vec{F}_2$ are directed along the axis and have magnitudes $F_1 = 4.0$ N and $F_2 = 2.0$ N. Force $\vec{F}_3$ is directed at angle $\theta = 30°$ and has magnitude $F_3 = 1.0$ N. In each situation, what is the acceleration of the puck?

KEY IDEA

In each situation we can relate the acceleration $\vec{a}$ to the net force $\vec{F}_{net}$ acting on the puck with Newton's second law, $\vec{F}_{net} = m\vec{a}$. However, because the motion is along only the x

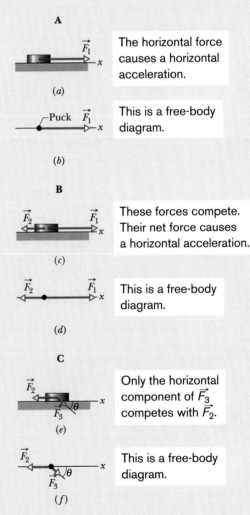

A

(a)

The horizontal force causes a horizontal acceleration.

(b)

This is a free-body diagram.

B

(c)

These forces compete. Their net force causes a horizontal acceleration.

(d)

This is a free-body diagram.

C

(e)

Only the horizontal component of $\vec{F}_3$ competes with $\vec{F}_2$.

(f)

This is a free-body diagram.

Fig. 5-3 In three situations, forces act on a puck that moves along an x axis. Free-body diagrams are also shown.

axis, we can simplify each situation by writing the second law for x components only:

$$F_{net,x} = ma_x. \qquad (5\text{-}4)$$

The free-body diagrams for the three situations are also given in Fig. 5-3, with the puck represented by a dot.

Situation A: For Fig. 5-3b, where only one horizontal force acts, Eq. 5-4 gives us

$$F_1 = ma_x,$$

which, with given data, yields

$$a_x = \frac{F_1}{m} = \frac{4.0 \text{ N}}{0.20 \text{ kg}} = 20 \text{ m/s}^2. \qquad \text{(Answer)}$$

The positive answer indicates that the acceleration is in the positive direction of the x axis.

Situation B: In Fig. 5-3d, two horizontal forces act on the puck, $\vec{F}_1$ in the positive direction of x and $\vec{F}_2$ in the negative direction. Now Eq. 5-4 gives us

$$F_1 - F_2 = ma_x,$$

which, with given data, yields

$$a_x = \frac{F_1 - F_2}{m} = \frac{4.0 \text{ N} - 2.0 \text{ N}}{0.20 \text{ kg}} = 10 \text{ m/s}^2.$$
$$\text{(Answer)}$$

Thus, the net force accelerates the puck in the positive direction of the x axis.

Situation C: In Fig. 5-3f, force $\vec{F}_3$ is not directed along the direction of the puck's acceleration; only x component $F_{3,x}$ is. (Force $\vec{F}_3$ is two-dimensional but the motion is only one-dimensional.) Thus, we write Eq. 5-4 as

$$F_{3,x} - F_2 = ma_x. \qquad (5\text{-}5)$$

From the figure, we see that $F_{3,x} = F_3 \cos \theta$. Solving for the acceleration and substituting for $F_{3,x}$ yield

$$a_x = \frac{F_{3,x} - F_2}{m} = \frac{F_3 \cos \theta - F_2}{m}$$

$$= \frac{(1.0 \text{ N})(\cos 30°) - 2.0 \text{ N}}{0.20 \text{ kg}} = -5.7 \text{ m/s}^2.$$
$$\text{(Answer)}$$

Thus, the net force accelerates the puck in the negative direction of the x axis.

Sample Problem

Two-dimensional forces, cookie tin

In the overhead view of Fig. 5-4a, a 2.0 kg cookie tin is accelerated at 3.0 m/s² in the direction shown by $\vec{a}$, over a frictionless horizontal surface. The acceleration is caused by three horizontal forces, only two of which are shown: $\vec{F_1}$ of magnitude 10 N and $\vec{F_2}$ of magnitude 20 N. What is the third force $\vec{F_3}$ in unit-vector notation and in magnitude-angle notation?

KEY IDEA

The net force $\vec{F}_{net}$ on the tin is the sum of the three forces and is related to the acceleration $\vec{a}$ via Newton's second law ($\vec{F}_{net} = m\vec{a}$). Thus,

$$\vec{F_1} + \vec{F_2} + \vec{F_3} = m\vec{a}, \tag{5-6}$$

which gives us

$$\vec{F_3} = m\vec{a} - \vec{F_1} - \vec{F_2}. \tag{5-7}$$

Calculations: Because this is a two-dimensional problem, we *cannot* find $\vec{F_3}$ merely by substituting the magnitudes for the vector quantities on the right side of Eq. 5-7. Instead, we must vectorially add $m\vec{a}$, $-\vec{F_1}$ (the reverse of $\vec{F_1}$), and $-\vec{F_2}$ (the reverse of $\vec{F_2}$), as shown in Fig. 5-4b. This addition can be done directly on a vector-capable calculator because we know both magnitude and angle for all three vectors. However, here we shall evaluate the right side of Eq. 5-7 in terms of components, first along the x axis and then along the y axis.

***x* components:** Along the x axis we have

$$F_{3,x} = ma_x - F_{1,x} - F_{2,x}$$
$$= m(a\cos 50°) - F_1\cos(-150°) - F_2\cos 90°.$$

Then, substituting known data, we find

$$F_{3,x} = (2.0 \text{ kg})(3.0 \text{ m/s}^2)\cos 50° - (10 \text{ N})\cos(-150°)$$
$$\quad - (20 \text{ N})\cos 90°$$
$$= 12.5 \text{ N.}$$

***y* components:** Similarly, along the y axis we find

$$F_{3,y} = ma_y - F_{1,y} - F_{2,y}$$
$$= m(a\sin 50°) - F_1\sin(-150°) - F_2\sin 90°$$
$$= (2.0 \text{ kg})(3.0 \text{ m/s}^2)\sin 50° - (10 \text{ N})\sin(-150°)$$
$$\quad - (20 \text{ N})\sin 90°$$
$$= -10.4 \text{ N.}$$

Vector: In unit-vector notation, we can write

$$\vec{F_3} = F_{3,x}\hat{i} + F_{3,y}\hat{j} = (12.5 \text{ N})\hat{i} - (10.4 \text{ N})\hat{j}$$
$$\approx (13 \text{ N})\hat{i} - (10 \text{ N})\hat{j}. \quad \text{(Answer)}$$

We can now use a vector-capable calculator to get the magnitude and the angle of $\vec{F_3}$. We can also use Eq. 3-6 to obtain the magnitude and the angle (from the positive direction of the x axis) as

$$F_3 = \sqrt{F_{3,x}^2 + F_{3,y}^2} = 16 \text{ N}$$

and

$$\theta = \tan^{-1}\frac{F_{3,y}}{F_{3,x}} = -40°. \quad \text{(Answer)}$$

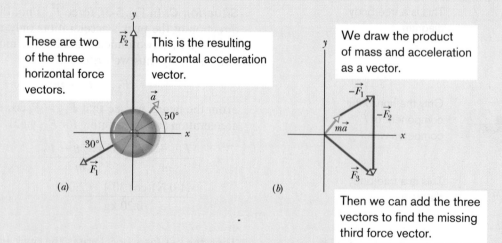

These are two of the three horizontal force vectors.

This is the resulting horizontal acceleration vector.

We draw the product of mass and acceleration as a vector.

Then we can add the three vectors to find the missing third force vector.

(a)   (b)

Fig. 5-4 (a) An overhead view of two of three horizontal forces that act on a cookie tin, resulting in acceleration $\vec{a}$. $\vec{F_3}$ is not shown. (b) An arrangement of vectors $m\vec{a}$, $-\vec{F_1}$, and $-\vec{F_2}$ to find force $\vec{F_3}$.

5-7 Some Particular Forces

The Gravitational Force

A **gravitational force** $\vec{F}_g$ on a body is a certain type of pull that is directed toward a second body. In these early chapters, we do not discuss the nature of this force and usually consider situations in which the second body is Earth. Thus, when we speak of *the* gravitational force $\vec{F}_g$ on a body, we usually mean a force that pulls on it directly toward the center of Earth—that is, directly down toward the ground. We shall assume that the ground is an inertial frame.

Suppose a body of mass m is in free fall with the free-fall acceleration of magnitude g. Then, if we neglect the effects of the air, the only force acting on the body is the gravitational force $\vec{F}_g$. We can relate this downward force and downward acceleration with Newton's second law ($\vec{F} = m\vec{a}$). We place a vertical y axis along the body's path, with the positive direction upward. For this axis, Newton's second law can be written in the form $F_{\text{net},y} = ma_y$, which, in our situation, becomes

$$-F_g = m(-g)$$

or
$$F_g = mg. \qquad (5\text{-}8)$$

In words, the magnitude of the gravitational force is equal to the product mg.

This same gravitational force, with the same magnitude, still acts on the body even when the body is not in free fall but is, say, at rest on a pool table or moving across the table. (For the gravitational force to disappear, Earth would have to disappear.)

We can write Newton's second law for the gravitational force in these vector forms:

$$\vec{F}_g = -F_g\hat{j} = -mg\hat{j} = m\vec{g}, \qquad (5\text{-}9)$$

where $\hat{j}$ is the unit vector that points upward along a y axis, directly away from the ground, and $\vec{g}$ is the free-fall acceleration (written as a vector), directed downward.

Weight

The **weight** W of a body is the magnitude of the net force required to prevent the body from falling freely, as measured by someone on the ground. For example, to keep a ball at rest in your hand while you stand on the ground, you must provide an upward force to balance the gravitational force on the ball from Earth. Suppose the magnitude of the gravitational force is 2.0 N. Then the magnitude of your upward force must be 2.0 N, and thus the weight W of the ball is 2.0 N. We also say that the ball *weighs* 2.0 N and speak about the ball *weighing* 2.0 N.

A ball with a weight of 3.0 N would require a greater force from you—namely, a 3.0 N force—to keep it at rest. The reason is that the gravitational force you must balance has a greater magnitude—namely, 3.0 N. We say that this second ball is *heavier* than the first ball.

Now let us generalize the situation. Consider a body that has an acceleration $\vec{a}$ of zero relative to the ground, which we again assume to be an inertial frame. Two forces act on the body: a downward gravitational force $\vec{F}_g$ and a balancing upward force of magnitude W. We can write Newton's second law for a vertical y axis, with the positive direction upward, as

$$F_{\text{net},y} = ma_y.$$

In our situation, this becomes

$$W - F_g = m(0) \qquad (5\text{-}10)$$

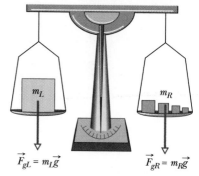

$$\vec{F}_{gL} = m_L \vec{g} \qquad \vec{F}_{gR} = m_R \vec{g}$$

Fig. 5-5 An equal-arm balance. When the device is in balance, the gravitational force $\vec{F}_{gL}$ on the body being weighed (on the left pan) and the total gravitational force $\vec{F}_{gR}$ on the reference bodies (on the right pan) are equal. Thus, the mass m_L of the body being weighed is equal to the total mass m_R of the reference bodies.

or $\qquad\qquad W = F_g \qquad$ (weight, with ground as inertial frame). $\qquad$ (5-11)

This equation tells us (assuming the ground is an inertial frame) that

> The weight W of a body is equal to the magnitude F_g of the gravitational force on the body.

Substituting mg for F_g from Eq. 5-8, we find

$$W = mg \qquad \text{(weight)}, \qquad (5\text{-}12)$$

which relates a body's weight to its mass.

To *weigh* a body means to measure its weight. One way to do this is to place the body on one of the pans of an equal-arm balance (Fig. 5-5) and then place reference bodies (whose masses are known) on the other pan until we strike a balance (so that the gravitational forces on the two sides match). The masses on the pans then match, and we know the mass of the body. If we know the value of g for the location of the balance, we can also find the weight of the body with Eq. 5-12.

We can also weigh a body with a spring scale (Fig. 5-6). The body stretches a spring, moving a pointer along a scale that has been calibrated and marked in either mass or weight units. (Most bathroom scales in the United States work this way and are marked in the force unit pounds.) If the scale is marked in mass units, it is accurate only where the value of g is the same as where the scale was calibrated.

The weight of a body must be measured when the body is not accelerating vertically relative to the ground. For example, you can measure your weight on a scale in your bathroom or on a fast train. However, if you repeat the measurement with the scale in an accelerating elevator, the reading differs from your weight because of the acceleration. Such a measurement is called an *apparent weight*.

Caution: A body's weight is not its mass. Weight is the magnitude of a force and is related to mass by Eq. 5-12. If you move a body to a point where the value of g is different, the body's mass (an intrinsic property) is not different but the weight is. For example, the weight of a bowling ball having a mass of 7.2 kg is 71 N on Earth but only 12 N on the Moon. The mass is the same on Earth and Moon, but the free-fall acceleration on the Moon is only 1.6 m/s².

The Normal Force

If you stand on a mattress, Earth pulls you downward, but you remain stationary. The reason is that the mattress, because it deforms downward due to you, pushes up on you. Similarly, if you stand on a floor, it deforms (it is compressed, bent, or buckled ever so slightly) and pushes up on you. Even a seemingly rigid concrete floor does this (if it is not sitting directly on the ground, enough people on the floor could break it).

The push on you from the mattress or floor is a **normal force** $\vec{F}_N$. The name comes from the mathematical term *normal*, meaning perpendicular: The force on you from, say, the floor is perpendicular to the floor.

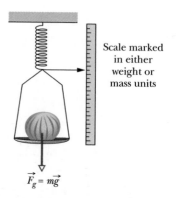

Scale marked in either weight or mass units

$$\vec{F}_g = m\vec{g}$$

Fig. 5-6 A spring scale. The reading is proportional to the *weight* of the object on the pan, and the scale gives that weight if marked in weight units. If, instead, it is marked in mass units, the reading is the object's weight only if the value of g at the location where the scale is being used is the same as the value of g at the location where the scale was calibrated.

> When a body presses against a surface, the surface (even a seemingly rigid one) deforms and pushes on the body with a normal force $\vec{F}_N$ that is perpendicular to the surface.

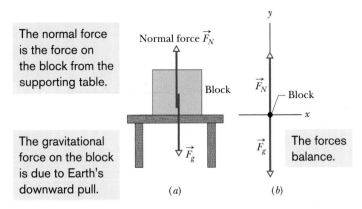

The normal force is the force on the block from the supporting table.

Normal force $\vec{F}_N$

Block

The gravitational force on the block is due to Earth's downward pull.

$\vec{F}_g$

$\vec{F}_N$

Block

x

$\vec{F}_g$

The forces balance.

(a)   (b)

Fig. 5-7 (a) A block resting on a table experiences a normal force $\vec{F}_N$ perpendicular to the tabletop. (b) The free-body diagram for the block.

Figure 5-7a shows an example. A block of mass m presses down on a table, deforming it somewhat because of the gravitational force $\vec{F}_g$ on the block. The table pushes up on the block with normal force $\vec{F}_N$. The free-body diagram for the block is given in Fig. 5-7b. Forces $\vec{F}_g$ and $\vec{F}_N$ are the only two forces on the block and they are both vertical. Thus, for the block we can write Newton's second law for a positive-upward y axis ($F_{net,y} = ma_y$) as

$$F_N - F_g = ma_y.$$

From Eq. 5-8, we substitute mg for F_g, finding

$$F_N - mg = ma_y.$$

Then the magnitude of the normal force is

$$F_N = mg + ma_y = m(g + a_y) \qquad (5\text{-}13)$$

for any vertical acceleration a_y of the table and block (they might be in an accelerating elevator). If the table and block are not accelerating relative to the ground, then $a_y = 0$ and Eq. 5-13 yields

$$F_N = mg. \qquad (5\text{-}14)$$

CHECKPOINT 3

In Fig. 5-7, is the magnitude of the normal force $\vec{F}_N$ greater than, less than, or equal to mg if the block and table are in an elevator moving upward (a) at constant speed and (b) at increasing speed?

Friction

If we either slide or attempt to slide a body over a surface, the motion is resisted by a bonding between the body and the surface. (We discuss this bonding more in the next chapter.) The resistance is considered to be a single force $\vec{f}$, called either the **frictional force** or simply **friction.** This force is directed along the surface, opposite the direction of the intended motion (Fig. 5-8). Sometimes, to simplify a situation, friction is assumed to be negligible (the surface is *frictionless*).

Tension

When a cord (or a rope, cable, or other such object) is attached to a body and pulled taut, the cord pulls on the body with a force $\vec{T}$ directed away from the

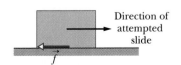

Direction of attempted slide

f

Fig. 5-8 A frictional force $\vec{f}$ opposes the attempted slide of a body over a surface.

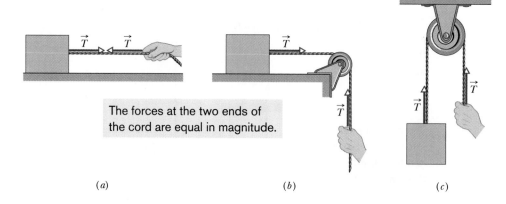

The forces at the two ends of the cord are equal in magnitude.

(a)   (b)   (c)

Fig. 5-9 (a) The cord, pulled taut, is under tension. If its mass is negligible, the cord pulls on the body and the hand with force $\vec{T}$, even if the cord runs around a massless, frictionless pulley as in (b) and (c).

body and along the cord (Fig. 5-9a). The force is often called a *tension force* because the cord is said to be in a state of *tension* (or to be *under tension*), which means that it is being pulled taut. The *tension in the cord* is the magnitude T of the force on the body. For example, if the force on the body from the cord has magnitude $T = 50$ N, the tension in the cord is 50 N.

A cord is often said to be *massless* (meaning its mass is negligible compared to the body's mass) and *unstretchable*. The cord then exists only as a connection between two bodies. It pulls on both bodies with the same force magnitude T, even if the bodies and the cord are accelerating and even if the cord runs around a *massless, frictionless pulley* (Figs. 5-9b and c). Such a pulley has negligible mass compared to the bodies and negligible friction on its axle opposing its rotation. If the cord wraps halfway around a pulley, as in Fig. 5-9c, the net force on the pulley from the cord has the magnitude $2T$.

CHECKPOINT 4

The suspended body in Fig. 5-9c weighs 75 N. Is T equal to, greater than, or less than 75 N when the body is moving upward (a) at constant speed, (b) at increasing speed, and (c) at decreasing speed?

5-8 Newton's Third Law

Two bodies are said to *interact* when they push or pull on each other—that is, when a force acts on each body due to the other body. For example, suppose you position a book B so it leans against a crate C (Fig. 5-10a). Then the book and crate interact: There is a horizontal force $\vec{F}_{BC}$ on the book from the crate (or due to the crate) and a horizontal force $\vec{F}_{CB}$ on the crate from the book (or due to the book). This pair of forces is shown in Fig. 5-10b. Newton's third law states that

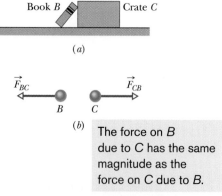

(a)

(b)

The force on B due to C has the same magnitude as the force on C due to B.

Fig. 5-10 (a) Book B leans against crate C. (b) Forces $\vec{F}_{BC}$ (the force on the book from the crate) and $\vec{F}_{CB}$ (the force on the crate from the book) have the same magnitude and are opposite in direction.

Newton's Third Law: When two bodies interact, the forces on the bodies from each other are always equal in magnitude and opposite in direction.

For the book and crate, we can write this law as the scalar relation

$$F_{BC} = F_{CB} \quad \text{(equal magnitudes)}$$

or as the vector relation

$$\vec{F}_{BC} = -\vec{F}_{CB} \quad \text{(equal magnitudes and opposite directions)}, \quad (5\text{-}15)$$

where the minus sign means that these two forces are in opposite directions. We can call the forces between two interacting bodies a **third-law force pair.** When

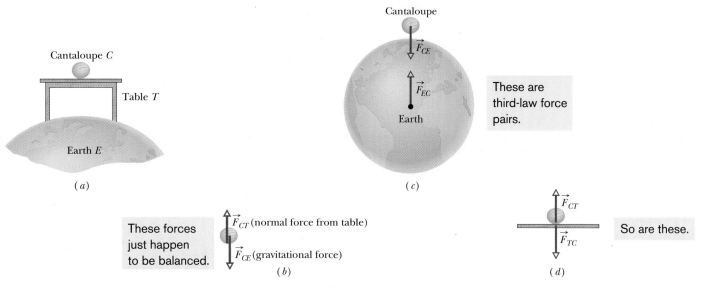

Fig. 5-11 (*a*) A cantaloupe lies on a table that stands on Earth. (*b*) The forces *on the cantaloupe* are $\vec{F}_{CT}$ and $\vec{F}_{CE}$. (*c*) The third-law force pair for the cantaloupe–Earth interaction. (*d*) The third-law force pair for the cantaloupe–table interaction.

any two bodies interact in any situation, a third-law force pair is present. The book and crate in Fig. 5-10*a* are stationary, but the third law would still hold if they were moving and even if they were accelerating.

As another example, let us find the third-law force pairs involving the cantaloupe in Fig. 5-11*a*, which lies on a table that stands on Earth. The cantaloupe interacts with the table and with Earth (this time, there are three bodies whose interactions we must sort out).

Let's first focus on the forces acting on the cantaloupe (Fig. 5-11*b*). Force $\vec{F}_{CT}$ is the normal force on the cantaloupe from the table, and force $\vec{F}_{CE}$ is the gravitational force on the cantaloupe due to Earth. Are they a third-law force pair? No, because they are forces on a single body, the cantaloupe, and not on two interacting bodies.

To find a third-law pair, we must focus not on the cantaloupe but on the interaction between the cantaloupe and one other body. In the cantaloupe–Earth interaction (Fig. 5-11*c*), Earth pulls on the cantaloupe with a gravitational force $\vec{F}_{CE}$ and the cantaloupe pulls on Earth with a gravitational force $\vec{F}_{EC}$. Are these forces a third-law force pair? Yes, because they are forces on two interacting bodies, the force on each due to the other. Thus, by Newton's third law,

$$\vec{F}_{CE} = -\vec{F}_{EC} \qquad \text{(cantaloupe – Earth interaction)}.$$

Next, in the cantaloupe–table interaction, the force on the cantaloupe from the table is $\vec{F}_{CT}$ and, conversely, the force on the table from the cantaloupe is $\vec{F}_{TC}$ (Fig. 5-11*d*). These forces are also a third-law force pair, and so

$$\vec{F}_{CT} = -\vec{F}_{TC} \qquad \text{(cantaloupe–table interaction)}.$$

✔ **CHECKPOINT 5**

Suppose that the cantaloupe and table of Fig. 5-11 are in an elevator cab that begins to accelerate upward. (a) Do the magnitudes of $\vec{F}_{TC}$ and $\vec{F}_{CT}$ increase, decrease, or stay the same? (b) Are those two forces still equal in magnitude and opposite in direction? (c) Do the magnitudes of $\vec{F}_{CE}$ and $\vec{F}_{EC}$ increase, decrease, or stay the same? (d) Are those two forces still equal in magnitude and opposite in direction?

5-9 Applying Newton's Laws

The rest of this chapter consists of sample problems. You should pore over them, learning their procedures for attacking a problem. Especially important is knowing how to translate a sketch of a situation into a free-body diagram with appropriate axes, so that Newton's laws can be applied.

Sample Problem

Block on table, block hanging

Figure 5-12 shows a block S (the *sliding block*) with mass $M = 3.3$ kg. The block is free to move along a horizontal frictionless surface and connected, by a cord that wraps over a frictionless pulley, to a second block H (the *hanging block*), with mass $m = 2.1$ kg. The cord and pulley have negligible masses compared to the blocks (they are "massless"). The hanging block H falls as the sliding block S accelerates to the right. Find (a) the acceleration of block S, (b) the acceleration of block H, and (c) the tension in the cord.

Q *What is this problem all about?*

You are given two bodies—sliding block and hanging block—but must also consider *Earth*, which pulls on both bodies. (Without Earth, nothing would happen here.) A total of five forces act on the blocks, as shown in Fig. 5-13:

1. The cord pulls to the right on sliding block S with a force of magnitude T.

2. The cord pulls upward on hanging block H with a force of the same magnitude T. This upward force keeps block H from falling freely.

3. Earth pulls down on block S with the gravitational force $\vec{F}_{gS}$, which has a magnitude equal to Mg.

4. Earth pulls down on block H with the gravitational force $\vec{F}_{gH}$, which has a magnitude equal to mg.

5. The table pushes up on block S with a normal force $\vec{F}_N$.

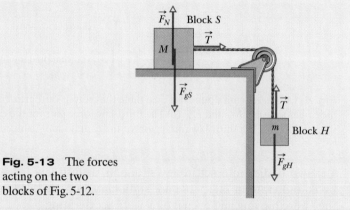

Fig. 5-13 The forces acting on the two blocks of Fig. 5-12.

There is another thing you should note. We assume that the cord does not stretch, so that if block H falls 1 mm in a certain time, block S moves 1 mm to the right in that same time. This means that the blocks move together and their accelerations have the same magnitude a.

Q *How do I classify this problem? Should it suggest a particular law of physics to me?*

Yes. Forces, masses, and accelerations are involved, and they should suggest Newton's second law of motion, $\vec{F}_{net} = m\vec{a}$. That is our starting **Key Idea**.

Q *If I apply Newton's second law to this problem, to which body should I apply it?*

We focus on two bodies, the sliding block and the hanging block. Although they are *extended objects* (they are not points), we can still treat each block as a particle because every part of it moves in exactly the same way. A second **Key Idea** is to apply Newton's second law separately to each block.

Q *What about the pulley?*

We cannot represent the pulley as a particle because different parts of it move in different ways. When we discuss rotation, we shall deal with pulleys in detail. Meanwhile, we eliminate the pulley from consideration by assuming its mass to be negligible compared with the masses of the two blocks. Its only function is to change the cord's orientation.

Q *OK. Now how do I apply $\vec{F}_{net} = m\vec{a}$ to the sliding block?*

Represent block S as a particle of mass M and draw *all* the forces that act *on* it, as in Fig. 5-14a. This is the block's

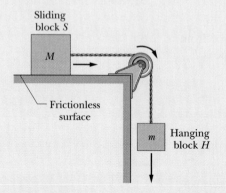

Fig. 5-12 A block S of mass M is connected to a block H of mass m by a cord that wraps over a pulley.

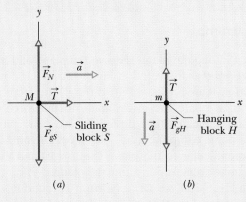

Fig. 5-14 (*a*) A free-body diagram for block *S* of Fig. 5-12. (*b*) A free-body diagram for block *H* of Fig. 5-12.

free-body diagram. Next, draw a set of axes. It makes sense to draw the *x* axis parallel to the table, in the direction in which the block moves.

Q *Thanks, but you still haven't told me how to apply* $\vec{F}_{net} = m\vec{a}$ *to the sliding block. All you've done is explain how to draw a free-body diagram.*

You are right, and here's the third **Key Idea:** The expression $\vec{F}_{net} = M\vec{a}$ is a vector equation, so we can write it as three component equations:

$$F_{net,x} = Ma_x \quad F_{net,y} = Ma_y \quad F_{net,z} = Ma_z \quad (5\text{-}16)$$

in which $F_{net,x}$, $F_{net,y}$, and $F_{net,z}$ are the components of the net force along the three axes. Now we apply each component equation to its corresponding direction. Because block *S* does not accelerate vertically, $F_{net,y} = Ma_y$ becomes

$$F_N - F_{gS} = 0 \quad \text{or} \quad F_N = F_{gS}. \quad (5\text{-}17)$$

Thus in the *y* direction, the magnitude of the normal force is equal to the magnitude of the gravitational force.

No force acts in the *z* direction, which is perpendicular to the page.

In the *x* direction, there is only one force component, which is *T*. Thus, $F_{net,x} = Ma_x$ becomes

$$T = Ma. \quad (5\text{-}18)$$

This equation contains two unknowns, *T* and *a*; so we cannot yet solve it. Recall, however, that we have not said anything about the hanging block.

Q *I agree. How do I apply* $\vec{F}_{net} = m\vec{a}$ *to the hanging block?*

We apply it just as we did for block *S*: Draw a free-body diagram for block *H*, as in Fig. 5-14*b*. Then apply $\vec{F}_{net} = m\vec{a}$ in component form. This time, because the acceleration is along the *y* axis, we use the *y* part of Eq. 5-16 ($F_{net,y} = ma_y$) to write

$$T - F_{gH} = ma_y. \quad (5\text{-}19)$$

We can now substitute *mg* for F_{gH} and $-a$ for a_y (negative because block *H* accelerates in the negative direction of the *y* axis). We find

$$T - mg = -ma. \quad (5\text{-}20)$$

Now note that Eqs. 5-18 and 5-20 are simultaneous equations with the same two unknowns, *T* and *a*. Subtracting these equations eliminates *T*. Then solving for *a* yields

$$a = \frac{m}{M + m} g. \quad (5\text{-}21)$$

Substituting this result into Eq. 5-18 yields

$$T = \frac{Mm}{M + m} g. \quad (5\text{-}22)$$

Putting in the numbers gives, for these two quantities,

$$a = \frac{m}{M + m} g = \frac{2.1 \text{ kg}}{3.3 \text{ kg} + 2.1 \text{ kg}} (9.8 \text{ m/s}^2)$$
$$= 3.8 \text{ m/s}^2 \quad \text{(Answer)}$$

and $\quad T = \frac{Mm}{M + m} g = \frac{(3.3 \text{ kg})(2.1 \text{ kg})}{3.3 \text{ kg} + 2.1 \text{ kg}} (9.8 \text{ m/s}^2)$
$$= 13 \text{ N.} \quad \text{(Answer)}$$

Q *The problem is now solved, right?*

That's a fair question, but the problem is not really finished until we have examined the results to see whether they make sense. (If you made these calculations on the job, wouldn't you want to see whether they made sense before you turned them in?)

Look first at Eq. 5-21. Note that it is dimensionally correct and that the acceleration *a* will always be less than *g*. This is as it must be, because the hanging block is not in free fall. The cord pulls upward on it.

Look now at Eq. 5-22, which we can rewrite in the form

$$T = \frac{M}{M + m} mg. \quad (5\text{-}23)$$

In this form, it is easier to see that this equation is also dimensionally correct, because both *T* and *mg* have dimensions of forces. Equation 5-23 also lets us see that the tension in the cord is always less than *mg*, and thus is always less than the gravitational force on the hanging block. That is a comforting thought because, if *T* were *greater* than *mg*, the hanging block would accelerate upward.

We can also check the results by studying special cases, in which we can guess what the answers must be. A simple example is to put *g* = 0, as if the experiment were carried out in interstellar space. We know that in that case, the blocks would not move from rest, there would be no forces on the ends of the cord, and so there would be no tension in the cord. Do the formulas predict this? Yes, they do. If you put *g* = 0 in Eqs. 5-21 and 5-22, you find *a* = 0 and *T* = 0. Two more special cases you might try are *M* = 0 and *m* → ∞.

 Additional examples, video, and practice available at *WileyPLUS*

Cord accelerates block up a ramp

In Fig. 5-15a, a cord pulls on a box of sea biscuits up along a frictionless plane inclined at $\theta = 30°$. The box has mass $m = 5.00$ kg, and the force from the cord has magnitude $T = 25.0$ N. What is the box's acceleration component a along the inclined plane?

The acceleration along the plane is set by the force components along the plane (not by force components perpendicular to the plane), as expressed by Newton's second law (Eq. 5-1).

Calculation: For convenience, we draw a coordinate system and a free-body diagram as shown in Fig. 5-15b. The positive direction of the x axis is up the plane. Force $\vec{T}$ from the cord is up the plane and has magnitude $T = 25.0$ N. The gravitational force $\vec{F}_g$ is downward and has magnitude $mg = (5.00 \text{ kg})(9.8 \text{ m/s}^2) = 49.0$ N. More important, its

component along the plane is down the plane and has magnitude $mg \sin \theta$ as indicated in Fig. 5-15g. (To see why that trig function is involved, we go through the steps of Figs. 5-15c to h to relate the given angle to the force components.) To indicate the direction, we can write the down-the-plane component as $-mg \sin \theta$. The normal force $\vec{F}_N$ is perpendicular to the plane (Fig. 5-15i) and thus does not determine acceleration along the plane.

From Fig. 5-15h, we write Newton's second law ($\vec{F}_{net} = m\vec{a}$) for motion along the x axis as

$$T - mg \sin \theta = ma. \tag{5-24}$$

Substituting data and solving for a, we find

$$a = 0.100 \text{ m/s}^2, \tag{Answer}$$

where the positive result indicates that the box accelerates up the plane.

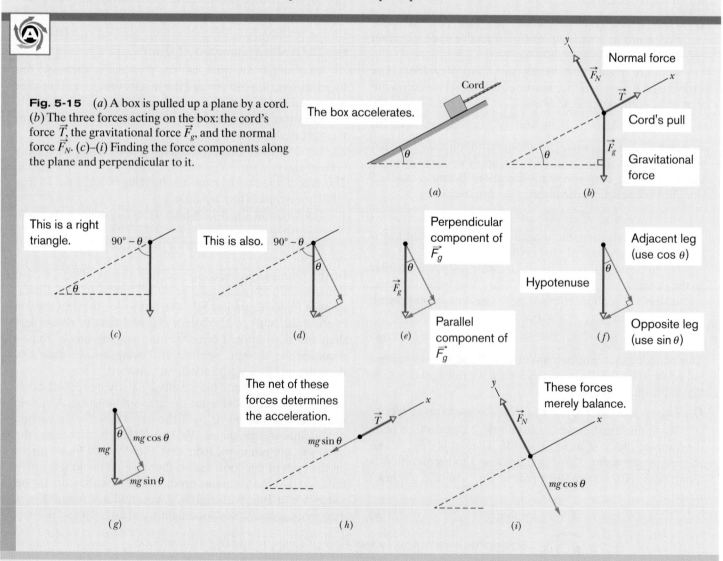

Fig. 5-15 (a) A box is pulled up a plane by a cord. (b) The three forces acting on the box: the cord's force $\vec{T}$, the gravitational force $\vec{F}_g$, and the normal force $\vec{F}_N$. (c)–(i) Finding the force components along the plane and perpendicular to it.

Sample Problem

Reading a force graph

Figure 5-16a shows the general arrangement in which two forces are applied to a 4.00 kg block on a frictionless floor, but only force $\vec{F_1}$ is indicated. That force has a fixed magnitude but can be applied at an adjustable angle θ to the positive direction of the x axis. Force $\vec{F_2}$ is horizontal and fixed in both magnitude and angle. Figure 5-16b gives the horizontal acceleration a_x of the block for any given value of θ from 0° to 90°. What is the value of a_x for $\theta = 180°$?

KEY IDEAS

(1) The horizontal acceleration a_x depends on the net horizontal force $F_{net,x}$, as given by Newton's second law. (2) The net horizontal force is the sum of the horizontal components of forces $\vec{F_1}$ and $\vec{F_2}$.

Calculations: The x component of $\vec{F_2}$ is F_2 because the vector is horizontal. The x component of $\vec{F_1}$ is $F_1 \cos \theta$. Using these expressions and a mass m of 4.00 kg, we can write Newton's second law ($\vec{F}_{net} = m\vec{a}$) for motion along the x axis as

$$F_1 \cos \theta + F_2 = 4.00a_x. \qquad (5\text{-}25)$$

From this equation we see that when $\theta = 90°$, $F_1 \cos \theta$ is zero and $F_2 = 4.00a_x$. From the graph we see that the corresponding acceleration is 0.50 m/s². Thus, $F_2 = 2.00$ N and $\vec{F_2}$ must be in the positive direction of the x axis.

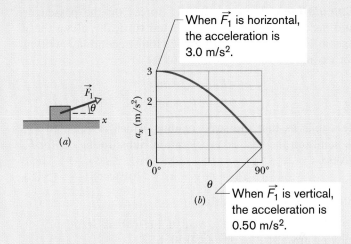

When $\vec{F_1}$ is horizontal, the acceleration is 3.0 m/s².

When $\vec{F_1}$ is vertical, the acceleration is 0.50 m/s².

Fig. 5-16 (a) One of the two forces applied to a block is shown. Its angle θ can be varied. (b) The block's acceleration component a_x versus θ.

From Eq. 5-25, we find that when $\theta = 0°$,

$$F_1 \cos 0° + 2.00 = 4.00a_x. \qquad (5\text{-}26)$$

From the graph we see that the corresponding acceleration is 3.0 m/s². From Eq. 5-26, we then find that $F_1 = 10$ N. Substituting $F_1 = 10$ N, $F_2 = 2.00$ N, and $\theta = 180°$ into Eq. 5-25 leads to

$$a_x = -2.00 \text{ m/s}^2. \qquad \text{(Answer)}$$

Sample Problem

Forces within an elevator cab

In Fig. 5-17a, a passenger of mass $m = 72.2$ kg stands on a platform scale in an elevator cab. We are concerned with the scale readings when the cab is stationary and when it is moving up or down.

(a) Find a general solution for the scale reading, whatever the vertical motion of the cab.

KEY IDEAS

(1) The reading is equal to the magnitude of the normal force $\vec{F_N}$ on the passenger from the scale. The only other force acting on the passenger is the gravitational force $\vec{F_g}$, as shown in the free-body diagram of Fig. 5-17b. (2) We can relate the forces on the passenger to his acceleration $\vec{a}$ by using Newton's second law ($\vec{F}_{net} = m\vec{a}$). However, recall that we can use this law only in an inertial frame. If the cab accelerates, then it is *not* an inertial frame. So we choose the ground

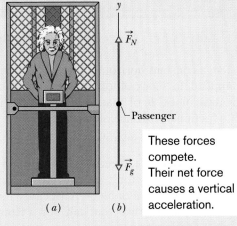

These forces compete. Their net force causes a vertical acceleration.

Fig. 5-17 (a) A passenger stands on a platform scale that indicates either his weight or his apparent weight. (b) The free-body diagram for the passenger, showing the normal force $\vec{F_N}$ on him from the scale and the gravitational force $\vec{F_g}$.

to be our inertial frame and make any measure of the passenger's acceleration relative to it.

Calculations: Because the two forces on the passenger and his acceleration are all directed vertically, along the y axis in Fig. 5-17b, we can use Newton's second law written for y components ($F_{net,y} = ma_y$) to get

$$F_N - F_g = ma$$

or

$$F_N = F_g + ma. \qquad (5\text{-}27)$$

This tells us that the scale reading, which is equal to F_N, depends on the vertical acceleration. Substituting mg for F_g gives us

$$F_N = m(g + a) \quad \text{(Answer)} \qquad (5\text{-}28)$$

for any choice of acceleration a.

(b) What does the scale read if the cab is stationary or moving upward at a constant 0.50 m/s?

KEY IDEA

For any constant velocity (zero or otherwise), the acceleration a of the passenger is zero.

Calculation: Substituting this and other known values into Eq. 5-28, we find

$$F_N = (72.2 \text{ kg})(9.8 \text{ m/s}^2 + 0) = 708 \text{ N}.$$

(Answer)

This is the weight of the passenger and is equal to the magnitude F_g of the gravitational force on him.

(c) What does the scale read if the cab accelerates upward at 3.20 m/s² and downward at 3.20 m/s²?

Calculations: For $a = 3.20$ m/s², Eq. 5-28 gives

$$F_N = (72.2 \text{ kg})(9.8 \text{ m/s}^2 + 3.20 \text{ m/s}^2)$$
$$= 939 \text{ N}, \qquad \text{(Answer)}$$

and for $a = -3.20$ m/s², it gives

$$F_N = (72.2 \text{ kg})(9.8 \text{ m/s}^2 - 3.20 \text{ m/s}^2)$$
$$= 477 \text{ N}. \qquad \text{(Answer)}$$

For an upward acceleration (either the cab's upward speed is increasing or its downward speed is decreasing), the scale reading is greater than the passenger's weight. That reading is a measurement of an apparent weight, because it is made in a noninertial frame. For a downward acceleration (either decreasing upward speed or increasing downward speed), the scale reading is less than the passenger's weight.

(d) During the upward acceleration in part (c), what is the magnitude F_{net} of the net force on the passenger, and what is the magnitude $a_{p,cab}$ of his acceleration as measured in the frame of the cab? Does $\vec{F}_{net} = m\vec{a}_{p,cab}$?

Calculation: The magnitude F_g of the gravitational force on the passenger does not depend on the motion of the passenger or the cab; so, from part (b), F_g is 708 N. From part (c), the magnitude F_N of the normal force on the passenger during the upward acceleration is the 939 N reading on the scale. Thus, the net force on the passenger is

$$F_{net} = F_N - F_g = 939 \text{ N} - 708 \text{ N} = 231 \text{ N}, \quad \text{(Answer)}$$

during the upward acceleration. However, his acceleration $a_{p,cab}$ relative to the frame of the cab is zero. Thus, in the noninertial frame of the accelerating cab, F_{net} is not equal to $ma_{p,cab}$, and Newton's second law does not hold.

Sample Problem

Acceleration of block pushing on block

In Fig. 5-18a, a constant horizontal force $\vec{F}_{app}$ of magnitude 20 N is applied to block A of mass $m_A = 4.0$ kg, which pushes against block B of mass $m_B = 6.0$ kg. The blocks slide over a frictionless surface, along an x axis.

(a) What is the acceleration of the blocks?

Serious Error: Because force $\vec{F}_{app}$ is applied directly to block A, we use Newton's second law to relate that force to the acceleration $\vec{a}$ of block A. Because the motion is along the x axis, we use that law for x components ($F_{net,x} = ma_x$), writing it as

$$F_{app} = m_A a.$$

However, this is seriously wrong because $\vec{F}_{app}$ is not the

only horizontal force acting on block A. There is also the force $\vec{F}_{AB}$ from block B (Fig. 5-18b).

Dead-End Solution: Let us now include force $\vec{F}_{AB}$ by writing, again for the x axis,

$$F_{app} - F_{AB} = m_A a.$$

(We use the minus sign to include the direction of $\vec{F}_{AB}$.) Because F_{AB} is a second unknown, we cannot solve this equation for a.

Successful Solution: Because of the direction in which force $\vec{F}_{app}$ is applied, the two blocks form a rigidly connected system. We can relate the net force *on the system* to the accel-

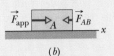

This force causes the acceleration of the full two-block system.

(a)

These are the two forces acting on just block A. Their net force causes its acceleration.

(b)

This is the only force causing the acceleration of block B.

(c)

Fig. 5-18 (a) A constant horizontal force $\vec{F}_{app}$ is applied to block A, which pushes against block B. (b) Two horizontal forces act on block A. (c) Only one horizontal force acts on block B.

eration *of the system* with Newton's second law. Here, once again for the x axis, we can write that law as

$$F_{app} = (m_A + m_B)a,$$

where now we properly apply $\vec{F}_{app}$ to the system with

total mass $m_A + m_B$. Solving for a and substituting known values, we find

$$a = \frac{F_{app}}{m_A + m_B} = \frac{20 \text{ N}}{4.0 \text{ kg} + 6.0 \text{ kg}} = 2.0 \text{ m/s}^2.$$

(Answer)

Thus, the acceleration of the system and of each block is in the positive direction of the x axis and has the magnitude 2.0 m/s^2.

(b) What is the (horizontal) force $\vec{F}_{BA}$ on block B from block A (Fig. 5-18c)?

KEY IDEA

We can relate the net force on block B to the block's acceleration with Newton's second law.

Calculation: Here we can write that law, still for components along the x axis, as

$$F_{BA} = m_B a,$$

which, with known values, gives

$$F_{BA} = (6.0 \text{ kg})(2.0 \text{ m/s}^2) = 12 \text{ N}. \quad \text{(Answer)}$$

Thus, force $\vec{F}_{BA}$ is in the positive direction of the x axis and has a magnitude of 12 N.

PLUS Additional examples, video, and practice available at *WileyPLUS*

REVIEW & SUMMARY

Newtonian Mechanics The velocity of an object can change (the object can accelerate) when the object is acted on by one or more **forces** (pushes or pulls) from other objects. *Newtonian mechanics* relates accelerations and forces.

Force Forces are vector quantities. Their magnitudes are defined in terms of the acceleration they would give the standard kilogram. A force that accelerates that standard body by exactly 1 m/s^2 is defined to have a magnitude of 1 N. The direction of a force is the direction of the acceleration it causes. Forces are combined according to the rules of vector algebra. The **net force** on a body is the vector sum of all the forces acting on the body.

Newton's First Law If there is no net force on a body, the body remains at rest if it is initially at rest or moves in a straight line at constant speed if it is in motion.

Inertial Reference Frames Reference frames in which Newtonian mechanics holds are called *inertial reference frames* or *inertial frames*. Reference frames in which Newtonian mechanics does not hold are called *noninertial reference frames* or *noninertial frames*.

Mass The **mass** of a body is the characteristic of that body that

relates the body's acceleration to the net force causing the acceleration. Masses are scalar quantities.

Newton's Second Law The net force $\vec{F}_{net}$ on a body with mass m is related to the body's acceleration $\vec{a}$ by

$$\vec{F}_{net} = m\vec{a}, \quad (5\text{-}1)$$

which may be written in the component versions

$$F_{net,x} = ma_x \quad F_{net,y} = ma_y \quad \text{and} \quad F_{net,z} = ma_z. \quad (5\text{-}2)$$

The second law indicates that in SI units

$$1 \text{ N} = 1 \text{ kg} \cdot \text{m/s}^2. \quad (5\text{-}3)$$

A **free-body diagram** is a stripped-down diagram in which only *one* body is considered. That body is represented by either a sketch or a dot. The external forces on the body are drawn, and a coordinate system is superimposed, oriented so as to simplify the solution.

Some Particular Forces A **gravitational force** $\vec{F}_g$ on a body is a pull by another body. In most situations in this book, the other body is Earth or some other astronomical body. For Earth, the force is directed down toward the ground, which is assumed to be an inertial frame. With that assumption, the magnitude of $\vec{F}_g$ is

$$F_g = mg, \qquad (5\text{-}8)$$

where m is the body's mass and g is the magnitude of the free-fall acceleration.

The **weight** W of a body is the magnitude of the upward force needed to balance the gravitational force on the body. A body's weight is related to the body's mass by

$$W = mg. \qquad (5\text{-}12)$$

A **normal force** $\vec{F}_N$ is the force on a body from a surface against which the body presses. The normal force is always perpendicular to the surface.

A **frictional force** $\vec{f}$ is the force on a body when the body slides or attempts to slide along a surface. The force is always parallel to the surface and directed so as to oppose the sliding. On a *frictionless surface*, the frictional force is negligible.

When a cord is under **tension**, each end of the cord pulls on a body. The pull is directed along the cord, away from the point of attachment to the body. For a *massless* cord (a cord with negligible mass), the pulls at both ends of the cord have the same magnitude T, even if the cord runs around a *massless, frictionless pulley* (a pulley with negligible mass and negligible friction on its axle to oppose its rotation).

Newton's Third Law   If a force $\vec{F}_{BC}$ acts on body B due to body C, then there is a force $\vec{F}_{CB}$ on body C due to body B:

$$\vec{F}_{BC} = -\vec{F}_{CB}.$$

QUESTIONS

1 Figure 5-19 gives the free-body diagram for four situations in which an object is pulled by several forces across a frictionless floor, as seen from overhead. In which situations does the object's acceleration $\vec{a}$ have (a) an x component and (b) a y component? (c) In each situation, give the direction of $\vec{a}$ by naming either a quadrant or a direction along an axis. (This can be done with a few mental calculations.)

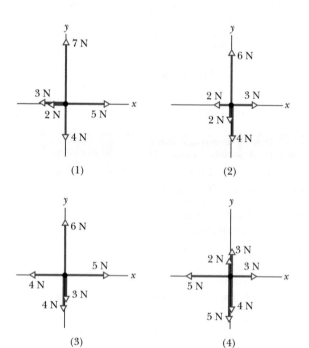

Fig. 5-19   Question 1.

2 Two horizontal forces,

$$\vec{F}_1 = (3\,\text{N})\hat{i} - (4\,\text{N})\hat{j} \quad \text{and} \quad \vec{F}_2 = -(1\,\text{N})\hat{i} - (2\,\text{N})\hat{j}$$

pull a banana split across a frictionless lunch counter. Without using a calculator, determine which of the vectors in the free-body diagram of Fig. 5-20 best represent (a) $\vec{F}_1$ and (b) $\vec{F}_2$. What is the net-force component along (c) the x axis and (d) the y axis? Into which quadrants do (e) the net-force vector and (f) the split's acceleration vector point?

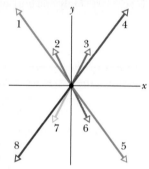

Fig. 5-20   Question 2.

3 In Fig. 5-21, forces $\vec{F}_1$ and $\vec{F}_2$ are applied to a lunchbox as it slides at constant velocity over a frictionless floor. We are to decrease angle θ without changing the magnitude of $\vec{F}_1$. For constant velocity, should we increase, decrease, or maintain the magnitude of $\vec{F}_2$?

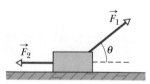

Fig. 5-21   Question 3.

4 At time $t = 0$, constant $\vec{F}$ begins to act on a rock moving through deep space in the $+x$ direction. (a) For time $t > 0$, which are possible functions $x(t)$ for the rock's position: (1) $x = 4t - 3$, (2) $x = -4t^2 + 6t - 3$, (3) $x = 4t^2 + 6t - 3$? (b) For which function is $\vec{F}$ directed opposite the rock's initial direction of motion?

5 Figure 5-22 shows overhead views of four situations in which forces act on a block that lies on a frictionless floor. If the force

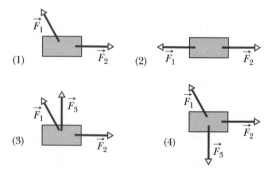

Fig. 5-22   Question 5.

magnitudes are chosen properly, in which situations is it possible that the block is (a) stationary and (b) moving with a constant velocity?

6 Figure 5-23 shows the same breadbox in four situations where horizontal forces are applied. Rank the situations according to the magnitude of the box's acceleration, greatest first.

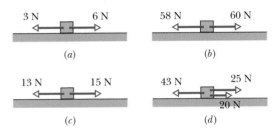

Fig. 5-23 Question 6.

7 ✈ July 17, 1981, Kansas City: The newly opened Hyatt Regency is packed with people listening and dancing to a band playing favorites from the 1940s. Many of the people are crowded onto the walkways that hang like bridges across the wide atrium. Suddenly two of the walkways collapse, falling onto the merrymakers on the main floor.

The walkways were suspended one above another on vertical rods and held in place by nuts threaded onto the rods. In the original design, only two long rods were to be used, each extending through all three walkways (Fig. 5-24a). If each walkway and the merrymakers on it have a combined mass of M, what is the total mass supported by the threads and two nuts on (a) the lowest walkway and (b) the highest walkway?

Threading nuts on a rod is impossible except at the ends, so the design was changed: Instead, six rods were used, each connecting two walkways (Fig. 5-24b). What now is the total mass supported by the threads and two nuts on (c) the lowest walkway, (d) the upper side of the highest walkway, and (e) the lower side of the highest walkway? It was this design that failed.

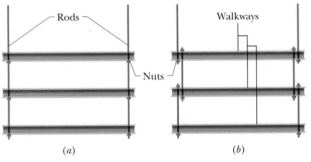

Fig. 5-24 Question 7.

8 Figure 5-25 gives three graphs of velocity component $v_x(t)$ and three graphs of velocity component $v_y(t)$. The graphs are not to scale. Which $v_x(t)$ graph and which $v_y(t)$ graph best correspond to each of the four situations in Question 1 and Fig. 5-19?

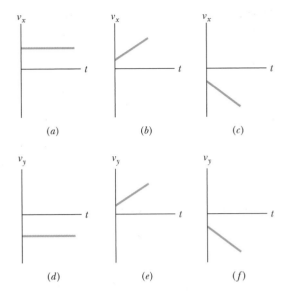

Fig. 5-25 Question 8.

9 Figure 5-26 shows a train of four blocks being pulled across a frictionless floor by force $\vec{F}$. What total mass is accelerated to the right by (a) force $\vec{F}$, (b) cord 3, and (c) cord 1? (d) Rank the blocks according to their accelerations, greatest first. (e) Rank the cords according to their tension, greatest first.

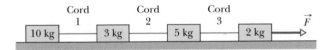

Fig. 5-26 Question 9.

10 Figure 5-27 shows three blocks being pushed across a frictionless floor by horizontal force $\vec{F}$. What total mass is accelerated to the right by (a) force $\vec{F}$, (b) force $\vec{F}_{21}$ on block 2 from block 1, and (c) force $\vec{F}_{32}$ on block 3 from block 2? (d) Rank the blocks according to their acceleration magnitudes, greatest first. (e) Rank forces $\vec{F}, \vec{F}_{21}$, and $\vec{F}_{32}$ according to magnitude, greatest first.

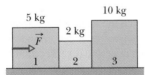

Fig. 5-27 Question 10.

11 A vertical force $\vec{F}$ is applied to a block of mass m that lies on a floor. What happens to the magnitude of the normal force $\vec{F}_N$ on the block from the floor as magnitude F is increased from zero if force $\vec{F}$ is (a) downward and (b) upward?

12 Figure 5-28 shows four choices for the direction of a force of magnitude F to be applied to a block on an inclined plane. The directions are either horizontal or vertical. (For choice b, the force is not enough to lift the block off the plane.) Rank the choices according to the magnitude of the normal force acting on the block from the plane, greatest first.

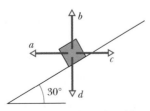

Fig. 5-28 Question 12.

GO   Tutoring problem available (at instructor's discretion) in *WileyPLUS* and WebAssign
SSM   Worked-out solution available in Student Solutions Manual   WWW Worked-out solution is at
• — •••   Number of dots indicates level of problem difficulty   ILW Interactive solution is at   http://www.wiley.com/college/halliday
       Additional information available in *The Flying Circus of Physics* and at flyingcircusofphysics.com

sec. 5-6   Newton's Second Law

•1   Only two horizontal forces act on a 3.0 kg body that can move over a frictionless floor. One force is 9.0 N, acting due east, and the other is 8.0 N, acting 62° north of west. What is the magnitude of the body's acceleration?

•2   Two horizontal forces act on a 2.0 kg chopping block that can slide over a frictionless kitchen counter, which lies in an xy plane. One force is $\vec{F}_1 = (3.0\text{ N})\hat{i} + (4.0\text{ N})\hat{j}$. Find the acceleration of the chopping block in unit-vector notation when the other force is (a) $\vec{F}_2 = (-3.0\text{ N})\hat{i} + (-4.0\text{ N})\hat{j}$, (b) $\vec{F}_2 = (-3.0\text{ N})\hat{i} + (4.0\text{ N})\hat{j}$, and (c) $\vec{F}_2 = (3.0\text{ N})\hat{i} + (-4.0\text{ N})\hat{j}$.

•3   If the 1 kg standard body has an acceleration of 2.00 m/s² at 20.0° to the positive direction of an x axis, what are (a) the x component and (b) the y component of the net force acting on the body, and (c) what is the net force in unit-vector notation?

••4   While two forces act on it, a particle is to move at the constant velocity $\vec{v} = (3\text{ m/s})\hat{i} - (4\text{ m/s})\hat{j}$. One of the forces is $\vec{F}_1 = (2\text{ N})\hat{i} + (-6\text{ N})\hat{j}$. What is the other force?

••5   GO   Three astronauts, propelled by jet backpacks, push and guide a 120 kg asteroid toward a processing dock, exerting the forces shown in Fig. 5-29, with $F_1 = 32$ N, $F_2 = 55$ N, $F_3 = 41$ N, $\theta_1 = 30°$, and $\theta_3 = 60°$. What is the asteroid's acceleration (a) in unit-vector notation and as (b) a magnitude and (c) a direction relative to the positive direction of the x axis?

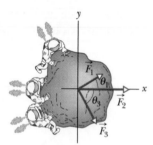

Fig. 5-29 Problem 5.

••6   In a two-dimensional tug-of-war, Alex, Betty, and Charles pull horizontally on an automobile tire at the angles shown in the overhead view of Fig. 5-30. The tire remains stationary in spite of the three pulls. Alex pulls with force $\vec{F}_A$ of magnitude 220 N, and Charles pulls with force $\vec{F}_C$ of magnitude 170 N. Note that the direction of $\vec{F}_C$ is not given. What is the magnitude of Betty's force $\vec{F}_B$?

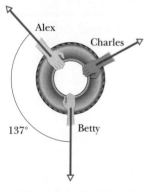

Fig. 5-30 Problem 6.

••7   SSM   There are two forces on the 2.00 kg box in the overhead view of Fig. 5-31, but only one is shown. For $F_1 = 20.0$ N, $a = 12.0$ m/s², and $\theta = 30.0°$, find the second force (a) in unit-vector notation and as (b) a magnitude and (c) an angle relative to the positive direction of the x axis.

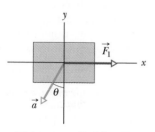

Fig. 5-31 Problem 7.

••8   A 2.00 kg object is subjected to three forces that give it an acceleration $\vec{a} = -(8.00\text{ m/s}^2)\hat{i} + (6.00\text{ m/s}^2)\hat{j}$. If two of the three forces are $\vec{F}_1 = (30.0\text{ N})\hat{i} + (16.0\text{ N})\hat{j}$ and $\vec{F}_2 = -(12.0\text{ N})\hat{i} + (8.00\text{ N})\hat{j}$, find the third force.

••9   A 0.340 kg particle moves in an xy plane according to $x(t) = -15.00 + 2.00t - 4.00t^3$ and $y(t) = 25.00 + 7.00t - 9.00t^2$, with x and y in meters and t in seconds. At $t = 0.700$ s, what are (a) the magnitude and (b) the angle (relative to the positive direction of the x axis) of the net force on the particle, and (c) what is the angle of the particle's direction of travel?

••10   A 0.150 kg particle moves along an x axis according to $x(t) = -13.00 + 2.00t + 4.00t^2 - 3.00t^3$, with x in meters and t in seconds. In unit-vector notation, what is the net force acting on the particle at $t = 3.40$ s?

••11   A 2.0 kg particle moves along an x axis, being propelled by a variable force directed along that axis. Its position is given by $x = 3.0$ m $+ (4.0$ m/s$)t + ct^2 - (2.0$ m/s³$)t^3$, with x in meters and t in seconds. The factor c is a constant. At $t = 3.0$ s, the force on the particle has a magnitude of 36 N and is in the negative direction of the axis. What is c?

•••12   GO   Two horizontal forces $\vec{F}_1$ and $\vec{F}_2$ act on a 4.0 kg disk that slides over frictionless ice, on which an xy coordinate system is laid out. Force $\vec{F}_1$ is in the positive direction of the x axis and has a magnitude of 7.0 N. Force $\vec{F}_2$ has a magnitude of 9.0 N. Figure 5-32 gives the x component v_x of the velocity of the disk as a function of time t during the sliding. What is the angle between the constant directions of forces $\vec{F}_1$ and $\vec{F}_2$?

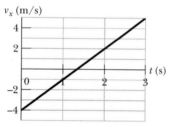

Fig. 5-32 Problem 12.

sec. 5-7   Some Particular Forces

•13   Figure 5-33 shows an arrangement in which four disks are suspended by cords. The longer, top cord loops over a frictionless pulley and pulls with a force of magnitude 98 N on the wall to which it is attached. The tensions in the three shorter cords are $T_1 = 58.8$ N, $T_2 = 49.0$ N, and $T_3 = 9.8$ N. What are the masses of (a) disk A, (b) disk B, (c) disk C, and (d) disk D?

•14   A block with a weight of 3.0 N is at rest on a horizontal surface. A 1.0 N upward force is applied to the block by means of an attached vertical string. What are the (a) magnitude and (b) direction of the force of the block on the horizontal surface?

•15   SSM   (a) An 11.0 kg salami is supported by a cord that runs to a spring scale,

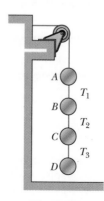

Fig. 5-33 Problem 13.

which is supported by a cord hung from the ceiling (Fig. 5-34a). What is the reading on the scale, which is marked in weight units? (b) In Fig. 5-34b the salami is supported by a cord that runs around a pulley and to a scale. The opposite end of the scale is attached by a cord to a wall. What is the reading on the scale? (c) In Fig. 5-34c the wall has been replaced with a second 11.0 kg salami, and the assembly is stationary. What is the reading on the scale?

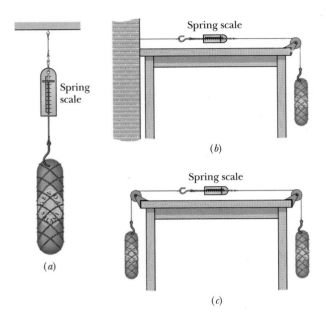

Fig. 5-34   Problem 15.

••16 Some insects can walk below a thin rod (such as a twig) by hanging from it. Suppose that such an insect has mass m and hangs from a horizontal rod as shown in Fig. 5-35, with angle $\theta = 40°$. Its six legs are all under the same tension, and the leg sections nearest the body are horizontal. (a) What is the ratio of the tension in each tibia (forepart of a leg) to the insect's weight? (b) If the insect straightens out its legs somewhat, does the tension in each tibia increase, decrease, or stay the same?

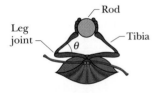

Fig. 5-35   Problem 16.

sec. 5-9   Applying Newton's Laws

•17 SSM WWW In Fig. 5-36, let the mass of the block be 8.5 kg and the angle θ be 30°. Find (a) the tension in the cord and (b) the normal force acting on the block. (c) If the cord is cut, find the magnitude of the resulting acceleration of the block.

•18 In April 1974, John Massis of Belgium managed to move two passenger railroad cars. He did so by clamping his teeth down on a bit that was attached to the cars with a rope and then leaning backward while pressing his feet against the railway ties. The cars together weighed 700 kN (about 80 tons). Assume

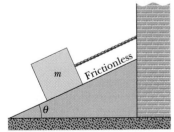

Fig. 5-36   Problem 17.

that he pulled with a constant force that was 2.5 times his body weight, at an upward angle θ of 30° from the horizontal. His mass was 80 kg, and he moved the cars by 1.0 m. Neglecting any retarding force from the wheel rotation, find the speed of the cars at the end of the pull.

•19 SSM A 500 kg rocket sled can be accelerated at a constant rate from rest to 1600 km/h in 1.8 s. What is the magnitude of the required net force?

•20 A car traveling at 53 km/h hits a bridge abutment. A passenger in the car moves forward a distance of 65 cm (with respect to the road) while being brought to rest by an inflated air bag. What magnitude of force (assumed constant) acts on the passenger's upper torso, which has a mass of 41 kg?

•21 A constant horizontal force $\vec{F}_a$ pushes a 2.00 kg FedEx package across a frictionless floor on which an xy coordinate system has been drawn. Figure 5-37 gives the package's x and y velocity components versus time t. What are the (a) magnitude and (b) direction of $\vec{F}_a$?

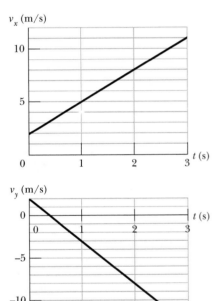

Fig. 5-37   Problem 21.

•22 A customer sits in an amusement park ride in which the compartment is to be pulled downward in the negative direction of a y axis with an acceleration magnitude of $1.24g$, with $g = 9.80$ m/s². A 0.567 g coin rests on the customer's knee. Once the motion begins and in unit-vector notation, what is the coin's acceleration relative to (a) the ground and (b) the customer? (c) How long does the coin take to reach the compartment ceiling, 2.20 m above the knee? In unit-vector notation, what are (d) the actual force on the coin and (e) the apparent force according to the customer's measure of the coin's acceleration?

•23 Tarzan, who weighs 820 N, swings from a cliff at the end of a 20.0 m vine that hangs from a high tree limb and initially makes an angle of 22.0° with the vertical. Assume that an x axis extends horizontally away from the cliff edge and a y axis extends upward.

Immediately after Tarzan steps off the cliff, the tension in the vine is 760 N. Just then, what are (a) the force on him from the vine in unit-vector notation and the net force on him (b) in unit-vector notation and as (c) a magnitude and (d) an angle relative to the positive direction of the x axis? What are the (e) magnitude and (f) angle of Tarzan's acceleration just then?

•24 There are two horizontal forces on the 2.0 kg box in the overhead view of Fig. 5-38 but only one (of magnitude $F_1 = 20$ N) is shown. The box moves along the x axis. For

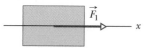

Fig. 5-38 Problem 24.

each of the following values for the acceleration a_x of the box, find the second force in unit-vector notation: (a) 10 m/s², (b) 20 m/s², (c) 0, (d) −10 m/s², and (e) −20 m/s².

•25 *Sunjamming.* A "sun yacht" is a spacecraft with a large sail that is pushed by sunlight. Although such a push is tiny in everyday circumstances, it can be large enough to send the spacecraft outward from the Sun on a cost-free but slow trip. Suppose that the spacecraft has a mass of 900 kg and receives a push of 20 N. (a) What is the magnitude of the resulting acceleration? If the craft starts from rest, (b) how far will it travel in 1 day and (c) how fast will it then be moving?

•26 The tension at which a fishing line snaps is commonly called the line's "strength." What minimum strength is needed for a line that is to stop a salmon of weight 85 N in 11 cm if the fish is initially drifting at 2.8 m/s? Assume a constant deceleration.

•27 SSM An electron with a speed of 1.2×10^7 m/s moves horizontally into a region where a constant vertical force of 4.5×10^{-16} N acts on it. The mass of the electron is 9.11×10^{-31} kg. Determine the vertical distance the electron is deflected during the time it has moved 30 mm horizontally.

•28 A car that weighs 1.30×10^4 N is initially moving at 40 km/h when the brakes are applied and the car is brought to a stop in 15 m. Assuming the force that stops the car is constant, find (a) the magnitude of that force and (b) the time required for the change in speed. If the initial speed is doubled, and the car experiences the same force during the braking, by what factors are (c) the stopping distance and (d) the stopping time multiplied? (There could be a lesson here about the danger of driving at high speeds.)

•29 A firefighter who weighs 712 N slides down a vertical pole with an acceleration of 3.00 m/s², directed downward. What are the (a) magnitude and (b) direction (up or down) of the vertical force on the firefighter from the pole and the (c) magnitude and (d) direction of the vertical force on the pole from the firefighter?

•30 The high-speed winds around a tornado can drive projectiles into trees, building walls, and even metal traffic signs. In a laboratory simulation, a standard wood toothpick was shot by pneumatic gun into an oak branch. The toothpick's mass was 0.13 g, its speed before entering the branch was 220 m/s, and its penetration depth was 15 mm. If its speed was decreased at a uniform rate, what was the magnitude of the force of the branch on the toothpick?

••31 SSM WWW A block is projected up a frictionless inclined plane with initial speed $v_0 = 3.50$ m/s. The angle of incline is $\theta = 32.0°$. (a) How far up the plane does the block go? (b) How long does it take to get there? (c) What is its speed when it gets back to the bottom?

••32 Figure 5-39 shows an overhead view of a 0.0250 kg lemon half and two of the three horizontal forces that act on it as it is on a frictionless table. Force $\vec{F}_1$ has a magnitude of 6.00 N and is at $\theta_1 = 30.0°$. Force $\vec{F}_2$ has a magnitude of 7.00 N and is at $\theta_2 = 30.0°$. In unit-vector notation, what is the third force if the lemon half (a) is stationary, (b) has the constant velocity $\vec{v} = (13.0\hat{i} - 14.0\hat{j})$ m/s, and (c) has the varying velocity $\vec{v} = (13.0t\hat{i} - 14.0t\hat{j})$ m/s², where t is time?

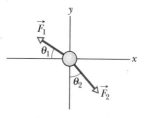

Fig. 5-39 Problem 32.

••33 An elevator cab and its load have a combined mass of 1600 kg. Find the tension in the supporting cable when the cab, originally moving downward at 12 m/s, is brought to rest with constant acceleration in a distance of 42 m.

••34 GO In Fig. 5-40, a crate of mass $m = 100$ kg is pushed at constant speed up a frictionless ramp ($\theta = 30.0°$) by a horizontal force $\vec{F}$. What are the magnitudes of (a) $\vec{F}$ and (b) the force on the crate from the ramp?

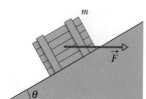

Fig. 5-40 Problem 34.

••35 The velocity of a 3.00 kg particle is given by $\vec{v} = (8.00t\hat{i} + 3.00t^2\hat{j})$ m/s, with time t in seconds. At the instant the net force on the particle has a magnitude of 35.0 N, what are the direction (relative to the positive direction of the x axis) of (a) the net force and (b) the particle's direction of travel?

••36 Holding on to a towrope moving parallel to a frictionless ski slope, a 50 kg skier is pulled up the slope, which is at an angle of 8.0° with the horizontal. What is the magnitude F_{rope} of the force on the skier from the rope when (a) the magnitude v of the skier's velocity is constant at 2.0 m/s and (b) $v = 2.0$ m/s as v increases at a rate of 0.10 m/s²?

••37 A 40 kg girl and an 8.4 kg sled are on the frictionless ice of a frozen lake, 15 m apart but connected by a rope of negligible mass. The girl exerts a horizontal 5.2 N force on the rope. What are the acceleration magnitudes of (a) the sled and (b) the girl? (c) How far from the girl's initial position do they meet?

••38 A 40 kg skier skis directly down a frictionless slope angled at 10° to the horizontal. Assume the skier moves in the negative direction of an x axis along the slope. A wind force with component F_x acts on the skier. What is F_x if the magnitude of the skier's velocity is (a) constant, (b) increasing at a rate of 1.0 m/s², and (c) increasing at a rate of 2.0 m/s²?

••39 ILW A sphere of mass 3.0×10^{-4} kg is suspended from a cord. A steady horizontal breeze pushes the sphere so that the cord makes a constant angle of 37° with the vertical. Find (a) the push magnitude and (b) the tension in the cord.

••40 GO A dated box of dates, of mass 5.00 kg, is sent sliding up a frictionless ramp at an angle of θ to the horizontal.

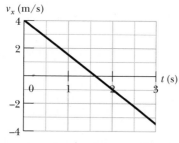

Fig. 5-41 Problem 40.

Figure 5-41 gives, as a function of time t, the component v_x of the box's velocity along an x axis that extends directly up the ramp. What is the magnitude of the normal force on the box from the ramp?

••41 Using a rope that will snap if the tension in it exceeds 387 N, you need to lower a bundle of old roofing material weighing 449 N from a point 6.1 m above the ground. (a) What magnitude of the bundle's acceleration will put the rope on the verge of snapping? (b) At that acceleration, with what speed would the bundle hit the ground?

••42 **GO** In earlier days, horses pulled barges down canals in the manner shown in Fig. 5-42. Suppose the horse pulls on the rope with a force of 7900 N at an angle of $\theta = 18°$ to the direction of motion of the barge, which is headed straight along the positive direction of an x axis. The mass of the barge is 9500 kg, and the magnitude of its acceleration is 0.12 m/s². What are the (a) magnitude and (b) direction (relative to positive x) of the force on the barge from the water?

Fig. 5-42   Problem 42.

••43 **SSM** In Fig. 5-43, a chain consisting of five links, each of mass 0.100 kg, is lifted vertically with constant acceleration of magnitude $a = 2.50$ m/s². Find the magnitudes of (a) the force on link 1 from link 2, (b) the force on link 2 from link 3, (c) the force on link 3 from link 4, and (d) the force on link 4 from link 5. Then find the magnitudes of (e) the force $\vec{F}$ on the top link from the person lifting the chain and (f) the *net* force accelerating each link.

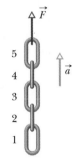

Fig. 5-43
Problem 43.

••44 A lamp hangs vertically from a cord in a descending elevator that decelerates at 2.4 m/s². (a) If the tension in the cord is 89 N, what is the lamp's mass? (b) What is the cord's tension when the elevator ascends with an upward acceleration of 2.4 m/s²?

••45 An elevator cab that weighs 27.8 kN moves upward. What is the tension in the cable if the cab's speed is (a) increasing at a rate of 1.22 m/s² and (b) decreasing at a rate of 1.22 m/s²?

••46 An elevator cab is pulled upward by a cable. The cab and its single occupant have a combined mass of 2000 kg. When that occupant drops a coin, its acceleration relative to the cab is 8.00 m/s² downward. What is the tension in the cable?

••47 The Zacchini family was renowned for their human-cannonball act in which a family member was shot from a cannon using either elastic bands or compressed air. In one version of the act, Emanuel Zacchini was shot over three Ferris wheels to land in a net at the same height as the open end of the cannon and at a range of 69 m. He was propelled inside the barrel for 5.2 m and launched at an angle of 53°. If his mass was 85 kg and he underwent constant acceleration inside the barrel, what was the magnitude of the force propelling him? (*Hint:* Treat the launch as though it were along a ramp at 53°. Neglect air drag.)

••48 **GO** In Fig. 5-44, elevator cabs A and B are connected by a short cable and can be pulled upward or lowered by the cable

above cab A. Cab A has mass 1700 kg; cab B has mass 1300 kg. A 12.0 kg box of catnip lies on the floor of cab A. The tension in the cable connecting the cabs is 1.91×10^4 N. What is the magnitude of the normal force on the box from the floor?

••49 In Fig. 5-45, a block of mass $m = 5.00$ kg is pulled along a horizontal frictionless floor by a cord that exerts a force of magnitude $F = 12.0$ N at an angle $\theta = 25.0°$. (a) What is the magnitude of the block's acceleration? (b) The force magnitude F is slowly increased. What is its value just before the block is lifted (completely) off the floor? (c) What is the magnitude of the block's acceleration just before it is lifted (completely) off the floor?

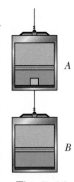

Fig. 5-44
Problem 48.

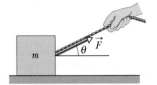

Fig. 5-45
Problems 49 and 60.

••50 **GO** In Fig. 5-46, three ballot boxes are connected by cords, one of which wraps over a pulley having negligible friction on its axle and negligible mass. The three masses are $m_A = 30.0$ kg, $m_B = 40.0$ kg, and $m_C = 10.0$ kg. When the assembly is released from rest, (a) what is the tension in the cord connecting B and C, and (b) how far does A move in the first 0.250 s (assuming it does not reach the pulley)?

Fig. 5-46   Problem 50.

••51 **GO** Figure 5-47 shows two blocks connected by a cord (of negligible mass) that passes over a frictionless pulley (also of negligible mass). The arrangement is known as *Atwood's machine*. One block has mass $m_1 = 1.30$ kg; the other has mass $m_2 = 2.80$ kg. What are (a) the magnitude of the blocks' acceleration and (b) the tension in the cord?

••52 An 85 kg man lowers himself to the ground from a height of 10.0 m by holding onto a rope that runs over a frictionless pulley to a 65 kg sandbag. With what speed does the man hit the ground if he started from rest?

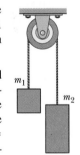

Fig. 5-47
Problems 51
and 65.

••53 In Fig. 5-48, three connected blocks are pulled to the right on a horizontal frictionless table by a force of magnitude $T_3 = 65.0$ N. If $m_1 = 12.0$ kg, $m_2 = 24.0$ kg, and $m_3 = 31.0$ kg, calculate (a) the magnitude of the system's acceleration, (b) the tension T_1, and (c) the tension T_2.

Fig. 5-48   Problem 53.

••54 GO Figure 5-49 shows four penguins that are being playfully pulled along very slippery (frictionless) ice by a curator. The masses of three penguins and the tension in two of the cords are $m_1 = 12$ kg, $m_3 = 15$ kg, $m_4 = 20$ kg, $T_2 = 111$ N, and $T_4 = 222$ N. Find the penguin mass m_2 that is not given.

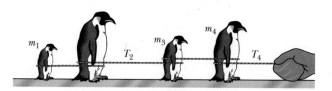

Fig. 5-49   Problem 54.

••55 SSM ILW WWW Two blocks are in contact on a frictionless table. A horizontal force is applied to the larger block, as shown in Fig. 5-50. (a) If $m_1 = 2.3$ kg, $m_2 = 1.2$ kg, and $F = 3.2$ N, find the magnitude of the force between the two blocks. (b) Show that if a force of the same magnitude F is applied to the smaller block but in the opposite direction, the magnitude of the force between the blocks is 2.1 N, which is not the same value calculated in (a). (c) Explain the difference.

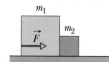

Fig. 5-50
Problem 55.

••56 In Fig. 5-51a, a constant horizontal force $\vec{F}_a$ is applied to block A, which pushes against block B with a 20.0 N force directed horizontally to the right. In Fig. 5-51b, the same force $\vec{F}_a$ is applied to block B; now block A pushes on block B with a 10.0 N force directed horizontally to the left. The blocks have a combined mass of 12.0 kg. What are the magnitudes of (a) their acceleration in Fig. 5-51a and (b) force $\vec{F}_a$?

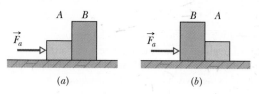

Fig. 5-51   Problem 56.

••57 ILW A block of mass $m_1 = 3.70$ kg on a frictionless plane inclined at angle $\theta = 30.0°$ is connected by a cord over a massless, frictionless pulley to a second block of mass $m_2 = 2.30$ kg (Fig. 5-52). What are (a) the magnitude of the acceleration of each block, (b) the direction of the acceleration of the hanging block, and (c) the tension in the cord?

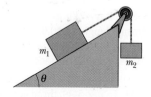

Fig. 5-52   Problem 57.

••58 Figure 5-53 shows a man sitting in a bosun's chair that dangles from a massless rope, which runs over a massless, frictionless pulley and back down to the man's hand. The combined mass of man and chair is 95.0 kg. With what force magnitude must the man pull on the rope if he is to rise (a) with a constant velocity and (b) with an upward acceleration of 1.30 m/s²? (*Hint:* A free-body diagram can really help.) If the rope on the right extends to the ground and is pulled by a co-worker, with what force magnitude must the co-worker pull for the man to rise (c) with a constant velocity and (d) with an upward acceleration of 1.30 m/s²? What is the magnitude of the force on the ceiling from the pulley system in (e) part a, (f) part b, (g) part c, and (h) part d?

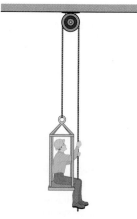

Fig. 5-53   Problem 58.

••59 SSM A 10 kg monkey climbs up a massless rope that runs over a frictionless tree limb and back down to a 15 kg package on the ground (Fig. 5-54). (a) What is the magnitude of the least acceleration the monkey must have if it is to lift the package off the ground? If, after the package has been lifted, the monkey stops its climb and holds onto the rope, what are the (b) magnitude and (c) direction of the monkey's acceleration and (d) the tension in the rope?

••60 Figure 5-45 shows a 5.00 kg block being pulled along a frictionless floor by a cord that applies a force of constant magnitude 20.0 N but with an angle $\theta(t)$ that varies with time. When angle $\theta = 25.0°$, at what rate is the acceleration of the block changing if (a) $\theta(t) = (2.00 \times 10^{-2}$ deg/s$)t$ and (b) $\theta(t) = -(2.00 \times 10^{-2}$ deg/s$)t$? (*Hint:* The angle should be in radians.)

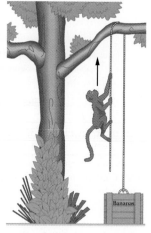

Fig. 5-54   Problem 59.

••61 SSM ILW A hot-air balloon of mass M is descending vertically with downward acceleration of magnitude a. How much mass (ballast) must be thrown out to give the balloon an upward acceleration of magnitude a? Assume that the upward force from the air (the lift) does not change because of the decrease in mass.

•••62 In shot putting, many athletes elect to launch the shot at an angle that is smaller than the theoretical one (about 42°) at which the distance of a projected ball at the same speed and height is greatest. One reason has to do with the speed the athlete can give the shot during the acceleration phase of the throw. Assume that a 7.260 kg shot is accelerated along a straight path of length 1.650 m by a constant applied force of magnitude 380.0 N, starting with an initial speed of 2.500 m/s (due to the athlete's preliminary motion). What is the shot's speed at the end of the acceleration phase if the angle between the path and the horizontal is (a) 30.00° and (b) 42.00°? (*Hint:* Treat the motion as though it were along a ramp at the given angle.) (c) By what percent is the launch speed decreased if the athlete increases the angle from 30.00° to 42.00°?

•••63 Figure 5-55 gives, as a function of time t, the force component F_x that acts on a 3.00 kg ice block that can move only along the x axis. At $t = 0$, the block is moving in the positive direction of the axis, with a speed of 3.0 m/s. What are its (a) speed and (b) direction of travel at $t = 11$ s?

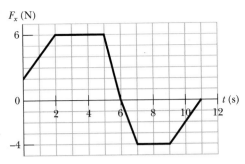

Fig. 5-55 Problem 63.

•••64 Figure 5-56 shows a box of mass $m_2 = 1.0$ kg on a frictionless plane inclined at angle $\theta = 30°$. It is connected by a cord of negligible mass to a box of mass $m_1 = 3.0$ kg on a horizontal frictionless surface. The pulley is frictionless and massless. (a) If the magnitude of horizontal force $\vec{F}$ is 2.3 N, what is the tension in the connecting cord? (b) What is the largest value the magnitude of $\vec{F}$ may have without the cord becoming slack?

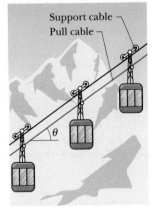

Fig. 5-56 Problem 64.

•••65 Figure 5-47 shows *Atwood's machine,* in which two containers are connected by a cord (of negligible mass) passing over a frictionless pulley (also of negligible mass). At time $t = 0$, container 1 has mass 1.30 kg and container 2 has mass 2.80 kg, but container 1 is losing mass (through a leak) at the constant rate of 0.200 kg/s. At what rate is the acceleration magnitude of the containers changing at (a) $t = 0$ and (b) $t = 3.00$ s? (c) When does the acceleration reach its maximum value?

•••66 Figure 5-57 shows a section of a cable-car system. The maximum permissible mass of each car with occupants is 2800 kg. The cars, riding on a support cable, are pulled by a second cable attached to the support tower on each car. Assume that the cables are taut and inclined at angle $\theta = 35°$. What is the difference in tension between adjacent sections of pull cable if the cars are at the maximum permissible mass and are being accelerated up the incline at 0.81 m/s²?

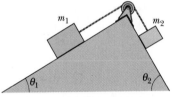

Fig. 5-57 Problem 66.

•••67 Figure 5-58 shows three blocks attached by cords that loop over frictionless pulleys. Block *B* lies on a frictionless table; the masses are $m_A = 6.00$ kg, $m_B = 8.00$ kg, and $m_C = 10.0$ kg. When the blocks are released, what is the tension in the cord at the right?

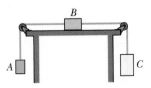

Fig. 5-58 Problem 67.

•••68 A shot putter launches a 7.260 kg shot by pushing it along a straight line of length 1.650 m and at an angle of 34.10° from the horizontal, accelerating the shot to the launch speed from its initial speed of 2.500 m/s (which is due to the athlete's preliminary motion). The shot leaves the hand at a height of 2.110 m and at an angle of 34.10°, and it lands at a horizontal distance of 15.90 m. What is the magnitude of the athlete's average force on the shot during the acceleration phase? (*Hint:* Treat the motion during the acceleration phase as though it were along a ramp at the given angle.)

Additional Problems

69 In Fig. 5-59, 4.0 kg block *A* and 6.0 kg block *B* are connected by a string of negligible mass. Force $\vec{F}_A = (12 \text{ N})\hat{i}$ acts on block *A*; force $\vec{F}_B = (24 \text{ N})\hat{i}$ acts on block *B*. What is the tension in the string?

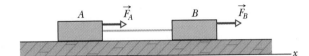

Fig. 5-59 Problem 69.

70 An 80 kg man drops to a concrete patio from a window 0.50 m above the patio. He neglects to bend his knees on landing, taking 2.0 cm to stop. (a) What is his average acceleration from when his feet first touch the patio to when he stops? (b) What is the magnitude of the average stopping force exerted on him by the patio?

71 SSM Figure 5-60 shows a box of dirty money (mass $m_1 = 3.0$ kg) on a frictionless plane inclined at angle $\theta_1 = 30°$. The box is connected via a cord of negligible mass to a box of laundered money (mass $m_2 = 2.0$ kg) on a frictionless plane inclined at angle $\theta_2 = 60°$. The pulley is frictionless and has negligible mass. What is the tension in the cord?

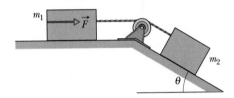

Fig. 5-60 Problem 71.

72 Three forces act on a particle that moves with unchanging velocity $\vec{v} = (2 \text{ m/s})\hat{i} - (7 \text{ m/s})\hat{j}$. Two of the forces are $\vec{F}_1 = (2 \text{ N})\hat{i} + (3 \text{ N})\hat{j} + (-2 \text{ N})\hat{k}$ and $\vec{F}_2 = (-5 \text{ N})\hat{i} + (8 \text{ N})\hat{j} + (-2 \text{ N})\hat{k}$. What is the third force?

73 SSM In Fig. 5-61, a tin of antioxidants ($m_1 = 1.0$ kg) on a frictionless inclined surface is connected to a tin of corned beef ($m_2 = 2.0$ kg). The pulley is massless and frictionless. An upward force of magnitude $F = 6.0$ N acts on the corned beef tin, which has a downward acceleration of 5.5 m/s². What are (a) the tension in the connecting cord and (b) angle β?

74 The only two forces acting on a body have magnitudes of 20 N and 35 N and directions that differ by 80°. The resulting acceleration has a magnitude of 20 m/s². What is the mass of the body?

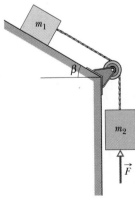

Fig. 5-61 Problem 73.

75 Figure 5-62 is an overhead view of a 12 kg tire that is to be pulled by three horizontal ropes. One rope's force (F_1 = 50 N) is indicated. The forces from the other ropes are to be oriented such that the tire's acceleration magnitude a is least. What is that least a if (a) F_2 = 30 N, F_3 = 20 N; (b) F_2 = 30 N, F_3 = 10 N; and (c) F_2 = F_3 = 30 N?

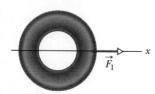

Fig. 5-62 Problem 75.

76 A block of mass M is pulled along a horizontal frictionless surface by a rope of mass m, as shown in Fig. 5-63. A horizontal force $\vec{F}$ acts on one end of the rope. (a)

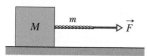

Fig. 5-63 Problem 76.

Show that the rope *must* sag, even if only by an imperceptible amount. Then, assuming that the sag is negligible, find (b) the acceleration of rope and block, (c) the force on the block from the rope, and (d) the tension in the rope at its midpoint.

77 SSM A worker drags a crate across a factory floor by pulling on a rope tied to the crate. The worker exerts a force of magnitude F = 450 N on the rope, which is inclined at an upward angle θ = 38° to the horizontal, and the floor exerts a horizontal force of magnitude f = 125 N that opposes the motion. Calculate the magnitude of the acceleration of the crate if (a) its mass is 310 kg and (b) its weight is 310 N.

78 In Fig. 5-64, a force $\vec{F}$ of magnitude 12 N is applied to a FedEx box of mass m_2 = 1.0 kg. The force is directed up a plane tilted by θ = 37°. The box is connected by a cord to a UPS box of mass m_1 = 3.0 kg on the floor. The floor, plane, and

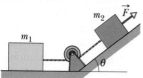

Fig. 5-64 Problem 78.

pulley are frictionless, and the masses of the pulley and cord are negligible. What is the tension in the cord?

79 A certain particle has a weight of 22 N at a point where g = 9.8 m/s². What are its (a) weight and (b) mass at a point where g = 4.9 m/s²? What are its (c) weight and (d) mass if it is moved to a point in space where g = 0?

80 An 80 kg person is parachuting and experiencing a downward acceleration of 2.5 m/s². The mass of the parachute is 5.0 kg. (a) What is the upward force on the open parachute from the air? (b) What is the downward force on the parachute from the person?

81 A spaceship lifts off vertically from the Moon, where g = 1.6 m/s². If the ship has an upward acceleration of 1.0 m/s² as it lifts off, what is the magnitude of the force exerted by the ship on its pilot, who weighs 735 N on Earth?

82 In the overhead view of Fig. 5-65, five forces pull on a box of mass m = 4.0 kg. The force magnitudes are F_1 = 11 N, F_2 = 17 N, F_3 = 3.0 N, F_4 = 14 N, and F_5 = 5.0 N, and angle θ_4 is 30°. Find the box's acceleration (a) in unit-vector notation and as (b) a magnitude and (c) an angle relative to the positive direction of the x axis.

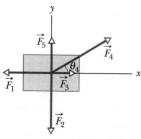

Fig. 5-65 Problem 82.

83 SSM A certain force gives an object of mass m_1 an acceleration of 12.0 m/s² and an object of mass m_2 an acceleration of 3.30 m/s². What acceleration would the force give to an object of mass (a) $m_2 - m_1$ and (b) $m_2 + m_1$?

84 You pull a short refrigerator with a constant force $\vec{F}$ across a greased (frictionless) floor, either with $\vec{F}$ horizontal (case 1) or with $\vec{F}$ tilted upward at an angle θ (case 2). (a) What is the ratio of the refrigerator's speed in case 2 to its speed in case 1 if you pull for a certain time t? (b) What is this ratio if you pull for a certain distance d?

85 A 52 kg circus performer is to slide down a rope that will break if the tension exceeds 425 N. (a) What happens if the performer hangs stationary on the rope? (b) At what magnitude of acceleration does the performer just avoid breaking the rope?

86 Compute the weight of a 75 kg space ranger (a) on Earth, (b) on Mars, where g = 3.7 m/s², and (c) in interplanetary space, where g = 0. (d) What is the ranger's mass at each location?

87 An object is hung from a spring balance attached to the ceiling of an elevator cab. The balance reads 65 N when the cab is standing still. What is the reading when the cab is moving upward (a) with a constant speed of 7.6 m/s and (b) with a speed of 7.6 m/s while decelerating at a rate of 2.4 m/s²?

88 Imagine a landing craft approaching the surface of Callisto, one of Jupiter's moons. If the engine provides an upward force (thrust) of 3260 N, the craft descends at constant speed; if the engine provides only 2200 N, the craft accelerates downward at 0.39 m/s². (a) What is the weight of the landing craft in the vicinity of Callisto's surface? (b) What is the mass of the craft? (c) What is the magnitude of the free-fall acceleration near the surface of Callisto?

89 A 1400 kg jet engine is fastened to the fuselage of a passenger jet by just three bolts (this is the usual practice). Assume that each bolt supports one-third of the load. (a) Calculate the force on each bolt as the plane waits in line for clearance to take off. (b) During flight, the plane encounters turbulence, which suddenly imparts an upward vertical acceleration of 2.6 m/s² to the plane. Calculate the force on each bolt now.

90 An interstellar ship has a mass of 1.20×10^6 kg and is initially at rest relative to a star system. (a) What constant acceleration is needed to bring the ship up to a speed of $0.10c$ (where c is the speed of light, 3.0×10^8 m/s) relative to the star system in 3.0 days? (b) What is that acceleration in g units? (c) What force is required for the acceleration? (d) If the engines are shut down when $0.10c$ is reached (the speed then remains constant), how long does the ship take (start to finish) to journey 5.0 light-months, the distance that light travels in 5.0 months?

91 SSM A motorcycle and 60.0 kg rider accelerate at 3.0 m/s² up a ramp inclined 10° above the horizontal. What are the magnitudes of (a) the net force on the rider and (b) the force on the rider from the motorcycle?

92 Compute the initial upward acceleration of a rocket of mass 1.3×10^4 kg if the initial upward force produced by its engine (the thrust) is 2.6×10^5 N. Do not neglect the gravitational force on the rocket.

93 SSM Figure 5-66a shows a mobile hanging from a ceiling; it consists of two metal pieces (m_1 = 3.5 kg and m_2 = 4.5 kg) that are strung together by cords of negligible mass. What is the tension

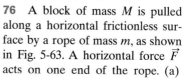

in (a) the bottom cord and (b) the top cord? Figure 5-66b shows a mobile consisting of three metal pieces. Two of the masses are $m_3 = 4.8$ kg and $m_5 = 5.5$ kg. The tension in the top cord is 199 N. What is the tension in (c) the lowest cord and (d) the middle cord?

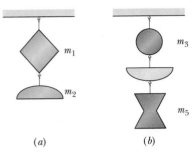

Fig. 5-66   Problem 93.

94   For sport, a 12 kg armadillo runs onto a large pond of level, frictionless ice. The armadillo's initial velocity is 5.0 m/s along the positive direction of an x axis. Take its initial position on the ice as being the origin. It slips over the ice while being pushed by a wind with a force of 17 N in the positive direction of the y axis. In unit-vector notation, what are the animal's (a) velocity and (b) position vector when it has slid for 3.0 s?

95   Suppose that in Fig. 5-12, the masses of the blocks are 2.0 kg and 4.0 kg. (a) Which mass should the hanging block have if the magnitude of the acceleration is to be as large as possible? What then are (b) the magnitude of the acceleration and (c) the tension in the cord?

96   A nucleus that captures a stray neutron must bring the neutron to a stop within the diameter of the nucleus by means of the *strong force*. That force, which "glues" the nucleus together, is approximately zero outside the nucleus. Suppose that a stray neutron with an initial speed of 1.4×10^7 m/s is just barely captured by a nucleus with diameter $d = 1.0 \times 10^{-14}$ m. Assuming the strong force on the neutron is constant, find the magnitude of that force. The neutron's mass is 1.67×10^{-27} kg.

6

FORCE AND MOTION-II

6-1 WHAT IS PHYSICS?

In this chapter we focus on the physics of three common types of force: frictional force, drag force, and centripetal force. An engineer preparing a car for the Indianapolis 500 must consider all three types. Frictional forces acting on the tires are crucial to the car's acceleration out of the pit and out of a curve (if the car hits an oil slick, the friction is lost and so is the car). Drag forces acting on the car from the passing air must be minimized or else the car will consume too much fuel and have to pit too early (even one 14 s pit stop can cost a driver the race). Centripetal forces are crucial in the turns (if there is insufficient centripetal force, the car slides into the wall). We start our discussion with frictional forces.

6-2 Friction

Frictional forces are unavoidable in our daily lives. If we were not able to counteract them, they would stop every moving object and bring to a halt every rotating shaft. About 20% of the gasoline used in an automobile is needed to counteract friction in the engine and in the drive train. On the other hand, if friction were totally absent, we could not get an automobile to go anywhere, and we could not walk or ride a bicycle. We could not hold a pencil, and, if we could, it would not write. Nails and screws would be useless, woven cloth would fall apart, and knots would untie.

Here we deal with the frictional forces that exist between dry solid surfaces, either stationary relative to each other or moving across each other at slow speeds. Consider three simple thought experiments:

1. Send a book sliding across a long horizontal counter. As expected, the book slows and then stops. This means the book must have an acceleration parallel to the counter surface, in the direction opposite the book's velocity. From Newton's second law, then, a force must act on the book parallel to the counter surface, in the direction opposite its velocity. That force is a frictional force.

2. Push horizontally on the book to make it travel at constant velocity along the counter. Can the force from you be the only horizontal force on the book? No, because then the book would accelerate. From Newton's second law, there must be a second force, directed opposite your force but with the same magnitude, so that the two forces balance. That second force is a frictional force, directed parallel to the counter.

3. Push horizontally on a heavy crate. The crate does not move. From Newton's second law, a second force must also be acting on the crate to counteract your force. Moreover, this second force must be directed opposite your force and have the same magnitude as your force, so that the two forces balance. That second force is a frictional force. Push even harder. The crate still does not move. Apparently the frictional force can change in magnitude so that the two

forces still balance. Now push with all your strength. The crate begins to slide. Evidently, there is a maximum magnitude of the frictional force. When you exceed that maximum magnitude, the crate slides.

Figure 6-1 shows a similar situation. In Fig. 6-1a, a block rests on a tabletop, with the gravitational force $\vec{F}_g$ balanced by a normal force $\vec{F}_N$. In Fig. 6-1b, you exert a force $\vec{F}$ on the block, attempting to pull it to the left. In response, a

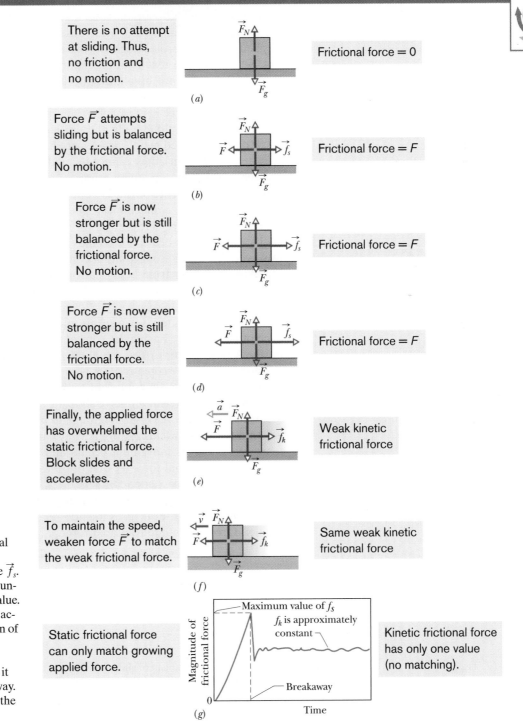

There is no attempt at sliding. Thus, no friction and no motion.

Frictional force = 0

(a)

Force $\vec{F}$ attempts sliding but is balanced by the frictional force. No motion.

Frictional force = F

(b)

Force $\vec{F}$ is now stronger but is still balanced by the frictional force. No motion.

Frictional force = F

(c)

Force $\vec{F}$ is now even stronger but is still balanced by the frictional force. No motion.

Frictional force = F

(d)

Finally, the applied force has overwhelmed the static frictional force. Block slides and accelerates.

Weak kinetic frictional force

(e)

To maintain the speed, weaken force $\vec{F}$ to match the weak frictional force.

Same weak kinetic frictional force

(f)

Static frictional force can only match growing applied force.

Kinetic frictional force has only one value (no matching).

(g)

Fig. 6-1 (a) The forces on a stationary block. (b–d) An external force $\vec{F}$, applied to the block, is balanced by a static frictional force $\vec{f}_s$. As F is increased, f_s also increases, until f_s reaches a certain maximum value. (e) The block then "breaks away," accelerating suddenly in the direction of $\vec{F}$. (f) If the block is now to move with constant velocity, F must be reduced from the maximum value it had just before the block broke away. (g) Some experimental results for the sequence (a) through (f).

frictional force $\vec{f}_s$ is directed to the right, exactly balancing your force. The force $\vec{f}_s$ is called the **static frictional force.** The block does not move.

Figures 6-1c and 6-1d show that as you increase the magnitude of your applied force, the magnitude of the static frictional force $\vec{f}_s$ also increases and the block remains at rest. When the applied force reaches a certain magnitude, however, the block "breaks away" from its intimate contact with the tabletop and accelerates leftward (Fig. 6-1e). The frictional force that then opposes the motion is called the **kinetic frictional force** $\vec{f}_k$.

Usually, the magnitude of the kinetic frictional force, which acts when there is motion, is less than the maximum magnitude of the static frictional force, which acts when there is no motion. Thus, if you wish the block to move across the surface with a constant speed, you must usually decrease the magnitude of the applied force once the block begins to move, as in Fig. 6-1f. As an example, Fig. 6-1g shows the results of an experiment in which the force on a block was slowly increased until breakaway occurred. Note the reduced force needed to keep the block moving at constant speed after breakaway.

A frictional force is, in essence, the vector sum of many forces acting between the surface atoms of one body and those of another body. If two highly polished and carefully cleaned metal surfaces are brought together in a very good vacuum (to keep them clean), they cannot be made to slide over each other. Because the surfaces are so smooth, many atoms of one surface contact many atoms of the other surface, and the surfaces *cold-weld* together instantly, forming a single piece of metal. If a machinist's specially polished gage blocks are brought together in air, there is less atom-to-atom contact, but the blocks stick firmly to each other and can be separated only by means of a wrenching motion. Usually, however, this much atom-to-atom contact is not possible. Even a highly polished metal surface is far from being flat on the atomic scale. Moreover, the surfaces of everyday objects have layers of oxides and other contaminants that reduce cold-welding.

When two ordinary surfaces are placed together, only the high points touch each other. (It is like having the Alps of Switzerland turned over and placed down on the Alps of Austria.) The actual *micro*scopic area of contact is much less than the apparent *macro*scopic contact area, perhaps by a factor of 10^4. Nonetheless, many contact points do cold-weld together. These welds produce static friction when an applied force attempts to slide the surfaces relative to each other.

If the applied force is great enough to pull one surface across the other, there is first a tearing of welds (at breakaway) and then a continuous re-forming and tearing of welds as movement occurs and chance contacts are made (Fig. 6-2). The kinetic frictional force $\vec{f}_k$ that opposes the motion is the vector sum of the forces at those many chance contacts.

If the two surfaces are pressed together harder, many more points cold-weld. Now getting the surfaces to slide relative to each other requires a greater applied force: The static frictional force $\vec{f}_s$ has a greater maximum value. Once

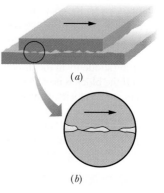

(a)

Fig. 6-2 The mechanism of sliding friction. (a) The upper surface is sliding to the right over the lower surface in this enlarged view. (b) A detail, showing two spots where cold-welding has occurred. Force is required to break the welds and maintain the motion.

(b)

the surfaces are sliding, there are many more points of momentary cold-welding, so the kinetic frictional force $\vec{f}_k$ also has a greater magnitude.

Often, the sliding motion of one surface over another is "jerky" because the two surfaces alternately stick together and then slip. Such repetitive *stick-and-slip* can produce squeaking or squealing, as when tires skid on dry pavement, fingernails scratch along a chalkboard, or a rusty hinge is opened. It can also produce beautiful and captivating sounds, as in music when a bow is drawn properly across a violin string.

6-3 Properties of Friction

Experiment shows that when a dry and unlubricated body presses against a surface in the same condition and a force $\vec{F}$ attempts to slide the body along the surface, the resulting frictional force has three properties:

Property 1.   If the body does not move, then the static frictional force $\vec{f}_s$ and the component of $\vec{F}$ that is parallel to the surface balance each other. They are equal in magnitude, and $\vec{f}_s$ is directed opposite that component of $\vec{F}$.

Property 2.   The magnitude of $\vec{f}_s$ has a maximum value $f_{s,\text{max}}$ that is given by

$$f_{s,\text{max}} = \mu_s F_N, \tag{6-1}$$

where μ_s is the **coefficient of static friction** and F_N is the magnitude of the normal force on the body from the surface. If the magnitude of the component of $\vec{F}$ that is parallel to the surface exceeds $f_{s,\text{max}}$, then the body begins to slide along the surface.

Property 3.   If the body begins to slide along the surface, the magnitude of the frictional force rapidly decreases to a value f_k given by

$$f_k = \mu_k F_N, \tag{6-2}$$

where μ_k is the **coefficient of kinetic friction.** Thereafter, during the sliding, a kinetic frictional force $\vec{f}_k$ with magnitude given by Eq. 6-2 opposes the motion.

The magnitude F_N of the normal force appears in properties 2 and 3 as a measure of how firmly the body presses against the surface. If the body presses harder, then, by Newton's third law, F_N is greater. Properties 1 and 2 are worded in terms of a single applied force $\vec{F}$, but they also hold for the net force of several applied forces acting on the body. Equations 6-1 and 6-2 are *not* vector equations; the direction of $\vec{f}_s$ or $\vec{f}_k$ is always parallel to the surface and opposed to the attempted sliding, and the normal force $\vec{F}_N$ is perpendicular to the surface.

The coefficients μ_s and μ_k are dimensionless and must be determined experimentally. Their values depend on certain properties of both the body and the surface; hence, they are usually referred to with the preposition "between," as in "the value of μ_s *between* an egg and a Teflon-coated skillet is 0.04, but that *between* rock-climbing shoes and rock is as much as 1.2." We assume that the value of μ_k does not depend on the speed at which the body slides along the surface.

✔ CHECKPOINT 1

A block lies on a floor. (a) What is the magnitude of the frictional force on it from the floor? (b) If a horizontal force of 5 N is now applied to the block, but the block does not move, what is the magnitude of the frictional force on it? (c) If the maximum value $f_{s,\text{max}}$ of the static frictional force on the block is 10 N, will the block move if the magnitude of the horizontally applied force is 8 N? (d) If it is 12 N? (e) What is the magnitude of the frictional force in part (c)?

Kinetic friction, constant acceleration, locked wheels

If a car's wheels are "locked" (kept from rolling) during emergency braking, the car slides along the road. Ripped-off bits of tire and small melted sections of road form the "skid marks" that reveal that cold-welding occurred during the slide. The record for the longest skid marks on a public road was reportedly set in 1960 by a Jaguar on the M1 highway in England (Fig. 6-3a)—the marks were 290 m long! Assuming that $\mu_k = 0.60$ and the car's acceleration was constant during the braking, how fast was the car going when the wheels became locked?

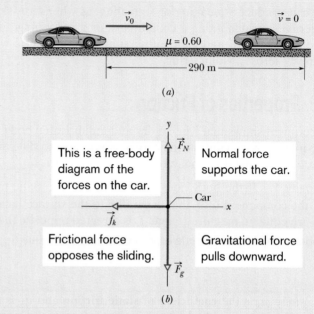

(a)

(b)

This is a free-body diagram of the forces on the car.

Normal force supports the car.

Frictional force opposes the sliding.

Gravitational force pulls downward.

Fig. 6-3 (a) A car sliding to the right and finally stopping after a displacement of 290 m. (b) A free-body diagram for the car.

KEY IDEAS

(1) Because the acceleration a is assumed constant, we can use the constant-acceleration equations of Table 2-1 to find the car's initial speed v_0. (2) If we neglect the effects of the air on the car, acceleration a was due only to a kinetic frictional force $\vec{f}_k$ on the car from the road, directed opposite the direction of the car's motion, assumed to be in the positive direction of an x axis (Fig. 6-3b). We can relate this force to the acceleration by writing Newton's second law for x components ($F_{net,x} = ma_x$) as

$$-f_k = ma, \tag{6-3}$$

where m is the car's mass. The minus sign indicates the direction of the kinetic frictional force.

Calculations: From Eq. 6-2, the frictional force has the magnitude $f_k = \mu_k F_N$, where F_N is the magnitude of the normal force on the car from the road. Because the car is not accelerating vertically, we know from Fig. 6-3b and Newton's second law that the magnitude of $\vec{F}_N$ is equal to the magnitude of the gravitational force $\vec{F}_g$ on the car, which is mg. Thus, $F_N = mg$.

Now solving Eq. 6-3 for a and substituting $f_k = \mu_k F_N = \mu_k mg$ for f_k yield

$$a = -\frac{f_k}{m} = -\frac{\mu_k mg}{m} = -\mu_k g, \tag{6-4}$$

where the minus sign indicates that the acceleration is in the negative direction of the x axis, opposite the direction of the velocity. Next, let's use Eq. 2-16,

$$v^2 = v_0^2 + 2a(x - x_0), \tag{6-5}$$

from the constant-acceleration equations of Chapter 2. We know that the displacement $x - x_0$ was 290 m and assume that the final speed v was 0. Substituting for a from Eq. 6-4 and solving for v_0 give

$$\begin{aligned} v_0 &= \sqrt{2\mu_k g(x - x_0)} \\ &= \sqrt{(2)(0.60)(9.8 \text{ m/s}^2)(290 \text{ m})} \\ &= 58 \text{ m/s} = 210 \text{ km/h}. \end{aligned} \tag{6-6}$$

We assumed that $v = 0$ at the far end of the skid marks. Actually, the marks ended only because the Jaguar left the road after 290 m. So v_0 was at *least* 210 km/h.

Friction, applied force at an angle

In Fig. 6-4a, a block of mass $m = 3.0$ kg slides along a floor while a force $\vec{F}$ of magnitude 12.0 N is applied to it at an upward angle θ. The coefficient of kinetic friction between the block and the floor is $\mu_k = 0.40$. We can vary θ from 0 to 90° (the block remains on the floor). What θ gives the maximum value of the block's acceleration magnitude a?

KEY IDEAS

Because the block is moving, a *kinetic* frictional force acts on it. The magnitude is given by Eq. 6-2 ($f_k = \mu_k F_N$, where F_N is the normal force). The direction is opposite the motion (the friction opposes the sliding).

Calculating F_N: Because we need the magnitude f_k of the frictional force, we first must calculate the magnitude F_N of the normal force. Figure 6-4b is a free-body diagram showing the forces along the vertical y axis. The normal force is upward, the gravitational force $\vec{F}_g$ with magnitude mg is downward, and (note) the vertical component F_y of the applied force is upward. That component is shown in Fig. 6-4c, where we can see that $F_y = F \sin \theta$. We can write Newton's second law ($\vec{F}_{net} = m\vec{a}$) for those forces along the y axis as

$$F_N + F \sin \theta - mg = m(0), \tag{6-7}$$

where we substituted zero for the acceleration along the y axis (the block does not even move along that axis). Thus,

$$F_N = mg - F \sin \theta. \tag{6-8}$$

Calculating acceleration a: Figure 6-4d is a free-body diagram for motion along the x axis. The horizontal component F_x of the applied force is rightward; from Fig. 6-4c, we see that $F_x = F \cos \theta$. The frictional force has magnitude f_k ($= \mu_k F_N$) and is leftward. Writing Newton's second law for motion along the x axis gives us

$$F \cos \theta - \mu_k F_N = ma. \tag{6-9}$$

Substituting for F_N from Eq. 6-8 and solving for a lead to

$$a = \frac{F}{m} \cos \theta - \mu_k \left(g - \frac{F}{m} \sin \theta \right). \tag{6-10}$$

Finding a maximum: To find the value of θ that maximizes a, we take the derivative of a with respect to θ and set the result equal to zero:

$$\frac{da}{d\theta} = -\frac{F}{m} \sin \theta + \mu_k \frac{F}{m} \cos \theta = 0. \tag{6-11}$$

Rearranging and using the identity $(\sin \theta)/(\cos \theta) = \tan \theta$ give us

$$\tan \theta = \mu_k. \tag{6-12}$$

Solving for θ and substituting the given $\mu_k = 0.40$, we find that the acceleration will be maximum if

$$\theta = \tan^{-1} \mu_k \tag{6-13}$$
$$= 21.8° \approx 22°. \qquad \text{(Answer)}$$

Comment: As we increase θ from 0, the acceleration tends to change in two opposing ways. First, more of the applied force $\vec{F}$ is upward, relieving the normal force. The decrease in the normal force causes a decrease in the frictional force, which opposes the block's motion. Thus, with the increase in θ, the block's acceleration tends to increase. However, second, the increase in θ also decreases the horizontal component of $\vec{F}$, and so the block's acceleration tends to decrease. These opposing tendencies produce a maximum acceleration at $\theta = 22°$.

This applied force accelerates block and helps support it.

These vertical forces balance.

Fig. 6-4 (a) A force is applied to a moving block. (b) The vertical forces. (c) The components of the applied force. (d) The horizontal forces and acceleration.

The applied force has these components.

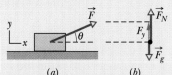

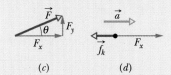

These two horizontal forces determine the acceleration.

 PLUS Additional examples, video, and practice available at *WileyPLUS*

6-4 The Drag Force and Terminal Speed

A **fluid** is anything that can flow—generally either a gas or a liquid. When there is a relative velocity between a fluid and a body (either because the body moves through the fluid or because the fluid moves past the body), the body experiences a **drag force** $\vec{D}$ that opposes the relative motion and points in the direction in which the fluid flows relative to the body.

Here we examine only cases in which air is the fluid, the body is blunt (like a baseball) rather than slender (like a javelin), and the relative motion is fast enough so that the air becomes turbulent (breaks up into swirls) behind the body.

Fig. 6-5 This skier crouches in an "egg position" so as to minimize her effective cross-sectional area and thus minimize the air drag acting on her. *(Karl-Josef Hildenbrand/dpa/Landov LLC)*

Table 6-1		
Some Terminal Speeds in Air		
Object	Terminal Speed (m/s)	95% Distance^a (m)
Shot (from shot put)	145	2500
Sky diver (typical)	60	430
Baseball	42	210
Tennis ball	31	115
Basketball	20	47
Ping-Pong ball	9	10
Raindrop (radius = 1.5 mm)	7	6
Parachutist (typical)	5	3

^{a}This is the distance through which the body must fall from rest to reach 95% of its terminal speed.
Source: Adapted from Peter J. Brancazio, *Sport Science,* 1984, Simon & Schuster, New York.

In such cases, the magnitude of the drag force $\vec{D}$ is related to the relative speed v by an experimentally determined **drag coefficient** C according to

$$D = \tfrac{1}{2}C\rho A v^2, \qquad (6\text{-}14)$$

where ρ is the air density (mass per volume) and A is the **effective cross-sectional area** of the body (the area of a cross section taken perpendicular to the velocity $\vec{v}$). The drag coefficient C (typical values range from 0.4 to 1.0) is not truly a constant for a given body because if v varies significantly, the value of C can vary as well. Here, we ignore such complications.

Downhill speed skiers know well that drag depends on A and v^2. To reach high speeds a skier must reduce D as much as possible by, for example, riding the skis in the "egg position" (Fig. 6-5) to minimize A.

When a blunt body falls from rest through air, the drag force $\vec{D}$ is directed upward; its magnitude gradually increases from zero as the speed of the body increases. This upward force $\vec{D}$ opposes the downward gravitational force $\vec{F}_g$ on the body. We can relate these forces to the body's acceleration by writing Newton's second law for a vertical y axis ($F_{\text{net},y} = ma_y$) as

$$D - F_g = ma, \qquad (6\text{-}15)$$

where m is the mass of the body. As suggested in Fig. 6-6, if the body falls long enough, D eventually equals F_g. From Eq. 6-15, this means that $a = 0$, and so the body's speed no longer increases. The body then falls at a constant speed, called the **terminal speed** v_t.

To find v_t, we set $a = 0$ in Eq. 6-15 and substitute for D from Eq. 6-14, obtaining

$$\tfrac{1}{2}C\rho A v_t^2 - F_g = 0,$$

which gives
$$v_t = \sqrt{\frac{2F_g}{C\rho A}}. \qquad (6\text{-}16)$$

Table 6-1 gives values of v_t for some common objects.

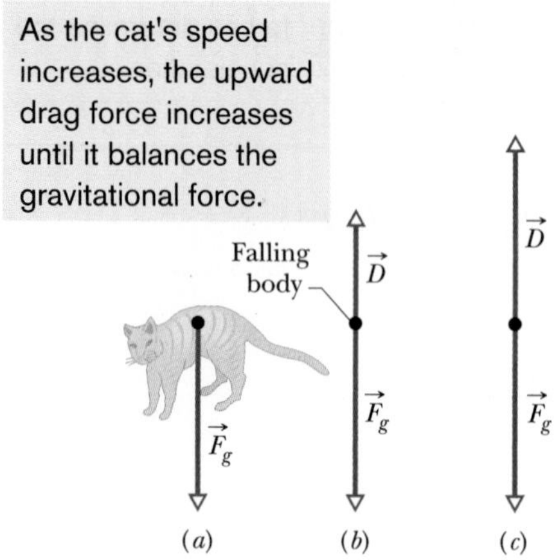

According to calculations* based on Eq. 6-14, a cat must fall about six floors to reach terminal speed. Until it does so, $F_g > D$ and the cat accelerates downward because of the net downward force. Recall from Chapter 2 that your body is an accelerometer, not a speedometer. Because the cat also senses the acceleration, it is frightened and keeps its feet underneath its body, its head tucked in, and its spine bent upward, making A small, v_t large, and injury likely.

*W. O. Whitney and C. J. Mehlhaff, "High-Rise Syndrome in Cats." *The Journal of the American Veterinary Medical Association,* 1987.

As the cat's speed increases, the upward drag force increases until it balances the gravitational force.

Falling body

(a)   (b)   (c)

Fig. 6-6 The forces that act on a body falling through air: (a) the body when it has just begun to fall and (b) the free-body diagram a little later, after a drag force has developed. (c) The drag force has increased until it balances the gravitational force on the body. The body now falls at its constant terminal speed.

However, if the cat does reach v_t during a longer fall, the acceleration vanishes and the cat relaxes somewhat, stretching its legs and neck horizontally outward and straightening its spine (it then resembles a flying squirrel). These actions increase area A and thus also, by Eq. 6-14, the drag D. The cat begins to slow because now $D > F_g$ (the net force is upward), until a new, smaller v_t is reached. The decrease in v_t reduces the possibility of serious injury on landing. Just before the end of the fall, when it sees it is nearing the ground, the cat pulls its legs back beneath its body to prepare for the landing.

Humans often fall from great heights for the fun of skydiving. However, in April 1987, during a jump, sky diver Gregory Robertson noticed that fellow sky diver Debbie Williams had been knocked unconscious in a collision with a third sky diver and was unable to open her parachute. Robertson, who was well above Williams at the time and who had not yet opened his parachute for the 4 km plunge, reoriented his body head-down so as to minimize A and maximize his downward speed. Reaching an estimated v_t of 320 km/h, he caught up with Williams and then went into a horizontal "spread eagle" (as in Fig. 6-7) to increase D so that he could grab her. He opened her parachute and then, after releasing her, his own, a scant 10 s before impact. Williams received extensive internal injuries due to her lack of control on landing but survived.

Fig. 6-7 Sky divers in a horizontal "spread eagle" maximize air drag. (*Steve Fitchett/Taxi/Getty Images*)

Sample Problem

Terminal speed of falling raindrop

A raindrop with radius $R = 1.5$ mm falls from a cloud that is at height $h = 1200$ m above the ground. The drag coefficient C for the drop is 0.60. Assume that the drop is spherical throughout its fall. The density of water ρ_w is 1000 kg/m³, and the density of air ρ_a is 1.2 kg/m³.

(a) As Table 6-1 indicates, the raindrop reaches terminal speed after falling just a few meters. What is the terminal speed?

KEY IDEA

The drop reaches a terminal speed v_t when the gravitational force on it is balanced by the air drag force on it, so its acceleration is zero. We could then apply Newton's second law and the drag force equation to find v_t, but Eq. 6-16 does all that for us.

Calculations: To use Eq. 6-16, we need the drop's effective cross-sectional area A and the magnitude F_g of the gravitational force. Because the drop is spherical, A is the area of a circle (πR^2) that has the same radius as the sphere. To find F_g, we use three facts: (1) $F_g = mg$, where m is the drop's mass; (2) the (spherical) drop's volume is $V = \frac{4}{3}\pi R^3$; and (3) the density of the water in the drop is the mass per volume, or $\rho_w = m/V$. Thus, we find

$$F_g = V\rho_w g = \tfrac{4}{3}\pi R^3 \rho_w g.$$

We next substitute this, the expression for A, and the given data into Eq. 6-16. Being careful to distinguish between the air den-

sity ρ_a and the water density ρ_w, we obtain

$$v_t = \sqrt{\frac{2F_g}{C\rho_a A}} = \sqrt{\frac{8\pi R^3 \rho_w g}{3C\rho_a \pi R^2}} = \sqrt{\frac{8R\rho_w g}{3C\rho_a}}$$

$$= \sqrt{\frac{(8)(1.5 \times 10^{-3}\ \text{m})(1000\ \text{kg/m}^3)(9.8\ \text{m/s}^2)}{(3)(0.60)(1.2\ \text{kg/m}^3)}}$$

$$= 7.4\ \text{m/s} \approx 27\ \text{km/h}. \qquad \text{(Answer)}$$

Note that the height of the cloud does not enter into the calculation.

(b) What would be the drop's speed just before impact if there were no drag force?

KEY IDEA

With no drag force to reduce the drop's speed during the fall, the drop would fall with the constant free-fall acceleration g, so the constant-acceleration equations of Table 2-1 apply.

Calculation: Because we know the acceleration is g, the initial velocity v_0 is 0, and the displacement $x - x_0$ is $-h$, we use Eq. 2-16 to find v:

$$v = \sqrt{2gh} = \sqrt{(2)(9.8\ \text{m/s}^2)(1200\ \text{m})}$$

$$= 153\ \text{m/s} \approx 550\ \text{km/h}. \qquad \text{(Answer)}$$

Had he known this, Shakespeare would scarcely have written, "it droppeth as the gentle rain from heaven, upon the place beneath." In fact, the speed is close to that of a bullet from a large-caliber handgun!

 Additional examples, video, and practice available at *WileyPLUS*

6-5 Uniform Circular Motion

From Section 4-7, recall that when a body moves in a circle (or a circular arc) at constant speed v, it is said to be in uniform circular motion. Also recall that the body has a centripetal acceleration (directed toward the center of the circle) of constant magnitude given by

$$a = \frac{v^2}{R} \quad \text{(centripetal acceleration)}, \quad (6\text{-}17)$$

where R is the radius of the circle.

Let us examine two examples of uniform circular motion:

1. *Rounding a curve in a car.* You are sitting in the center of the rear seat of a car moving at a constant high speed along a flat road. When the driver suddenly turns left, rounding a corner in a circular arc, you slide across the seat toward the right and then jam against the car wall for the rest of the turn. What is going on?

   While the car moves in the circular arc, it is in uniform circular motion; that is, it has an acceleration that is directed toward the center of the circle. By Newton's second law, a force must cause this acceleration. Moreover, the force must also be directed toward the center of the circle. Thus, it is a **centripetal force,** where the adjective indicates the direction. In this example, the centripetal force is a frictional force on the tires from the road; it makes the turn possible.

   If you are to move in uniform circular motion along with the car, there must also be a centripetal force on you. However, apparently the frictional force on you from the seat was not great enough to make you go in a circle with the car. Thus, the seat slid beneath you, until the right wall of the car jammed into you. Then its push on you provided the needed centripetal force on you, and you joined the car's uniform circular motion.

2. *Orbiting Earth.* This time you are a passenger in the space shuttle *Atlantis.* As it and you orbit Earth, you float through your cabin. What is going on?

   Both you and the shuttle are in uniform circular motion and have accelerations directed toward the center of the circle. Again by Newton's second law, centripetal forces must cause these accelerations. This time the centripetal forces are gravitational pulls (the pull on you and the pull on the shuttle) exerted by Earth and directed radially inward, toward the center of Earth.

In both car and shuttle you are in uniform circular motion, acted on by a centripetal force—yet your sensations in the two situations are quite different. In the car, jammed up against the wall, you are aware of being compressed by the wall. In the orbiting shuttle, however, you are floating around with no sensation of any force acting on you. Why this difference?

The difference is due to the nature of the two centripetal forces. In the car, the centripetal force is the push on the part of your body touching the car wall. You can sense the compression on that part of your body. In the shuttle, the centripetal force is Earth's gravitational pull on every atom of your body. Thus, there is no compression (or pull) on any one part of your body and no sensation of a force acting on you. (The sensation is said to be one of "weightlessness," but that description is tricky. The pull on you by Earth has certainly not disappeared and, in fact, is only a little less than it would be with you on the ground.)

Another example of a centripetal force is shown in Fig. 6-8. There a hockey puck moves around in a circle at constant speed v while tied to a string looped around a central peg. This time the centripetal force is the radially inward pull on the puck from the string. Without that force, the puck would slide off in a straight line instead of moving in a circle.

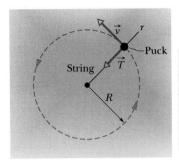

The puck moves in uniform circular motion only because of a toward-the-center force.

Fig. 6-8 An overhead view of a hockey puck moving with constant speed v in a circular path of radius R on a horizontal frictionless surface. The centripetal force on the puck is $\vec{T}$, the pull from the string, directed inward along the radial axis r extending through the puck.

Note again that a centripetal force is not a new kind of force. The name merely indicates the direction of the force. It can, in fact, be a frictional force, a gravitational force, the force from a car wall or a string, or any other force. For any situation:

➡️ A centripetal force accelerates a body by changing the direction of the body's velocity without changing the body's speed.

From Newton's second law and Eq. 6-17 ($a = v^2/R$), we can write the magnitude F of a centripetal force (or a net centripetal force) as

$$F = m\,\frac{v^2}{R} \qquad \text{(magnitude of centripetal force).} \qquad (6\text{-}18)$$

Because the speed v here is constant, the magnitudes of the acceleration and the force are also constant.

However, the directions of the centripetal acceleration and force are not constant; they vary continuously so as to always point toward the center of the circle. For this reason, the force and acceleration vectors are sometimes drawn along a radial axis r that moves with the body and always extends from the center of the circle to the body, as in Fig. 6-8. The positive direction of the axis is radially outward, but the acceleration and force vectors point radially inward.

✔️**CHECKPOINT 2**

When you ride in a Ferris wheel at constant speed, what are the directions of your acceleration $\vec{a}$ and the normal force $\vec{F}_N$ on you (from the always upright seat) as you pass through (a) the highest point and (b) the lowest point of the ride?

Sample Problem

Vertical circular loop, Diavolo

In a 1901 circus performance, Allo "Dare Devil" Diavolo introduced the stunt of riding a bicycle in a loop-the-loop (Fig. 6-9a). Assuming that the loop is a circle with radius $R = 2.7$ m, what is the least speed v that Diavolo and his bicycle could have at the top of the loop to remain in contact with it there?

KEY IDEA

We can assume that Diavolo and his bicycle travel through the top of the loop as a single particle in uniform circular motion. Thus, at the top, the acceleration $\vec{a}$ of this particle must have the magnitude $a = v^2/R$ given by Eq. 6-17 and be directed downward, toward the center of the circular loop.

(a)

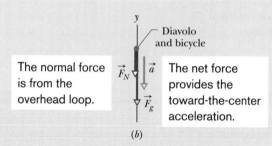

The normal force $\vec{F}_N$ is from the overhead loop.

The net force provides the toward-the-center acceleration.

(b)

Fig. 6-9 (a) Contemporary advertisement for Diavolo and (b) free-body diagram for the performer at the top of the loop. (Photograph in part a reproduced with permission of Circus World Museum)

Calculations: The forces on the particle when it is at the top of the loop are shown in the free-body diagram of Fig 6-9b. The gravitational force $\vec{F}_g$ is downward along a y axis; so is the normal force $\vec{F}_N$ on the particle from the loop; so also is the centripetal acceleration of the particle. Thus, Newton's second law for y components ($F_{\text{net},y} = ma_y$) gives us

$$-F_N - F_g = m(-a)$$

and

$$-F_N - mg = m\left(-\frac{v^2}{R}\right). \tag{6-19}$$

If the particle has the *least speed v* needed to remain in contact, then it is on the *verge of losing contact* with the loop (falling away from the loop), which means that $F_N = 0$ at the top of the loop (the particle and loop touch but without any normal force). Substituting 0 for F_N in Eq. 6-19, solving for v, and then substituting known values give us

$$v = \sqrt{gR} = \sqrt{(9.8 \text{ m/s}^2)(2.7 \text{ m})}$$
$$= 5.1 \text{ m/s}. \tag{Answer}$$

Comments: Diavolo made certain that his speed at the top of the loop was greater than 5.1 m/s so that he did not lose contact with the loop and fall away from it. Note that this speed requirement is independent of the mass of Diavolo and his bicycle. Had he feasted on, say, pierogies before his performance, he still would have had to exceed only 5.1 m/s to maintain contact as he passed through the top of the loop.

Sample Problem

Car in flat circular turn

Upside-down racing: A modern race car is designed so that the passing air pushes down on it, allowing the car to travel much faster through a flat turn in a Grand Prix without friction failing. This downward push is called *negative lift*. Can a race car have so much negative lift that it could be driven upside down on a long ceiling, as done fictionally by a sedan in the first *Men in Black* movie?

Figure 6-10a represents a Grand Prix race car of mass $m = 600$ kg as it travels on a flat track in a circular arc of radius $R = 100$ m. Because of the shape of the car and the wings on it, the passing air exerts a negative lift $\vec{F}_L$ downward on the car. The coefficient of static friction between the tires and the track is 0.75. (Assume that the forces on the four tires are identical.)

(a) If the car is on the verge of sliding out of the turn when its speed is 28.6 m/s, what is the magnitude of the negative lift $\vec{F}_L$ acting downward on the car?

KEY IDEAS

1. A centripetal force must act on the car because the car is moving around a circular arc; that force must be directed toward the center of curvature of the arc (here, that is horizontally).

2. The only horizontal force acting on the car is a frictional force on the tires from the road. So the required centripetal force is a frictional force.

3. Because the car is not sliding, the frictional force must be a *static* frictional force $\vec{f}_s$ (Fig. 6-10a).

4. Because the car is on the verge of sliding, the magnitude f_s is equal to the maximum value $f_{s,\text{max}} = \mu_s F_N$, where F_N is the magnitude of the normal force $\vec{F}_N$ acting on the car from the track.

Radial calculations: The frictional force $\vec{f}_s$ is shown in the free-body diagram of Fig. 6-10b. It is in the negative direction of a radial axis r that always extends from the center of curvature through the car as the car moves. The force produces a centripetal acceleration of magnitude v^2/R. We can relate the force and acceleration by writing Newton's second law for components along the r axis ($F_{net,r} = ma_r$) as

$$-f_s = m\left(-\frac{v^2}{R}\right). \qquad (6\text{-}20)$$

Substituting $f_{s,max} = \mu_s F_N$ for f_s leads us to

$$\mu_s F_N = m\left(\frac{v^2}{R}\right). \qquad (6\text{-}21)$$

Vertical calculations: Next, let's consider the vertical forces on the car. The normal force $\vec{F}_N$ is directed up, in the positive direction of the y axis in Fig. 6-10b. The gravitational force $\vec{F}_g = m\vec{g}$ and the negative lift $\vec{F}_L$ are directed down. The acceleration of the car along the y axis is zero. Thus we can write Newton's second law for components along the y axis ($F_{net,y} = ma_y$) as

$$F_N - mg - F_L = 0,$$

or $\qquad F_N = mg + F_L. \qquad (6\text{-}22)$

Combining results: Now we can combine our results along the two axes by substituting Eq. 6-22 for F_N in Eq. 6-21. Doing so and then solving for F_L lead to

$$F_L = m\left(\frac{v^2}{\mu_s R} - g\right)$$

$$= (600 \text{ kg})\left(\frac{(28.6 \text{ m/s})^2}{(0.75)(100 \text{ m})} - 9.8 \text{ m/s}^2\right)$$

$$= 663.7 \text{ N} \approx 660 \text{ N}. \qquad \text{(Answer)}$$

(b) The magnitude F_L of the negative lift on a car depends on the square of the car's speed v^2, just as the drag force does (Eq. 6-14). Thus, the negative lift on the car here is greater when the car travels faster, as it does on a straight section of track. What is the magnitude of the negative lift for a speed of 90 m/s?

KEY IDEA

F_L is proportional to v^2.

Calculations: Thus we can write a ratio of the negative lift $F_{L,90}$ at $v = 90$ m/s to our result for the negative lift F_L at $v = 28.6$ m/s as

$$\frac{F_{L,90}}{F_L} = \frac{(90 \text{ m/s})^2}{(28.6 \text{ m/s})^2}.$$

Substituting our known negative lift of $F_L = 663.7$ N and solving for $F_{L,90}$ give us

$$F_{L,90} = 6572 \text{ N} \approx 6600 \text{ N}. \qquad \text{(Answer)}$$

Upside-down racing: The gravitational force is, of course, the force to beat if there is a chance of racing upside down:

$$F_g = mg = (600 \text{ kg})(9.8 \text{ m/s}^2)$$

$$= 5880 \text{ N}.$$

With the car upside down, the negative lift is an *upward* force of 6600 N, which exceeds the downward 5880 N. Thus, the car could run on a long ceiling *provided* that it moves at about 90 m/s ($= 324$ km/h $= 201$ mi/h). However, moving that fast while right side up on a horizontal track is dangerous enough, so you are not likely to see upside-down racing except in the movies.

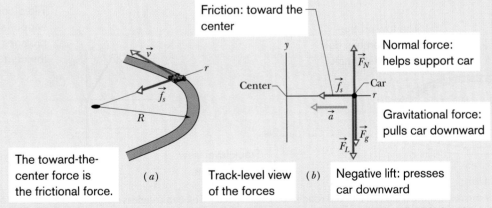

Friction: toward the center

Normal force: helps support car

Center

Car

Gravitational force: pulls car downward

The toward-the-center force is the frictional force.

(a)

Track-level view of the forces

(b)

Negative lift: presses car downward

Fig. 6-10 (a) A race car moves around a flat curved track at constant speed v. The frictional force $\vec{f}_s$ provides the necessary centripetal force along a radial axis r. (b) A free-body diagram (not to scale) for the car, in the vertical plane containing r.

 Additional examples, video, and practice available at *WileyPLUS*

Sample Problem

Car in banked circular turn

Curved portions of highways are always banked (tilted) to prevent cars from sliding off the highway. When a highway is dry, the frictional force between the tires and the road surface may be enough to prevent sliding. When the highway is wet, however, the frictional force may be negligible, and banking is then essential. Figure 6-11a represents a car of mass m as it moves at a constant speed v of 20 m/s around a banked circular track of radius $R = 190$ m. (It is a normal car, rather than a race car, which means any vertical force from the passing air is negligible.) If the frictional force from the track is negligible, what bank angle θ prevents sliding?

KEY IDEAS

Here the track is banked so as to tilt the normal force $\vec{F}_N$ on the car toward the center of the circle (Fig. 6-11b). Thus, $\vec{F}_N$ now has a centripetal component of magnitude F_{Nr}, directed inward along a radial axis r. We want to find the value of the bank angle θ such that this centripetal component keeps the car on the circular track without need of friction.

Radial calculation: As Fig. 6-11b shows (and as you should verify), the angle that force $\vec{F}_N$ makes with the vertical is equal to the bank angle θ of the track. Thus, the radial component F_{Nr} is equal to $F_N \sin \theta$. We can now write Newton's second law for components along the r axis ($F_{net,r} = ma_r$) as

$$-F_N \sin \theta = m\left(-\frac{v^2}{R}\right). \qquad (6\text{-}23)$$

We cannot solve this equation for the value of θ because it also contains the unknowns F_N and m.

Vertical calculations: We next consider the forces and acceleration along the y axis in Fig. 6-11b. The vertical component of the normal force is $F_{Ny} = F_N \cos \theta$, the gravitational force $\vec{F}_g$ on the car has the magnitude mg, and the acceleration of the car along the y axis is zero. Thus we can write Newton's second law for components along the y axis ($F_{net,y} = ma_y$) as

$$F_N \cos \theta - mg = m(0),$$

from which

$$F_N \cos \theta = mg. \qquad (6\text{-}24)$$

Combining results: Equation 6-24 also contains the unknowns F_N and m, but note that dividing Eq. 6-23 by Eq. 6-24 neatly eliminates both those unknowns. Doing so, replacing (sin θ)/(cos θ) with tan θ, and solving for θ then yield

$$\theta = \tan^{-1}\frac{v^2}{gR}$$

$$= \tan^{-1}\frac{(20 \text{ m/s})^2}{(9.8 \text{ m/s}^2)(190 \text{ m})} = 12°. \qquad \text{(Answer)}$$

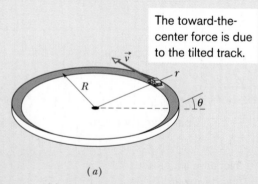

The toward-the-center force is due to the tilted track.

Track-level view of the forces

(a)

Tilted normal force supports car and provides the toward-the-center force.

The gravitational force pulls car downward.

(b)

Fig. 6-11 (a) A car moves around a curved banked road at constant speed v. The bank angle is exaggerated for clarity. (b) A free-body diagram for the car, assuming that friction between tires and road is zero and that the car lacks negative lift. The radially inward component F_{Nr} of the normal force (along radial axis r) provides the necessary centripetal force and radial acceleration.

 Additional examples, video, and practice available at *WileyPLUS*

REVIEW & SUMMARY

Friction   When a force $\vec{F}$ tends to slide a body along a surface, a **frictional force** from the surface acts on the body. The frictional force is parallel to the surface and directed so as to oppose the sliding. It is due to bonding between the body and the surface.

If the body does not slide, the frictional force is a **static frictional force** $\vec{f}_s$. If there is sliding, the frictional force is a **kinetic frictional force** $\vec{f}_k$.

1. If a body does not move, the static frictional force $\vec{f}_s$ and the component of $\vec{F}$ parallel to the surface are equal in magnitude, and $\vec{f}_s$ is directed opposite that component. If the component increases, f_s also increases.

2. The magnitude of $\vec{f}_s$ has a maximum value $f_{s,max}$ given by

$$f_{s,max} = \mu_s F_N, \tag{6-1}$$

where μ_s is the **coefficient of static friction** and F_N is the magnitude of the normal force. If the component of $\vec{F}$ parallel to the surface exceeds $f_{s,max}$, the body slides on the surface.

3. If the body begins to slide on the surface, the magnitude of the frictional force rapidly decreases to a constant value f_k given by

$$f_k = \mu_k F_N, \tag{6-2}$$

where μ_k is the **coefficient of kinetic friction.**

Drag Force   When there is relative motion between air (or some other fluid) and a body, the body experiences a **drag force** $\vec{D}$ that opposes the relative motion and points in the direction in which the fluid flows relative to the body. The magnitude of $\vec{D}$ is

related to the relative speed v by an experimentally determined **drag coefficient** C according to

$$D = \tfrac{1}{2}C\rho A v^2, \tag{6-14}$$

where ρ is the fluid density (mass per unit volume) and A is the **effective cross-sectional area** of the body (the area of a cross section taken perpendicular to the relative velocity $\vec{v}$).

Terminal Speed   When a blunt object has fallen far enough through air, the magnitudes of the drag force $\vec{D}$ and the gravitational force $\vec{F}_g$ on the body become equal. The body then falls at a constant **terminal speed** v_t given by

$$v_t = \sqrt{\frac{2F_g}{C\rho A}}. \tag{6-16}$$

Uniform Circular Motion   If a particle moves in a circle or a circular arc of radius R at constant speed v, the particle is said to be in **uniform circular motion.** It then has a **centripetal acceleration** $\vec{a}$ with magnitude given by

$$a = \frac{v^2}{R}. \tag{6-17}$$

This acceleration is due to a net **centripetal force** on the particle, with magnitude given by

$$F = \frac{mv^2}{R}, \tag{6-18}$$

where m is the particle's mass. The vector quantities $\vec{a}$ and $\vec{F}$ are directed toward the center of curvature of the particle's path.

QUESTIONS

1   In Fig. 6-12, if the box is stationary and the angle θ between the horizontal and force $\vec{F}$ is increased somewhat, do the following quantities increase, decrease, or remain the same: (a) F_x; (b) f_s; (c) F_N; (d) $f_{s,max}$? (e) If, instead, the box is sliding and θ is increased, does the magnitude of the frictional force on the box increase, decrease, or remain the same?

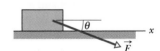

Fig. 6-12   Question 1.

2   Repeat Question 1 for force $\vec{F}$ angled upward instead of downward as drawn.

3   In Fig. 6-13, horizontal force $\vec{F}_1$ of magnitude 10 N is applied to a box on a floor, but the box does not slide. Then, as the magnitude of vertical force $\vec{F}_2$ is increased from zero, do the following quantities increase, decrease, or stay the same: (a) the magnitude of the frictional force $\vec{f}_s$ on the box; (b) the magnitude of the normal force $\vec{F}_N$ on the box from the floor; (c) the maximum value $f_{s,max}$ of

the magnitude of the static frictional force on the box? (d) Does the box eventually slide?

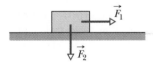

Fig. 6-13   Question 3.

4   In three experiments, three different horizontal forces are applied to the same block lying on the same countertop. The force magnitudes are $F_1 = 12$ N, $F_2 = 8$ N, and $F_3 = 4$ N. In each experiment, the block remains stationary in spite of the applied force. Rank the forces according to (a) the magnitude f_s of the static frictional force on the block from the countertop and (b) the maximum value $f_{s,max}$ of that force, greatest first.

5   If you press an apple crate against a wall so hard that the crate cannot slide down the wall, what is the direction of (a) the static frictional force $\vec{f}_s$ on the crate from the wall and (b) the normal force $\vec{F}_N$ on the crate from the wall? If you increase your push, what happens to (c) f_s, (d) F_N, and (e) $f_{s,max}$?

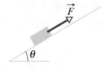

Fig. 6-14 Question 6.

6 In Fig. 6-14, a block of mass m is held stationary on a ramp by the frictional force on it from the ramp. A force $\vec{F}$, directed up the ramp, is then applied to the block and gradually increased in magnitude from zero. During the increase, what happens to the direction and magnitude of the frictional force on the block?

7 Reconsider Question 6 but with the force $\vec{F}$ now directed down the ramp. As the magnitude of $\vec{F}$ is increased from zero, what happens to the direction and magnitude of the frictional force on the block?

8 In Fig. 6-15, a horizontal force of 100 N is to be applied to a 10 kg slab that is initially stationary on a frictionless floor, to accelerate the slab. A 10 kg block lies on top of the slab; the coefficient of friction μ between the block and the slab is not known, and the block might slip. (a) Considering that possibility, what is the possible range of values for the magnitude of the slab's acceleration a_{slab}? (*Hint:* You don't need written calculations; just consider extreme values for μ.) (b) What is the possible range for the magnitude a_{block} of the block's acceleration?

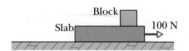

Fig. 6-15 Question 8.

9 Figure 6-16 shows the path of a park ride that travels at constant speed through five circular arcs of radii R_0, $2R_0$, and $3R_0$. Rank the arcs according to the magnitude of the centripetal force on a rider traveling in the arcs, greatest first.

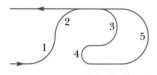

Fig. 6-16 Question 9.

10 In 1987, as a Halloween stunt, two sky divers passed a pumpkin back and forth between them while they were in free fall just west of Chicago. The stunt was great fun until the last sky diver with the pumpkin opened his parachute. The pumpkin broke free from his grip, plummeted about 0.5 km, ripped through the roof of a house, slammed into the kitchen floor, and splattered all over the newly remodeled kitchen. From the sky diver's viewpoint and from the pumpkin's viewpoint, why did the sky diver lose control of the pumpkin?

11 A person riding a Ferris wheel moves through positions at (1) the top, (2) the bottom, and (3) midheight. If the wheel rotates at a constant rate, rank these three positions according to (a) the magnitude of the person's centripetal acceleration, (b) the magnitude of the net centripetal force on the person, and (c) the magnitude of the normal force on the person, greatest first.

P R O B L E M S

GO	Tutoring problem available (at instructor's discretion) in *WileyPLUS* and WebAssign
SSM	Worked-out solution available in Student Solutions Manual **WWW** Worked-out solution is at
• – •••	Number of dots indicates level of problem difficulty **ILW** Interactive solution is at http://www.wiley.com/college/halliday
	Additional information available in *The Flying Circus of Physics* and at flyingcircusofphysics.com

sec. 6-3 Properties of Friction

•1 The floor of a railroad flatcar is loaded with loose crates having a coefficient of static friction of 0.25 with the floor. If the train is initially moving at a speed of 48 km/h, in how short a distance can the train be stopped at constant acceleration without causing the crates to slide over the floor?

•2 In a pickup game of dorm shuffleboard, students crazed by final exams use a broom to propel a calculus book along the dorm hallway. If the 3.5 kg book is pushed from rest through a distance of 0.90 m by the horizontal 25 N force from the broom and then has a speed of 1.60 m/s, what is the coefficient of kinetic friction between the book and floor?

•3 **SSM** **WWW** A bedroom bureau with a mass of 45 kg, including drawers and clothing, rests on the floor. (a) If the coefficient of static friction between the bureau and the floor is 0.45, what is the magnitude of the minimum horizontal force that a person must apply to start the bureau moving? (b) If the drawers and clothing, with 17 kg mass, are removed before the bureau is pushed, what is the new minimum magnitude?

•4 A slide-loving pig slides down a certain 35° slide in twice the time it would take to slide down a frictionless 35° slide. What is the coefficient of kinetic friction between the pig and the slide?

•5 **GO** A 2.5 kg block is initially at rest on a horizontal surface. A horizontal force $\vec{F}$ of magnitude 6.0 N and a vertical force $\vec{P}$ are then applied to the block (Fig. 6-17). The coefficients of friction for the block and surface are $\mu_s = 0.40$ and $\mu_k = 0.25$. Determine the magnitude of the frictional force acting on the block if the magnitude of $\vec{P}$ is (a) 8.0 N, (b) 10 N, and (c) 12 N.

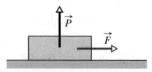

Fig. 6-17 Problem 5.

•6 A baseball player with mass $m = 79$ kg, sliding into second base, is retarded by a frictional force of magnitude 470 N. What is the coefficient of kinetic friction μ_k between the player and the ground?

•7 **SSM** **ILW** A person pushes horizontally with a force of 220 N on a 55 kg crate to move it across a level floor. The coefficient of kinetic friction is 0.35. What is the magnitude of (a) the frictional force and (b) the crate's acceleration?

•8 ~~━━~~ *The mysterious sliding stones.* Along the remote Racetrack Playa in Death Valley, California, stones sometimes gouge out prominent trails in the desert floor, as if the stones had been migrating (Fig. 6-18). For years curiosity mounted about why the stones moved. One explanation was that strong winds during occasional rainstorms would drag the rough stones over ground softened by rain. When the desert dried out, the trails behind the stones were hard-baked in place. According to measurements, the coefficient of kinetic friction between the stones and the wet playa ground is about 0.80. What horizontal force must act on a 20 kg stone (a typical mass) to maintain the stone's motion once a gust has started it moving? (Story continues with Problem 37.)

Fig. 6-18 Problem 8. What moved the stone? *(Jerry Schad/ Photo Researchers)*

•9 **GO** A 3.5 kg block is pushed along a horizontal floor by a force $\vec{F}$ of magnitude 15 N at an angle $\theta = 40°$ with the horizontal (Fig. 6-19). The coefficient of kinetic friction between the block and the floor is 0.25. Calculate the magnitudes of (a) the frictional force on the block from the floor and (b) the block's acceleration.

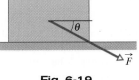

Fig. 6-19 Problems 9 and 32.

•10 Figure 6-20 shows an initially stationary block of mass m on a floor. A force of magnitude $0.500mg$ is then applied at upward angle $\theta = 20°$. What is the magnitude of the acceleration of the block across the floor if the friction coefficients are (a) $\mu_s = 0.600$ and $\mu_k = 0.500$ and (b) $\mu_s = 0.400$ and $\mu_k = 0.300$?

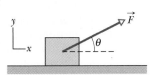

Fig. 6-20 Problem 10.

•11 **SSM** A 68 kg crate is dragged across a floor by pulling on a rope attached to the crate and inclined 15° above the horizontal. (a) If the coefficient of static friction is 0.50, what minimum force magnitude is required from the rope to start the crate moving? (b) If $\mu_k = 0.35$, what is the magnitude of the initial acceleration of the crate?

•12 In about 1915, Henry Sincosky of Philadelphia suspended himself from a rafter by gripping the rafter with the thumb of each hand on one side and the fingers on the opposite side (Fig. 6-21). Sincosky's mass was 79 kg. If the coefficient of static friction between hand and rafter was 0.70, what was the least magnitude of the normal force on the rafter from each thumb or opposite fingers? (After suspending himself, Sincosky chinned himself on the rafter and then moved hand-over-hand along the rafter. If you do not think Sincosky's grip was remarkable, try to repeat his stunt.)

Fig. 6-21 Problem 12.

•13 A worker pushes horizontally on a 35 kg crate with a force of magnitude 110 N. The coefficient of static friction between the crate and the floor is 0.37. (a) What is the value of $f_{s,max}$ under the circumstances? (b) Does the crate move? (c) What is the frictional force on the crate from the floor? (d) Suppose, next, that a second worker pulls directly upward on the crate to help out. What is the least vertical pull that will allow the first worker's 110 N push to move the crate? (e) If, instead, the second worker pulls horizontally to help out, what is the least pull that will get the crate moving?

•14 Figure 6-22 shows the cross section of a road cut into the side of a mountain. The solid line AA' represents a weak bedding plane along which sliding is possible. Block B directly above the highway is separated from uphill rock by a large crack (called a *joint*), so that only friction between the block and the bedding plane prevents sliding. The mass of the block is 1.8×10^7 kg, the *dip angle* θ of the bedding plane is 24°, and the coefficient of static friction between block and plane is 0.63. (a) Show that the block will not slide under these circumstances. (b) Next, water seeps into the joint and expands upon freezing, exerting on the block a force $\vec{F}$ parallel to AA'. What minimum value of force magnitude F will trigger a slide down the plane?

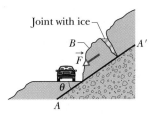

Fig. 6-22 Problem 14.

•15 The coefficient of static friction between Teflon and scrambled eggs is about 0.04. What is the smallest angle from the horizontal that will cause the eggs to slide across the bottom of a Teflon-coated skillet?

••16 A loaded penguin sled weighing 80 N rests on a plane inclined at angle $\theta = 20°$ to the horizontal (Fig. 6-23). Between the sled and the plane, the coefficient of static friction is 0.25, and the coefficient of kinetic friction is 0.15. (a) What is the least magnitude of the force $\vec{F}$, parallel to the plane, that will prevent the sled from slipping down the plane? (b) What is the minimum magnitude F that will start the sled moving up the plane? (c) What value of F is required to move the sled up the plane at constant velocity?

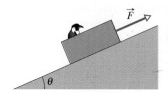

Fig. 6-23 Problems 16 and 22.

••17 In Fig. 6-24, a force $\vec{P}$ acts on a block weighing 45 N. The block is initially at rest on a plane inclined at angle $\theta = 15°$ to the horizontal. The positive direction of the x axis is up the plane. The coefficients of friction between block and plane are $\mu_s = 0.50$ and $\mu_k = 0.34$. In unit-vector notation, what is the frictional force on the block from the plane when $\vec{P}$ is (a) $(-5.0 \text{ N})\hat{i}$, (b) $(-8.0 \text{ N})\hat{i}$, and (c) $(-15 \text{ N})\hat{i}$?

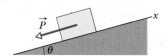

Fig. 6-24 Problem 17.

••18 You testify as an *expert witness* in a case involving an accident in which car A slid into the rear of car B, which was stopped at a red light along a road headed down a hill (Fig. 6-25). You find that the slope of the hill is $\theta = 12.0°$, that the cars were separated by distance $d = 24.0$ m when the driver of car A put the car into a slide (it lacked any automatic anti-brake-lock system), and that the speed of car A at the onset of braking was $v_0 = 18.0$ m/s. With what speed did car A hit car B if the coefficient of kinetic friction was (a) 0.60 (dry road surface) and (b) 0.10 (road surface covered with wet leaves)?

Fig. 6-25 Problem 18.

••19 A 12 N horizontal force $\vec{F}$ pushes a block weighing 5.0 N against a vertical wall (Fig. 6-26). The coefficient of static friction between the wall and the block is 0.60, and the coefficient of kinetic friction is 0.40. Assume that the block is not

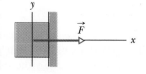

Fig. 6-26 Problem 19.

moving initially. (a) Will the block move? (b) In unit-vector notation, what is the force on the block from the wall?

••20 **GO** In Fig. 6-27, a box of Cheerios (mass $m_C = 1.0$ kg) and a box of Wheaties (mass $m_W = 3.0$ kg) are accelerated across a horizontal surface by a horizontal force $\vec{F}$ applied to the Cheerios box. The magnitude of the frictional force on the Cheerios box is 2.0 N, and the magnitude of the frictional force on the Wheaties box is 4.0 N. If the magnitude of $\vec{F}$ is 12 N, what is the magnitude of the force on the Wheaties box from the Cheerios box?

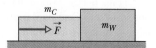

Fig. 6-27 Problem 20.

••21 An initially stationary box of sand is to be pulled across a floor by means of a cable in which the tension should not exceed 1100 N. The coefficient of static friction between the box and the floor is 0.35. (a) What should be the angle between the cable and the horizontal in order to pull the greatest possible amount of sand, and (b) what is the weight of the sand and box in that situation?

••22 **GO** In Fig. 6-23, a sled is held on an inclined plane by a cord pulling directly up the plane. The sled is to be on the verge of moving up the plane. In Fig. 6-28, the magnitude F required of the cord's force on the sled is plotted versus a range of values for the coefficient of static friction μ_s between sled and plane: $F_1 = 2.0$ N, $F_2 = 5.0$ N, and $\mu_2 = 0.50$. At what angle θ is the plane inclined?

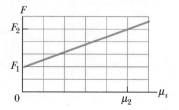

Fig. 6-28 Problem 22.

••23 When the three blocks in Fig. 6-29 are released from rest, they accelerate with a magnitude of 0.500 m/s². Block 1 has mass M, block 2 has $2M$, and block 3 has $2M$. What is the coefficient of kinetic friction between block 2 and the table?

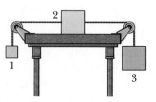

Fig. 6-29 Problem 23.

••24 A 4.10 kg block is pushed along a floor by a constant applied force that is horizontal and has a magnitude of 40.0 N. Figure 6-30 gives the block's speed v versus time t as the block moves along an x axis on the floor. The scale of the figure's vertical axis is set by $v_s = 5.0$ m/s. What is the coefficient of kinetic friction between the block and the floor?

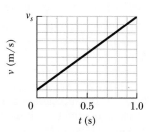

Fig. 6-30 Problem 24.

••25 SSM WWW Block B in Fig. 6-31 weighs 711 N. The coefficient of static friction between block and table is 0.25; angle θ is 30°; assume that the cord between B and the knot is horizontal. Find the maximum weight of block A for which the system will be stationary.

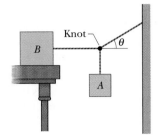

Fig. 6-31 Problem 25.

••26 GO Figure 6-32 shows three crates being pushed over a concrete floor by a horizontal force $\vec{F}$ of magnitude 440 N. The masses of the crates are $m_1 = 30.0$ kg, $m_2 = 10.0$ kg, and $m_3 = 20.0$ kg. The coefficient of kinetic friction between the floor and each of the crates is 0.700. (a) What is the magnitude F_{32} of the force on crate 3 from crate 2? (b) If the crates then slide onto a polished floor, where the coefficient of kinetic friction is less than 0.700, is magnitude F_{32} more than, less than, or the same as it was when the coefficient was 0.700?

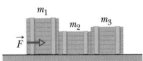

Fig. 6-32 Problem 26.

••27 Body A in Fig. 6-33 weighs 102 N, and body B weighs 32 N. The coefficients of friction between A and the incline are $\mu_s = 0.56$ and $\mu_k = 0.25$. Angle θ is 40°. Let the positive direction of an x axis be up the incline. In unit-vector notation, what is the acceleration of A if A is initially (a) at rest, (b) moving up the incline, and (c) moving down the incline?

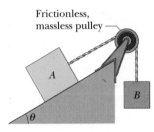

Fig. 6-33
Problems 27 and 28.

••28 In Fig. 6-33, two blocks are connected over a pulley. The mass of block A is 10 kg, and the coefficient of kinetic friction between A and the incline is 0.20. Angle θ of the incline is 30°. Block A slides down the incline at constant speed. What is the mass of block B?

••29 In Fig. 6-34, blocks A and B have weights of 44 N and 22 N, respectively. (a) Determine the minimum weight of block C to keep A from sliding if μ_s between A and the table is 0.20. (b) Block C suddenly is lifted off A. What is the acceleration of block A if μ_k between A and the table is 0.15?

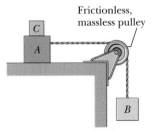

Fig. 6-34 Problem 29.

••30 A toy chest and its contents have a combined weight of 180 N. The coefficient of static friction between toy chest and floor is 0.42. The child in Fig. 6-35 attempts to move the chest across the floor by pulling on an attached rope. (a) If θ is 42°, what is the mag-

nitude of the force $\vec{F}$ that the child must exert on the rope to put the chest on the verge of moving? (b) Write an expression for the magnitude F required to put the chest on the verge of moving as a function of the angle θ. Determine (c) the value of θ for which F is a minimum and (d) that minimum magnitude.

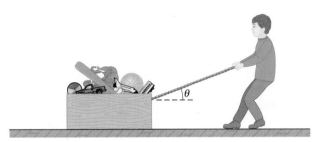

Fig. 6-35 Problem 30.

••31 SSM Two blocks, of weights 3.6 N and 7.2 N, are connected by a massless string and slide down a 30° inclined plane. The coefficient of kinetic friction between the lighter block and the plane is 0.10, and the coefficient between the heavier block and the plane is 0.20. Assuming that the lighter block leads, find (a) the magnitude of the acceleration of the blocks and (b) the tension in the taut string.

••32 GO A block is pushed across a floor by a constant force that is applied at downward angle θ (Fig. 6-19). Figure 6-36 gives the acceleration magnitude a versus a range of values for the coefficient of kinetic friction μ_k between block and floor: $a_1 = 3.0$ m/s^2, $\mu_{k2} = 0.20$, and $\mu_{k3} = 0.40$. What is the value of θ?

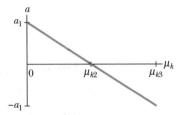

Fig. 6-36 Problem 32.

•••33 SSM A 1000 kg boat is traveling at 90 km/h when its engine is shut off. The magnitude of the frictional force $\vec{f}_k$ between boat and water is proportional to the speed v of the boat: $f_k = 70v$, where v is in meters per second and f_k is in newtons. Find the time required for the boat to slow to 45 km/h.

•••34 GO In Fig. 6-37, a slab of mass $m_1 = 40$ kg rests on a frictionless floor, and a block of mass $m_2 = 10$ kg rests on top of the slab. Between block and slab, the coefficient of static friction is 0.60, and the coefficient of kinetic friction is 0.40. A horizontal force $\vec{F}$ of magnitude 100 N begins to pull directly on the block, as shown. In unit-vector notation, what are the resulting accelerations of (a) the block and (b) the slab?

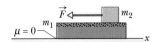

Fig. 6-37 Problem 34.

•••**35** ILW The two blocks (m = 16 kg and M = 88 kg) in Fig. 6-38 are not attached to each other. The coefficient of static friction between the blocks is μ_s = 0.38, but the surface beneath the larger block is frictionless. What is the minimum magnitude of the horizontal force $\vec{F}$ required to keep the smaller block from slipping down the larger block?

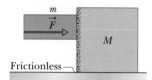

Fig. 6-38 Problem 35.

sec. 6-4 The Drag Force and Terminal Speed

•**36** The terminal speed of a sky diver is 160 km/h in the spread-eagle position and 310 km/h in the nosedive position. Assuming that the diver's drag coefficient C does not change from one position to the other, find the ratio of the effective cross-sectional area A in the slower position to that in the faster position.

••**37** *Continuation of Problem 8.* Now assume that Eq. 6-14 gives the magnitude of the air drag force on the typical 20 kg stone, which presents to the wind a vertical cross-sectional area of 0.040 m² and has a drag coefficient C of 0.80. Take the air density to be 1.21 kg/m³, and the coefficient of kinetic friction to be 0.80. (a) In kilometers per hour, what wind speed V along the ground is needed to maintain the stone's motion once it has started moving? Because winds along the ground are retarded by the ground, the wind speeds reported for storms are often measured at a height of 10 m. Assume wind speeds are 2.00 times those along the ground. (b) For your answer to (a), what wind speed would be reported for the storm? (c) Is that value reasonable for a high-speed wind in a storm? (Story continues with Problem 65.)

••**38** Assume Eq. 6-14 gives the drag force on a pilot plus ejection seat just after they are ejected from a plane traveling horizontally at 1300 km/h. Assume also that the mass of the seat is equal to the mass of the pilot and that the drag coefficient is that of a sky diver. Making a reasonable guess of the pilot's mass and using the appropriate v_t value from Table 6-1, estimate the magnitudes of (a) the drag force on the *pilot + seat* and (b) their horizontal deceleration (in terms of g), both just after ejection. (The result of (a) should indicate an engineering requirement: The seat must include a protective barrier to deflect the initial wind blast away from the pilot's head.)

••**39** Calculate the ratio of the drag force on a jet flying at 1000 km/h at an altitude of 10 km to the drag force on a prop-driven transport flying at half that speed and altitude. The density of air is 0.38 kg/m³ at 10 km and 0.67 kg/m³ at 5.0 km. Assume that the airplanes have the same effective cross-sectional area and drag coefficient C.

••**40** In downhill speed skiing a skier is retarded by both the air drag force on the body and the kinetic frictional force on the skis. (a) Suppose the slope angle is θ = 40.0°, the snow is dry snow with a coefficient of kinetic friction μ_k = 0.0400, the mass of the skier and equipment is m = 85.0 kg, the cross-sectional area of the (tucked) skier is A = 1.30 m², the drag coefficient is C = 0.150, and the air density is 1.20 kg/m³. (a) What is the terminal speed? (b) If a skier can vary C by a slight amount dC by adjusting, say, the hand positions, what is the corresponding variation in the terminal speed?

sec. 6-5 Uniform Circular Motion

•**41** A cat dozes on a stationary merry-go-round, at a radius of 5.4 m from the center of the ride. Then the operator turns on the ride

and brings it up to its proper turning rate of one complete rotation every 6.0 s. What is the least coefficient of static friction between the cat and the merry-go-round that will allow the cat to stay in place, without sliding?

•**42** Suppose the coefficient of static friction between the road and the tires on a car is 0.60 and the car has no negative lift. What speed will put the car on the verge of sliding as it rounds a level curve of 30.5 m radius?

•**43** ILW What is the smallest radius of an unbanked (flat) track around which a bicyclist can travel if her speed is 29 km/h and the μ_s between tires and track is 0.32?

•**44** During an Olympic bobsled run, the Jamaican team makes a turn of radius 7.6 m at a speed of 96.6 km/h. What is their acceleration in terms of g?

••**45** SSM ILW A student of weight 667 N rides a steadily rotating Ferris wheel (the student sits upright). At the highest point, the magnitude of the normal force $\vec{F}_N$ on the student from the seat is 556 N. (a) Does the student feel "light" or "heavy" there? (b) What is the magnitude of $\vec{F}_N$ at the lowest point? If the wheel's speed is doubled, what is the magnitude F_N at the (c) highest and (d) lowest point?

••**46** A police officer in hot pursuit drives her car through a circular turn of radius 300 m with a constant speed of 80.0 km/h. Her mass is 55.0 kg. What are (a) the magnitude and (b) the angle (relative to vertical) of the *net* force of the officer on the car seat? (*Hint:* Consider both horizontal and vertical forces.)

••**47** A circular-motion addict of mass 80 kg rides a Ferris wheel around in a vertical circle of radius 10 m at a constant speed of 6.1 m/s. (a) What is the period of the motion? What is the magnitude of the normal force on the addict from the seat when both go through (b) the highest point of the circular path and (c) the lowest point?

••**48** A roller-coaster car has a mass of 1200 kg when fully loaded with passengers. As the car passes over the top of a circular hill of radius 18 m, its speed is not changing. At the top of the hill, what are the (a) magnitude F_N and (b) direction (up or down) of the normal force on the car from the track if the car's speed is v = 11 m/s? What are (c) F_N and (d) the direction if v = 14 m/s?

••**49** In Fig. 6-39, a car is driven at constant speed over a circular hill and then into a circular valley with the same radius. At the top of the hill, the normal force on the driver from the car seat is 0. The driver's mass is 70.0 kg. What is the magnitude of the normal force on the driver from the seat when the car passes through the bottom of the valley?

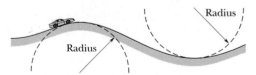

Fig. 6-39 Problem 49.

••**50** An 85.0 kg passenger is made to move along a circular path of radius r = 3.50 m in uniform circular motion. (a) Figure 6-40a is a plot of the required magnitude F of the net centripetal force for a range of possible values of the passenger's speed v. What is the plot's slope at $v = 8.30$ m/s? (b) Figure 6-40b is a plot of F for a range of possible values of T, the period of the motion. What is the plot's slope at $T = 2.50$ s?

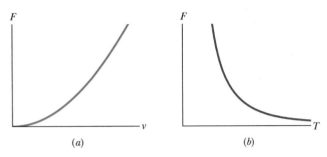

Fig. 6-40   Problem 50.

••**51** **SSM** **WWW** An airplane is flying in a horizontal circle at a speed of 480 km/h (Fig. 6-41). If its wings are tilted at angle $\theta = 40°$ to the horizontal, what is the radius of the circle in which the plane is flying? Assume that the required force is provided entirely by an "aerodynamic lift" that is perpendicular to the wing surface.

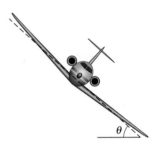

Fig. 6-41   Problem 51.

••**52** ✈ An amusement park ride consists of a car moving in a vertical circle on the end of a rigid boom of negligible mass. The combined weight of the car and riders is 5.0 kN, and the circle's radius is 10 m. At the top of the circle, what are the (a) magnitude F_B and (b) direction (up or down) of the force on the car from the boom if the car's speed is $v = 5.0$ m/s? What are (c) F_B and (d) the direction if $v = 12$ m/s?

••**53** An old streetcar rounds a flat corner of radius 9.1 m, at 16 km/h. What angle with the vertical will be made by the loosely hanging hand straps?

••**54** ✈ In designing circular rides for amusement parks, mechanical engineers must consider how small variations in certain parameters can alter the net force on a passenger. Consider a passenger of mass m riding around a horizontal circle of radius r at speed v. What is the variation dF in the net force magnitude for (a) a variation dr in the radius with v held constant, (b) a variation dv in the speed with r held constant, and (c) a variation dT in the period with r held constant?

••**55** A bolt is threaded onto one end of a thin horizontal rod, and the rod is then rotated horizontally about its other end. An engineer monitors the motion by flashing a strobe lamp onto the rod and bolt, adjusting the strobe rate until the bolt appears to be in the same eight places during each full rotation of the rod (Fig. 6-42). The strobe rate is 2000 flashes per second; the bolt has mass 30 g and is at radius 3.5 cm. What is the magnitude of the force on the bolt from the rod?

••**56** **GO** A banked circular highway curve is designed for traffic moving at 60 km/h. The radius of the curve is 200 m. Traffic is moving along the highway at 40 km/h on a rainy day. What is the minimum coefficient of friction between tires and road that will allow cars to take the turn without sliding off the road? (Assume the cars do not have negative lift.)

••**57** **GO** A puck of mass $m = 1.50$ kg slides in a circle of radius $r = 20.0$ cm on a frictionless table while attached to a hanging cylinder of mass $M = 2.50$ kg by means of a cord that extends through a hole in the table (Fig. 6-43). What speed keeps the cylinder at rest?

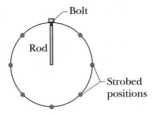

Fig. 6-42   Problem 55.

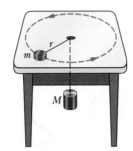

Fig. 6-43
Problem 57.

••**58** ✈ *Brake or turn?* Figure 6-44 depicts an overhead view of a car's path as the car travels toward a wall. Assume that the driver begins to brake the car when the distance to the wall is $d = 107$ m, and take the car's mass as $m = 1400$ kg, its initial speed as $v_0 = 35$ m/s, and the coefficient of static friction as $\mu_s = 0.50$. Assume that the car's weight is distributed evenly on the four wheels, even during braking. (a) What magnitude of static friction is needed (between tires and road) to stop the car just as it reaches the wall? (b) What is the maximum possible static friction $f_{s,\,max}$? (c) If the coefficient of kinetic friction between the (sliding) tires and the road is $\mu_k = 0.40$, at what speed will the car hit the wall? To avoid the crash, a driver could elect to turn the car so that it just barely misses the wall, as shown in the figure. (d) What magnitude of frictional force would be required to keep the car in a circular path of radius d and at the given speed v_0, so that the car moves in a quarter circle and then parallel to the wall? (e) Is the required force less than $f_{s,\,max}$ so that a circular path is possible?

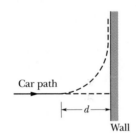

Fig. 6-44
Problem 58.

•••**59** **SSM** **ILW** In Fig. 6-45, a 1.34 kg ball is connected by means of two massless strings, each of length $L = 1.70$ m, to a vertical, rotating rod. The strings are tied to the rod with separation $d = 1.70$ m and are taut. The tension in the upper string is 35 N. What are the (a) tension in the lower string, (b) magnitude of the net force $\vec{F}_{net}$ on the ball, and (c) speed of the ball? (d) What is the direction of $\vec{F}_{net}$?

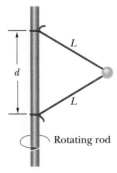

Fig. 6-45   Problem 59.

Additional Problems

60 In Fig. 6-46, a box of ant aunts (total mass $m_1 = 1.65$ kg) and a box of ant uncles (total mass $m_2 = 3.30$ kg) slide down an inclined plane while attached by a massless rod parallel to the plane. The angle of incline is $\theta = 30.0°$. The coefficient of kinetic friction between the aunt box and the incline is $\mu_1 = 0.226$; that between the uncle box and the incline is $\mu_2 = 0.113$. Compute (a) the tension in the rod and (b) the magnitude of the common acceleration of the two boxes. (c) How would the answers to (a) and (b) change if the uncles trailed the aunts?

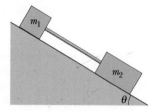

Fig. 6-46 Problem 60.

61 SSM A block of mass $m_t = 4.0$ kg is put on top of a block of mass $m_b = 5.0$ kg. To cause the top block to slip on the bottom one while the bottom one is held fixed, a horizontal force of at least 12 N must be applied to the top block. The assembly of blocks is now placed on a horizontal, frictionless table (Fig. 6-47). Find the magnitudes of (a) the maximum horizontal force $\vec{F}$ that can be applied to the lower block so that the blocks will move together and (b) the resulting acceleration of the blocks.

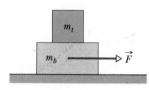

Fig. 6-47 Problem 61.

62 A 5.00 kg stone is rubbed across the horizontal ceiling of a cave passageway (Fig. 6-48). If the coefficient of kinetic friction is 0.65 and the force applied to the stone is angled at $\theta = 70.0°$, what must the magnitude of the force be for the stone to move at constant velocity?

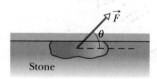

Fig. 6-48 Problem 62.

63 In Fig. 6-49, a 49 kg rock climber is climbing a "chimney." The coefficient of static friction between her shoes and the

Fig. 6-49 Problem 63.

rock is 1.2; between her back and the rock is 0.80. She has reduced her push against the rock until her back and her shoes are on the verge of slipping. (a) Draw a free-body diagram of her. (b) What is the magnitude of her push against the rock? (c) What fraction of her weight is supported by the frictional force on her shoes?

64 A high-speed railway car goes around a flat, horizontal circle of radius 470 m at a constant speed. The magnitudes of the horizontal and vertical components of the force of the car on a 51.0 kg passenger are 210 N and 500 N, respectively. (a) What is the magnitude of the net force (of *all* the forces) on the passenger? (b) What is the speed of the car?

65 *Continuation of Problems 8 and 37.* Another explanation is that the stones move only when the water dumped on the playa during a storm freezes into a large, thin sheet of ice. The stones are trapped in place in the ice. Then, as air flows across the ice during a wind, the air-drag forces on the ice and stones move them both, with the stones gouging out the trails. The magnitude of the air-drag force on this horizontal "ice sail" is given by $D_{ice} = 4C_{ice}\rho A_{ice}v^2$, where C_{ice} is the drag coefficient (2.0×10^{-3}), ρ is the air density (1.21 kg/m³), A_{ice} is the horizontal area of the ice, and v is the wind speed along the ice.

Assume the following: The ice sheet measures 400 m by 500 m by 4.0 mm and has a coefficient of kinetic friction of 0.10 with the ground and a density of 917 kg/m³. Also assume that 100 stones identical to the one in Problem 8 are trapped in the ice. To maintain the motion of the sheet, what are the required wind speeds (a) near the sheet and (b) at a height of 10 m? (c) Are these reasonable values for high-speed winds in a storm?

66 GO In Fig. 6-50, block 1 of mass $m_1 = 2.0$ kg and block 2 of mass $m_2 = 3.0$ kg are connected by a string of negligible mass and are initially held in place. Block 2 is on a frictionless surface tilted at $\theta = 30°$. The coefficient of kinetic friction between block 1 and the horizontal surface is 0.25. The pulley has negligible mass and friction. Once they are released, the blocks move. What then is the tension in the string?

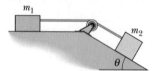

Fig. 6-50 Problem 66.

67 In Fig. 6-51, a crate slides down an inclined right-angled trough. The coefficient of kinetic friction between the crate and the trough is μ_k. What is the acceleration of the crate in terms of μ_k, θ, and g?

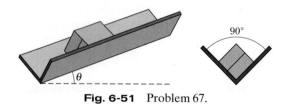

Fig. 6-51 Problem 67.

68 *Engineering a highway curve.* If a car goes through a curve too fast, the car tends to slide out of the curve. For a banked curve with friction, a frictional force acts on a fast car to oppose the tendency to slide out of the curve; the force is directed down the bank (in the direction water would drain). Consider a circular curve of radius $R = 200$ m and bank angle θ, where the coefficient of static friction between tires and pavement is μ_s. A car (without negative lift) is driven around the curve as shown in Fig. 6-11. (a) Find an expression for the car speed v_{max} that puts the car on the verge of sliding out. (b) On the same graph, plot v_{max} versus angle θ for the range $0°$ to $50°$, first for $\mu_s = 0.60$ (dry pavement) and then for $\mu_s = 0.050$ (wet or icy pavement). In kilometers per hour, evaluate v_{max} for a bank angle of $\theta = 10°$ and for (c) $\mu_s = 0.60$ and (d) $\mu_s = 0.050$. (Now you can see why accidents occur in highway curves when icy conditions are not obvious to drivers, who tend to drive at normal speeds.)

69 A student, crazed by final exams, uses a force $\vec{P}$ of magnitude 80 N and angle $\theta = 70°$ to push a 5.0 kg block across the ceiling of his room (Fig. 6-52). If the coefficient of kinetic friction between the block and the ceiling is 0.40, what is the magnitude of the block's acceleration?

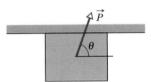

Fig. 6-52 Problem 69.

70 Figure 6-53 shows a *conical pendulum*, in which the bob (the small object at the lower end of the cord) moves in a horizontal circle at constant speed. (The cord sweeps out a cone as the bob rotates.) The bob has a mass of 0.040 kg, the string has length $L = 0.90$ m and negligible mass, and the bob follows a circular path of circumference 0.94 m. What are (a) the tension in the string and (b) the period of the motion?

71 An 8.00 kg block of steel is at rest on a horizontal table. The coefficient of static friction between the block and the table is 0.450. A force is to be applied to the block. To three significant figures, what is the magnitude of that applied force if it puts the block on the verge of sliding when the force is directed (a) horizontally, (b) upward at 60.0° from the horizontal, and (c) downward at 60.0° from the horizontal?

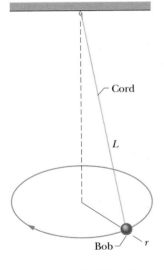

Fig. 6-53 Problem 70.

72 A box of canned goods slides down a ramp from street level into the basement of a grocery store with acceleration 0.75 m/s² directed down the ramp. The ramp makes an angle of 40° with the horizontal. What is the coefficient of kinetic friction between the box and the ramp?

73 In Fig. 6-54, the coefficient of kinetic friction between the block and inclined plane is 0.20, and angle θ is 60°. What are the (a) magnitude a and (b) direction (up or down the plane) of the

block's acceleration if the block is sliding down the plane? What are (c) a and (d) the direction if the block is sent sliding up the plane?

74 A 110 g hockey puck sent sliding over ice is stopped in 15 m by the frictional force on it from the ice. (a) If its initial speed is 6.0 m/s, what is the magnitude of the frictional force? (b) What is the coefficient of friction between the puck and the ice?

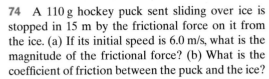

Fig. 6-54 Problem 73.

75 A locomotive accelerates a 25-car train along a level track. Every car has a mass of 5.0×10^4 kg and is subject to a frictional force $f = 250v$, where the speed v is in meters per second and the force f is in newtons. At the instant when the speed of the train is 30 km/h, the magnitude of its acceleration is 0.20 m/s². (a) What is the tension in the coupling between the first car and the locomotive? (b) If this tension is equal to the maximum force the locomotive can exert on the train, what is the steepest grade up which the locomotive can pull the train at 30 km/h?

76 A house is built on the top of a hill with a nearby slope at angle $\theta = 45°$ (Fig. 6-55). An engineering study indicates that the slope angle should be reduced because the top layers of soil along the slope might slip past the lower layers. If the coefficient of static friction between two such layers is 0.5, what is the least angle ϕ through which the present slope should be reduced to prevent slippage?

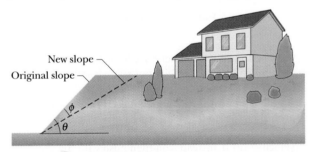

Fig. 6-55 Problem 76.

77 What is the terminal speed of a 6.00 kg spherical ball that has a radius of 3.00 cm and a drag coefficient of 1.60? The density of the air through which the ball falls is 1.20 kg/m³.

78 A student wants to determine the coefficients of static friction and kinetic friction between a box and a plank. She places the box on the plank and gradually raises one end of the plank. When the angle of inclination with the horizontal reaches 30°, the box starts to slip, and it then slides 2.5 m down the plank in 4.0 s at constant acceleration. What are (a) the coefficient of static friction and (b) the coefficient of kinetic friction between the box and the plank?

79 SSM Block A in Fig. 6-56 has mass $m_A = 4.0$ kg, and block B has mass $m_B = 2.0$ kg. The coefficient of kinetic friction between block B

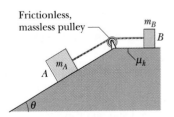

Fig. 6-56 Problem 79.

and the horizontal plane is $\mu_k = 0.50$. The inclined plane is frictionless and at angle $\theta = 30°$. The pulley serves only to change the direction of the cord connecting the blocks. The cord has negligible mass. Find (a) the tension in the cord and (b) the magnitude of the acceleration of the blocks.

80 Calculate the magnitude of the drag force on a missile 53 cm in diameter cruising at 250 m/s at low altitude, where the density of air is 1.2 kg/m³. Assume $C = 0.75$.

81 SSM A bicyclist travels in a circle of radius 25.0 m at a constant speed of 9.00 m/s. The bicycle–rider mass is 85.0 kg. Calculate the magnitudes of (a) the force of friction on the bicycle from the road and (b) the *net* force on the bicycle from the road.

82 In Fig. 6-57, a stuntman drives a car (without negative lift) over the top of a hill, the cross section of which can be approximated by a circle of radius $R = 250$ m. What is the greatest speed at which he can drive without the car leaving the road at the top of the hill?

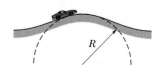

Fig. 6-57 Problem 82.

83 You must push a crate across a floor to a docking bay. The crate weighs 165 N. The coefficient of static friction between crate and floor is 0.510, and the coefficient of kinetic friction is 0.32. Your force on the crate is directed horizontally. (a) What magnitude of your push puts the crate on the verge of sliding? (b) With what magnitude must you then push to keep the crate moving at a constant velocity? (c) If, instead, you then push with the same magnitude as the answer to (a), what is the magnitude of the crate's acceleration?

84 In Fig. 6-58, force $\vec{F}$ is applied to a crate of mass m on a floor where the coefficient of static friction between crate and floor is μ_s. Angle θ is initially 0° but is gradually increased so that the force vector rotates clockwise in the figure. During the rotation, the magnitude F of the force is continuously adjusted so that the crate is always on the verge of sliding. For $\mu_s = 0.70$, (a) plot the ratio F/mg versus θ and (b) determine the angle θ_{inf} at which the ratio approaches an infinite value. (c) Does lubricating the floor increase or decrease θ_{inf}, or is the value unchanged? (d) What is θ_{inf} for $\mu_s = 0.60$?

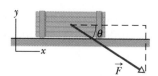

Fig. 6-58 Problem 84.

85 In the early afternoon, a car is parked on a street that runs down a steep hill, at an angle of 35.0° relative to the horizontal. Just then the coefficient of static friction between the tires and the street surface is 0.725. Later, after nightfall, a sleet storm hits the

area, and the coefficient decreases due to both the ice and a chemical change in the road surface because of the temperature decrease. By what percentage must the coefficient decrease if the car is to be in danger of sliding down the street?

86 A sling-thrower puts a stone (0.250 kg) in the sling's pouch (0.010 kg) and then begins to make the stone and pouch move in a vertical circle of radius 0.650 m. The cord between the pouch and the person's hand has negligible mass and will break when the tension in the cord is 33.0 N or more. Suppose the sling-thrower could gradually increase the speed of the stone. (a) Will the breaking occur at the lowest point of the circle or at the highest point? (b) At what speed of the stone will that breaking occur?

87 SSM A car weighing 10.7 kN and traveling at 13.4 m/s without negative lift attempts to round an unbanked curve with a radius of 61.0 m. (a) What magnitude of the frictional force on the tires is required to keep the car on its circular path? (b) If the coefficient of static friction between the tires and the road is 0.350, is the attempt at taking the curve successful?

88 In Fig. 6-59, block 1 of mass $m_1 = 2.0$ kg and block 2 of mass $m_2 = 1.0$ kg are connected by a string of negligible mass. Block 2 is pushed by force $\vec{F}$ of magnitude 20 N and angle $\theta = 35°$. The coefficient of kinetic friction between each block and the horizontal surface is 0.20. What is the tension in the string?

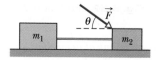

Fig. 6-59 Problem 88.

89 SSM A filing cabinet weighing 556 N rests on the floor. The coefficient of static friction between it and the floor is 0.68, and the coefficient of kinetic friction is 0.56. In four different attempts to move it, it is pushed with horizontal forces of magnitudes (a) 222 N, (b) 334 N, (c) 445 N, and (d) 556 N. For each attempt, calculate the magnitude of the frictional force on it from the floor. (The cabinet is initially at rest.) (e) In which of the attempts does the cabinet move?

90 In Fig. 6-60, a block weighing 22 N is held at rest against a vertical wall by a horizontal force $\vec{F}$ of magnitude 60 N. The coefficient of static friction between the wall and the block is 0.55, and the coefficient of kinetic friction between them is 0.38. In six experiments, a second force $\vec{P}$ is applied to the block and directed parallel to the wall with these magnitudes and directions: (a) 34 N, up, (b) 12 N, up, (c) 48 N, up, (d) 62 N, up, (e) 10 N, down, and (f) 18 N, down. In each experiment, what is the magnitude of the frictional force on the block? In which does the block move (g) up the wall and (h) down the wall? (i) In which is the frictional force directed down the wall?

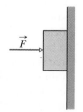

Fig. 6-60 Problem 90.

91 SSM A block slides with constant velocity down an inclined plane that has slope angle θ. The block is then projected up the same plane with an initial speed v_0. (a) How far up the plane will it move before coming to rest? (b) After the block comes to rest, will it slide down the plane again? Give an argument to back your answer.

92 A circular curve of highway is designed for traffic moving at 60 km/h. Assume the traffic consists of cars without negative lift. (a) If the radius of the curve is 150 m, what is the correct angle of banking of the road? (b) If the curve were not banked, what would be the minimum coefficient of friction between tires and road that would keep traffic from skidding out of the turn when traveling at 60 km/h?

93 A 1.5 kg box is initially at rest on a horizontal surface when at $t = 0$ a horizontal force $\vec{F} = (1.8t)\hat{i}$ N (with t in seconds) is applied to the box. The acceleration of the box as a function of time t is given by $\vec{a} = 0$ for $0 \le t \le 2.8$ s and $\vec{a} = (1.2t - 2.4)\hat{i}$ m/s^2 for $t > 2.8$ s. (a) What is the coefficient of static friction between the box and the surface? (b) What is the coefficient of kinetic friction between the box and the surface?

94 A child weighing 140 N sits at rest at the top of a playground slide that makes an angle of 25° with the horizontal. The child keeps from sliding by holding onto the sides of the slide. After letting go of the sides, the child has a constant acceleration of 0.86 m/s^2 (down the slide, of course). (a) What is the coefficient of kinetic friction between the child and the slide? (b) What maximum and minimum values for the coefficient of static friction between the child and the slide are consistent with the information given here?

95 In Fig. 6-61 a fastidious worker pushes directly along the handle of a mop with a force $\vec{F}$. The handle is at an angle θ with the vertical, and μ_s and μ_k are the coefficients of static and kinetic friction between the head of the mop and the floor. Ignore the mass of the handle and assume that all the mop's mass m is

Fig. 6-61 Problem 95.

in its head. (a) If the mop head moves along the floor with a constant velocity, then what is F? (b) Show that if θ is less than a certain value θ_0, then $\vec{F}$ (still directed along the handle) is unable to move the mop head. Find θ_0.

96 A child places a picnic basket on the outer rim of a merry-go-round that has a radius of 4.6 m and revolves once every 30 s. (a) What is the speed of a point on that rim? (b) What is the lowest value of the coefficient of static friction between basket and merry-go-round that allows the basket to stay on the ride?

97 SSM A warehouse worker exerts a constant horizontal force of magnitude 85 N on a 40 kg box that is initially at rest on the horizontal floor of the warehouse. When the box has moved a distance of 1.4 m, its speed is 1.0 m/s. What is the coefficient of kinetic friction between the box and the floor?

98 In Fig. 6-62, a 5.0 kg block is sent sliding up a plane inclined at $\theta = 37°$ while a horizontal force $\vec{F}$ of magnitude 50 N acts on it. The coefficient of kinetic friction between block and plane is 0.30. What are the (a) magnitude and (b) direction (up or down the plane) of the block's acceleration? The block's initial speed is 4.0 m/s. (c) How far up the plane does the block go? (d) When it reaches its highest point, does it remain at rest or slide back down the plane?

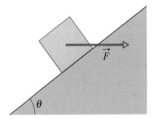

Fig. 6-62 Problem 98.

KINETIC ENERGY AND WORK

7-1 WHAT IS PHYSICS?

One of the fundamental goals of physics is to investigate something that everyone talks about: energy. The topic is obviously important. Indeed, our civilization is based on acquiring and effectively using energy.

For example, everyone knows that any type of motion requires energy: Flying across the Pacific Ocean requires it. Lifting material to the top floor of an office building or to an orbiting space station requires it. Throwing a fastball requires it. We spend a tremendous amount of money to acquire and use energy. Wars have been started because of energy resources. Wars have been ended because of a sudden, overpowering use of energy by one side. Everyone knows many examples of energy and its use, but what does the term *energy* really mean?

7-2 What Is Energy?

The term *energy* is so broad that a clear definition is difficult to write. Technically, energy is a scalar quantity associated with the state (or condition) of one or more objects. However, this definition is too vague to be of help to us now.

A looser definition might at least get us started. Energy is a number that we associate with a system of one or more objects. If a force changes one of the objects by, say, making it move, then the energy number changes. After countless experiments, scientists and engineers realized that if the scheme by which we assign energy numbers is planned carefully, the numbers can be used to predict the outcomes of experiments and, even more important, to build machines, such as flying machines. This success is based on a wonderful property of our universe: Energy can be transformed from one type to another and transferred from one object to another, but the total amount is always the same (energy is *conserved*). No exception to this *principle of energy conservation* has ever been found.

Think of the many types of energy as being numbers representing money in many types of bank accounts. Rules have been made about what such money numbers mean and how they can be changed. You can transfer money numbers from one account to another or from one system to another, perhaps electronically with nothing material actually moving. However, the total amount (the total of all the money numbers) can always be accounted for: It is always conserved.

In this chapter we focus on only one type of energy (*kinetic energy*) and on only one way in which energy can be transferred (*work*). In the next chapter we examine a few other types of energy and how the principle of energy conservation can be written as equations to be solved.

7-3 Kinetic Energy

Kinetic energy K is energy associated with the *state of motion* of an object. The faster the object moves, the greater is its kinetic energy. When the object is stationary, its kinetic energy is zero.

For an object of mass m whose speed v is well below the speed of light,

$$K = \tfrac{1}{2}mv^2 \qquad \text{(kinetic energy).} \qquad (7\text{-}1)$$

For example, a 3.0 kg duck flying past us at 2.0 m/s has a kinetic energy of $6.0 \ \text{kg} \cdot \text{m}^2/\text{s}^2$; that is, we associate that number with the duck's motion.

The SI unit of kinetic energy (and every other type of energy) is the **joule** (J), named for James Prescott Joule, an English scientist of the 1800s. It is defined directly from Eq. 7-1 in terms of the units for mass and velocity:

$$1 \text{ joule} = 1 \text{ J} = 1 \ \text{kg} \cdot \text{m}^2/\text{s}^2. \qquad (7\text{-}2)$$

Thus, the flying duck has a kinetic energy of 6.0 J.

Sample Problem

Kinetic energy, train crash

In 1896 in Waco, Texas, William Crush parked two locomotives at opposite ends of a 6.4-km-long track, fired them up, tied their throttles open, and then allowed them to crash head-on at full speed (Fig. 7-1) in front of 30,000 spectators. Hundreds of people were hurt by flying debris; several were killed. Assuming each locomotive weighed 1.2×10^6 N and its acceleration was a constant $0.26 \ \text{m/s}^2$, what was the total kinetic energy of the two locomotives just before the collision?

KEY IDEAS

(1) We need to find the kinetic energy of each locomotive with Eq. 7-1, but that means we need each locomotive's speed just before the collision and its mass. (2) Because we can assume each locomotive had constant acceleration, we can use the equations in Table 2-1 to find its speed v just before the collision.

Calculations: We choose Eq. 2-16 because we know values for all the variables except v:

$$v^2 = v_0^2 + 2a(x - x_0).$$

With $v_0 = 0$ and $x - x_0 = 3.2 \times 10^3$ m (half the initial separation), this yields

$$v^2 = 0 + 2(0.26 \ \text{m/s}^2)(3.2 \times 10^3 \ \text{m}),$$

or

$$v = 40.8 \ \text{m/s}$$

(about 150 km/h).

Fig. 7-1 The aftermath of an 1896 crash of two locomotives. *(Courtesy Library of Congress)*

We can find the mass of each locomotive by dividing its given weight by g:

$$m = \frac{1.2 \times 10^6 \ \text{N}}{9.8 \ \text{m/s}^2} = 1.22 \times 10^5 \ \text{kg}.$$

Now, using Eq. 7-1, we find the total kinetic energy of the two locomotives just before the collision as

$$K = 2(\tfrac{1}{2}mv^2) = (1.22 \times 10^5 \ \text{kg})(40.8 \ \text{m/s})^2$$
$$= 2.0 \times 10^8 \ \text{J}. \qquad \text{(Answer)}$$

This collision was like an exploding bomb.

WILEY PLUS   Additional examples, video, and practice available at *WileyPLUS*

7-4 Work

If you accelerate an object to a greater speed by applying a force to the object, you increase the kinetic energy $K (= \frac{1}{2}mv^2)$ of the object. Similarly, if you decelerate the object to a lesser speed by applying a force, you decrease the kinetic energy of the object. We account for these changes in kinetic energy by saying that your force has transferred energy *to* the object from yourself or *from* the object to yourself. In such a transfer of energy via a force, **work** W is said to be *done on the object by the force.* More formally, we define work as follows:

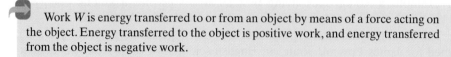

> Work W is energy transferred to or from an object by means of a force acting on the object. Energy transferred to the object is positive work, and energy transferred from the object is negative work.

"Work," then, is transferred energy; "doing work" is the act of transferring the energy. Work has the same units as energy and is a scalar quantity.

The term *transfer* can be misleading. It does not mean that anything material flows into or out of the object; that is, the transfer is not like a flow of water. Rather, it is like the electronic transfer of money between two bank accounts: The number in one account goes up while the number in the other account goes down, with nothing material passing between the two accounts.

Note that we are not concerned here with the common meaning of the word "work," which implies that *any* physical or mental labor is work. For example, if you push hard against a wall, you tire because of the continuously repeated muscle contractions that are required, and you are, in the common sense, working. However, such effort does not cause an energy transfer to or from the wall and thus is not work done on the wall as defined here.

To avoid confusion in this chapter, we shall use the symbol W only for work and shall represent a weight with its equivalent mg.

7-5 Work and Kinetic Energy

Finding an Expression for Work

Let us find an expression for work by considering a bead that can slide along a frictionless wire that is stretched along a horizontal x axis (Fig. 7-2). A constant force $\vec{F}$, directed at an angle ϕ to the wire, accelerates the bead along the wire. We can relate the force and the acceleration with Newton's second law, written for components along the x axis:

$$F_x = ma_x, \tag{7-3}$$

where m is the bead's mass. As the bead moves through a displacement $\vec{d}$, the force changes the bead's velocity from an initial value $\vec{v}_0$ to some other value $\vec{v}$. Because the force is constant, we know that the acceleration is also constant. Thus, we can use Eq. 2-16 to write, for components along the x axis,

$$v^2 = v_0^2 + 2a_x d. \tag{7-4}$$

Solving this equation for a_x, substituting into Eq. 7-3, and rearranging then give us

$$\tfrac{1}{2}mv^2 - \tfrac{1}{2}mv_0^2 = F_x d. \tag{7-5}$$

The first term on the left side of the equation is the kinetic energy K_f of the bead at the end of the displacement d, and the second term is the kinetic energy K_i of the bead at the start of the displacement. Thus, the left side of Eq. 7-5 tells us the kinetic energy has been changed by the force, and the right side tells us the change is equal to $F_x d$. Therefore, the work W done on the bead by the force (the energy transfer due to the force) is

$$W = F_x d. \qquad (7\text{-}6)$$

If we know values for F_x and d, we can use this equation to calculate the work W done on the bead by the force.

> To calculate the work a force does on an object as the object moves through some displacement, we use only the force component along the object's displacement. The force component perpendicular to the displacement does zero work.

From Fig. 7-2, we see that we can write F_x as $F \cos \phi$, where ϕ is the angle between the directions of the displacement $\vec{d}$ and the force $\vec{F}$. Thus,

$$W = Fd \cos \phi \qquad \text{(work done by a constant force)}. \qquad (7\text{-}7)$$

Because the right side of this equation is equivalent to the scalar (dot) product $\vec{F} \cdot \vec{d}$, we can also write

$$W = \vec{F} \cdot \vec{d} \qquad \text{(work done by a constant force)}, \qquad (7\text{-}8)$$

where F is the magnitude of $\vec{F}$. (You may wish to review the discussion of scalar products in Section 3-8.) Equation 7-8 is especially useful for calculating the work when $\vec{F}$ and $\vec{d}$ are given in unit-vector notation.

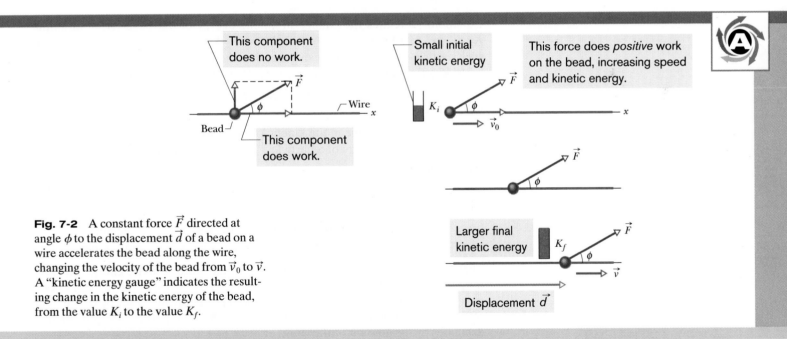

Fig. 7-2 A constant force $\vec{F}$ directed at angle ϕ to the displacement $\vec{d}$ of a bead on a wire accelerates the bead along the wire, changing the velocity of the bead from $\vec{v}_0$ to $\vec{v}$. A "kinetic energy gauge" indicates the resulting change in the kinetic energy of the bead, from the value K_i to the value K_f.

Cautions: There are two restrictions to using Eqs. 7-6 through 7-8 to calculate work done on an object by a force. First, the force must be a *constant force;* that is, it must not change in magnitude or direction as the object moves. (Later, we shall discuss what to do with a *variable force* that changes in magnitude.) Second, the object must be *particle-like*. This means that the object must be *rigid;* all parts of it must move together, in the same direction. In this chapter we consider only particle-like objects, such as the bed and its occupant being pushed in Fig. 7-3.

Signs for work. The work done on an object by a force can be either positive work or negative work. For example, if angle ϕ in Eq. 7-7 is less than 90°, then cos ϕ is positive and thus so is the work. If ϕ is greater than 90° (up to 180°), then cos ϕ is

Fig. 7-3 A contestant in a bed race. We can approximate the bed and its occupant as being a particle for the purpose of calculating the work done on them by the force applied by the student.

negative and thus so is the work. (Can you see that the work is zero when $\phi = 90°$?) These results lead to a simple rule. To find the sign of the work done by a force, consider the force vector component that is parallel to the displacement:

 A force does positive work when it has a vector component in the same direction as the displacement, and it does negative work when it has a vector component in the opposite direction. It does zero work when it has no such vector component.

Units for work. Work has the SI unit of the joule, the same as kinetic energy. However, from Eqs. 7-6 and 7-7 we can see that an equivalent unit is the newton-meter (N·m). The corresponding unit in the British system is the foot-pound (ft·lb). Extending Eq. 7-2, we have

$$1 \text{ J} = 1 \text{ kg} \cdot \text{m}^2/\text{s}^2 = 1 \text{ N} \cdot \text{m} = 0.738 \text{ ft} \cdot \text{lb}. \tag{7-9}$$

Net work done by several forces. When two or more forces act on an object, the **net work** done on the object is the sum of the works done by the individual forces. We can calculate the net work in two ways. (1) We can find the work done by each force and then sum those works. (2) Alternatively, we can first find the net force $\vec{F}_{\text{net}}$ of those forces. Then we can use Eq. 7-7, substituting the magnitude F_{net} for F and also the angle between the directions of $\vec{F}_{\text{net}}$ and $\vec{d}$ for ϕ. Similarly, we can use Eq. 7-8 with $\vec{F}_{\text{net}}$ substituted for $\vec{F}$.

Work–Kinetic Energy Theorem

Equation 7-5 relates the change in kinetic energy of the bead (from an initial $K_i = \frac{1}{2}mv_0^2$ to a later $K_f = \frac{1}{2}mv^2$) to the work W ($= F_x d$) done on the bead. For such particle-like objects, we can generalize that equation. Let ΔK be the change in the kinetic energy of the object, and let W be the net work done on it. Then

$$\Delta K = K_f - K_i = W, \tag{7-10}$$

which says that

$$\begin{pmatrix} \text{change in the kinetic} \\ \text{energy of a particle} \end{pmatrix} = \begin{pmatrix} \text{net work done on} \\ \text{the particle} \end{pmatrix}.$$

We can also write

$$K_f = K_i + W, \tag{7-11}$$

which says that

$$\begin{pmatrix} \text{kinetic energy after} \\ \text{the net work is done} \end{pmatrix} = \begin{pmatrix} \text{kinetic energy} \\ \text{before the net work} \end{pmatrix} + \begin{pmatrix} \text{the net} \\ \text{work done} \end{pmatrix}.$$

These statements are known traditionally as the **work–kinetic energy theorem** for particles. They hold for both positive and negative work: If the net work done on a particle is positive, then the particle's kinetic energy increases by the amount of the work. If the net work done is negative, then the particle's kinetic energy decreases by the amount of the work.

For example, if the kinetic energy of a particle is initially 5 J and there is a net transfer of 2 J to the particle (positive net work), the final kinetic energy is 7 J. If, instead, there is a net transfer of 2 J from the particle (negative net work), the final kinetic energy is 3 J.

✔ **CHECKPOINT 1**

A particle moves along an x axis. Does the kinetic energy of the particle increase, decrease, or remain the same if the particle's velocity changes (a) from -3 m/s to -2 m/s and (b) from -2 m/s to 2 m/s? (c) In each situation, is the work done on the particle positive, negative, or zero?

Sample Problem

Work done by two constant forces, industrial spies

Figure 7-4a shows two industrial spies sliding an initially stationary 225 kg floor safe a displacement $\vec{d}$ of magnitude 8.50 m, straight toward their truck. The push $\vec{F_1}$ of spy 001 is 12.0 N, directed at an angle of 30.0° downward from the horizontal; the pull $\vec{F_2}$ of spy 002 is 10.0 N, directed at 40.0° above the horizontal. The magnitudes and directions of these forces do not change as the safe moves, and the floor and safe make frictionless contact.

(a) What is the net work done on the safe by forces $\vec{F_1}$ and $\vec{F_2}$ during the displacement $\vec{d}$?

KEY IDEAS

(1) The net work W done on the safe by the two forces is the sum of the works they do individually. (2) Because we can treat the safe as a particle and the forces are constant in both magnitude and direction, we can use either Eq. 7-7 ($W = Fd \cos \phi$) or Eq. 7-8 ($W = \vec{F} \cdot \vec{d}$) to calculate those works. Since we know the magnitudes and directions of the forces, we choose Eq. 7-7.

Calculations: From Eq. 7-7 and the free-body diagram for the safe in Fig. 7-4b, the work done by $\vec{F_1}$ is

$$W_1 = F_1 d \cos \phi_1 = (12.0 \text{ N})(8.50 \text{ m})(\cos 30.0°)$$
$$= 88.33 \text{ J},$$

and the work done by $\vec{F_2}$ is

$$W_2 = F_2 d \cos \phi_2 = (10.0 \text{ N})(8.50 \text{ m})(\cos 40.0°)$$
$$= 65.11 \text{ J}.$$

Thus, the net work W is

$$W = W_1 + W_2 = 88.33 \text{ J} + 65.11 \text{ J}$$
$$= 153.4 \text{ J} \approx 153 \text{ J}. \qquad \text{(Answer)}$$

During the 8.50 m displacement, therefore, the spies transfer 153 J of energy to the kinetic energy of the safe.

(b) During the displacement, what is the work W_g done on the safe by the gravitational force $\vec{F_g}$ and what is the work W_N done on the safe by the normal force $\vec{F_N}$ from the floor?

KEY IDEA

Because these forces are constant in both magnitude and direction, we can find the work they do with Eq. 7-7.

Calculations: Thus, with mg as the magnitude of the gravitational force, we write

$$W_g = mgd \cos 90° = mgd(0) = 0 \qquad \text{(Answer)}$$

and $$W_N = F_N d \cos 90° = F_N d(0) = 0. \qquad \text{(Answer)}$$

We should have known this result. Because these forces are perpendicular to the displacement of the safe, they do zero work on the safe and do not transfer any energy to or from it.

(c) The safe is initially stationary. What is its speed v_f at the end of the 8.50 m displacement?

KEY IDEA

The speed of the safe changes because its kinetic energy is changed when energy is transferred to it by $\vec{F_1}$ and $\vec{F_2}$.

Calculations: We relate the speed to the work done by combining Eqs. 7-10 and 7-1:

$$W = K_f - K_i = \tfrac{1}{2}mv_f^2 - \tfrac{1}{2}mv_i^2.$$

The initial speed v_i is zero, and we now know that the work done is 153.4 J. Solving for v_f and then substituting known data, we find that

$$v_f = \sqrt{\frac{2W}{m}} = \sqrt{\frac{2(153.4 \text{ J})}{225 \text{ kg}}}$$
$$= 1.17 \text{ m/s}. \qquad \text{(Answer)}$$

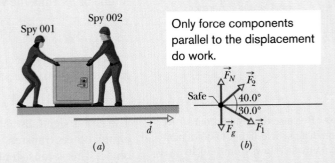

Spy 002

Spy 001

Only force components parallel to the displacement do work.

$\vec{F_N}$ $\vec{F_2}$

Safe 40.0°

30.0°

$\vec{F_g}$ $\vec{F_1}$

$\vec{d}$

(a)

(b)

Fig. 7-4 (a) Two spies move a floor safe through a displacement $\vec{d}$. (b) A free-body diagram for the safe.

 Additional examples, video, and practice available at *WileyPLUS*

Sample Problem

Work done by a constant force in unit-vector notation

During a storm, a crate of crepe is sliding across a slick, oily parking lot through a displacement $\vec{d} = (-3.0\text{ m})\hat{i}$ while a steady wind pushes against the crate with a force $\vec{F} = (2.0\text{ N})\hat{i} + (-6.0\text{ N})\hat{j}$. The situation and coordinate axes are shown in Fig. 7-5.

(a) How much work does this force do on the crate during the displacement?

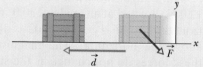

The parallel force component does *negative* work, slowing the crate.

Fig. 7-5 Force $\vec{F}$ slows a crate during displacement $\vec{d}$.

KEY IDEA

Because we can treat the crate as a particle and because the wind force is constant ("steady") in both magnitude and direction during the displacement, we can use either Eq. 7-7 ($W = Fd\cos\phi$) or Eq. 7-8 ($W = \vec{F}\cdot\vec{d}$) to calculate the work. Since we know $\vec{F}$ and $\vec{d}$ in unit-vector notation, we choose Eq. 7-8.

Calculations: We write

$$W = \vec{F}\cdot\vec{d} = [(2.0\text{ N})\hat{i} + (-6.0\text{ N})\hat{j}]\cdot[(-3.0\text{ m})\hat{i}].$$

Of the possible unit-vector dot products, only $\hat{i}\cdot\hat{i}$, $\hat{j}\cdot\hat{j}$, and $\hat{k}\cdot\hat{k}$ are nonzero (see Appendix E). Here we obtain

$$W = (2.0\text{ N})(-3.0\text{ m})\hat{i}\cdot\hat{i} + (-6.0\text{ N})(-3.0\text{ m})\hat{j}\cdot\hat{i}$$
$$= (-6.0\text{ J})(1) + 0 = -6.0\text{ J}. \qquad \text{(Answer)}$$

Thus, the force does a negative 6.0 J of work on the crate, transferring 6.0 J of energy from the kinetic energy of the crate.

(b) If the crate has a kinetic energy of 10 J at the beginning of displacement $\vec{d}$, what is its kinetic energy at the end of $\vec{d}$?

KEY IDEA

Because the force does negative work on the crate, it reduces the crate's kinetic energy.

Calculation: Using the work–kinetic energy theorem in the form of Eq. 7-11, we have

$$K_f = K_i + W = 10\text{ J} + (-6.0\text{ J}) = 4.0\text{ J}. \qquad \text{(Answer)}$$

Less kinetic energy means that the crate has been slowed.

WILEY PLUS   Additional examples, video, and practice available at *WileyPLUS*

7-6 Work Done by the Gravitational Force

We next examine the work done on an object by the gravitational force acting on it. Figure 7-6 shows a particle-like tomato of mass m that is thrown upward with initial speed v_0 and thus with initial kinetic energy $K_i = \frac{1}{2}mv_0^2$. As the tomato rises, it is slowed by a gravitational force $\vec{F}_g$; that is, the tomato's kinetic energy decreases because $\vec{F}_g$ does work on the tomato as it rises. Because we can treat the tomato as a particle, we can use Eq. 7-7 ($W = Fd\cos\phi$) to express the work done during a displacement $\vec{d}$. For the force magnitude F, we use mg as the magnitude of $\vec{F}_g$. Thus, the work W_g done by the gravitational force $\vec{F}_g$ is

$$W_g = mgd\cos\phi \qquad \text{(work done by gravitational force).} \qquad (7\text{-}12)$$

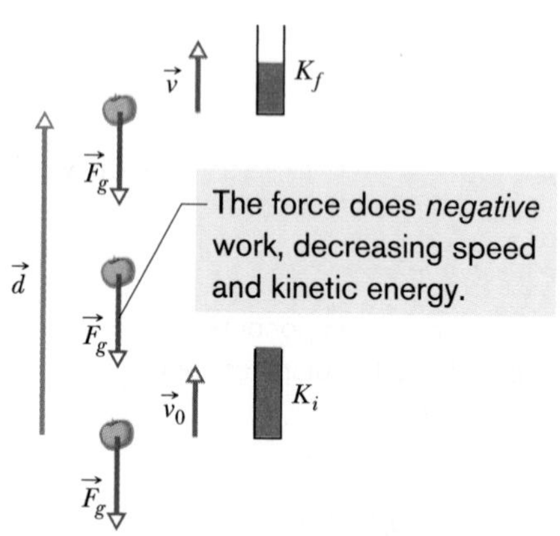

For a rising object, force $\vec{F}_g$ is directed opposite the displacement $\vec{d}$, as indicated in Fig. 7-6. Thus, $\phi = 180°$ and

$$W_g = mgd\cos 180° = mgd(-1) = -mgd. \qquad (7\text{-}13)$$

The minus sign tells us that during the object's rise, the gravitational force acting on the object transfers energy in the amount mgd from the kinetic energy of the object. This is consistent with the slowing of the object as it rises.

After the object has reached its maximum height and is falling back down, the angle ϕ between force $\vec{F}_g$ and displacement $\vec{d}$ is zero. Thus,

$$W_g = mgd\cos 0° = mgd(+1) = +mgd. \qquad (7\text{-}14)$$

The force does *negative* work, decreasing speed and kinetic energy.

Fig. 7-6 Because the gravitational force $\vec{F}_g$ acts on it, a particle-like tomato of mass m thrown upward slows from velocity $\vec{v}_0$ to velocity $\vec{v}$ during displacement $\vec{d}$. A kinetic energy gauge indicates the resulting change in the kinetic energy of the tomato, from $K_i (= \frac{1}{2}mv_0^2)$ to $K_f (= \frac{1}{2}mv^2)$.

The plus sign tells us that the gravitational force now transfers energy in the amount mgd to the kinetic energy of the object. This is consistent with the speeding up of the object as it falls. (Actually, as we shall see in Chapter 8, energy transfers associated with lifting and lowering an object involve the full object–Earth system.)

Work Done in Lifting and Lowering an Object

Now suppose we lift a particle-like object by applying a vertical force $\vec{F}$ to it. During the upward displacement, our applied force does positive work W_a on the object while the gravitational force does negative work W_g on it. Our applied force tends to transfer energy to the object while the gravitational force tends to transfer energy from it. By Eq. 7-10, the change ΔK in the kinetic energy of the object due to these two energy transfers is

$$\Delta K = K_f - K_i = W_a + W_g, \qquad (7\text{-}15)$$

in which K_f is the kinetic energy at the end of the displacement and K_i is that at the start of the displacement. This equation also applies if we lower the object, but then the gravitational force tends to transfer energy *to* the object while our force tends to transfer energy *from* it.

In one common situation, the object is stationary before and after the lift—for example, when you lift a book from the floor to a shelf. Then K_f and K_i are both zero, and Eq. 7-15 reduces to

$$W_a + W_g = 0$$

or

$$W_a = -W_g. \qquad (7\text{-}16)$$

Note that we get the same result if K_f and K_i are not zero but are still equal. Either way, the result means that the work done by the applied force is the negative of the work done by the gravitational force; that is, the applied force transfers the same amount of energy to the object as the gravitational force transfers from the object. Using Eq. 7-12, we can rewrite Eq. 7-16 as

$$W_a = -mgd \cos \phi \qquad \text{(work done in lifting and lowering; } K_f = K_i), \qquad (7\text{-}17)$$

with ϕ being the angle between $\vec{F}_g$ and $\vec{d}$. If the displacement is vertically upward (Fig. 7-7a), then $\phi = 180°$ and the work done by the applied force equals mgd. If the displacement is vertically downward (Fig. 7-7b), then $\phi = 0°$ and the work done by the applied force equals $-mgd$.

Equations 7-16 and 7-17 apply to any situation in which an object is lifted or lowered, with the object stationary before and after the lift. They are independent of the magnitude of the force used. For example, if you lift a mug from the floor to over your head, your force on the mug varies considerably during the lift. Still, because the mug is stationary before and after the lift, the work your force does on the mug is given by Eqs. 7-16 and 7-17, where, in Eq. 7-17, mg is the weight of the mug and d is the distance you lift it.

Fig. 7-7 (a) An applied force $\vec{F}$ lifts an object. The object's displacement $\vec{d}$ makes an angle $\phi = 180°$ with the gravitational force $\vec{F}_g$ on the object. The applied force does positive work on the object. (b) An applied force $\vec{F}$ lowers an object. The displacement $\vec{d}$ of the object makes an angle $\phi = 0°$ with the gravitational force $\vec{F}_g$. The applied force does negative work on the object.

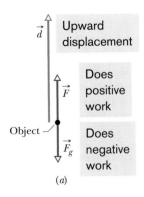

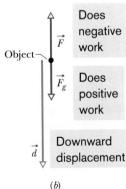

Work done on an accelerating elevator cab

An elevator cab of mass $m = 500$ kg is descending with speed $v_i = 4.0$ m/s when its supporting cable begins to slip, allowing it to fall with constant acceleration $\vec{a} = \vec{g}/5$ (Fig. 7-8a).

(a) During the fall through a distance $d = 12$ m, what is the work W_g done on the cab by the gravitational force $\vec{F}_g$?

KEY IDEA

We can treat the cab as a particle and thus use Eq. 7-12 ($W_g = mgd \cos \phi$) to find the work W_g.

Calculation: From Fig. 7-8b, we see that the angle between the directions of $\vec{F}_g$ and the cab's displacement $\vec{d}$ is 0°. Then, from Eq. 7-12, we find

$$W_g = mgd \cos 0° = (500 \text{ kg})(9.8 \text{ m/s}^2)(12 \text{ m})(1)$$
$$= 5.88 \times 10^4 \text{ J} \approx 59 \text{ kJ}. \qquad \text{(Answer)}$$

(b) During the 12 m fall, what is the work W_T done on the cab by the upward pull $\vec{T}$ of the elevator cable?

KEY IDEAS

(1) We can calculate work W_T with Eq. 7-7 ($W = Fd \cos \phi$) if we first find an expression for the magnitude T of the cable's pull. (2) We can find that expression by writing Newton's second law for components along the y axis in Fig. 7-8b ($F_{net,y} = ma_y$).

Calculations: We get

$$T - F_g = ma. \qquad (7\text{-}18)$$

Solving for T, substituting mg for F_g, and then substituting the result in Eq. 7-7, we obtain

$$W_T = Td \cos \phi = m(a + g)d \cos \phi. \qquad (7\text{-}19)$$

Next, substituting $-g/5$ for the (downward) acceleration a and then 180° for the angle ϕ between the directions of forces $\vec{T}$ and $m\vec{g}$, we find

$$W_T = m\left(-\frac{g}{5} + g\right)d \cos \phi = \frac{4}{5} mgd \cos \phi$$

$$= \frac{4}{5}(500 \text{ kg})(9.8 \text{ m/s}^2)(12 \text{ m}) \cos 180°$$

$$= -4.70 \times 10^4 \text{ J} \approx -47 \text{ kJ}. \qquad \text{(Answer)}$$

Caution: Note that W_T is not simply the negative of W_g. The reason is that, because the cab accelerates during the

fall, its speed changes during the fall, and thus its kinetic energy also changes. Therefore, Eq. 7-16 (which assumes that the initial and final kinetic energies are equal) does *not* apply here.

(c) What is the net work W done on the cab during the fall?

Calculation: The net work is the sum of the works done by the forces acting on the cab:

$$W = W_g + W_T = 5.88 \times 10^4 \text{ J} - 4.70 \times 10^4 \text{ J}$$
$$= 1.18 \times 10^4 \text{ J} \approx 12 \text{ kJ}. \qquad \text{(Answer)}$$

(d) What is the cab's kinetic energy at the end of the 12 m fall?

KEY IDEA

The kinetic energy changes *because* of the net work done on the cab, according to Eq. 7-11 ($K_f = K_i + W$).

Calculation: From Eq. 7-1, we can write the kinetic energy at the start of the fall as $K_i = \frac{1}{2}mv_i^2$. We can then write Eq. 7-11 as

$$K_f = K_i + W = \frac{1}{2}mv_i^2 + W$$
$$= \frac{1}{2}(500 \text{ kg})(4.0 \text{ m/s})^2 + 1.18 \times 10^4 \text{ J}$$
$$= 1.58 \times 10^4 \text{ J} \approx 16 \text{ kJ}. \qquad \text{(Answer)}$$

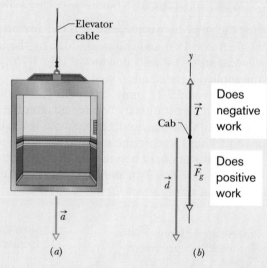

Fig. 7-8 An elevator cab, descending with speed v_i, suddenly begins to accelerate downward. (a) It moves through a displacement $\vec{d}$ with constant acceleration $\vec{a} = \vec{g}/5$. (b) A free-body diagram for the cab, displacement included.

7-7 Work Done by a Spring Force

We next want to examine the work done on a particle-like object by a particular type of *variable force*—namely, a **spring force**, the force from a spring. Many forces in nature have the same mathematical form as the spring force. Thus, by examining this one force, you can gain an understanding of many others.

The Spring Force

Figure 7-9a shows a spring in its **relaxed state**—that is, neither compressed nor extended. One end is fixed, and a particle-like object—a block, say—is attached to the other, free end. If we stretch the spring by pulling the block to the right as in Fig. 7-9b, the spring pulls on the block toward the left. (Because a spring force acts to restore the relaxed state, it is sometimes said to be a *restoring force*.) If we compress the spring by pushing the block to the left as in Fig. 7-9c, the spring now pushes on the block toward the right.

To a good approximation for many springs, the force $\vec{F}_s$ from a spring is proportional to the displacement $\vec{d}$ of the free end from its position when the spring is in the relaxed state. The *spring force* is given by

$$\vec{F}_s = -k\vec{d} \qquad \text{(Hooke's law)}, \qquad (7\text{-}20)$$

which is known as **Hooke's law** after Robert Hooke, an English scientist of the late 1600s. The minus sign in Eq. 7-20 indicates that the direction of the spring force is always opposite the direction of the displacement of the spring's free end. The constant k is called the **spring constant** (or **force constant**) and is a measure of the stiffness of the spring. The larger k is, the stiffer the spring; that is, the larger k is, the stronger the spring's pull or push for a given displacement. The SI unit for k is the newton per meter.

In Fig. 7-9 an x axis has been placed parallel to the length of the spring, with the origin ($x = 0$) at the position of the free end when the spring is in its relaxed state. For this common arrangement, we can write Eq. 7-20 as

$$F_x = -kx \qquad \text{(Hooke's law)}, \qquad (7\text{-}21)$$

where we have changed the subscript. If x is positive (the spring is stretched toward the right on the x axis), then F_x is negative (it is a pull toward the left). If x is negative (the spring is compressed toward the left), then F_x is positive (it is a push toward the right). Note that a spring force is a *variable force* because it is a function of x, the position of the free end. Thus F_x can be symbolized as $F(x)$. Also note that Hooke's law is a *linear* relationship between F_x and x.

The Work Done by a Spring Force

To find the work done by the spring force as the block in Fig. 7-9a moves, let us make two simplifying assumptions about the spring. (1) It is *massless;* that is, its mass is negligible relative to the block's mass. (2) It is an *ideal spring;* that is, it obeys Hooke's law exactly. Let us also assume that the contact between the block and the floor is frictionless and that the block is particle-like.

We give the block a rightward jerk to get it moving and then leave it alone. As the block moves rightward, the spring force F_x does work on the block, decreasing the kinetic energy and slowing the block. However, we *cannot* find this work by using Eq. 7-7 ($W = Fd \cos \phi$) because that equation assumes a constant force. The spring force is a variable force.

To find the work done by the spring, we use calculus. Let the block's initial position be x_i and its later position x_f. Then divide the distance between those two

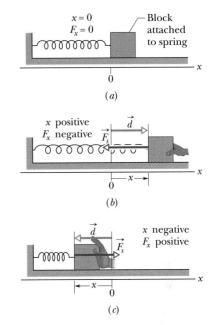

Fig. 7-9 (a) A spring in its relaxed state. The origin of an x axis has been placed at the end of the spring that is attached to a block. (b) The block is displaced by $\vec{d}$, and the spring is stretched by a positive amount x. Note the restoring force $\vec{F}_s$ exerted by the spring. (c) The spring is compressed by a negative amount x. Again, note the restoring force.

positions into many segments, each of tiny length Δx. Label these segments, starting from x_i, as segments 1, 2, and so on. As the block moves through a segment, the spring force hardly varies because the segment is so short that x hardly varies. Thus, we can approximate the force magnitude as being constant within the segment. Label these magnitudes as F_{x1} in segment 1, F_{x2} in segment 2, and so on.

With the force now constant in each segment, we *can* find the work done within each segment by using Eq. 7-7. Here $\phi = 180°$, and so $\cos \phi = -1$. Then the work done is $-F_{x1} \Delta x$ in segment 1, $-F_{x2} \Delta x$ in segment 2, and so on. The net work W_s done by the spring, from x_i to x_f, is the sum of all these works:

$$W_s = \sum -F_{xj} \, \Delta x, \tag{7-22}$$

where j labels the segments. In the limit as Δx goes to zero, Eq. 7-22 becomes

$$W_s = \int_{x_i}^{x_f} -F_x \, dx. \tag{7-23}$$

From Eq. 7-21, the force magnitude F_x is kx. Thus, substitution leads to

$$W_s = \int_{x_i}^{x_f} -kx \, dx = -k \int_{x_i}^{x_f} x \, dx$$

$$= (-\tfrac{1}{2}k)[x^2]_{x_i}^{x_f} = (-\tfrac{1}{2}k)(x_f^2 - x_i^2). \tag{7-24}$$

Multiplied out, this yields

$$W_s = \tfrac{1}{2}kx_i^2 - \tfrac{1}{2}kx_f^2 \qquad \text{(work by a spring force).} \tag{7-25}$$

This work W_s done by the spring force can have a positive or negative value, depending on whether the *net* transfer of energy is to or from the block as the block moves from x_i to x_f. *Caution:* The final position x_f appears in the *second* term on the right side of Eq. 7-25. Therefore, Eq. 7-25 tells us:

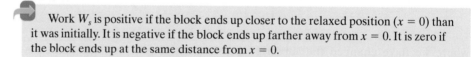

Work W_s is positive if the block ends up closer to the relaxed position ($x = 0$) than it was initially. It is negative if the block ends up farther away from $x = 0$. It is zero if the block ends up at the same distance from $x = 0$.

If $x_i = 0$ and if we call the final position x, then Eq. 7-25 becomes

$$W_s = -\tfrac{1}{2}kx^2 \qquad \text{(work by a spring force).} \tag{7-26}$$

The Work Done by an Applied Force

Now suppose that we displace the block along the x axis while continuing to apply a force $\vec{F}_a$ to it. During the displacement, our applied force does work W_a on the block while the spring force does work W_s. By Eq. 7-10, the change ΔK in the kinetic energy of the block due to these two energy transfers is

$$\Delta K = K_f - K_i = W_a + W_s, \tag{7-27}$$

in which K_f is the kinetic energy at the end of the displacement and K_i is that at the start of the displacement. If the block is stationary before and after the displacement, then K_f and K_i are both zero and Eq. 7-27 reduces to

$$W_a = -W_s. \tag{7-28}$$

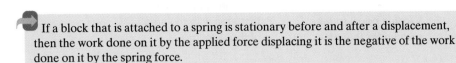
 If a block that is attached to a spring is stationary before and after a displacement, then the work done on it by the applied force displacing it is the negative of the work done on it by the spring force.

Caution: If the block is not stationary before and after the displacement, then this statement is *not* true.

✓ CHECKPOINT 2

For three situations, the initial and final positions, respectively, along the x axis for the block in Fig. 7-9 are (a) -3 cm, 2 cm; (b) 2 cm, 3 cm; and (c) -2 cm, 2 cm. In each situation, is the work done by the spring force on the block positive, negative, or zero?

Sample Problem

Work done by spring to change kinetic energy

In Fig. 7-10, a cumin canister of mass $m = 0.40$ kg slides across a horizontal frictionless counter with speed $v = 0.50$ m/s. It then runs into and compresses a spring of spring constant $k = 750$ N/m. When the canister is momentarily stopped by the spring, by what distance d is the spring compressed?

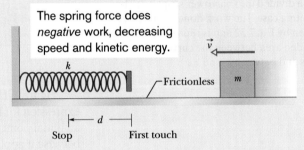

The spring force does *negative* work, decreasing speed and kinetic energy.

Fig. 7-10 A canister of mass m moves at velocity $\vec{v}$ toward a spring that has spring constant k.

KEY IDEAS

1. The work W_s done on the canister by the spring force is related to the requested distance d by Eq. 7-26 ($W_s = -\frac{1}{2}kx^2$), with d replacing x.

2. The work W_s is also related to the kinetic energy of the canister by Eq. 7-10 ($K_f - K_i = W$).

3. The canister's kinetic energy has an initial value of $K = \frac{1}{2}mv^2$ and a value of zero when the canister is momentarily at rest.

Calculations: Putting the first two of these ideas together, we write the work–kinetic energy theorem for the canister as

$$K_f - K_i = -\tfrac{1}{2}kd^2.$$

Substituting according to the third key idea gives us this expression

$$0 - \tfrac{1}{2}mv^2 = -\tfrac{1}{2}kd^2.$$

Simplifying, solving for d, and substituting known data then give us

$$d = v\sqrt{\frac{m}{k}} = (0.50 \text{ m/s})\sqrt{\frac{0.40 \text{ kg}}{750 \text{ N/m}}}$$

$$= 1.2 \times 10^{-2} \text{ m} = 1.2 \text{ cm}. \qquad \text{(Answer)}$$

 Additional examples, video, and practice available at *WileyPLUS*

7-8 Work Done by a General Variable Force

One-Dimensional Analysis

Let us return to the situation of Fig. 7-2 but now consider the force to be in the positive direction of the x axis and the force magnitude to vary with position x. Thus, as the bead (particle) moves, the magnitude $F(x)$ of the force doing work on it changes. Only the magnitude of this variable force changes, not its direction, and the magnitude at any position does not change with time.

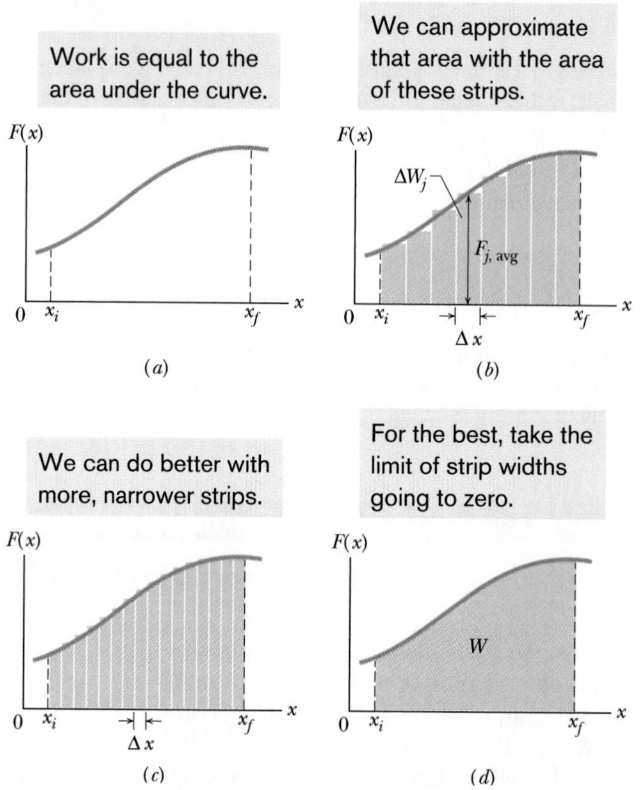

We can approximate
that area with the area
of these strips.

Work is equal to the
area under the curve.

(a)

(b)

Fig. 7-11 (a) A one-dimensional force $\vec{F}(x)$ plotted against the displacement x of a particle on which it acts. The particle moves from x_i to x_f. (b) Same as (a) but with the area under the curve divided into narrow strips. (c) Same as (b) but with the area divided into narrower strips. (d) The limiting case. The work done by the force is given by Eq. 7-32 and is represented by the shaded area between the curve and the x axis and between x_i and x_f.

We can do better with
more, narrower strips.

For the best, take the
limit of strip widths
going to zero.

(c)

(d)

Figure 7-11a shows a plot of such a *one-dimensional variable force.* We want an expression for the work done on the particle by this force as the particle moves from an initial point x_i to a final point x_f. However, we *cannot* use Eq. 7-7 ($W = Fd \cos \phi$) because it applies only for a constant force $\vec{F}$. Here, again, we shall use calculus. We divide the area under the curve of Fig. 7-11a into a number of narrow strips of width Δx (Fig. 7-11b). We choose Δx small enough to permit us to take the force $F(x)$ as being reasonably constant over that interval. We let $F_{j,\text{avg}}$ be the average value of $F(x)$ within the jth interval. Then in Fig. 7-11b, $F_{j,\text{avg}}$ is the height of the jth strip.

With $F_{j,\text{avg}}$ considered constant, the increment (small amount) of work ΔW_j done by the force in the jth interval is now approximately given by Eq. 7-7 and is

$$\Delta W_j = F_{j,\text{avg}} \Delta x. \tag{7-29}$$

In Fig. 7-11b, ΔW_j is then equal to the area of the jth rectangular, shaded strip.

To approximate the total work W done by the force as the particle moves from x_i to x_f, we add the areas of all the strips between x_i and x_f in Fig. 7-11b:

$$W = \sum \Delta W_j = \sum F_{j,\text{avg}} \Delta x. \tag{7-30}$$

Equation 7-30 is an approximation because the broken "skyline" formed by the tops of the rectangular strips in Fig. 7-11b only approximates the actual curve of $F(x)$.

We can make the approximation better by reducing the strip width Δx and using more strips (Fig. 7-11c). In the limit, we let the strip width approach zero; the number of strips then becomes infinitely large and we have, as an exact result,

$$W = \lim_{\Delta x \to 0} \sum F_{j,\text{avg}} \Delta x. \tag{7-31}$$

This limit is exactly what we mean by the integral of the function $F(x)$ between the limits x_i and x_f. Thus, Eq. 7-31 becomes

$$W = \int_{x_i}^{x_f} F(x)\, dx \qquad \text{(work: variable force).} \tag{7-32}$$

If we know the function $F(x)$, we can substitute it into Eq. 7-32, introduce the proper limits of integration, carry out the integration, and thus find the work.

(Appendix E contains a list of common integrals.) Geometrically, the work is equal to the area between the $F(x)$ curve and the x axis, between the limits x_i and x_f (shaded in Fig. 7-11d).

Three-Dimensional Analysis

Consider now a particle that is acted on by a three-dimensional force

$$\vec{F} = F_x\hat{i} + F_y\hat{j} + F_z\hat{k}, \tag{7-33}$$

in which the components F_x, F_y, and F_z can depend on the position of the particle; that is, they can be functions of that position. However, we make three simplifications: F_x may depend on x but not on y or z, F_y may depend on y but not on x or z, and F_z may depend on z but not on x or y. Now let the particle move through an incremental displacement

$$d\vec{r} = dx\hat{i} + dy\hat{j} + dz\hat{k}. \tag{7-34}$$

The increment of work dW done on the particle by $\vec{F}$ during the displacement $d\vec{r}$ is, by Eq. 7-8,

$$dW = \vec{F}\cdot d\vec{r} = F_x\,dx + F_y\,dy + F_z\,dz. \tag{7-35}$$

The work W done by $\vec{F}$ while the particle moves from an initial position r_i having coordinates (x_i, y_i, z_i) to a final position r_f having coordinates (x_f, y_f, z_f) is then

$$W = \int_{r_i}^{r_f} dW = \int_{x_i}^{x_f} F_x\,dx + \int_{y_i}^{y_f} F_y\,dy + \int_{z_i}^{z_f} F_z\,dz. \tag{7-36}$$

If $\vec{F}$ has only an x component, then the y and z terms in Eq. 7-36 are zero and the equation reduces to Eq. 7-32.

Work–Kinetic Energy Theorem with a Variable Force

Equation 7-32 gives the work done by a variable force on a particle in a one-dimensional situation. Let us now make certain that the work is equal to the change in kinetic energy, as the work–kinetic energy theorem states.

Consider a particle of mass m, moving along an x axis and acted on by a net force $F(x)$ that is directed along that axis. The work done on the particle by this force as the particle moves from position x_i to position x_f is given by Eq. 7-32 as

$$W = \int_{x_i}^{x_f} F(x)\,dx = \int_{x_i}^{x_f} ma\,dx, \tag{7-37}$$

in which we use Newton's second law to replace $F(x)$ with ma. We can write the quantity $ma\,dx$ in Eq. 7-37 as

$$ma\,dx = m\frac{dv}{dt}\,dx. \tag{7-38}$$

From the chain rule of calculus, we have

$$\frac{dv}{dt} = \frac{dv}{dx}\frac{dx}{dt} = \frac{dv}{dx}v, \tag{7-39}$$

and Eq. 7-38 becomes

$$ma\,dx = m\frac{dv}{dx}v\,dx = mv\,dv. \tag{7-40}$$

Substituting Eq. 7-40 into Eq. 7-37 yields

$$W = \int_{v_i}^{v_f} mv\,dv = m\int_{v_i}^{v_f} v\,dv$$

$$= \tfrac{1}{2}mv_f^2 - \tfrac{1}{2}mv_i^2. \tag{7-41}$$

Note that when we change the variable from x to v we are required to express the limits on the integral in terms of the new variable. Note also that because the mass m is a constant, we are able to move it outside the integral.

Recognizing the terms on the right side of Eq. 7-41 as kinetic energies allows us to write this equation as

$$W = K_f - K_i = \Delta K,$$

which is the work–kinetic energy theorem.

Sample Problem

Work calculated by graphical integration

In an epidural procedure, as used in childbirth, a surgeon or an anesthetist must run a needle through the skin on the patient's back, through various tissue layers and into a narrow region called the epidural space that lies within the spinal canal surrounding the spinal cord. The needle is intended to deliver an anesthetic fluid. This tricky procedure requires much practice so that the doctor knows when the needle has reached the epidural space and not overshot it, a mistake that could result in serious complications.

The feel a doctor has for the needle's penetration is the variable force that must be applied to advance the needle through the tissues. Figure 7-12a is a graph of the force magnitude F versus displacement x of the needle tip in a typical epidural procedure. (The line segments have been straightened somewhat from the original data.) As x increases from 0, the skin resists the needle, but at $x = 8.0$ mm the force is finally great enough to pierce the skin, and then the required force decreases. Similarly, the needle finally pierces the interspinous ligament at $x = 18$ mm and the relatively tough ligamentum flavum at $x = 30$ mm. The needle then enters the epidural space (where it is to deliver the anesthetic fluid), and the force drops sharply. A new doctor must learn this pattern of force versus displacement to recognize when to stop pushing on the needle. (This is the pattern to be programmed into a virtual-reality simulation of an epidural procedure.) How much work W is done by the force exerted on the needle to get the needle to the epidural space at $x = 30$ mm?

KEY IDEAS

(1) We can calculate the work W done by a variable force $F(x)$ by integrating the force versus position x. Equation 7-32 tells us that

$$W = \int_{x_i}^{x_f} F(x)\, dx.$$

We want the work done by the force during the displacement from $x_i = 0$ to $x_f = 0.030$ m. (2) We can evaluate the integral by finding the area under the curve on the graph of Fig. 7-12a.

$$W = \begin{pmatrix} \text{area between force curve} \\ \text{and } x \text{ axis, from } x_i \text{ to } x_f \end{pmatrix}.$$

Calculations: Because our graph consists of straight-line segments, we can find the area by splitting the region below the curve into rectangular and triangular regions, as shown in Fig. 7-12b. For example, the area in triangular region A is

$$\text{area}_A = \tfrac{1}{2}(0.0080 \text{ m})(12 \text{ N}) = 0.048 \text{ N·m} = 0.048 \text{ J}.$$

Once we've calculated the areas for all the labeled regions in Fig. 7-12b, we find that the total work is

$W = $ (sum of the areas of regions A through K)

$= 0.048 + 0.024 + 0.012 + 0.036 + 0.009 + 0.001$

$\quad + 0.016 + 0.048 + 0.016 + 0.004 + 0.024$

$= 0.238 \text{ J}.$ (Answer)

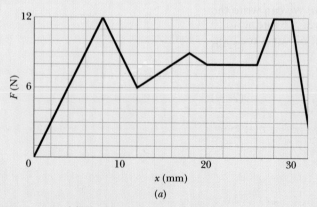

(a)

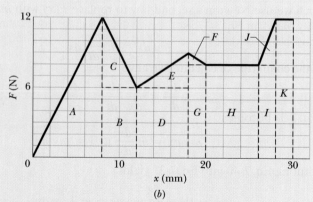

(b)

Fig. 7-12 (a) The force magnitude F versus the displacement x of the needle in an epidural procedure. (b) Breaking up the region between the plotted curve and the displacement axis to calculate the area.

 Additional examples, video, and practice available at *WileyPLUS*

Sample Problem

Work, two-dimensional integration

Force $\vec{F} = (3x^2 \text{ N})\hat{i} + (4 \text{ N})\hat{j}$, with x in meters, acts on a particle, changing only the kinetic energy of the particle. How much work is done on the particle as it moves from coordinates (2 m, 3 m) to (3 m, 0 m)? Does the speed of the particle increase, decrease, or remain the same?

KEY IDEA

The force is a variable force because its x component depends on the value of x. Thus, we cannot use Eqs. 7-7 and 7-8 to find the work done. Instead, we must use Eq. 7-36 to integrate the force.

Calculation: We set up two integrals, one along each axis:

$$W = \int_2^3 3x^2 \, dx + \int_3^0 4 \, dy = 3\int_2^3 x^2 \, dx + 4\int_3^0 dy$$

$$= 3[\tfrac{1}{3}x^3]_2^3 + 4[y]_3^0 = [3^3 - 2^3] + 4[0 - 3]$$

$$= 7.0 \text{ J}. \qquad \text{(Answer)}$$

The positive result means that energy is transferred to the particle by force $\vec{F}$. Thus, the kinetic energy of the particle increases and, because $K = \tfrac{1}{2}mv^2$, its speed must also increase. If the work had come out negative, the kinetic energy and speed would have decreased.

 Additional examples, video, and practice available at *WileyPLUS*

7-9 Power

The time rate at which work is done by a force is said to be the **power** due to the force. If a force does an amount of work W in an amount of time Δt, the **average power** due to the force during that time interval is

$$P_{\text{avg}} = \frac{W}{\Delta t} \qquad \text{(average power)}. \qquad (7\text{-}42)$$

The **instantaneous power** P is the instantaneous time rate of doing work, which we can write as

$$P = \frac{dW}{dt} \qquad \text{(instantaneous power)}. \qquad (7\text{-}43)$$

Suppose we know the work $W(t)$ done by a force as a function of time. Then to get the instantaneous power P at, say, time $t = 3.0$ s during the work, we would first take the time derivative of $W(t)$ and then evaluate the result for $t = 3.0$ s.

The SI unit of power is the joule per second. This unit is used so often that it has a special name, the **watt** (W), after James Watt, who greatly improved the rate at which steam engines could do work. In the British system, the unit of power is the foot-pound per second. Often the horsepower is used. These are related by

$$1 \text{ watt} = 1 \text{ W} = 1 \text{ J/s} = 0.738 \text{ ft} \cdot \text{lb/s} \qquad (7\text{-}44)$$

and

$$1 \text{ horsepower} = 1 \text{ hp} = 550 \text{ ft} \cdot \text{lb/s} = 746 \text{ W}. \qquad (7\text{-}45)$$

Inspection of Eq. 7-42 shows that work can be expressed as power multiplied by time, as in the common unit kilowatt-hour. Thus,

$$1 \text{ kilowatt-hour} = 1 \text{ kW} \cdot \text{h} = (10^3 \text{ W})(3600 \text{ s})$$

$$= 3.60 \times 10^6 \text{ J} = 3.60 \text{ MJ}. \qquad (7\text{-}46)$$

Perhaps because they appear on our utility bills, the watt and the kilowatt-hour have become identified as electrical units. They can be used equally well as units for other examples of power and energy. Thus, if you pick up a book from the floor and put it on a tabletop, you are free to report the work that you have done as, say, $4 \times 10^{-6} \text{ kW} \cdot \text{h}$ (or more conveniently as 4 mW · h).

Fig. 7-13 The power due to the truck's applied force on the trailing load is the rate at which that force does work on the load. *(REGLAIN FREDERIC/Gamma-Presse, Inc.)*

We can also express the rate at which a force does work on a particle (or particle-like object) in terms of that force and the particle's velocity. For a particle that is moving along a straight line (say, an x axis) and is acted on by a constant force $\vec{F}$ directed at some angle ϕ to that line, Eq. 7-43 becomes

$$P = \frac{dW}{dt} = \frac{F \cos \phi \, dx}{dt} = F \cos \phi \left(\frac{dx}{dt}\right),$$

or

$$P = Fv \cos \phi. \qquad (7\text{-}47)$$

Reorganizing the right side of Eq. 7-47 as the dot product $\vec{F} \cdot \vec{v}$, we may also write the equation as

$$P = \vec{F} \cdot \vec{v} \qquad \text{(instantaneous power)}. \qquad (7\text{-}48)$$

For example, the truck in Fig. 7-13 exerts a force $\vec{F}$ on the trailing load, which has velocity $\vec{v}$ at some instant. The instantaneous power due to $\vec{F}$ is the rate at which $\vec{F}$ does work on the load at that instant and is given by Eqs. 7-47 and 7-48. Saying that this power is "the power of the truck" is often acceptable, but keep in mind what is meant: Power is the rate at which the applied *force* does work.

✔️**CHECKPOINT 3**

A block moves with uniform circular motion because a cord tied to the block is anchored at the center of a circle. Is the power due to the force on the block from the cord positive, negative, or zero?

Sample Problem

Power, force, and velocity

Figure 7-14 shows constant forces $\vec{F}_1$ and $\vec{F}_2$ acting on a box as the box slides rightward across a frictionless floor. Force $\vec{F}_1$ is horizontal, with magnitude 2.0 N; force $\vec{F}_2$ is angled upward by 60° to the floor and has magnitude 4.0 N. The speed v of the box at a certain instant is 3.0 m/s. What is the power due to each force acting on the box at that instant, and what is the net power? Is the net power changing at that instant?

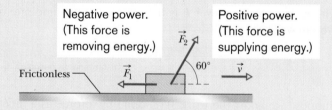

Fig. 7-14 Two forces $\vec{F}_1$ and $\vec{F}_2$ act on a box that slides rightward across a frictionless floor. The velocity of the box is $\vec{v}$.

KEY IDEA

We want an instantaneous power, not an average power over a time period. Also, we know the box's velocity (rather than the work done on it).

Calculation: We use Eq. 7-47 for each force. For force $\vec{F}_1$, at angle $\phi_1 = 180°$ to velocity $\vec{v}$, we have

$$P_1 = F_1 v \cos \phi_1 = (2.0\text{ N})(3.0\text{ m/s}) \cos 180°$$
$$= -6.0 \text{ W.} \qquad \text{(Answer)}$$

This negative result tells us that force $\vec{F}_1$ is transferring energy *from* the box at the rate of 6.0 J/s.

For force $\vec{F}_2$, at angle $\phi_2 = 60°$ to velocity $\vec{v}$, we have

$$P_2 = F_2 v \cos \phi_2 = (4.0\text{ N})(3.0\text{ m/s}) \cos 60°$$
$$= 6.0 \text{ W.} \qquad \text{(Answer)}$$

This positive result tells us that force $\vec{F}_2$ is transferring energy *to* the box at the rate of 6.0 J/s.

The net power is the sum of the individual powers:

$$P_{\text{net}} = P_1 + P_2$$
$$= -6.0 \text{ W} + 6.0 \text{ W} = 0, \qquad \text{(Answer)}$$

which tells us that the net rate of transfer of energy to or from the box is zero. Thus, the kinetic energy ($K = \frac{1}{2}mv^2$) of the box is not changing, and so the speed of the box will remain at 3.0 m/s. With neither the forces $\vec{F}_1$ and $\vec{F}_2$ nor the velocity $\vec{v}$ changing, we see from Eq. 7-48 that P_1 and P_2 are constant and thus so is P_{net}.

 Additional examples, video, and practice available at *WileyPLUS*

REVIEW & SUMMARY

Kinetic Energy The **kinetic energy** K associated with the motion of a particle of mass m and speed v, where v is well below the speed of light, is

$$K = \tfrac{1}{2}mv^2 \qquad \text{(kinetic energy)}. \qquad (7\text{-}1)$$

Work Work W is energy transferred to or from an object via a force acting on the object. Energy transferred to the object is positive work, and from the object, negative work.

Work Done by a Constant Force The work done on a particle by a constant force $\vec{F}$ during displacement $\vec{d}$ is

$$W = Fd \cos \phi = \vec{F} \cdot \vec{d} \qquad \text{(work, constant force)}, \qquad (7\text{-}7, 7\text{-}8)$$

in which ϕ is the constant angle between the directions of $\vec{F}$ and $\vec{d}$. Only the component of $\vec{F}$ that is along the displacement $\vec{d}$ can do work on the object. When two or more forces act on an object, their **net work** is the sum of the individual works done by the forces, which is also equal to the work that would be done on the object by the net force $\vec{F}_{net}$ of those forces.

Work and Kinetic Energy For a particle, a change ΔK in the kinetic energy equals the net work W done on the particle:

$$\Delta K = K_f - K_i = W \quad \text{(work–kinetic energy theorem)}, \qquad (7\text{-}10)$$

in which K_i is the initial kinetic energy of the particle and K_f is the kinetic energy after the work is done. Equation 7-10 rearranged gives us

$$K_f = K_i + W. \qquad (7\text{-}11)$$

Work Done by the Gravitational Force The work W_g done by the gravitational force $\vec{F}_g$ on a particle-like object of mass m as the object moves through a displacement $\vec{d}$ is given by

$$W_g = mgd \cos \phi, \qquad (7\text{-}12)$$

in which ϕ is the angle between $\vec{F}_g$ and $\vec{d}$.

Work Done in Lifting and Lowering an Object The work W_a done by an applied force as a particle-like object is either lifted or lowered is related to the work W_g done by the gravitational force and the change ΔK in the object's kinetic energy by

$$\Delta K = K_f - K_i = W_a + W_g. \qquad (7\text{-}15)$$

If $K_f = K_i$, then Eq. 7-15 reduces to

$$W_a = -W_g, \qquad (7\text{-}16)$$

which tells us that the applied force transfers as much energy to the object as the gravitational force transfers from it.

Spring Force The force $\vec{F}_s$ from a spring is

$$\vec{F}_s = -k\vec{d} \qquad \text{(Hooke's law)}, \qquad (7\text{-}20)$$

where $\vec{d}$ is the displacement of the spring's free end from its position when the spring is in its **relaxed state** (neither compressed nor extended), and k is the **spring constant** (a measure of the spring's stiffness). If an x axis lies along the spring, with the origin at the location of the spring's free end when the spring is in its relaxed state, Eq. 7-20 can be written as

$$F_x = -kx \qquad \text{(Hooke's law)}. \qquad (7\text{-}21)$$

A spring force is thus a variable force: It varies with the displacement of the spring's free end.

Work Done by a Spring Force If an object is attached to the spring's free end, the work W_s done on the object by the spring force when the object is moved from an initial position x_i to a final position x_f is

$$W_s = \tfrac{1}{2}kx_i^2 - \tfrac{1}{2}kx_f^2. \qquad (7\text{-}25)$$

If $x_i = 0$ and $x_f = x$, then Eq. 7-25 becomes

$$W_s = -\tfrac{1}{2}kx^2. \qquad (7\text{-}26)$$

Work Done by a Variable Force When the force $\vec{F}$ on a particle-like object depends on the position of the object, the work done by $\vec{F}$ on the object while the object moves from an initial position r_i with coordinates (x_i, y_i, z_i) to a final position r_f with coordinates (x_f, y_f, z_f) must be found by integrating the force. If we assume that component F_x may depend on x but not on y or z, component F_y may depend on y but not on x or z, and component F_z may depend on z but not on x or y, then the work is

$$W = \int_{x_i}^{x_f} F_x \, dx + \int_{y_i}^{y_f} F_y \, dy + \int_{z_i}^{z_f} F_z \, dz. \qquad (7\text{-}36)$$

If $\vec{F}$ has only an x component, then Eq. 7-36 reduces to

$$W = \int_{x_i}^{x_f} F(x) \, dx. \qquad (7\text{-}32)$$

Power The **power** due to a force is the *rate* at which that force does work on an object. If the force does work W during a time interval Δt, the *average power* due to the force over that time interval is

$$P_{avg} = \frac{W}{\Delta t}. \qquad (7\text{-}42)$$

Instantaneous power is the instantaneous rate of doing work:

$$P = \frac{dW}{dt}. \qquad (7\text{-}43)$$

For a force $\vec{F}$ at an angle ϕ to the direction of travel of the instantaneous velocity $\vec{v}$, the instantaneous power is

$$P = Fv \cos \phi = \vec{F} \cdot \vec{v}. \qquad (7\text{-}47, 7\text{-}48)$$

QUESTIONS

1 Rank the following velocities according to the kinetic energy a particle will have with each velocity, greatest first: (a) $\vec{v} = 4\hat{i} + 3\hat{j}$, (b) $\vec{v} = -4\hat{i} + 3\hat{j}$, (c) $\vec{v} = -3\hat{i} + 4\hat{j}$, (d) $\vec{v} = 3\hat{i} - 4\hat{j}$, (e) $\vec{v} = 5\hat{i}$, and (f) $v = 5$ m/s at $30°$ to the horizontal.

2 Figure 7-15a shows two horizontal forces that act on a block that is sliding to the right across a frictionless floor. Figure 7-15b shows three plots of the block's kinetic energy K versus time t.

Fig. 7-15
Question 2.

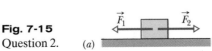
(a)

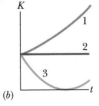
(b)

Which of the plots best corresponds to the following three situations: (a) $F_1 = F_2$, (b) $F_1 > F_2$, (c) $F_1 < F_2$?

3   Is positive or negative work done by a constant force $\vec{F}$ on a particle during a straight-line displacement $\vec{d}$ if (a) the angle between $\vec{F}$ and $\vec{d}$ is 30°; (b) the angle is 100°; (c) $\vec{F} = 2\hat{i} - 3\hat{j}$ and $\vec{d} = -4\hat{i}$?

4   In three situations, a briefly applied horizontal force changes the velocity of a hockey puck that slides over frictionless ice. The overhead views of Fig. 7-16 indicate, for each situation, the puck's initial speed v_i, its final speed v_f, and the directions of the corresponding velocity vectors. Rank the situations according to the work done on the puck by the applied force, most positive first and most negative last.

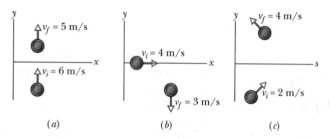

Fig. 7-16   Question 4.

5   Figure 7-17 shows four graphs (drawn to the same scale) of the x component F_x of a variable force (directed along an x axis) versus the position x of a particle on which the force acts. Rank the graphs according to the work done by the force on the particle from $x = 0$ to $x = x_1$, from most positive work first to most negative work last.

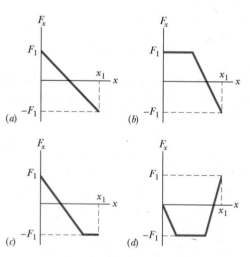

Fig. 7-17   Question 5.

6   Figure 7-18 gives the x component F_x of a force that can act on a particle. If the particle begins at rest at $x = 0$, what is its coordinate when it has (a) its greatest kinetic energy, (b) its greatest speed, and (c) zero speed? (d) What is the particle's direction of travel after it reaches $x = 6$ m?

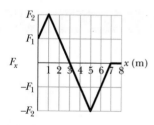

Fig. 7-18   Question 6.

7   In Fig. 7-19, a greased pig has a choice of three frictionless slides along which to slide to the ground. Rank the slides according to how much work the gravitational force does on the pig during the descent, greatest first.

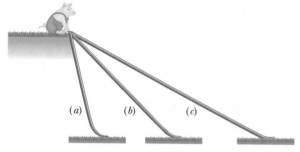

Fig. 7-19   Question 7.

8   Figure 7-20a shows four situations in which a horizontal force acts on the same block, which is initially at rest. The force magnitudes are $F_2 = F_4 = 2F_1 = 2F_3$. The horizontal component v_x of the block's velocity is shown in Fig. 7-20b for the four situations. (a) Which plot in Fig. 7-20b best corresponds to which force in Fig. 7-20a? (b) Which plot in Fig. 7-20c (for kinetic energy K versus time t) best corresponds to which plot in Fig. 7-20b?

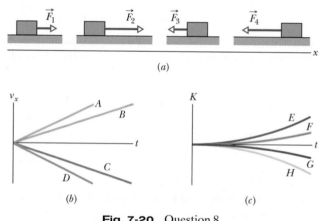

Fig. 7-20   Question 8.

9   Spring A is stiffer than spring B ($k_A > k_B$). The spring force of which spring does more work if the springs are compressed (a) the same distance and (b) by the same applied force?

10   A glob of slime is launched or dropped from the edge of a cliff. Which of the graphs in Fig. 7-21 could possibly show how the kinetic energy of the glob changes during its flight?

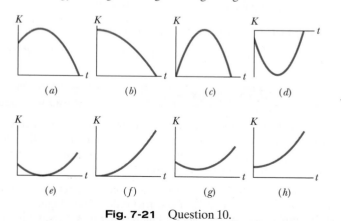

Fig. 7-21   Question 10.

GO	Tutoring problem available (at instructor's discretion) in *WileyPLUS* and WebAssign
SSM	Worked-out solution available in Student Solutions Manual
• – •••	Number of dots indicates level of problem difficulty
✈	Additional information available in *The Flying Circus of Physics* and at flyingcircusofphysics.com

WWW Worked-out solution is at
ILW Interactive solution is at
http://www.wiley.com/college/halliday

sec. 7-3 Kinetic Energy

•1 SSM A proton (mass $m = 1.67 \times 10^{-27}$ kg) is being accelerated along a straight line at 3.6×10^{15} m/s^2 in a machine. If the proton has an initial speed of 2.4×10^7 m/s and travels 3.5 cm, what then is (a) its speed and (b) the increase in its kinetic energy?

•2 If a Saturn V rocket with an Apollo spacecraft attached had a combined mass of 2.9×10^5 kg and reached a speed of 11.2 km/s, how much kinetic energy would it then have?

•3 ✈ On August 10, 1972, a large meteorite skipped across the atmosphere above the western United States and western Canada, much like a stone skipped across water. The accompanying fireball was so bright that it could be seen in the daytime sky and was brighter than the usual meteorite trail. The meteorite's mass was about 4×10^6 kg; its speed was about 15 km/s. Had it entered the atmosphere vertically, it would have hit Earth's surface with about the same speed. (a) Calculate the meteorite's loss of kinetic energy (in joules) that would have been associated with the vertical impact. (b) Express the energy as a multiple of the explosive energy of 1 megaton of TNT, which is 4.2×10^{15} J. (c) The energy associated with the atomic bomb explosion over Hiroshima was equivalent to 13 kilotons of TNT. To how many Hiroshima bombs would the meteorite impact have been equivalent?

••4 A bead with mass 1.8×10^{-2} kg is moving along a wire in the positive direction of an x axis. Beginning at time $t = 0$, when the bead passes through $x = 0$ with speed 12 m/s, a constant force acts on the bead. Figure 7-22 indicates the bead's position at these four times: $t_0 = 0$, $t_1 = 1.0$ s, $t_2 = 2.0$ s, and $t_3 = 3.0$ s. The bead momentarily stops at $t = 3.0$ s. What is the kinetic energy of the bead at $t = 10$ s?

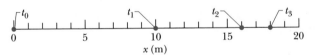

Fig. 7-22 Problem 4.

••5 A father racing his son has half the kinetic energy of the son, who has half the mass of the father. The father speeds up by 1.0 m/s and then has the same kinetic energy as the son. What are the original speeds of (a) the father and (b) the son?

••6 A force $\vec{F}_a$ is applied to a bead as the bead is moved along a straight wire through displacement +5.0 cm. The magnitude of $\vec{F}_a$ is set at a certain value, but the angle ϕ between $\vec{F}_a$ and the bead's displacement can be chosen. Figure 7-23 gives the work W done by $\vec{F}_a$ on the bead for a range of ϕ values; $W_0 = 25$ J. How much work is done by $\vec{F}_a$ if ϕ is (a) 64° and (b) 147°?

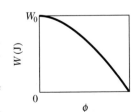

Fig. 7-23 Problem 6.

sec. 7-5 Work and Kinetic Energy

•7 A 3.0 kg body is at rest on a frictionless horizontal air track when a constant horizontal force $\vec{F}$ acting in the positive direction of an x axis along the track is applied to the body. A stroboscopic graph of the position of the body as it slides to the right is shown in Fig. 7-24. The force $\vec{F}$ is applied to the body at $t = 0$, and the graph records the position of the body at 0.50 s intervals. How much work is done on the body by the applied force $\vec{F}$ between $t = 0$ and $t = 2.0$ s?

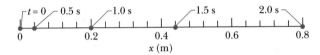

Fig. 7-24 Problem 7.

•8 A ice block floating in a river is pushed through a displacement $\vec{d} = (15$ m$)\hat{i} - (12$ m$)\hat{j}$ along a straight embankment by rushing water, which exerts a force $\vec{F} = (210$ N$)\hat{i} - (150$ N$)\hat{j}$ on the block. How much work does the force do on the block during the displacement?

•9 The only force acting on a 2.0 kg canister that is moving in an xy plane has a magnitude of 5.0 N. The canister initially has a velocity of 4.0 m/s in the positive x direction and some time later has a velocity of 6.0 m/s in the positive y direction. How much work is done on the canister by the 5.0 N force during this time?

•10 A coin slides over a frictionless plane and across an xy coordinate system from the origin to a point with xy coordinates (3.0 m, 4.0 m) while a constant force acts on it. The force has magnitude 2.0 N and is directed at a counterclockwise angle of 100° from the positive direction of the x axis. How much work is done by the force on the coin during the displacement?

••11 A 12.0 N force with a fixed orientation does work on a particle as the particle moves through the three-dimensional displacement $\vec{d} = (2.00\hat{i} - 4.00\hat{j} + 3.00\hat{k})$ m. What is the angle between the force and the displacement if the change in the particle's kinetic energy is (a) +30.0 J and (b) −30.0 J?

••12 A can of bolts and nuts is pushed 2.00 m along an x axis by a broom along the greasy (frictionless) floor of a car repair shop in a version of shuffleboard. Figure 7-25 gives the work W done on the

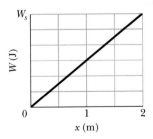

Fig. 7-25 Problem 12.

can by the constant horizontal force from the broom, versus the can's position x. The scale of the figure's vertical axis is set by $W_s = 6.0$ J. (a) What is the magnitude of that force? (b) If the can had an initial kinetic energy of 3.00 J, moving in the positive direction of the x axis, what is its kinetic energy at the end of the 2.00 m?

••13 A luge and its rider, with a total mass of 85 kg, emerge from a downhill track onto a horizontal straight track with an initial speed of 37 m/s. If a force slows them to a stop at a constant rate of 2.0 m/s², (a) what magnitude F is required for the force, (b) what distance d do they travel while slowing, and (c) what work W is done on them by the force? What are (d) F, (e) d, and (f) W if they, instead, slow at 4.0 m/s²?

••14 **GO** Figure 7-26 shows an overhead view of three horizontal forces acting on a cargo canister that was initially stationary but now moves across a frictionless floor. The force magnitudes are $F_1 = 3.00$ N, $F_2 = 4.00$ N, and $F_3 = 10.0$ N, and the indicated angles are $\theta_2 = 50.0°$ and $\theta_3 = 35.0°$. What is the net work done on the canister by the three forces during the first 4.00 m of displacement?

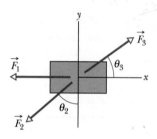

Fig. 7-26 Problem 14.

••15 **GO** Figure 7-27 shows three forces applied to a trunk that moves leftward by 3.00 m over a frictionless floor. The force magnitudes are $F_1 = 5.00$ N, $F_2 = 9.00$ N, and $F_3 = 3.00$ N, and the indicated angle is $\theta = 60.0°$. During the displacement, (a) what is the net work done on the trunk by the three forces and (b) does the kinetic energy of the trunk increase or decrease?

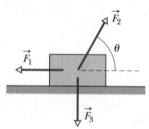

Fig. 7-27 Problem 15.

••16 **GO** An 8.0 kg object is moving in the positive direction of an x axis. When it passes through $x = 0$, a constant force directed along the axis begins to act on it. Figure 7-28 gives its kinetic energy K versus position x as it moves from $x = 0$ to $x = 5.0$ m; $K_0 = 30.0$ J. The force continues to act. What is v when the object moves back through $x = -3.0$ m?

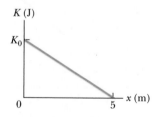

Fig. 7-28 Problem 16.

sec. 7-6 Work Done by the Gravitational Force

•17 **SSM** **WWW** A helicopter lifts a 72 kg astronaut 15 m vertically from the ocean by means of a cable. The acceleration of the astronaut is $g/10$. How much work is done on the astronaut by (a) the force from the helicopter and (b) the gravitational force on her? Just before she reaches the helicopter, what are her (c) kinetic energy and (d) speed?

•18 (a) In 1975 the roof of Montreal's Velodrome, with a weight of 360 kN, was lifted by 10 cm so that it could be centered. How much work was done on the roof by the forces making the lift? (b) In 1960 a Tampa, Florida, mother reportedly raised one end of a car that had fallen onto her son when a jack failed. If her panic lift effectively raised 4000 N (about $\frac{1}{4}$ of the car's weight) by 5.0 cm, how much work did her force do on the car?

••19 **GO** In Fig. 7-29, a block of ice slides down a frictionless ramp at angle $\theta = 50°$ while an ice worker pulls on the block (via a rope) with a force $\vec{F}_r$ that has a magnitude of 50 N and is directed up the ramp. As the block slides through distance $d = 0.50$ m along the ramp, its kinetic energy increases by 80 J. How much greater would its kinetic energy have been if the rope had not been attached to the block?

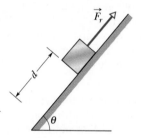

Fig. 7-29 Problem 19.

••20 A block is sent up a frictionless ramp along which an x axis extends upward. Figure 7-30 gives the kinetic energy of the block as a function of position x; the scale of the figure's vertical axis is set by $K_s = 40.0$ J. If the block's initial speed is 4.00 m/s, what is the normal force on the block?

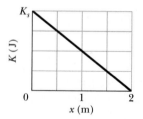

Fig. 7-30 Problem 20.

••21 **SSM** A cord is used to vertically lower an initially stationary block of mass M at a constant downward acceleration of $g/4$. When the block has fallen a distance d, find (a) the work done by the cord's force on the block, (b) the work done by the gravitational force on the block, (c) the kinetic energy of the block, and (d) the speed of the block.

••22 A cave rescue team lifts an injured spelunker directly upward and out of a sinkhole by means of a motor-driven cable. The lift is performed in three stages, each requiring a vertical distance of 10.0 m: (a) the initially stationary spelunker is accelerated to a speed of 5.00 m/s; (b) he is then lifted at the constant speed of 5.00 m/s; (c) finally he is decelerated to zero speed. How much work is done on the 80.0 kg rescuee by the force lifting him during each stage?

••23 In Fig. 7-31, a constant force $\vec{F}_a$ of magnitude 82.0 N is applied to a 3.00 kg shoe box at angle $\phi = 53.0°$, causing the box to move up a frictionless ramp at constant speed. How much work is done on the box by $\vec{F}_a$ when the box has moved through vertical distance $h = 0.150$ m?

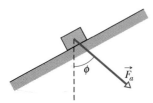

Fig. 7-31 Problem 23.

••24 GO In Fig. 7-32, a horizontal force $\vec{F}_a$ of magnitude 20.0 N is applied to a 3.00 kg psychology book as the book slides a distance $d = 0.500$ m up a frictionless ramp at angle $\theta = 30.0°$. (a) During the displacement, what is the net work done on the book by $\vec{F}_a$, the gravitational force on the book, and the normal force on the book? (b) If the book has zero kinetic energy at the start of the displacement, what is its speed at the end of the displacement?

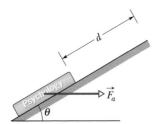

Fig. 7-32 Problem 24.

•••25 GO In Fig. 7-33, a 0.250 kg block of cheese lies on the floor of a 900 kg elevator cab that is being pulled upward by a cable through distance $d_1 = 2.40$ m and then through distance $d_2 = 10.5$ m. (a) Through d_1, if the normal force on the block from the floor has constant magnitude $F_N = 3.00$ N, how much work is done on the cab by the force from the cable? (b) Through d_2, if the work done on the cab by the (constant) force from the cable is 92.61 kJ, what is the magnitude of F_N?

Fig. 7-33
Problem 25.

sec. 7-7 Work Done by a Spring Force

•26 In Fig. 7-9, we must apply a force of magnitude 80 N to hold the block stationary at $x = -2.0$ cm. From that position, we then slowly move the block so that our force does +4.0 J of work on the spring–block system; the block is then again stationary. What is the block's position? (*Hint:* There are two answers.)

•27 A spring and block are in the arrangement of Fig. 7-9. When the block is pulled out to $x = +4.0$ cm, we must apply a force of magnitude 360 N to hold it there. We pull the block to $x = 11$ cm and then release

it. How much work does the spring do on the block as the block moves from $x_i = +5.0$ cm to (a) $x = +3.0$ cm, (b) $x = -3.0$ cm, (c) $x = -5.0$ cm, and (d) $x = -9.0$ cm?

•28 During spring semester at MIT, residents of the parallel buildings of the East Campus dorms battle one another with large catapults that are made with surgical hose mounted on a window frame. A balloon filled with dyed water is placed in a pouch attached to the hose, which is then stretched through the width of the room. Assume that the stretching of the hose obeys Hooke's law with a spring constant of 100 N/m. If the hose is stretched by 5.00 m and then released, how much work does the force from the hose do on the balloon in the pouch by the time the hose reaches its relaxed length?

••29 In the arrangement of Fig. 7-9, we gradually pull the block from $x = 0$ to $x = +3.0$ cm, where it is stationary. Figure 7-34 gives the work that our force does on the block. The scale of the figure's vertical axis is set by $W_s = 1.0$ J. We then pull the block out to $x = +5.0$ cm and release it from rest. How much work does the spring do on the block when the block moves from $x_i = +5.0$ cm to (a) $x = +4.0$ cm, (b) $x = -2.0$ cm, and (c) $x = -5.0$ cm?

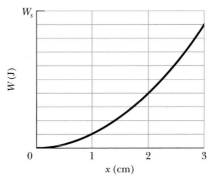

Fig. 7-34 Problem 29.

••30 In Fig. 7-9a, a block of mass m lies on a horizontal frictionless surface and is attached to one end of a horizontal spring (spring constant k) whose other end is fixed. The block is initially at rest at the position where the spring is unstretched ($x = 0$) when a constant horizontal force $\vec{F}$ in the positive direction of the x axis is applied to it. A plot of the resulting kinetic energy of the block versus its position x is shown in Fig. 7-35. The scale of the figure's vertical axis is set by $K_s = 4.0$ J. (a) What is the magnitude of $\vec{F}$? (b) What is the value of k?

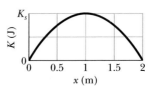

Fig. 7-35 Problem 30.

••31 SSM WWW The only force acting on a 2.0 kg body as it moves along a positive x axis has an x component $F_x = -6x$ N, with x in meters. The velocity at $x = 3.0$ m is 8.0 m/s. (a) What is the velocity of the body at $x = 4.0$ m? (b) At what positive value of x will the body have a velocity of 5.0 m/s?

••32 Figure 7-36 gives spring force F_x versus position x for the spring–block arrangement of Fig. 7-9. The scale is set by $F_s = 160.0$ N. We release the block at $x = 12$ cm. How much work does the spring do on the block when the block moves from $x_i = +8.0$ cm to (a) $x = +5.0$ cm, (b) $x = -5.0$ cm, (c) $x = -8.0$ cm, and (d) $x = -10.0$ cm?

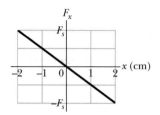

Fig. 7-36 Problem 32.

•••33 The block in Fig. 7-9a lies on a horizontal frictionless surface, and the spring constant is 50 N/m. Initially, the spring is at its relaxed length and the block is stationary at position $x = 0$. Then an applied force with a constant magnitude of 3.0 N pulls the block in the positive direction of the x axis, stretching the spring until the block stops. When that stopping point is reached, what are (a) the position of the block, (b) the work that has been done on the block by the applied force, and (c) the work that has been done on the block by the spring force? During the block's displacement, what are (d) the block's position when its kinetic energy is maximum and (e) the value of that maximum kinetic energy?

sec. 7-8   Work Done by a General Variable Force

•34 ILW A 10 kg brick moves along an x axis. Its acceleration as a function of its position is shown in Fig. 7-37. The scale of the figure's vertical axis is set by $a_s = 20.0$ m/s². What is the net work performed on the brick by the force causing the acceleration as the brick moves from $x = 0$ to $x = 8.0$ m?

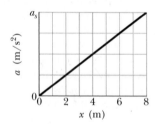

Fig. 7-37 Problem 34.

•35 SSM WWW The force on a particle is directed along an x axis and given by $F = F_0(x/x_0 - 1)$. Find the work done by the force in moving the particle from $x = 0$ to $x = 2x_0$ by (a) plotting $F(x)$ and measuring the work from the graph and (b) integrating $F(x)$.

•36 A 5.0 kg block moves in a straight line on a horizontal frictionless surface under the influence of a force that varies with position as shown in Fig. 7-38.

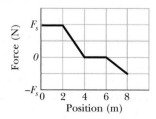

Fig. 7-38 Problem 36.

The scale of the figure's vertical axis is set by $F_s = 10.0$ N. How much work is done by the force as the block moves from the origin to $x = 8.0$ m?

••37 Figure 7-39 gives the acceleration of a 2.00 kg particle as an applied force $\vec{F}_a$ moves it from rest along an x axis from $x = 0$ to $x = 9.0$ m. The scale of the figure's vertical axis is set by $a_s = 6.0$ m/s². How much work has the force done on the particle when the particle reaches (a) $x = 4.0$ m, (b) $x = 7.0$ m, and (c) $x = 9.0$ m? What is the particle's speed and direction of travel when it reaches (d) $x = 4.0$ m, (e) $x = 7.0$ m, and (f) $x = 9.0$ m?

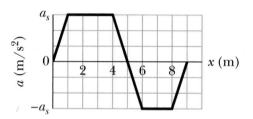

Fig. 7-39 Problem 37.

••38 A 1.5 kg block is initially at rest on a horizontal frictionless surface when a horizontal force along an x axis is applied to the block. The force is given by $\vec{F}(x) = (2.5 - x^2)\hat{i}$ N, where x is in meters and the initial position of the block is $x = 0$. (a) What is the kinetic energy of the block as it passes through $x = 2.0$ m? (b) What is the maximum kinetic energy of the block between $x = 0$ and $x = 2.0$ m?

••39 GO A force $\vec{F} = (cx - 3.00x^2)\hat{i}$ acts on a particle as the particle moves along an x axis, with $\vec{F}$ in newtons, x in meters, and c a constant. At $x = 0$, the particle's kinetic energy is 20.0 J; at $x = 3.00$ m, it is 11.0 J. Find c.

••40 A can of sardines is made to move along an x axis from $x = 0.25$ m to $x = 1.25$ m by a force with a magnitude given by $F = \exp(-4x^2)$, with x in meters and F in newtons. (Here exp is the exponential function.) How much work is done on the can by the force?

••41 A single force acts on a 3.0 kg particle-like object whose position is given by $x = 3.0t - 4.0t^2 + 1.0t^3$, with x in meters and t in seconds. Find the work done on the object by the force from $t = 0$ to $t = 4.0$ s.

•••42 Figure 7-40 shows a cord attached to a cart that can slide along a frictionless horizontal rail aligned along an x axis. The left end of the cord is pulled over a pulley, of negligible mass and friction and at cord height $h = 1.20$ m, so the cart slides from $x_1 = 3.00$ m to $x_2 = 1.00$ m. During the move, the tension in the cord is a constant 25.0 N. What is the change in the kinetic energy of the cart during the move?

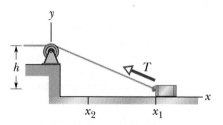

Fig. 7-40 Problem 42.

sec. 7-9   **Power**

•**43** SSM A force of 5.0 N acts on a 15 kg body initially at rest. Compute the work done by the force in (a) the first, (b) the second, and (c) the third seconds and (d) the instantaneous power due to the force at the end of the third second.

•**44** A skier is pulled by a towrope up a frictionless ski slope that makes an angle of 12° with the horizontal. The rope moves parallel to the slope with a constant speed of 1.0 m/s. The force of the rope does 900 J of work on the skier as the skier moves a distance of 8.0 m up the incline. (a) If the rope moved with a constant speed of 2.0 m/s, how much work would the force of the rope do on the skier as the skier moved a distance of 8.0 m up the incline? At what rate is the force of the rope doing work on the skier when the rope moves with a speed of (b) 1.0 m/s and (c) 2.0 m/s?

•**45** SSM ILW A 100 kg block is pulled at a constant speed of 5.0 m/s across a horizontal floor by an applied force of 122 N directed 37° above the horizontal. What is the rate at which the force does work on the block?

•**46** The loaded cab of an elevator has a mass of 3.0×10^3 kg and moves 210 m up the shaft in 23 s at constant speed. At what average rate does the force from the cable do work on the cab?

••**47** A machine carries a 4.0 kg package from an initial position of $\vec{d}_i = (0.50\ \text{m})\hat{i} + (0.75\ \text{m})\hat{j} + (0.20\ \text{m})\hat{k}$ at $t = 0$ to a final position of $\vec{d}_f = (7.50\ \text{m})\hat{i} + (12.0\ \text{m})\hat{j} + (7.20\ \text{m})\hat{k}$ at $t = 12$ s. The constant force applied by the machine on the package is $\vec{F} = (2.00\ \text{N})\hat{i} + (4.00\ \text{N})\hat{j} + (6.00\ \text{N})\hat{k}$. For that displacement, find (a) the work done on the package by the machine's force and (b) the average power of the machine's force on the package.

••**48** A 0.30 kg ladle sliding on a horizontal frictionless surface is attached to one end of a horizontal spring ($k = 500$ N/m) whose other end is fixed. The ladle has a kinetic energy of 10 J as it passes through its equilibrium position (the point at which the spring force is zero). (a) At what rate is the spring doing work on the ladle as the ladle passes through its equilibrium position? (b) At what rate is the spring doing work on the ladle when the spring is compressed 0.10 m and the ladle is moving away from the equilibrium position?

••**49** SSM A fully loaded, slow-moving freight elevator has a cab with a total mass of 1200 kg, which is required to travel upward 54 m in 3.0 min, starting and ending at rest. The elevator's counterweight has a mass of only 950 kg, and so the elevator motor must help. What average power is required of the force the motor exerts on the cab via the cable?

••**50** (a) At a certain instant, a particle-like object is acted on by a force $\vec{F} = (4.0\ \text{N})\hat{i} - (2.0\ \text{N})\hat{j} + (9.0\ \text{N})\hat{k}$ while the object's velocity is $\vec{v} = -(2.0\ \text{m/s})\hat{i} + (4.0\ \text{m/s})\hat{k}$. What is the instantaneous rate at which the force does work on the object? (b) At some other time, the velocity consists of only a y component. If the force is unchanged and the instantaneous power is -12 W, what is the velocity of the object?

••**51** A force $\vec{F} = (3.00\ \text{N})\hat{i} + (7.00\ \text{N})\hat{j} + (7.00\ \text{N})\hat{k}$ acts on a 2.00 kg mobile object that moves from an initial position of $\vec{d}_i = (3.00\ \text{m})\hat{i} - (2.00\ \text{m})\hat{j} + (5.00\ \text{m})\hat{k}$ to a final position of $\vec{d}_f = -(5.00\ \text{m})\hat{i} + (4.00\ \text{m})\hat{j} + (7.00\ \text{m})\hat{k}$ in 4.00 s. Find (a) the work done on the object by the force in the 4.00 s interval, (b) the average power due to the force during that interval, and (c) the angle between vectors $\vec{d}_i$ and $\vec{d}_f$.

•••**52** A funny car accelerates from rest through a measured track distance in time T with the engine operating at a constant power P. If the track crew can increase the engine power by a differential amount dP, what is the change in the time required for the run?

Additional Problems

53 Figure 7-41 shows a cold package of hot dogs sliding rightward across a frictionless floor through a distance $d = 20.0$ cm while three forces act on the package. Two of them are horizontal and have the magnitudes $F_1 = 5.00$ N and $F_2 = 1.00$ N; the third is angled down by $\theta = 60.0°$ and has the magnitude $F_3 = 4.00$ N. (a) For the 20.0 cm displacement, what is the *net* work done on the package by the three applied forces, the gravitational force on the package, and the normal force on the package? (b) If the package has a mass of 2.0 kg and an initial kinetic energy of 0, what is its speed at the end of the displacement?

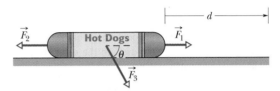

Fig. 7-41   Problem 53.

54 The only force acting on a 2.0 kg body as the body moves along an x axis varies as shown in Fig. 7-42. The scale of the figure's vertical axis is set by $F_s = 4.0$ N. The velocity of the body at $x = 0$ is 4.0 m/s. (a) What is the kinetic energy of the body at $x = 3.0$ m? (b) At what value of x will the body have a kinetic energy of 8.0 J? (c) What is the maximum kinetic energy of the body between $x = 0$ and $x = 5.0$ m?

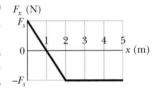

Fig. 7-42   Problem 54.

55 SSM A horse pulls a cart with a force of 40 lb at an angle of 30° above the horizontal and moves along at a speed of 6.0 mi/h. (a) How much work does the force do in 10 min? (b) What is the average power (in horsepower) of the force?

56 An initially stationary 2.0 kg object accelerates horizontally and uniformly to a speed of 10 m/s in 3.0 s. (a) In that 3.0 s interval, how much work is done on the object by the force accelerating it? What is the instantaneous power due to that force (b) at the end of the interval and (c) at the end of the first half of the interval?

57 A 230 kg crate hangs from the end of a rope of length $L = 12.0$ m. You push horizontally on the crate with a varying force $\vec{F}$ to move it distance $d = 4.00$ m to the side (Fig. 7-43). (a) What is the magnitude of $\vec{F}$ when the crate is in this final position? During the crate's displacement, what are (b) the total work done on it, (c) the work done by the gravitational force on the crate, and (d) the work done by the pull on the crate from the rope? (e) Knowing that the crate is motionless before and after its displacement, use the answers to (b), (c), and (d) to find the work your

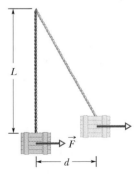

Fig. 7-43   Problem 57.

force $\vec{F}$ does on the crate. (f) Why is the work of your force not equal to the product of the horizontal displacement and the answer to (a)?

58 To pull a 50 kg crate across a horizontal frictionless floor, a worker applies a force of 210 N, directed 20° above the horizontal. As the crate moves 3.0 m, what work is done on the crate by (a) the worker's force, (b) the gravitational force on the crate, and (c) the normal force on the crate from the floor? (d) What is the total work done on the crate?

59 An explosion at ground level leaves a crater with a diameter that is proportional to the energy of the explosion raised to the $\frac{1}{3}$ power; an explosion of 1 megaton of TNT leaves a crater with a 1 km diameter. Below Lake Huron in Michigan there appears to be an ancient impact crater with a 50 km diameter. What was the kinetic energy associated with that impact, in terms of (a) megatons of TNT (1 megaton yields 4.2×10^{15} J) and (b) Hiroshima bomb equivalents (13 kilotons of TNT each)? (Ancient meteorite or comet impacts may have significantly altered Earth's climate and contributed to the extinction of the dinosaurs and other life-forms.)

60 A frightened child is restrained by her mother as the child slides down a frictionless playground slide. If the force on the child from the mother is 100 N up the slide, the child's kinetic energy increases by 30 J as she moves down the slide a distance of 1.8 m. (a) How much work is done on the child by the gravitational force during the 1.8 m descent? (b) If the child is not restrained by her mother, how much will the child's kinetic energy increase as she comes down the slide that same distance of 1.8 m?

61 How much work is done by a force $\vec{F} = (2x \text{ N})\hat{i} + (3 \text{ N})\hat{j}$, with x in meters, that moves a particle from a position $\vec{r}_i = (2 \text{ m})\hat{i} + (3 \text{ m})\hat{j}$ to a position $\vec{r}_f = -(4 \text{ m})\hat{i} - (3 \text{ m})\hat{j}$?

62 A 250 g block is dropped onto a relaxed vertical spring that has a spring constant of $k = 2.5$ N/cm (Fig. 7-44). The block becomes attached to the spring and compresses the spring 12 cm before momentarily stopping. While the spring is being compressed, what work is done on the block by (a) the gravitational force on it and (b) the spring force? (c) What is the speed of the block just before it hits the spring? (Assume that friction is negligible.) (d) If the speed at impact is doubled, what is the maximum compression of the spring?

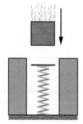

Fig. 7-44
Problem 62.

63 SSM To push a 25.0 kg crate up a frictionless incline, angled at 25.0° to the horizontal, a worker exerts a force of 209 N parallel to the incline. As the crate slides 1.50 m, how much work is done on the crate by (a) the worker's applied force, (b) the gravitational force on the crate, and (c) the normal force exerted by the incline on the crate? (d) What is the total work done on the crate?

64 Boxes are transported from one location to another in a warehouse by means of a conveyor belt that moves with a constant speed of 0.50 m/s. At a certain location the conveyor belt moves for 2.0 m up an incline that makes an angle of 10° with the horizontal, then for 2.0 m horizontally, and finally for 2.0 m down an incline that makes an angle of 10° with the horizontal. Assume that a 2.0 kg box rides on the belt without slipping. At what rate is the force of the conveyor belt doing work on the box as the box moves (a) up the 10° incline, (b) horizontally, and (c) down the 10° incline?

65 In Fig. 7-45, a cord runs around two massless, frictionless pulleys. A canister with mass $m = 20$ kg hangs from one pulley, and you exert a force $\vec{F}$ on the free end of the cord. (a) What must be the magnitude of $\vec{F}$ if you are to lift the canister at a constant speed? (b) To lift the canister by 2.0 cm, how far must you pull the free end of the cord? During that lift, what is the work done on the canister by (c) your force (via the cord) and (d) the gravitational force? (*Hint:* When a cord loops around a pulley as shown, it pulls on the pulley with a net force that is twice the tension in the cord.)

Fig. 7-45 Problem 65.

66 If a car of mass 1200 kg is moving along a highway at 120 km/h, what is the car's kinetic energy as determined by someone standing alongside the highway?

67 SSM A spring with a pointer attached is hanging next to a scale marked in millimeters. Three different packages are hung from the spring, in turn, as shown in Fig. 7-46. (a) Which mark on the scale will the pointer indicate when no package is hung from the spring? (b) What is the weight W of the third package?

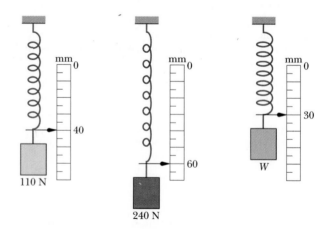

Fig. 7-46 Problem 67.

68 An iceboat is at rest on a frictionless frozen lake when a sudden wind exerts a constant force of 200 N, toward the east, on the boat. Due to the angle of the sail, the wind causes the boat to slide in a straight line for a distance of 8.0 m in a direction 20° north of east. What is the kinetic energy of the iceboat at the end of that 8.0 m?

69 If a ski lift raises 100 passengers averaging 660 N in weight to a height of 150 m in 60.0 s, at constant speed, what average power is required of the force making the lift?

70 A force $\vec{F} = (4.0 \text{ N})\hat{i} + c\hat{j}$ acts on a particle as the particle goes through displacement $\vec{d} = (3.0 \text{ m})\hat{i} - (2.0 \text{ m})\hat{j}$. (Other forces also act on the particle.) What is c if the work done on the particle by force $\vec{F}$ is (a) 0, (b) 17 J, and (c) −18 J?

71 A constant force of magnitude 10 N makes an angle of 150° (measured counterclockwise) with the positive x direction as it acts on a 2.0 kg object moving in an xy plane. How much work is done on the object by the force as the object moves from the origin to the point having position vector $(2.0 \text{ m})\hat{i} - (4.0 \text{ m})\hat{j}$?

72 In Fig. 7-47a, a 2.0 N force is applied to a 4.0 kg block at a downward angle θ as the block moves rightward through 1.0 m across a frictionless floor. Find an expression for the speed v_f of the block at the end of that distance if the block's initial velocity is (a) 0 and (b) 1.0 m/s to the right. (c) The situation in Fig. 7-47b is similar in that the block is initially moving at 1.0 m/s to the right, but now the 2.0 N force is directed downward to the left. Find an expression for the speed v_f of the block at the end of the 1.0 m distance. (d) Graph all three expressions for v_f versus downward angle θ for $\theta = 0°$ to $\theta = 90°$. Interpret the graphs.

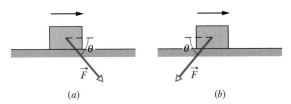

(a)   (b)

Fig. 7-47   Problem 72.

73 A force $\vec{F}$ in the positive direction of an x axis acts on an object moving along the axis. If the magnitude of the force is $F = 10e^{-x/2.0}$ N, with x in meters, find the work done by $\vec{F}$ as the object moves from $x = 0$ to $x = 2.0$ m by (a) plotting $F(x)$ and estimating the area under the curve and (b) integrating to find the work analytically.

74 A particle moves along a straight path through displacement $\vec{d} = (8 \text{ m})\hat{i} + c\hat{j}$ while force $\vec{F} = (2 \text{ N})\hat{i} - (4 \text{ N})\hat{j}$ acts on it. (Other forces also act on the particle.) What is the value of c if the work done by $\vec{F}$ on the particle is (a) zero, (b) positive, and (c) negative?

75 SSM An elevator cab has a mass of 4500 kg and can carry a maximum load of 1800 kg. If the cab is moving upward at full load at 3.80 m/s, what power is required of the force moving the cab to maintain that speed?

76 A 45 kg block of ice slides down a frictionless incline 1.5 m long and 0.91 m high. A worker pushes up against the ice, parallel to the incline, so that the block slides down at constant speed. (a) Find the magnitude of the worker's force. How much work is done on the block by (b) the worker's force, (c) the gravitational force on the block, (d) the normal force on the block from the surface of the incline, and (e) the net force on the block?

77 As a particle moves along an x axis, a force in the positive direction of the axis acts on it. Figure 7-48 shows the magnitude F of the force versus position x of the particle. The curve is given by $F = a/x^2$, with $a = 9.0 \text{ N} \cdot \text{m}^2$. Find the work done on the particle by the force as the particle moves from $x = 1.0$ m to $x = 3.0$ m by (a) estimating the work from the graph and (b) integrating the force function.

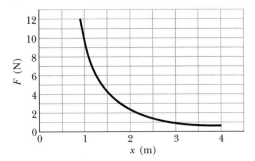

Fig. 7-48   Problem 77.

78 A CD case slides along a floor in the positive direction of an x axis while an applied force $\vec{F}_a$ acts on the case. The force is directed along the x axis and has the x component $F_{ax} = 9x - 3x^2$, with x in meters and F_{ax} in newtons. The case starts at rest at the position $x = 0$, and it moves until it is again at rest. (a) Plot the work $\vec{F}_a$ does on the case as a function of x. (b) At what position is the work maximum, and (c) what is that maximum value? (d) At what position has the work decreased to zero? (e) At what position is the case again at rest?

79 SSM A 2.0 kg lunchbox is sent sliding over a frictionless surface, in the positive direction of an x axis along the surface. Beginning at time $t = 0$, a steady wind pushes on the lunchbox in the negative direction of the x axis. Figure 7-49 shows the position x of the lunchbox as a function of time t as the wind pushes on the lunchbox. From the graph, estimate the kinetic energy of the lunchbox at (a) $t = 1.0$ s and (b) $t = 5.0$ s. (c) How much work does the force from the wind do on the lunchbox from $t = 1.0$ s to $t = 5.0$ s?

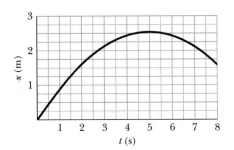

Fig. 7-49   Problem 79.

80 *Numerical integration.* A breadbox is made to move along an x axis from $x = 0.15$ m to $x = 1.20$ m by a force with a magnitude given by $F = \exp(-2x^2)$, with x in meters and F in newtons. (Here exp is the exponential function.) How much work is done on the breadbox by the force?

8 POTENTIAL ENERGY AND CONSERVATION OF ENERGY

8-1 WHAT IS PHYSICS?

One job of physics is to identify the different types of energy in the world, especially those that are of common importance. One general type of energy is **potential energy** U. Technically, potential energy is energy that can be associated with the configuration (arrangement) of a system of objects that exert forces on one another.

This is a pretty formal definition of something that is actually familiar to you. An example might help better than the definition: A bungee-cord jumper plunges from a staging platform (Fig. 8-1). The system of objects consists of Earth and the jumper. The force between the objects is the gravitational force. The configuration of the system changes (the separation between the jumper and Earth decreases—that is, of course, the thrill of the jump). We can account for the jumper's motion and increase in kinetic energy by defining a **gravitational potential energy** U. This

Fig. 8-1 The kinetic energy of a bungee-cord jumper increases during the free fall, and then the cord begins to stretch, slowing the jumper. *(KOFUJIWARA/amana images/ Getty Images News and Sport Services)*

is the energy associated with the state of separation between two objects that attract each other by the gravitational force, here the jumper and Earth.

When the jumper begins to stretch the bungee cord near the end of the plunge, the system of objects consists of the cord and the jumper. The force between the objects is an elastic (spring-like) force. The configuration of the system changes (the cord stretches). We can account for the jumper's decrease in kinetic energy and the cord's increase in length by defining an **elastic potential energy** U. This is the energy associated with the state of compression or extension of an elastic object, here the bungee cord.

Physics determines how the potential energy of a system can be calculated so that energy might be stored or put to use. For example, before any particular bungee-cord jumper takes the plunge, someone (probably a mechanical engineer) must determine the correct cord to be used by calculating the gravitational and elastic potential energies that can be expected. Then the jump is only thrilling and not fatal.

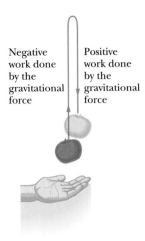

Fig. 8-2 A tomato is thrown upward. As it rises, the gravitational force does negative work on it, decreasing its kinetic energy. As the tomato descends, the gravitational force does positive work on it, increasing its kinetic energy.

8-2 Work and Potential Energy

In Chapter 7 we discussed the relation between work and a change in kinetic energy. Here we discuss the relation between work and a change in potential energy.

Let us throw a tomato upward (Fig. 8-2). We already know that as the tomato rises, the work W_g done on the tomato by the gravitational force is negative because the force transfers energy *from* the kinetic energy of the tomato. We can now finish the story by saying that this energy is transferred by the gravitational force *to* the gravitational potential energy of the tomato–Earth system.

The tomato slows, stops, and then begins to fall back down because of the gravitational force. During the fall, the transfer is reversed: The work W_g done on the tomato by the gravitational force is now positive—that force transfers energy *from* the gravitational potential energy of the tomato–Earth system *to* the kinetic energy of the tomato.

For either rise or fall, the change ΔU in gravitational potential energy is defined as being equal to the negative of the work done on the tomato by the gravitational force. Using the general symbol W for work, we write this as

$$\Delta U = -W. \tag{8-1}$$

This equation also applies to a block–spring system, as in Fig. 8-3. If we abruptly shove the block to send it moving rightward, the spring force acts leftward and thus does negative work on the block, transferring energy from the kinetic energy of the block to the elastic potential energy of the spring–block system. The block slows and eventually stops, and then begins to move leftward because the spring force is still leftward. The transfer of energy is then reversed—it is from potential energy of the spring–block system to kinetic energy of the block.

Conservative and Nonconservative Forces

Let us list the key elements of the two situations we just discussed:

1. The *system* consists of two or more objects.
2. A *force* acts between a particle-like object (tomato or block) in the system and the rest of the system.
3. When the system configuration changes, the force does *work* (call it W_1) on the particle-like object, transferring energy between the kinetic energy K of the object and some other type of energy of the system.

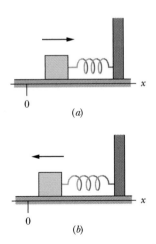

Fig. 8-3 A block, attached to a spring and initially at rest at $x = 0$, is set in motion toward the right. (*a*) As the block moves rightward (as indicated by the arrow), the spring force does negative work on it. (*b*) Then, as the block moves back toward $x = 0$, the spring force does positive work on it.

4. When the configuration change is reversed, the force reverses the energy transfer, doing work W_2 in the process.

In a situation in which $W_1 = -W_2$ is always true, the other type of energy is a potential energy and the force is said to be a **conservative force.** As you might suspect, the gravitational force and the spring force are both conservative (since otherwise we could not have spoken of gravitational potential energy and elastic potential energy, as we did previously).

A force that is not conservative is called a **nonconservative force.** The kinetic frictional force and drag force are nonconservative. For an example, let us send a block sliding across a floor that is not frictionless. During the sliding, a kinetic frictional force from the floor slows the block by transferring energy from its kinetic energy to a type of energy called *thermal energy* (which has to do with the random motions of atoms and molecules). We know from experiment that this energy transfer cannot be reversed (thermal energy cannot be transferred back to kinetic energy of the block by the kinetic frictional force). Thus, although we have a system (made up of the block and the floor), a force that acts between parts of the system, and a transfer of energy by the force, the force is not conservative. Therefore, thermal energy is not a potential energy.

When only conservative forces act on a particle-like object, we can greatly simplify otherwise difficult problems involving motion of the object. The next section, in which we develop a test for identifying conservative forces, provides one means for simplifying such problems.

8-3 Path Independence of Conservative Forces

The primary test for determining whether a force is conservative or nonconservative is this: Let the force act on a particle that moves along any *closed path,* beginning at some initial position and eventually returning to that position (so that the particle makes a *round trip* beginning and ending at the initial position). The force is conservative only if the total energy it transfers to and from the particle during the round trip along this and any other closed path is zero. In other words:

The net work done by a conservative force on a particle moving around any closed path is zero.

We know from experiment that the gravitational force passes this *closed-path test.* An example is the tossed tomato of Fig. 8-2. The tomato leaves the launch point with speed v_0 and kinetic energy $\frac{1}{2}mv_0^2$. The gravitational force acting on the tomato slows it, stops it, and then causes it to fall back down. When the tomato returns to the launch point, it again has speed v_0 and kinetic energy $\frac{1}{2}mv_0^2$. Thus, the gravitational force transfers as much energy *from* the tomato during the ascent as it transfers *to* the tomato during the descent back to the launch point. The net work done on the tomato by the gravitational force during the round trip is zero.

An important result of the closed-path test is that:

The work done by a conservative force on a particle moving between two points does not depend on the path taken by the particle.

For example, suppose that a particle moves from point *a* to point *b* in Fig. 8-4a along either path 1 or path 2. If only a conservative force acts on the particle, then the work done on the particle is the same along the two paths. In symbols, we can

The force is conservative. Any choice of path between the points gives the same amount of work.

(a)

And a round trip gives a total work of zero.

(b)

Fig. 8-4 (a) As a conservative force acts on it, a particle can move from point a to point b along either path 1 or path 2. (b) The particle moves in a round trip, from point a to point b along path 1 and then back to point a along path 2.

write this result as

$$W_{ab,1} = W_{ab,2}, \tag{8-2}$$

where the subscript ab indicates the initial and final points, respectively, and the subscripts 1 and 2 indicate the path.

This result is powerful because it allows us to simplify difficult problems when only a conservative force is involved. Suppose you need to calculate the work done by a conservative force along a given path between two points, and the calculation is difficult or even impossible without additional information. You can find the work by substituting some other path between those two points for which the calculation is easier and possible.

Proof of Equation 8-2

Figure 8-4b shows an arbitrary round trip for a particle that is acted upon by a single force. The particle moves from an initial point a to point b along path 1 and then back to point a along path 2. The force does work on the particle as the particle moves along each path. Without worrying about where positive work is done and where negative work is done, let us just represent the work done from a to b along path 1 as $W_{ab,1}$ and the work done from b back to a along path 2 as $W_{ba,2}$. If the force is conservative, then the net work done during the round trip must be zero:

$$W_{ab,1} + W_{ba,2} = 0,$$

and thus

$$W_{ab,1} = -W_{ba,2}. \tag{8-3}$$

In words, the work done along the outward path must be the negative of the work done along the path back.

Let us now consider the work $W_{ab,2}$ done on the particle by the force when the particle moves from a to b along path 2, as indicated in Fig. 8-4a. If the force is conservative, that work is the negative of $W_{ba,2}$:

$$W_{ab,2} = -W_{ba,2}. \tag{8-4}$$

Substituting $W_{ab,2}$ for $-W_{ba,2}$ in Eq. 8-3, we obtain

$$W_{ab,1} = W_{ab,2},$$

which is what we set out to prove.

CHECKPOINT 1

The figure shows three paths connecting points a and b. A single force $\vec{F}$ does the indicated work on a particle moving along each path in the indicated direction. On the basis of this information, is force $\vec{F}$ conservative?

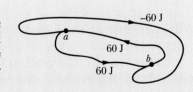

Sample Problem

Equivalent paths for calculating work, slippery cheese

Figure 8-5*a* shows a 2.0 kg block of slippery cheese that slides along a frictionless track from point *a* to point *b*. The cheese travels through a total distance of 2.0 m along the track, and a net vertical distance of 0.80 m. How much work is done on the cheese by the gravitational force during the slide?

KEY IDEAS

(1) We *cannot* calculate the work by using Eq. 7-12 ($W_g = mgd \cos \phi$). The reason is that the angle ϕ between the direc-

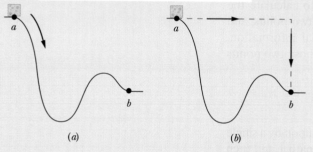

> The gravitational force is conservative. Any choice of path between the points gives the same amount of work.

Fig. 8-5 (*a*) A block of cheese slides along a frictionless track from point *a* to point *b*. (*b*) Finding the work done on the cheese by the gravitational force is easier along the dashed path than along the actual path taken by the cheese; the result is the same for both paths.

tions of the gravitational force $\vec{F}_g$ and the displacement $\vec{d}$ varies along the track in an unknown way. (Even if we did know the shape of the track and could calculate ϕ along it, the calculation could be very difficult.) (2) Because $\vec{F}_g$ is a conservative force, we can find the work by choosing some other path between *a* and *b*—one that makes the calculation easy.

Calculations: Let us choose the dashed path in Fig. 8-5*b*; it consists of two straight segments. Along the horizontal segment, the angle ϕ is a constant 90°. Even though we do not know the displacement along that horizontal segment, Eq. 7-12 tells us that the work W_h done there is

$$W_h = mgd \cos 90° = 0.$$

Along the vertical segment, the displacement *d* is 0.80 m and, with $\vec{F}_g$ and $\vec{d}$ both downward, the angle ϕ is a constant 0°. Thus, Eq. 7-12 gives us, for the work W_v done along the vertical part of the dashed path,

$$W_v = mgd \cos 0°$$

$$= (2.0 \text{ kg})(9.8 \text{ m/s}^2)(0.80 \text{ m})(1) = 15.7 \text{ J}.$$

The total work done on the cheese by $\vec{F}_g$ as the cheese moves from point *a* to point *b* along the dashed path is then

$$W = W_h + W_v = 0 + 15.7 \text{ J} \approx 16 \text{ J}. \quad \text{(Answer)}$$

This is also the work done as the cheese slides along the track from *a* to *b*.

WILEY PLUS Additional examples, video, and practice available at *WileyPLUS*

8-4 Determining Potential Energy Values

Here we find equations that give the value of the two types of potential energy discussed in this chapter: gravitational potential energy and elastic potential energy. However, first we must find a general relation between a conservative force and the associated potential energy.

Consider a particle-like object that is part of a system in which a conservative force $\vec{F}$ acts. When that force does work W on the object, the change ΔU in the potential energy associated with the system is the negative of the work done. We wrote this fact as Eq. 8-1 ($\Delta U = -W$). For the most general case, in which the force may vary with position, we may write the work W as in Eq. 7-32:

$$W = \int_{x_i}^{x_f} F(x) \, dx. \quad (8-5)$$

This equation gives the work done by the force when the object moves from point x_i to point x_f, changing the configuration of the system. (Because the force is conservative, the work is the same for all paths between those two points.)

Substituting Eq. 8-5 into Eq. 8-1, we find that the change in potential energy due to the change in configuration is, in general notation,

$$\Delta U = -\int_{x_i}^{x_f} F(x) \, dx. \quad (8-6)$$

Gravitational Potential Energy

We first consider a particle with mass m moving vertically along a y axis (the positive direction is upward). As the particle moves from point y_i to point y_f, the gravitational force $\vec{F}_g$ does work on it. To find the corresponding change in the gravitational potential energy of the particle–Earth system, we use Eq. 8-6 with two changes: (1) We integrate along the y axis instead of the x axis, because the gravitational force acts vertically. (2) We substitute $-mg$ for the force symbol F, because $\vec{F}_g$ has the magnitude mg and is directed down the y axis. We then have

$$\Delta U = -\int_{y_i}^{y_f} (-mg)\, dy = mg \int_{y_i}^{y_f} dy = mg\Big[\, y\, \Big]_{y_i}^{y_f},$$

which yields

$$\Delta U = mg(y_f - y_i) = mg\,\Delta y. \tag{8-7}$$

Only *changes* ΔU in gravitational potential energy (or any other type of potential energy) are physically meaningful. However, to simplify a calculation or a discussion, we sometimes would like to say that a certain gravitational potential value U is associated with a certain particle–Earth system when the particle is at a certain height y. To do so, we rewrite Eq. 8-7 as

$$U - U_i = mg(y - y_i). \tag{8-8}$$

Then we take U_i to be the gravitational potential energy of the system when it is in a **reference configuration** in which the particle is at a **reference point** y_i. Usually we take $U_i = 0$ and $y_i = 0$. Doing this changes Eq. 8-8 to

$$U(y) = mgy \qquad \text{(gravitational potential energy).} \tag{8-9}$$

This equation tells us:

The gravitational potential energy associated with a particle–Earth system depends only on the vertical position y (or height) of the particle relative to the reference position $y = 0$, not on the horizontal position.

Elastic Potential Energy

We next consider the block–spring system shown in Fig. 8-3, with the block moving on the end of a spring of spring constant k. As the block moves from point x_i to point x_f, the spring force $F_x = -kx$ does work on the block. To find the corresponding change in the elastic potential energy of the block–spring system, we substitute $-kx$ for $F(x)$ in Eq. 8-6. We then have

$$\Delta U = -\int_{x_i}^{x_f} (-kx)\, dx = k \int_{x_i}^{x_f} x\, dx = \tfrac{1}{2}k\Big[\, x^2\, \Big]_{x_i}^{x_f},$$

or

$$\Delta U = \tfrac{1}{2}kx_f^2 - \tfrac{1}{2}kx_i^2. \tag{8-10}$$

To associate a potential energy value U with the block at position x, we choose the reference configuration to be when the spring is at its relaxed length and the block is at $x_i = 0$. Then the elastic potential energy U_i is 0, and Eq. 8-10 becomes

$$U - 0 = \tfrac{1}{2}kx^2 - 0,$$

which gives us

$$U(x) = \tfrac{1}{2}kx^2 \qquad \text{(elastic potential energy).} \tag{8-11}$$

CHECKPOINT 2

A particle is to move along an x axis from $x = 0$ to x_1 while a conservative force, directed along the x axis, acts on the particle. The figure shows three situations in

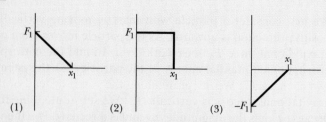

which the x component of that force varies with x. The force has the same maximum magnitude F_1 in all three situations. Rank the situations according to the change in the associated potential energy during the particle's motion, most positive first.

Sample Problem

Choosing reference level for gravitational potential energy, sloth

A 2.0 kg sloth hangs 5.0 m above the ground (Fig. 8-6).

(a) What is the gravitational potential energy U of the sloth–Earth system if we take the reference point $y = 0$ to be (1) at the ground, (2) at a balcony floor that is 3.0 m above the ground, (3) at the limb, and (4) 1.0 m above the limb? Take the gravitational potential energy to be zero at $y = 0$.

KEY IDEA

Once we have chosen the reference point for $y = 0$, we can calculate the gravitational potential energy U of the system *relative to that reference point* with Eq. 8-9.

Calculations: For choice (1) the sloth is at $y = 5.0$ m, and

$$U = mgy = (2.0 \text{ kg})(9.8 \text{ m/s}^2)(5.0 \text{ m})$$

$$= 98 \text{ J}. \qquad \text{(Answer)}$$

For the other choices, the values of U are

(2) $U = mgy = mg(2.0 \text{ m}) = 39 \text{ J}$,

(3) $U = mgy = mg(0) = 0 \text{ J}$,

(4) $U = mgy = mg(-1.0 \text{ m})$

$$= -19.6 \text{ J} \approx -20 \text{ J}. \qquad \text{(Answer)}$$

(b) The sloth drops to the ground. For each choice of reference point, what is the change ΔU in the potential energy of the sloth–Earth system due to the fall?

KEY IDEA

The *change* in potential energy does not depend on the choice of the reference point for $y = 0$; instead, it depends on the change in height Δy.

Fig. 8-6 Four choices of reference point $y = 0$. Each y axis is marked in units of meters. The choice affects the value of the potential energy U of the sloth–Earth system. However, it does not affect the change ΔU in potential energy of the system if the sloth moves by, say, falling.

Calculation: For all four situations, we have the same $\Delta y = -5.0$ m. Thus, for (1) to (4), Eq. 8-7 tells us that

$$\Delta U = mg \, \Delta y = (2.0 \text{ kg})(9.8 \text{ m/s}^2)(-5.0 \text{ m})$$

$$= -98 \text{ J}. \qquad \text{(Answer)}$$

 Additional examples, video, and practice available at *WileyPLUS*

8-5 Conservation of Mechanical Energy

The **mechanical energy** E_{mec} of a system is the sum of its potential energy U and the kinetic energy K of the objects within it:

$$E_{mec} = K + U \qquad \text{(mechanical energy).} \qquad (8\text{-}12)$$

In this section, we examine what happens to this mechanical energy when only conservative forces cause energy transfers within the system—that is, when frictional and drag forces do not act on the objects in the system. Also, we shall assume that the system is *isolated* from its environment; that is, no *external force* from an object outside the system causes energy changes inside the system.

    When a conservative force does work W on an object within the system, that force transfers energy between kinetic energy K of the object and potential energy U of the system. From Eq. 7-10, the change ΔK in kinetic energy is

$$\Delta K = W \qquad (8\text{-}13)$$

and from Eq. 8-1, the change ΔU in potential energy is

$$\Delta U = -W. \qquad (8\text{-}14)$$

Combining Eqs. 8-13 and 8-14, we find that

$$\Delta K = -\Delta U. \qquad (8\text{-}15)$$

In words, one of these energies increases exactly as much as the other decreases.
    We can rewrite Eq. 8-15 as

$$K_2 - K_1 = -(U_2 - U_1), \qquad (8\text{-}16)$$

where the subscripts refer to two different instants and thus to two different arrangements of the objects in the system. Rearranging Eq. 8-16 yields

$$K_2 + U_2 = K_1 + U_1 \qquad \text{(conservation of mechanical energy).} \qquad (8\text{-}17)$$

In words, this equation says:

$$\begin{pmatrix} \text{the sum of } K \text{ and } U \text{ for} \\ \text{any state of a system} \end{pmatrix} = \begin{pmatrix} \text{the sum of } K \text{ and } U \text{ for} \\ \text{any other state of the system} \end{pmatrix},$$

when the system is isolated and only conservative forces act on the objects in the system. In other words:

> In an isolated system where only conservative forces cause energy changes, the kinetic energy and potential energy can change, but their sum, the mechanical energy E_{mec} of the system, cannot change.

This result is called the **principle of conservation of mechanical energy.** (Now you can see where *conservative* forces got their name.) With the aid of Eq. 8-15, we can write this principle in one more form, as

$$\Delta E_{mec} = \Delta K + \Delta U = 0. \qquad (8\text{-}18)$$

    The principle of conservation of mechanical energy allows us to solve problems that would be quite difficult to solve using only Newton's laws:

> When the mechanical energy of a system is conserved, we can relate the sum of kinetic energy and potential energy at one instant to that at another instant *without considering the intermediate motion* and *without finding the work done by the forces involved.*

In olden days, a person would be tossed via a blanket to be able to see farther over the flat terrain. Nowadays, it is done just for fun. During the ascent of the person in the photograph, energy is transferred from kinetic energy to gravitational potential energy. The maximum height is reached when that transfer is complete. Then the transfer is reversed during the fall. *(©AP/Wide World Photos)*

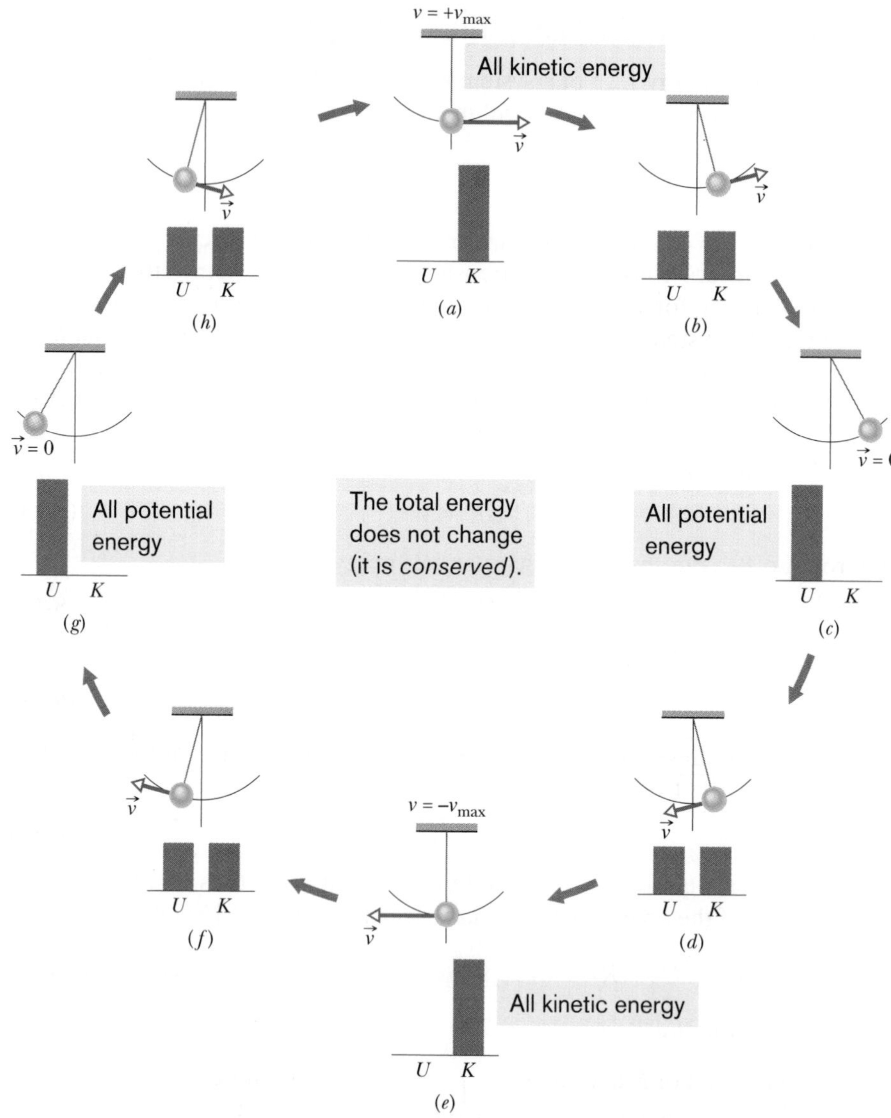

$v = +v_{max}$

All kinetic energy

U K
(a)

(h)

(b)

Fig. 8-7 A pendulum, with its mass concentrated in a bob at the lower end, swings back and forth. One full cycle of the motion is shown. During the cycle the values of the potential and kinetic energies of the pendulum–Earth system vary as the bob rises and falls, but the mechanical energy E_{mec} of the system remains constant. The energy E_{mec} can be described as continuously shifting between the kinetic and potential forms. In stages (a) and (e), all the energy is kinetic energy. The bob then has its greatest speed and is at its lowest point. In stages (c) and (g), all the energy is potential energy. The bob then has zero speed and is at its highest point. In stages (b), (d), (f), and (h), half the energy is kinetic energy and half is potential energy. If the swinging involved a frictional force at the point where the pendulum is attached to the ceiling, or a drag force due to the air, then E_{mec} would not be conserved, and eventually the pendulum would stop.

$\vec{v} = 0$

All potential energy

The total energy does not change (it is *conserved*).

All potential energy

$\vec{v} = 0$

U K
(g)

U K
(c)

$v = -v_{max}$

U K
(f)

U K
(d)

All kinetic energy

U K
(e)

Figure 8-7 shows an example in which the principle of conservation of mechanical energy can be applied: As a pendulum swings, the energy of the pendulum–Earth system is transferred back and forth between kinetic energy K and gravitational potential energy U, with the sum $K + U$ being constant. If we know the gravitational potential energy when the pendulum bob is at its highest point (Fig. 8-7c), Eq. 8-17 gives us the kinetic energy of the bob at the lowest point (Fig. 8-7e).

For example, let us choose the lowest point as the reference point, with the gravitational potential energy $U_2 = 0$. Suppose then that the potential energy at the highest point is $U_1 = 20$ J relative to the reference point. Because the bob momentarily stops at its highest point, the kinetic energy there is $K_1 = 0$. Putting these values into Eq. 8-17 gives us the kinetic energy K_2 at the lowest point:

$$K_2 + 0 = 0 + 20 \text{ J} \qquad \text{or} \qquad K_2 = 20 \text{ J}.$$

Note that we get this result without considering the motion between the highest and lowest points (such as in Fig. 8-7d) and without finding the work done by any forces involved in the motion.

✓ CHECKPOINT 3

The figure shows four situations—one in which an initially stationary block is dropped and three in which the block is allowed to slide down frictionless ramps. (a) Rank the situations according to the kinetic energy of the block at point B, greatest first. (b) Rank them according to the speed of the block at point B, greatest first.

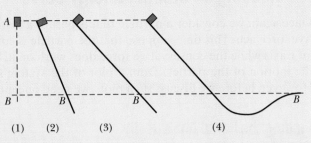

Sample Problem

Conservation of mechanical energy, water slide

In Fig. 8-8, a child of mass m is released from rest at the top of a water slide, at height $h = 8.5$ m above the bottom of the slide. Assuming that the slide is frictionless because of the water on it, find the child's speed at the bottom of the slide.

KEY IDEAS

(1) We cannot find her speed at the bottom by using her acceleration along the slide as we might have in earlier chapters because we do not know the slope (angle) of the slide. However, because that speed is related to her kinetic energy, perhaps we can use the principle of conservation of mechanical energy to get the speed. Then we would not need to know the slope. (2) Mechanical energy is conserved in a system *if* the system is isolated and *if* only conservative forces cause energy transfers within it. Let's check.

Forces: Two forces act on the child. The *gravitational force,* a conservative force, does work on her. The *normal force* on her from the slide does no work because its direction at any point during the descent is always perpendicular to the direction in which the child moves.

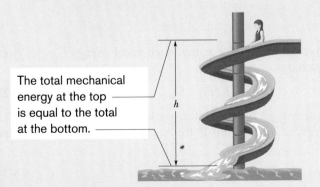

The total mechanical energy at the top is equal to the total at the bottom.

Fig. 8-8 A child slides down a water slide as she descends a height h.

System: Because the only force doing work on the child is the gravitational force, we choose the child–Earth system as our system, which we can take to be isolated.

Thus, we have only a conservative force doing work in an isolated system, so we *can* use the principle of conservation of mechanical energy.

Calculations: Let the mechanical energy be $E_{\text{mec},t}$ when the child is at the top of the slide and $E_{\text{mec},b}$ when she is at the bottom. Then the conservation principle tells us

$$E_{\text{mec},b} = E_{\text{mec},t}. \tag{8-19}$$

To show both kinds of mechanical energy, we have

$$K_b + U_b = K_t + U_t, \tag{8-20}$$

or

$$\tfrac{1}{2}mv_b^2 + mgy_b = \tfrac{1}{2}mv_t^2 + mgy_t.$$

Dividing by m and rearranging yield

$$v_b^2 = v_t^2 + 2g(y_t - y_b).$$

Putting $v_t = 0$ and $y_t - y_b = h$ leads to

$$v_b = \sqrt{2gh} = \sqrt{(2)(9.8 \text{ m/s}^2)(8.5 \text{ m})}$$
$$= 13 \text{ m/s.} \qquad \text{(Answer)}$$

This is the same speed that the child would reach if she fell 8.5 m vertically. On an actual slide, some frictional forces would act and the child would not be moving quite so fast.

Comments: Although this problem is hard to solve directly with Newton's laws, using conservation of mechanical energy makes the solution much easier. However, if we were asked to find the time taken for the child to reach the bottom of the slide, energy methods would be of no use; we would need to know the shape of the slide, and we would have a difficult problem.

8-6 Reading a Potential Energy Curve

Once again we consider a particle that is part of a system in which a conservative force acts. This time suppose that the particle is constrained to move along an x axis while the conservative force does work on it. We can learn a lot about the motion of the particle from a plot of the system's potential energy $U(x)$. However, before we discuss such plots, we need one more relationship.

Finding the Force Analytically

Equation 8-6 tells us how to find the change ΔU in potential energy between two points in a one-dimensional situation if we know the force $F(x)$. Now we want to go the other way; that is, we know the potential energy function $U(x)$ and want to find the force.

For one-dimensional motion, the work W done by a force that acts on a particle as the particle moves through a distance Δx is $F(x)\,\Delta x$. We can then write Eq. 8-1 as

$$\Delta U(x) = -W = -F(x)\,\Delta x. \tag{8-21}$$

Solving for $F(x)$ and passing to the differential limit yield

$$F(x) = -\frac{dU(x)}{dx} \quad \text{(one-dimensional motion)}, \tag{8-22}$$

which is the relation we sought.

We can check this result by putting $U(x) = \frac{1}{2}kx^2$, which is the elastic potential energy function for a spring force. Equation 8-22 then yields, as expected, $F(x) = -kx$, which is Hooke's law. Similarly, we can substitute $U(x) = mgx$, which is the gravitational potential energy function for a particle–Earth system, with a particle of mass m at height x above Earth's surface. Equation 8-22 then yields $F = -mg$, which is the gravitational force on the particle.

The Potential Energy Curve

Figure 8-9a is a plot of a potential energy function $U(x)$ for a system in which a particle is in one-dimensional motion while a conservative force $F(x)$ does work on it. We can easily find $F(x)$ by (graphically) taking the slope of the $U(x)$ curve at various points. (Equation 8-22 tells us that $F(x)$ is the negative of the slope of the $U(x)$ curve.) Figure 8-9b is a plot of $F(x)$ found in this way.

Turning Points

In the absence of a nonconservative force, the mechanical energy E of a system has a constant value given by

$$U(x) + K(x) = E_{\text{mec}}. \tag{8-23}$$

Here $K(x)$ is the *kinetic energy function* of a particle in the system (this $K(x)$ gives the kinetic energy as a function of the particle's location x). We may rewrite Eq. 8-23 as

$$K(x) = E_{\text{mec}} - U(x). \tag{8-24}$$

Suppose that E_{mec} (which has a constant value, remember) happens to be 5.0 J. It would be represented in Fig. 8-9c by a horizontal line that runs through the value 5.0 J on the energy axis. (It is, in fact, shown there.)

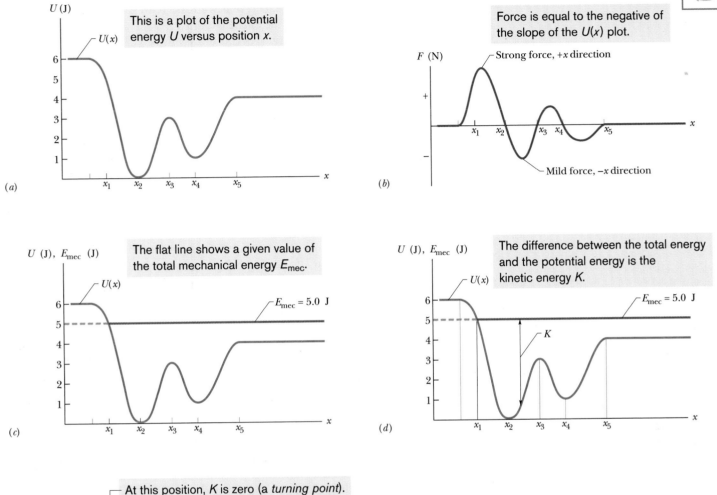

U (J)

This is a plot of the potential energy *U* versus position *x*.

U(x)

(a)

Force is equal to the negative of the slope of the *U(x)* plot.

F (N)

Strong force, +*x* direction

Mild force, −*x* direction

(b)

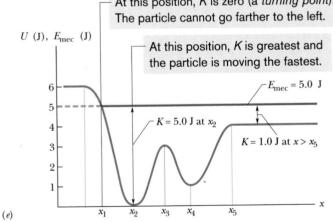

U (J), E_{mec} (J)

The flat line shows a given value of the total mechanical energy E_{mec}.

U(x)

E_{mec} = 5.0 J

(c)

The difference between the total energy and the potential energy is the kinetic energy *K*.

U (J), E_{mec} (J)

U(x)

E_{mec} = 5.0 J

K

(d)

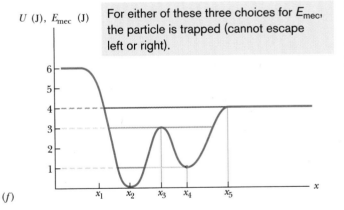

At this position, *K* is zero (a *turning point*). The particle cannot go farther to the left.

U (J), E_{mec} (J)

At this position, *K* is greatest and the particle is moving the fastest.

E_{mec} = 5.0 J

K = 5.0 J at x_2

K = 1.0 J at *x* > x_5

(e)

U (J), E_{mec} (J)

For either of these three choices for E_{mec}, the particle is trapped (cannot escape left or right).

(f)

Fig. 8-9 (*a*) A plot of *U(x)*, the potential energy function of a system containing a particle confined to move along an *x* axis. There is no friction, so mechanical energy is conserved. (*b*) A plot of the force *F(x)* acting on the particle, derived from the potential energy plot by taking its slope at various points. (*c*)–(*e*) How to determine the kinetic energy. (*f*) The *U(x)* plot of (*a*) with three possible values of E_{mec} shown.

Equation 8-24 and Fig. 8-9*d* tell us how to determine the kinetic energy *K* for any location *x* of the particle: On the *U*(*x*) curve, find *U* for that location *x* and then subtract *U* from E_{mec}. In Fig. 8-9*e* for example, if the particle is at any point to the right of x_5, then $K = 1.0$ J. The value of *K* is greatest (5.0 J) when the particle is at x_2 and least (0 J) when the particle is at x_1.

Since *K* can never be negative (because v^2 is always positive), the particle can never move to the left of x_1, where $E_{mec} - U$ is negative. Instead, as the particle moves toward x_1 from x_2, *K* decreases (the particle slows) until *K* = 0 at x_1 (the particle stops there).

Note that when the particle reaches x_1, the force on the particle, given by Eq. 8-22, is positive (because the slope *dU*/*dx* is negative). This means that the particle does not remain at x_1 but instead begins to move to the right, opposite its earlier motion. Hence x_1 is a **turning point,** a place where *K* = 0 (because *U* = *E*) and the particle changes direction. There is no turning point (where *K* = 0) on the right side of the graph. When the particle heads to the right, it will continue indefinitely.

Equilibrium Points

Figure 8-9*f* shows three different values for E_{mec} superposed on the plot of the potential energy function *U*(*x*) of Fig. 8-9*a*. Let us see how they change the situation. If $E_{mec} = 4.0$ J (purple line), the turning point shifts from x_1 to a point between x_1 and x_2. Also, at any point to the right of x_5, the system's mechanical energy is equal to its potential energy; thus, the particle has no kinetic energy and (by Eq. 8-22) no force acts on it, and so it must be stationary. A particle at such a position is said to be in **neutral equilibrium.** (A marble placed on a horizontal tabletop is in that state.)

If $E_{mec} = 3.0$ J (pink line), there are two turning points: One is between x_1 and x_2, and the other is between x_4 and x_5. In addition, x_3 is a point at which *K* = 0. If the particle is located exactly there, the force on it is also zero, and the particle remains stationary. However, if it is displaced even slightly in either direction, a nonzero force pushes it farther in the same direction, and the particle continues to move. A particle at such a position is said to be in **unstable equilibrium.** (A marble balanced on top of a bowling ball is an example.)

Next consider the particle's behavior if $E_{mec} = 1.0$ J (green line). If we place it at x_4, it is stuck there. It cannot move left or right on its own because to do so would require a negative kinetic energy. If we push it slightly left or right, a restoring force appears that moves it back to x_4. A particle at such a position is said to be in **stable equilibrium.** (A marble placed at the bottom of a hemispherical bowl is an example.) If we place the particle in the cup-like *potential well* centered at x_2, it is between two turning points. It can still move somewhat, but only partway to x_1 or x_3.

✔ **CHECKPOINT 4**

The figure gives the potential energy function *U*(*x*) for a system in which a particle is in one-dimensional motion. (a) Rank regions *AB*, *BC*, and *CD* according to the magnitude of the force on the particle, greatest first. (b) What is the direction of the force when the particle is in region *AB*?

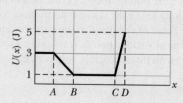

Sample Problem

Reading a potential energy graph

A 2.00 kg particle moves along an x axis in one-dimensional motion while a conservative force along that axis acts on it. The potential energy $U(x)$ associated with the force is plotted in Fig. 8-10a. That is, if the particle were placed at any position between $x = 0$ and $x = 7.00$ m, it would have the plotted value of U. At $x = 6.5$ m, the particle has velocity $v_0 = (-4.00 \text{ m/s})\hat{i}$.

(a) From Fig. 8-10a, determine the particle's speed at $x_1 = 4.5$ m.

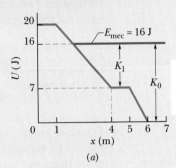

Kinetic energy is the difference between the total energy and the potential energy.

KEY IDEAS

(1) The particle's kinetic energy is given by Eq. 7-1 $(K = \frac{1}{2}mv^2)$. (2) Because only a conservative force acts on the particle, the mechanical energy $E_{mec} (= K + U)$ is conserved as the particle moves. (3) Therefore, on a plot of $U(x)$ such as Fig. 8-10a, the kinetic energy is equal to the difference between E_{mec} and U.

Calculations: At $x = 6.5$ m, the particle has kinetic energy

$$K_0 = \tfrac{1}{2}mv_0^2 = \tfrac{1}{2}(2.00 \text{ kg})(4.00 \text{ m/s})^2$$
$$= 16.0 \text{ J}.$$

Because the potential energy there is $U = 0$, the mechanical energy is

$$E_{mec} = K_0 + U_0 = 16.0 \text{ J} + 0 = 16.0 \text{ J}.$$

This value for E_{mec} is plotted as a horizontal line in Fig. 8-10a. From that figure we see that at $x = 4.5$ m, the potential energy is $U_1 = 7.0$ J. The kinetic energy K_1 is the difference between E_{mec} and U_1:

$$K_1 = E_{mec} - U_1 = 16.0 \text{ J} - 7.0 \text{ J} = 9.0 \text{ J}.$$

Because $K_1 = \frac{1}{2}mv_1^2$, we find

$$v_1 = 3.0 \text{ m/s}. \qquad \text{(Answer)}$$

(b) Where is the particle's turning point located?

KEY IDEA

The turning point is where the force momentarily stops and then reverses the particle's motion. That is, it is where the particle momentarily has $v = 0$ and thus $K = 0$.

Calculations: Because K is the difference between E_{mec} and U, we want the point in Fig. 8-10a where the plot of U rises to meet the horizontal line of E_{mec}, as shown in Fig. 8-10b. Because the plot of U is a straight line in Fig. 8-10b, we can draw nested right triangles as shown and then write

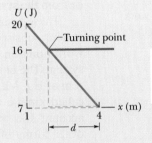

The kinetic energy is zero at the turning point (the particle speed is zero).

Fig. 8-10 (a) A plot of potential energy U versus position x. (b) A section of the plot used to find where the particle turns around.

the proportionality of distances

$$\frac{16 - 7.0}{d} = \frac{20 - 7.0}{4.0 - 1.0},$$

which gives us $d = 2.08$ m. Thus, the turning point is at

$$x = 4.0 \text{ m} - d = 1.9 \text{ m}. \qquad \text{(Answer)}$$

(c) Evaluate the force acting on the particle when it is in the region 1.9 m $< x <$ 4.0 m.

KEY IDEA

The force is given by Eq. 8-22 $(F(x) = -dU(x)/dx)$. The equation states that the force is equal to the negative of the slope on a graph of $U(x)$.

Calculations: For the graph of Fig. 8-10b, we see that for the range 1.0 m $< x <$ 4.0 m the force is

$$F = -\frac{20 \text{ J} - 7.0 \text{ J}}{1.0 \text{ m} - 4.0 \text{ m}} = 4.3 \text{ N}. \qquad \text{(Answer)}$$

Thus, the force has magnitude 4.3 N and is in the positive direction of the x axis. This result is consistent with the fact that the initially leftward-moving particle is stopped by the force and then sent rightward.

 Additional examples, video, and practice available at *WileyPLUS*

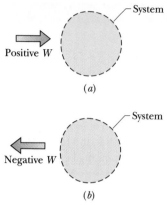

Fig. 8-11 (*a*) Positive work *W* done on an arbitrary system means a transfer of energy to the system. (*b*) Negative work *W* means a transfer of energy from the system.

8-7 Work Done on a System by an External Force

In Chapter 7, we defined work as being energy transferred to or from an object by means of a force acting on the object. We can now extend that definition to an external force acting on a system of objects.

> Work is energy transferred to or from a system by means of an external force acting on that system.

Figure 8-11*a* represents positive work (a transfer of energy *to* a system), and Fig. 8-11*b* represents negative work (a transfer of energy *from* a system). When more than one force acts on a system, their *net work* is the energy transferred to or from the system.

These transfers are like transfers of money to and from a bank account. If a system consists of a single particle or particle-like object, as in Chapter 7, the work done on the system by a force can change only the kinetic energy of the system. The energy statement for such transfers is the work–kinetic energy theorem of Eq. 7-10 ($\Delta K = W$); that is, a single particle has only one energy account, called kinetic energy. External forces can transfer energy into or out of that account. If a system is more complicated, however, an external force can change other forms of energy (such as potential energy); that is, a more complicated system can have multiple energy accounts.

Let us find energy statements for such systems by examining two basic situations, one that does not involve friction and one that does.

No Friction Involved

To compete in a bowling-ball-hurling contest, you first squat and cup your hands under the ball on the floor. Then you rapidly straighten up while also pulling your hands up sharply, launching the ball upward at about face level. During your upward motion, your applied force on the ball obviously does work; that is, it is an external force that transfers energy, but to what system?

To answer, we check to see which energies change. There is a change ΔK in the ball's kinetic energy and, because the ball and Earth become more separated, there is a change ΔU in the gravitational potential energy of the ball–Earth system. To include both changes, we need to consider the ball–Earth system. Then your force is an external force doing work on that system, and the work is

$$W = \Delta K + \Delta U, \tag{8-25}$$

or
$$W = \Delta E_{\text{mec}} \quad \text{(work done on system, no friction involved),} \tag{8-26}$$

where ΔE_{mec} is the change in the mechanical energy of the system. These two equations, which are represented in Fig. 8-12, are equivalent energy statements for work done on a system by an external force when friction is not involved.

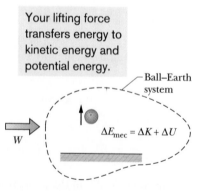

Your lifting force transfers energy to kinetic energy and potential energy.

Fig. 8-12 Positive work *W* is done on a system of a bowling ball and Earth, causing a change ΔE_{mec} in the mechanical energy of the system, a change ΔK in the ball's kinetic energy, and a change ΔU in the system's gravitational potential energy.

Friction Involved

We next consider the example in Fig. 8-13*a*. A constant horizontal force $\vec{F}$ pulls a block along an *x* axis and through a displacement of magnitude *d*, increasing the block's velocity from $\vec{v}_0$ to $\vec{v}$. During the motion, a constant kinetic frictional force $\vec{f}_k$ from the floor acts on the block. Let us first choose the block as our

The applied force supplies energy. The frictional force transfers some of it to thermal energy.

So, the work done by the applied force goes into kinetic energy and also thermal energy.

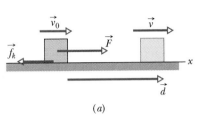

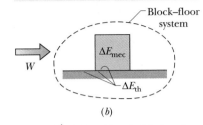

(a)

(b)

Fig. 8-13 (a) A block is pulled across a floor by force $\vec{F}$ while a kinetic frictional force $\vec{f}_k$ opposes the motion. The block has velocity $\vec{v}_0$ at the start of a displacement $\vec{d}$ and velocity $\vec{v}$ at the end of the displacement. (b) Positive work W is done on the block–floor system by force $\vec{F}$, resulting in a change ΔE_{mec} in the block's mechanical energy and a change ΔE_{th} in the thermal energy of the block and floor.

system and apply Newton's second law to it. We can write that law for components along the x axis ($F_{net,x} = ma_x$) as

$$F - f_k = ma. \tag{8-27}$$

Because the forces are constant, the acceleration $\vec{a}$ is also constant. Thus, we can use Eq. 2-16 to write

$$v^2 = v_0^2 + 2ad.$$

Solving this equation for a, substituting the result into Eq. 8-27, and rearranging then give us

$$Fd = \tfrac{1}{2}mv^2 - \tfrac{1}{2}mv_0^2 + f_k d \tag{8-28}$$

or, because $\tfrac{1}{2}mv^2 - \tfrac{1}{2}mv_0^2 = \Delta K$ for the block,

$$Fd = \Delta K + f_k d. \tag{8-29}$$

In a more general situation (say, one in which the block is moving up a ramp), there can be a change in potential energy. To include such a possible change, we generalize Eq. 8-29 by writing

$$Fd = \Delta E_{mec} + f_k d. \tag{8-30}$$

By experiment we find that the block and the portion of the floor along which it slides become warmer as the block slides. As we shall discuss in Chapter 18, the temperature of an object is related to the object's thermal energy E_{th} (the energy associated with the random motion of the atoms and molecules in the object). Here, the thermal energy of the block and floor increases because (1) there is friction between them and (2) there is sliding. Recall that friction is due to the cold-welding between two surfaces. As the block slides over the floor, the sliding causes repeated tearing and re-forming of the welds between the block and the floor, which makes the block and floor warmer. Thus, the sliding increases their thermal energy E_{th}.

Through experiment, we find that the increase ΔE_{th} in thermal energy is equal to the product of the magnitudes f_k and d:

$$\Delta E_{th} = f_k d \qquad \text{(increase in thermal energy by sliding).} \tag{8-31}$$

Thus, we can rewrite Eq. 8-30 as

$$Fd = \Delta E_{mec} + \Delta E_{th}. \tag{8-32}$$

Fd is the work W done by the external force $\vec{F}$ (the energy transferred by the force), but on which system is the work done (where are the energy transfers made)?

To answer, we check to see which energies change. The block's mechanical energy changes, and the thermal energies of the block and floor also change. Therefore, the work done by force $\vec{F}$ is done on the block–floor system. That work is

$$W = \Delta E_{mec} + \Delta E_{th} \quad \text{(work done on system, friction involved).} \quad (8\text{-}33)$$

This equation, which is represented in Fig. 8-13b, is the energy statement for the work done on a system by an external force when friction is involved.

 CHECKPOINT 5

In three trials, a block is pushed by a horizontal applied force across a floor that is not frictionless, as in Fig. 8-13a. The magnitudes F of the applied force and the results of the pushing on the block's speed are given in the table. In all three trials, the block is pushed through the same distance d. Rank the three trials according to the change in the thermal energy of the block and floor that occurs in that distance d, greatest first.

Trial	F	Result on Block's Speed
a	5.0 N	decreases
b	7.0 N	remains constant
c	8.0 N	increases

Sample Problem

Work, friction, change in thermal energy, cabbage heads

A food shipper pushes a wood crate of cabbage heads (total mass $m = 14$ kg) across a concrete floor with a constant horizontal force $\vec{F}$ of magnitude 40 N. In a straight-line displacement of magnitude $d = 0.50$ m, the speed of the crate decreases from $v_0 = 0.60$ m/s to $v = 0.20$ m/s.

(a) How much work is done by force $\vec{F}$, and on what system does it do the work?

KEY IDEA

Because the applied force $\vec{F}$ is constant, we can calculate the work it does by using Eq. 7-7 ($W = Fd \cos \phi$).

Calculation: Substituting given data, including the fact that force $\vec{F}$ and displacement $\vec{d}$ are in the same direction, we find

$$W = Fd \cos \phi = (40 \text{ N})(0.50 \text{ m}) \cos 0°$$
$$= 20 \text{ J}. \qquad \text{(Answer)}$$

Reasoning: We can determine the system on which the work is done to see which energies change. Because the crate's speed changes, there is certainly a change ΔK in the crate's kinetic energy. Is there friction between the floor and the crate, and thus a change in thermal energy? Note that $\vec{F}$ and the crate's velocity have the same direction.

Thus, if there is no friction, then $\vec{F}$ should be accelerating the crate to a *greater* speed. However, the crate is *slowing*, so there must be friction and a change ΔE_{th} in thermal energy of the crate and the floor. Therefore, the system on which the work is done is the crate–floor system, because both energy changes occur in that system.

(b) What is the increase ΔE_{th} in the thermal energy of the crate and floor?

KEY IDEA

We can relate ΔE_{th} to the work W done by $\vec{F}$ with the energy statement of Eq. 8-33 for a system that involves friction:

$$W = \Delta E_{mec} + \Delta E_{th}. \qquad (8\text{-}34)$$

Calculations: We know the value of W from (a). The change ΔE_{mec} in the crate's mechanical energy is just the change in its kinetic energy because no potential energy changes occur, so we have

$$\Delta E_{mec} = \Delta K = \tfrac{1}{2}mv^2 - \tfrac{1}{2}mv_0^2.$$

Substituting this into Eq. 8-34 and solving for ΔE_{th}, we find

$$\Delta E_{th} = W - (\tfrac{1}{2}mv^2 - \tfrac{1}{2}mv_0^2) = W - \tfrac{1}{2}m(v^2 - v_0^2)$$
$$= 20 \text{ J} - \tfrac{1}{2}(14 \text{ kg})[(0.20 \text{ m/s})^2 - (0.60 \text{ m/s})^2]$$
$$= 22.2 \text{ J} \approx 22 \text{ J}. \qquad \text{(Answer)}$$

 Additional examples, video, and practice available at *WileyPLUS*

8-8 Conservation of Energy

We now have discussed several situations in which energy is transferred to or from objects and systems, much like money is transferred between accounts. In each situation we assume that the energy that was involved could always be accounted for; that is, energy could not magically appear or disappear. In more formal language, we assumed (correctly) that energy obeys a law called the **law of conservation of energy,** which is concerned with the **total energy** E of a system. That total is the sum of the system's mechanical energy, thermal energy, and any type of *internal energy* in addition to thermal energy. (We have not yet discussed other types of internal energy.) The law states that

> The total energy E of a system can change only by amounts of energy that are transferred to or from the system.

The only type of energy transfer that we have considered is work W done on a system. Thus, for us at this point, this law states that

$$W = \Delta E = \Delta E_{mec} + \Delta E_{th} + \Delta E_{int}, \qquad (8\text{-}35)$$

where ΔE_{mec} is any change in the mechanical energy of the system, ΔE_{th} is any change in the thermal energy of the system, and ΔE_{int} is any change in any other type of internal energy of the system. Included in ΔE_{mec} are changes ΔK in kinetic energy and changes ΔU in potential energy (elastic, gravitational, or any other type we might find).

This law of conservation of energy is *not* something we have derived from basic physics principles. Rather, it is a law based on countless experiments. Scientists and engineers have never found an exception to it.

Isolated System

If a system is isolated from its environment, there can be no energy transfers to or from it. For that case, the law of conservation of energy states:

> The total energy E of an isolated system cannot change.

Fig. 8-14 To descend, the rock climber must transfer energy from the gravitational potential energy of a system consisting of him, his gear, and Earth. He has wrapped the rope around metal rings so that the rope rubs against the rings. This allows most of the transferred energy to go to the thermal energy of the rope and rings rather than to his kinetic energy. *(Tyler Stableford/The Image Bank/ Getty Images)*

Many energy transfers may be going on *within* an isolated system—between, say, kinetic energy and a potential energy or between kinetic energy and thermal energy. However, the total of all the types of energy in the system cannot change.

We can use the rock climber in Fig. 8-14 as an example, approximating him, his gear, and Earth as an isolated system. As he rappels down the rock face, changing the configuration of the system, he needs to control the transfer of energy from the gravitational potential energy of the system. (That energy cannot just disappear.) Some of it is transferred to his kinetic energy. However, he obviously does not want very much transferred to that type or he will be moving too quickly, so he has wrapped the rope around metal rings to produce friction between the rope and the rings as he moves down. The sliding of the rings on the rope then transfers the gravitational potential energy of the system to thermal energy of the rings and rope in a way that he can control. The total energy of the climber–gear–Earth system (the total of its gravitational potential energy, kinetic energy, and thermal energy) does not change during his descent.

For an isolated system, the law of conservation of energy can be written in two ways. First, by setting $W = 0$ in Eq. 8-35, we get

$$\Delta E_{mec} + \Delta E_{th} + \Delta E_{int} = 0 \qquad \text{(isolated system).} \qquad (8\text{-}36)$$

We can also let $\Delta E_{mec} = E_{mec,2} - E_{mec,1}$, where the subscripts 1 and 2 refer to two different instants—say, before and after a certain process has occurred. Then Eq. 8-36 becomes

$$E_{mec,2} = E_{mec,1} - \Delta E_{th} - \Delta E_{int}. \qquad (8\text{-}37)$$

Equation 8-37 tells us:

> In an isolated system, we can relate the total energy at one instant to the total energy at another instant *without considering the energies at intermediate times.*

This fact can be a very powerful tool in solving problems about isolated systems when you need to relate energies of a system before and after a certain process occurs in the system.

In Section 8-5, we discussed a special situation for isolated systems—namely, the situation in which nonconservative forces (such as a kinetic frictional force) do not act within them. In that special situation, ΔE_{th} and ΔE_{int} are both zero, and so Eq. 8-37 reduces to Eq. 8-18. In other words, the mechanical energy of an isolated system is conserved when nonconservative forces do not act in it.

External Forces and Internal Energy Transfers

An external force can change the kinetic energy or potential energy of an object without doing work on the object—that is, without transferring energy to the object. Instead, the force is responsible for transfers of energy from one type to another inside the object.

Figure 8-15 shows an example. An initially stationary ice-skater pushes away from a railing and then slides over the ice (Figs. 8-15a and b). Her kinetic energy increases because of an external force $\vec{F}$ on her from the rail. However, that force does not transfer energy from the rail to her. Thus, the force does no work on

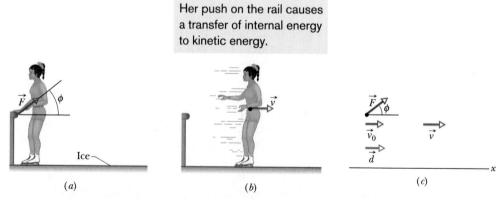

Her push on the rail causes a transfer of internal energy to kinetic energy.

Fig. 8-15 (a) As a skater pushes herself away from a railing, the force on her from the railing is $\vec{F}$. (b) After the skater leaves the railing, she has velocity $\vec{v}$. (c) External force $\vec{F}$ acts on the skater, at angle ϕ with a horizontal x axis. When the skater goes through displacement $\vec{d}$, her velocity is changed from $\vec{v}_0$ (= 0) to $\vec{v}$ by the horizontal component of $\vec{F}$.

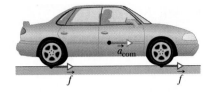

Fig. 8-16 A vehicle accelerates to the right using four-wheel drive. The road exerts four frictional forces (two of them shown) on the bottom surfaces of the tires. Taken together, these four forces make up the net external force $\vec{F}$ acting on the car.

her. Rather, her kinetic energy increases as a result of internal transfers from the biochemical energy in her muscles.

Figure 8-16 shows another example. An engine increases the speed of a car with four-wheel drive (all four wheels are made to turn by the engine). During the acceleration, the engine causes the tires to push backward on the road surface. This push produces frictional forces $\vec{f}$ that act on each tire in the forward direction. The net external force $\vec{F}$ from the road, which is the sum of these frictional forces, accelerates the car, increasing its kinetic energy. However, $\vec{F}$ does not transfer energy from the road to the car and so does no work on the car. Rather, the car's kinetic energy increases as a result of internal transfers from the energy stored in the fuel.

In situations like these two, we can sometimes relate the external force $\vec{F}$ on an object to the change in the object's mechanical energy if we can simplify the situation. Consider the ice-skater example. During her push through distance d in Fig. 8-15c, we can simplify by assuming that the acceleration is constant, her speed changing from $v_0 = 0$ to v. (That is, we assume $\vec{F}$ has constant magnitude F and angle ϕ.) After the push, we can simplify the skater as being a particle and neglect the fact that the exertions of her muscles have increased the thermal energy in her muscles and changed other physiological features. Then we can apply Eq. 7-5 ($\frac{1}{2}mv^2 - \frac{1}{2}mv_0^2 = F_x d$) to write

$$K - K_0 = (F \cos \phi)d,$$

or $$\Delta K = Fd \cos \phi. \tag{8-38}$$

If the situation also involves a change in the elevation of an object, we can include the change ΔU in gravitational potential energy by writing

$$\Delta U + \Delta K = Fd \cos \phi. \tag{8-39}$$

The force on the right side of this equation does no work on the object but is still responsible for the changes in energy shown on the left side.

Power

Now that you have seen how energy can be transferred from one type to another, we can expand the definition of power given in Section 7-9. There power is defined as the rate at which work is done by a force. In a more general sense, power P is the rate at which energy is transferred by a force from one type to another. If an amount of energy ΔE is transferred in an amount of time Δt, the **average power** due to the force is

$$P_{\text{avg}} = \frac{\Delta E}{\Delta t}. \tag{8-40}$$

Similarly, the **instantaneous power** due to the force is

$$P = \frac{dE}{dt}. \tag{8-41}$$

Sample Problem

Energy, friction, spring, and tamales

In Fig. 8-17, a 2.0 kg package of tamales slides along a floor with speed $v_1 = 4.0$ m/s. It then runs into and compresses a spring, until the package momentarily stops. Its path to the initially relaxed spring is frictionless, but as it compresses the spring, a kinetic frictional force from the floor, of magnitude 15 N, acts on the package. If $k = 10\,000$ N/m, by what distance d is the spring compressed when the package stops?

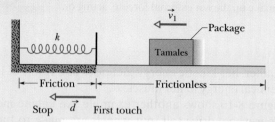

During the rubbing, kinetic energy is transferred to potential energy and thermal energy.

Fig. 8-17 A package slides across a frictionless floor with velocity $\vec{v}_1$ toward a spring of spring constant k. When the package reaches the spring, a frictional force from the floor acts on the package.

KEY IDEAS

We need to examine all the forces and then to determine whether we have an isolated system or a system on which an external force is doing work.

Forces: The normal force on the package from the floor does no work on the package because the direction of this force is always perpendicular to the direction of the package's displacement. For the same reason, the gravitational force on the package does no work. As the spring is compressed, however, a spring force does work on the package, transferring energy to elastic potential energy of the spring. The spring force also pushes against a rigid wall. Because there is friction between the package and the floor, the sliding of the package across the floor increases their thermal energies.

System: The package–spring–floor–wall system includes all these forces and energy transfers in one isolated system. Therefore, because the system is isolated, its total energy cannot change. We can then apply the law of conservation of energy in the form of Eq. 8-37 to the system:

$$E_{\text{mec},2} = E_{\text{mec},1} - \Delta E_{\text{th}}. \qquad (8\text{-}42)$$

Calculations: In Eq. 8-42, let subscript 1 correspond to the initial state of the sliding package and subscript 2 correspond to the state in which the package is momentarily stopped and the spring is compressed by distance d. For both states the mechanical energy of the system is the sum

of the package's kinetic energy ($K = \frac{1}{2}mv^2$) and the spring's potential energy ($U = \frac{1}{2}kx^2$). For state 1, $U = 0$ (because the spring is not compressed), and the package's speed is v_1. Thus, we have

$$E_{\text{mec},1} = K_1 + U_1 = \tfrac{1}{2}mv_1^2 + 0.$$

For state 2, $K = 0$ (because the package is stopped), and the compression distance is d. Therefore, we have

$$E_{\text{mec},2} = K_2 + U_2 = 0 + \tfrac{1}{2}kd^2.$$

Finally, by Eq. 8-31, we can substitute $f_k d$ for the change ΔE_{th} in the thermal energy of the package and the floor. We can now rewrite Eq. 8-42 as

$$\tfrac{1}{2}kd^2 = \tfrac{1}{2}mv_1^2 - f_k d.$$

Rearranging and substituting known data give us

$$5000d^2 + 15d - 16 = 0.$$

Solving this quadratic equation yields

$$d = 0.055 \text{ m} = 5.5 \text{ cm}. \qquad \text{(Answer)}$$

 Additional examples, video, and practice available at *WileyPLUS*

REVIEW & SUMMARY

Conservative Forces   A force is a **conservative force** if the net work it does on a particle moving around any closed path, from an initial point and then back to that point, is zero. Equivalently, a force is conservative if the net work it does on a particle moving between two points does not depend on the path taken by the particle. The gravitational force and the spring force are conservative forces; the kinetic frictional force is a **nonconservative force**.

Potential Energy   A **potential energy** is energy that is associated with the configuration of a system in which a conservative force acts. When the conservative force does work W on a particle within the system, the change ΔU in the potential energy of the system is

$$\Delta U = -W. \qquad (8\text{-}1)$$

If the particle moves from point x_i to point x_f, the change in the potential energy of the system is

$$\Delta U = -\int_{x_i}^{x_f} F(x)\, dx. \qquad (8\text{-}6)$$

Gravitational Potential Energy   The potential energy associated with a system consisting of Earth and a nearby particle is **gravitational potential energy**. If the particle moves from height y_i

to height y_f, the change in the gravitational potential energy of the particle–Earth system is

$$\Delta U = mg(y_f - y_i) = mg\,\Delta y. \qquad (8\text{-}7)$$

If the **reference point** of the particle is set as $y_i = 0$ and the corresponding gravitational potential energy of the system is set as $U_i = 0$, then the gravitational potential energy U when the particle is at any height y is

$$U(y) = mgy. \qquad (8\text{-}9)$$

Elastic Potential Energy **Elastic potential energy** is the energy associated with the state of compression or extension of an elastic object. For a spring that exerts a spring force $F = -kx$ when its free end has displacement x, the elastic potential energy is

$$U(x) = \tfrac{1}{2}kx^2. \qquad (8\text{-}11)$$

The **reference configuration** has the spring at its relaxed length, at which $x = 0$ and $U = 0$.

Mechanical Energy The **mechanical energy** E_{mec} of a system is the sum of its kinetic energy K and potential energy U:

$$E_{mec} = K + U. \qquad (8\text{-}12)$$

An *isolated system* is one in which no *external force* causes energy changes. If only conservative forces do work within an isolated system, then the mechanical energy E_{mec} of the system cannot change. This **principle of conservation of mechanical energy** is written as

$$K_2 + U_2 = K_1 + U_1, \qquad (8\text{-}17)$$

in which the subscripts refer to different instants during an energy transfer process. This conservation principle can also be written as

$$\Delta E_{mec} = \Delta K + \Delta U = 0. \qquad (8\text{-}18)$$

Potential Energy Curves If we know the potential energy function $U(x)$ for a system in which a one-dimensional force $F(x)$ acts on a particle, we can find the force as

$$F(x) = -\frac{dU(x)}{dx}. \qquad (8\text{-}22)$$

If $U(x)$ is given on a graph, then at any value of x, the force $F(x)$ is the negative of the slope of the curve there and the kinetic energy of the particle is given by

$$K(x) = E_{mec} - U(x), \qquad (8\text{-}24)$$

where E_{mec} is the mechanical energy of the system. A **turning point**

is a point x at which the particle reverses its motion (there, $K = 0$). The particle is in **equilibrium** at points where the slope of the $U(x)$ curve is zero (there, $F(x) = 0$).

Work Done on a System by an External Force Work W is energy transferred to or from a system by means of an external force acting on the system. When more than one force acts on a system, their *net work* is the transferred energy. When friction is not involved, the work done on the system and the change ΔE_{mec} in the mechanical energy of the system are equal:

$$W = \Delta E_{mec} = \Delta K + \Delta U. \qquad (8\text{-}26, 8\text{-}25)$$

When a kinetic frictional force acts within the system, then the thermal energy E_{th} of the system changes. (This energy is associated with the random motion of atoms and molecules in the system.) The work done on the system is then

$$W = \Delta E_{mec} + \Delta E_{th}. \qquad (8\text{-}33)$$

The change ΔE_{th} is related to the magnitude f_k of the frictional force and the magnitude d of the displacement caused by the external force by

$$\Delta E_{th} = f_k d. \qquad (8\text{-}31)$$

Conservation of Energy The **total energy** E of a system (the sum of its mechanical energy and its internal energies, including thermal energy) can change only by amounts of energy that are transferred to or from the system. This experimental fact is known as the **law of conservation of energy**. If work W is done on the system, then

$$W = \Delta E = \Delta E_{mec} + \Delta E_{th} + \Delta E_{int}. \qquad (8\text{-}35)$$

If the system is isolated ($W = 0$), this gives

$$\Delta E_{mec} + \Delta E_{th} + \Delta E_{int} = 0 \qquad (8\text{-}36)$$

and

$$E_{mec,2} = E_{mec,1} - \Delta E_{th} - \Delta E_{int}, \qquad (8\text{-}37)$$

where the subscripts 1 and 2 refer to two different instants.

Power The **power** due to a force is the *rate* at which that force transfers energy. If an amount of energy ΔE is transferred by a force in an amount of time Δt, the **average power** of the force is

$$P_{avg} = \frac{\Delta E}{\Delta t}. \qquad (8\text{-}40)$$

The **instantaneous power** due to a force is

$$P = \frac{dE}{dt}. \qquad (8\text{-}41)$$

QUESTIONS

1 In Fig. 8-18, a horizontally moving block can take three frictionless routes, differing only in elevation, to reach the dashed finish line.

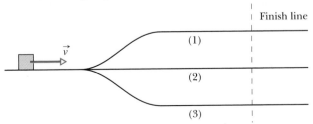

Fig. 8-18 Question 1.

Rank the routes according to (a) the speed of the block at the finish line and (b) the travel time of the block to the finish line, greatest first.

2 Figure 8-19 gives the potential energy function of a particle.

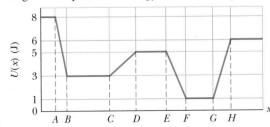

Fig. 8-19 Question 2.

(a) Rank regions *AB, BC, CD*, and *DE* according to the magnitude of the force on the particle, greatest first. What value must the mechanical energy E_{mec} of the particle not exceed if the particle is to be (b) trapped in the potential well at the left, (c) trapped in the potential well at the right, and (d) able to move between the two potential wells but not to the right of point *H*? For the situation of (d), in which of regions *BC, DE*, and *FG* will the particle have (e) the greatest kinetic energy and (f) the least speed?

3 Figure 8-20 shows one direct path and four indirect paths from point *i* to point *f*. Along the direct path and three of the indirect paths, only a conservative force F_c acts on a certain object. Along the fourth indirect path, both F_c and a nonconservative force F_{nc} act on the object. The change

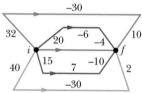

Fig. 8-20   Question 3.

ΔE_{mec} in the object's mechanical energy (in joules) in going from *i* to *f* is indicated along each straight-line segment of the indirect paths. What is ΔE_{mec} (a) from *i* to *f* along the direct path and (b) due to F_{nc} along the one path where it acts?

4 In Fig. 8-21, a small, initially stationary block is released on a frictionless ramp at a height of 3.0 m. Hill heights along the ramp are as shown. The hills have identical circular tops, and the block does not fly off any hill. (a) Which hill is the first the block cannot cross? (b) What does the block do after failing to cross that hill? On which hilltop is (c) the centripetal acceleration of the block greatest and (d) the normal force on the block least?

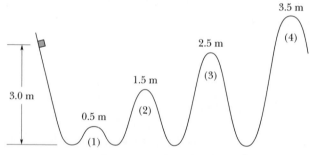

Fig. 8-21   Question 4.

5 In Fig. 8-22, a block slides from *A* to *C* along a frictionless ramp, and then it passes through horizontal region *CD*, where a frictional force acts on it. Is the block's kinetic energy increasing, decreasing, or constant in (a) region *AB*, (b) region *BC*, and (c) region *CD*? (d) Is the block's mechanical energy increasing, decreasing, or constant in those regions?

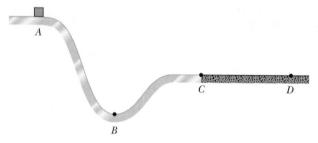

Fig. 8-22   Question 5.

6 In Fig. 8-23*a*, you pull upward on a rope that is attached to a cylinder on a vertical rod. Because the cylinder fits tightly on the rod, the cylinder slides along the rod with considerable friction. Your force

does work $W = +100$ J on the cylinder–rod–Earth system (Fig. 8-23*b*). An "energy statement" for the system is shown in Fig. 8-23*c*: the kinetic energy *K* increases by 50 J, and the gravitational potential energy U_g increases by 20 J. The only other change in energy within the system is for the thermal energy E_{th}. What is the change ΔE_{th}?

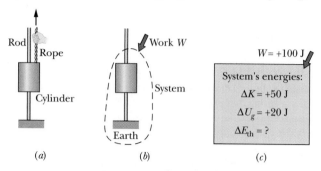

Fig. 8-23   Question 6.

7 The arrangement shown in Fig. 8-24 is similar to that in Question 6. Here you pull downward on the rope that is attached to the cylinder, which fits tightly on the rod. Also, as the cylinder descends, it pulls on a block via a second rope, and the block slides over a lab table. Again consider the cylinder–rod–Earth system, similar to that shown in Fig. 8-23*b*. Your work on the system is 200 J. The system does work of 60 J on the block. Within the system, the kinetic energy

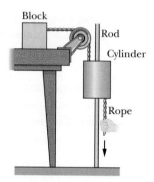

Fig. 8-24   Question 7.

increases by 130 J and the gravitational potential energy decreases by 20 J. (a) Draw an "energy statement" for the system, as in Fig. 8-23*c*. (b) What is the change in the thermal energy within the system?

8 In Fig. 8-25, a block slides along a track that descends through distance *h*. The track is frictionless except for the lower section. There the block slides to a stop in a certain distance *D* because of friction. (a) If we decrease *h*, will the block now slide to a stop in a distance that is greater than, less than, or equal to *D*? (b) If, instead, we increase the mass of the block, will the stopping distance now be greater than, less than, or equal to *D*?

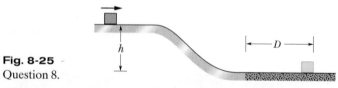

Fig. 8-25 Question 8.

9 Figure 8-26 shows three situations involving a plane that is not frictionless and a block sliding along the plane. The block begins with the same speed in all three situations and slides until the kinetic frictional force has stopped it. Rank the situations according to the increase in thermal energy due to the sliding, greatest first.

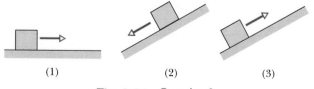

Fig. 8-26   Question 9.

sec. 8-4   Determining Potential Energy Values

•1   SSM   What is the spring constant of a spring that stores 25 J of elastic potential energy when compressed by 7.5 cm?

•2   In Fig. 8-27, a single frictionless roller-coaster car of mass $m = 825$ kg tops the first hill with speed $v_0 = 17.0$ m/s at height $h = 42.0$ m. How much work does the gravitational force do on the car from that point to (a) point A, (b) point B, and (c) point C? If the gravitational potential energy of the car–Earth system is taken to be zero at C, what is its value when the car is at (d) B and (e) A? (f) If mass m were doubled, would the change in the gravitational potential energy of the system between points A and B increase, decrease, or remain the same?

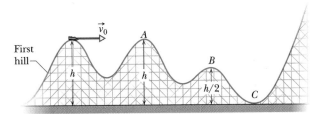

Fig. 8-27   Problems 2 and 9.

•3   You drop a 2.00 kg book to a friend who stands on the ground at distance $D = 10.0$ m below. If your friend's outstretched hands are at distance $d = 1.50$ m above the ground (Fig. 8-28), (a) how much work W_g does the gravitational force do on the book as it drops to her hands? (b) What is the change ΔU in the gravitational potential energy of the book–Earth system during the drop? If the gravitational potential energy U of that system is taken to be zero at ground level, what is U (c) when the book is released and (d) when it reaches her hands? Now take U to be 100 J at ground level and again find (e) W_g, (f) ΔU, (g) U at the release point, and (h) U at her hands.

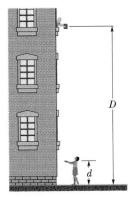

Fig. 8-28 Problems 3 and 10.

•4   Figure 8-29 shows a ball with mass $m = 0.341$ kg attached to the end of a thin rod with length $L = 0.452$ m and negligible mass. The other end of the rod is pivoted so that the ball can move in a vertical circle. The rod is held horizontally as shown and then given enough of a downward push to cause the ball to swing down and around and just reach the vertically up position, with zero speed there. How much work is done on the ball by the gravitational force from the initial point to (a) the lowest point, (b) the highest point, and (c) the point on the right level with the initial point? If the gravitational potential energy of the ball–Earth system is taken to be zero at the initial point, what is it when the ball reaches (d) the lowest point, (e) the highest point, and (f) the point on the right level with the initial point? (g) Suppose the rod were pushed harder so that the ball passed through the highest point with a nonzero speed. Would ΔU_g from the lowest point to the highest point then be greater than, less than, or the same as it was when the ball stopped at the highest point?

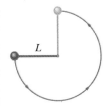

Fig. 8-29
Problems 4 and 14.

•5   SSM   In Fig. 8-30, a 2.00 g ice flake is released from the edge of a hemispherical bowl whose radius r is 22.0 cm. The flake–bowl contact is frictionless. (a) How much work is done on the flake by the gravitational force during the flake's descent to the bottom of the bowl? (b) What is the change in the potential energy of the flake–Earth system during that descent? (c) If that potential energy is taken to be zero at the bottom of the bowl, what is its value when the flake is released? (d) If, instead, the potential energy is taken to be zero at the release point, what is its value when the flake reaches the bottom of the bowl? (e) If the mass of the flake were doubled, would the magnitudes of the answers to (a) through (d) increase, decrease, or remain the same?

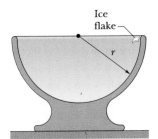

Fig. 8-30   Problems 5 and 11.

••6   In Fig. 8-31, a small block of mass $m = 0.032$ kg can slide along the frictionless loop-the-loop, with loop radius $R = 12$ cm. The block is released from rest at point P, at height $h = 5.0R$ above the bottom of the loop. How much work does the gravitational force do on the block as the block travels from point P to (a) point Q and (b) the top of the loop? If the gravitational potential energy of the block–Earth system is taken to be zero at the bottom of the loop, what is that potential energy when the block is (c) at point P, (d) at point Q, and (e) at the top of the loop? (f) If, instead of merely being released, the block is given some initial speed downward along the track, do the answers to (a) through (e) increase, decrease, or remain the same?

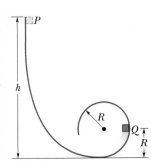

Fig. 8-31   Problems 6 and 17.

••7 Figure 8-32 shows a thin rod, of length $L = 2.00$ m and negligible mass, that can pivot about one end to rotate in a vertical circle. A ball of mass $m = 5.00$ kg is attached to the other end. The rod is pulled aside to angle $\theta_0 = 30.0°$ and released with initial velocity $\vec{v}_0 = 0$. As the ball descends to its lowest point, (a) how much work does the gravitational force do on it and (b) what is the change in the gravitational potential energy of the ball–Earth system? (c) If the gravitational potential energy is taken to be zero at the lowest point, what is its value just as the ball is released? (d) Do the magnitudes of the answers to (a) through (c) increase, decrease, or remain the same if angle θ_0 is increased?

••8 A 1.50 kg snowball is fired from a cliff 12.5 m high. The snowball's initial velocity is 14.0 m/s, directed 41.0° above the horizontal. (a) How much work is done on the snowball by the gravitational force during its flight to the flat ground below the cliff? (b) What is the change in the gravitational potential energy of the snowball–Earth system during the flight? (c) If that gravitational potential energy is taken to be zero at the height of the cliff, what is its value when the snowball reaches the ground?

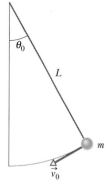

Fig. 8-32
Problems 7, 18, and 21.

sec. 8-5 Conservation of Mechanical Energy

•9 In Problem 2, what is the speed of the car at (a) point A, (b) point B, and (c) point C? (d) How high will the car go on the last hill, which is too high for it to cross? (e) If we substitute a second car with twice the mass, what then are the answers to (a) through (d)?

•10 (a) In Problem 3, what is the speed of the book when it reaches the hands? (b) If we substituted a second book with twice the mass, what would its speed be? (c) If, instead, the book were thrown down, would the answer to (a) increase, decrease, or remain the same?

•11 SSM WWW (a) In Problem 5, what is the speed of the flake when it reaches the bottom of the bowl? (b) If we substituted a second flake with twice the mass, what would its speed be? (c) If, instead, we gave the flake an initial downward speed along the bowl, would the answer to (a) increase, decrease, or remain the same?

•12 (a) In Problem 8, using energy techniques rather than the techniques of Chapter 4, find the speed of the snowball as it reaches the ground below the cliff. What is that speed (b) if the launch angle is changed to 41.0° *below* the horizontal and (c) if the mass is changed to 2.50 kg?

•13 SSM A 5.0 g marble is fired vertically upward using a spring gun. The spring must be compressed 8.0 cm if the marble is to just reach a target 20 m above the marble's position on the compressed spring. (a) What is the change ΔU_g in the gravitational potential energy of the marble–Earth system during the 20 m ascent? (b) What is the change ΔU_s in the elastic potential energy of the spring during its launch of the marble? (c) What is the spring constant of the spring?

•14 (a) In Problem 4, what initial speed must be given the ball so that it reaches the vertically upward position with zero speed? What then is its speed at (b) the lowest point and (c) the point on the right at which the ball is level with the initial point? (d) If the ball's mass

were doubled, would the answers to (a) through (c) increase, decrease, or remain the same?

•15 SSM In Fig. 8-33, a runaway truck with failed brakes is moving downgrade at 130 km/h just before the driver steers the truck up a frictionless emergency escape ramp with an inclination of $\theta = 15°$. The truck's mass is 1.2×10^4 kg. (a) What minimum length L must the ramp have if the truck is to stop (momentarily) along it? (Assume the truck is a particle, and justify that assumption.) Does the minimum length L increase, decrease, or remain the same if (b) the truck's mass is decreased and (c) its speed is decreased?

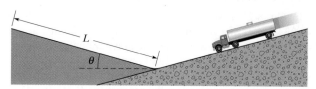

Fig. 8-33 Problem 15.

••16 A 700 g block is released from rest at height h_0 above a vertical spring with spring constant $k = 400$ N/m and negligible mass. The block sticks to the spring and momentarily stops after compressing the spring 19.0 cm. How much work is done (a) by the block on the spring and (b) by the spring on the block? (c) What is the value of h_0? (d) If the block were released from height $2.00h_0$ above the spring, what would be the maximum compression of the spring?

••17 In Problem 6, what are the magnitudes of (a) the horizontal component and (b) the vertical component of the *net* force acting on the block at point Q? (c) At what height h should the block be released from rest so that it is on the verge of losing contact with the track at the top of the loop? (*On the verge of losing contact* means that the normal force on the block from the track has just then become zero.) (d) Graph the magnitude of the normal force on the block at the top of the loop versus initial height h, for the range $h = 0$ to $h = 6R$.

••18 (a) In Problem 7, what is the speed of the ball at the lowest point? (b) Does the speed increase, decrease, or remain the same if the mass is increased?

••19 Figure 8-34 shows an 8.00 kg stone at rest on a spring. The spring is compressed 10.0 cm by the stone. (a) What is the spring constant? (b) The stone is pushed down an additional 30.0 cm and released. What is the elastic potential energy of the compressed spring just before that release? (c) What is the change in the gravitational potential energy of the stone–Earth system when the stone moves from the release point to its maximum height? (d) What is that maximum height, measured from the release point?

Fig. 8-34
Problem 19.

••20 A pendulum consists of a 2.0 kg stone swinging on a 4.0 m string of negligible mass. The stone has a speed of 8.0 m/s when it passes its lowest point. (a) What is the speed when the string is at 60° to the vertical? (b) What is the greatest angle with the vertical that the string will reach during the stone's motion? (c) If the potential energy of the pendulum–Earth system is taken to be zero at the stone's lowest point, what is the total mechanical energy of the system?

••21   Figure 8-32 shows a pendulum of length $L = 1.25$ m. Its bob (which effectively has all the mass) has speed v_0 when the cord makes an angle $\theta_0 = 40.0°$ with the vertical. (a) What is the speed of the bob when it is in its lowest position if $v_0 = 8.00$ m/s? What is the least value that v_0 can have if the pendulum is to swing down and then up (b) to a horizontal position, and (c) to a vertical position with the cord remaining straight? (d) Do the answers to (b) and (c) increase, decrease, or remain the same if θ_0 is increased by a few degrees?

••22 ✈ A 60 kg skier starts from rest at height $H = 20$ m above the end of a ski-jump ramp (Fig. 8-35) and leaves the ramp at angle $\theta = 28°$. Neglect the effects of air resistance and assume the ramp is frictionless. (a) What is the maximum height h of his jump above the end of the ramp? (b) If he increased his weight by putting on a backpack, would h then be greater, less, or the same?

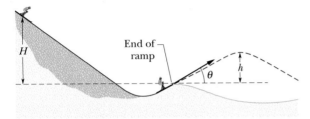

End of
ramp

H

θ

h

Fig. 8-35   Problem 22.

••23 ILW The string in Fig. 8-36 is $L = 120$ cm long, has a ball attached to one end, and is fixed at its other end. The distance d from the fixed end to a fixed peg at point P is 75.0 cm. When the initially stationary ball is released with the string horizontal as shown, it will swing along the dashed arc. What is its speed when it reaches (a) its lowest point and (b) its highest point after the string catches on the peg?

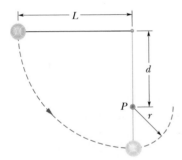

L

d

P

r

Fig. 8-36   Problems 23 and 70.

••24   A block of mass $m = 2.0$ kg is dropped from height $h = 40$ cm onto a spring of spring constant $k = 1960$ N/m (Fig. 8-37). Find the maximum distance the spring is compressed.

••25   At $t = 0$ a 1.0 kg ball is thrown from a tall tower with $\vec{v} = (18$ m/s$)\hat{i} + (24$ m/s$)\hat{j}$. What is ΔU of the ball–Earth system between $t = 0$ and $t = 6.0$ s (still free fall)?

••26   A conservative force $\vec{F} = (6.0x - 12)\hat{i}$ N, where x is in meters, acts on a particle moving along an x axis. The potential energy U associated with this force is assigned a value of 27 J at $x = 0$. (a) Write an expression for U as a function of x,

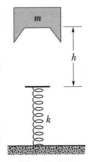

m

h

k

Fig. 8-37
Problem 24.

with U in joules and x in meters. (b) What is the maximum positive potential energy? At what (c) negative value and (d) positive value of x is the potential energy equal to zero?

••27   Tarzan, who weighs 688 N, swings from a cliff at the end of a vine 18 m long (Fig. 8-38). From the top of the cliff to the bottom of the swing, he descends by 3.2 m. The vine will break if the force on it exceeds 950 N. (a) Does the vine break? (b) If no, what is the greatest force on it during the swing? If yes, at what angle with the vertical does it break?

Fig. 8-38   Problem 27.

••28   Figure 8-39a applies to the spring in a cork gun (Fig. 8-39b); it shows the spring force as a function of the stretch or compression of the spring. The spring is compressed by 5.5 cm and used to propel a 3.8 g cork from the gun. (a) What is the speed of the cork if it is released as the spring passes through its relaxed position? (b) Suppose, instead, that the cork sticks to the spring and stretches it 1.5 cm before separation occurs. What now is the speed of the cork at the time of release?

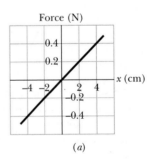

Force (N)

0.4

0.2

x (cm)

−4 −2

2 4

−0.2

−0.4

(a)

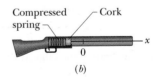

Compressed
spring

Cork

x

0

(b)

Fig. 8-39   Problem 28.

••29 SSM WWW In Fig. 8-40, a block of mass $m = 12$ kg is released from rest on a frictionless incline of angle $\theta = 30°$. Below the block is a spring that can be compressed 2.0 cm by a force of 270 N. The block momentarily stops when it compresses the spring by 5.5 cm. (a) How far does the block move down the incline from its rest position to this stopping point? (b) What is the speed of the block just as it touches the spring?

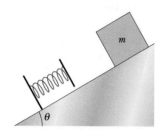

Fig. 8-40 Problems 29 and 35.

••30 GO A 2.0 kg breadbox on a frictionless incline of angle $\theta =$ 40° is connected, by a cord that runs over a pulley, to a light spring of spring constant $k = 120$ N/m, as shown in Fig. 8-41. The box is released from rest when the spring is unstretched. Assume that the pulley is massless and frictionless. (a) What is the speed of the box when it has moved 10 cm down the incline? (b) How far down the incline from its point of release does the box slide before momentarily stopping, and what are the (c) magnitude and (d) direction (up or down the incline) of the box's acceleration at the instant the box momentarily stops?

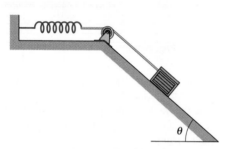

Fig. 8-41 Problem 30.

••31 ILW A block with mass $m = 2.00$ kg is placed against a spring on a frictionless incline with angle $\theta = 30.0°$ (Fig. 8-42). (The block is not attached to the spring.) The spring, with spring constant $k = 19.6$ N/cm, is compressed 20.0 cm and then released. (a) What is the elastic potential energy of the compressed spring? (b) What is the change in the gravitational potential energy of the block–Earth system as the block moves from the release point to its highest point on the incline? (c) How far along the incline is the highest point from the release point?

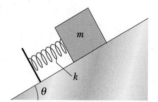

Fig. 8-42 Problem 31.

••32 In Fig. 8-43, a chain is held on a frictionless table with one-fourth of its length hanging over the edge. If the chain has length

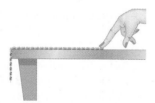

Fig. 8-43 Problem 32.

$L = 28$ cm and mass $m = 0.012$ kg, how much work is required to pull the hanging part back onto the table?

•••33 GO In Fig. 8-44, a spring with $k = 170$ N/m is at the top of a frictionless incline of angle $\theta = 37.0°$. The lower end of the incline is distance $D = 1.00$ m from the end of the spring, which is at its relaxed length. A 2.00 kg canister is pushed against the spring until the spring is compressed 0.200 m and released from rest. (a) What is the speed of the canister at the instant the spring returns to its relaxed length (which is when the canister loses contact with the spring)? (b) What is the speed of the canister when it reaches the lower end of the incline?

Fig. 8-44 Problem 33.

•••34 GO A boy is initially seated on the top of a hemispherical ice mound of radius $R = 13.8$ m. He begins to slide down the ice, with a negligible initial speed (Fig. 8-45). Approximate the ice as being frictionless. At what height does the boy lose contact with the ice?

Fig. 8-45 Problem 34.

•••35 In Fig. 8-40, a block of mass $m = 3.20$ kg slides from rest a distance d down a frictionless incline at angle $\theta = 30.0°$ where it runs into a spring of spring constant 431 N/m. When the block momentarily stops, it has compressed the spring by 21.0 cm. What are (a) distance d and (b) the distance between the point of the first block–spring contact and the point where the block's speed is greatest?

•••36 GO Two children are playing a game in which they try to hit a small box on the floor with a marble fired from a spring-loaded gun that is mounted on a table. The target box is horizontal distance $D = 2.20$ m from the edge of the table; see Fig. 8-46. Bobby

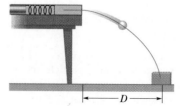

Fig. 8-46 Problem 36.

compresses the spring 1.10 cm, but the center of the marble falls 27.0 cm short of the center of the box. How far should Rhoda compress the spring to score a direct hit? Assume that neither the spring nor the ball encounters friction in the gun.

•••37   A uniform cord of length 25 cm and mass 15 g is initially stuck to a ceiling. Later, it hangs vertically from the ceiling with only one end still stuck. What is the change in the gravitational potential energy of the cord with this change in orientation? (*Hint:* Consider a differential slice of the cord and then use integral calculus.)

sec. 8-6   Reading a Potential Energy Curve

••38   Figure 8-47 shows a plot of potential energy U versus position x of a 0.200 kg particle that can travel only along an x axis under the influence of a conservative force. The graph has these values: $U_A = 9.00$ J, $U_C = 20.00$ J, and $U_D = 24.00$ J. The particle is released at the point where U forms a "potential hill" of "height" $U_B = 12.00$ J, with kinetic energy 4.00 J. What is the speed of the particle at (a) $x = 3.5$ m and (b) $x = 6.5$ m? What is the position of the turning point on (c) the right side and (d) the left side?

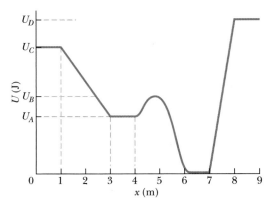

Fig. 8-47   Problem 38.

••39   GO   Figure 8-48 shows a plot of potential energy U versus position x of a 0.90 kg particle that can travel only along an x axis. (Nonconservative forces are not involved.) Three values are $U_A = 15.0$ J, $U_B = 35.0$ J, and $U_C = 45.0$ J. The particle is released at $x = 4.5$ m with an initial speed of 7.0 m/s, headed in the negative x direction. (a) If the particle can reach $x = 1.0$ m, what is its speed there, and if it cannot, what is its turning point? What are the (b) magnitude and (c) direction of the force on the particle as it begins to move to the left of $x = 4.0$ m? Suppose, instead, the particle is headed in the positive x direction when it is released at $x = 4.5$ m at speed 7.0 m/s. (d) If the particle can reach $x = 7.0$ m, what is its speed there, and if it cannot, what is its turning point? What are the (e) magnitude and (f) direction of the force on the particle as it begins to move to the right of $x = 5.0$ m?

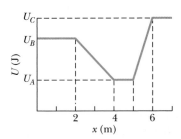

Fig. 8-48   Problem 39.

••40   The potential energy of a diatomic molecule (a two-atom system like H_2 or O_2) is given by

$$U = \frac{A}{r^{12}} - \frac{B}{r^6},$$

where r is the separation of the two atoms of the molecule and A and B are positive constants. This potential energy is associated with the force that binds the two atoms together. (a) Find the *equilibrium separation*—that is, the distance between the atoms at which the force on each atom is zero. Is the force repulsive (the atoms are pushed apart) or attractive (they are pulled together) if their separation is (b) smaller and (c) larger than the equilibrium separation?

•••41   A single conservative force $F(x)$ acts on a 1.0 kg particle that moves along an x axis. The potential energy $U(x)$ associated with $F(x)$ is given by

$$U(x) = -4x\,e^{-x/4} \text{ J},$$

where x is in meters. At $x = 5.0$ m the particle has a kinetic energy of 2.0 J. (a) What is the mechanical energy of the system? (b) Make a plot of $U(x)$ as a function of x for $0 \le x \le 10$ m, and on the same graph draw the line that represents the mechanical energy of the system. Use part (b) to determine (c) the least value of x the particle can reach and (d) the greatest value of x the particle can reach. Use part (b) to determine (e) the maximum kinetic energy of the particle and (f) the value of x at which it occurs. (g) Determine an expression in newtons and meters for $F(x)$ as a function of x. (h) For what (finite) value of x does $F(x) = 0$?

sec. 8-7   Work Done on a System by an External Force

•42   A worker pushed a 27 kg block 9.2 m along a level floor at constant speed with a force directed 32° below the horizontal. If the coefficient of kinetic friction between block and floor was 0.20, what were (a) the work done by the worker's force and (b) the increase in thermal energy of the block–floor system?

•43   A collie drags its bed box across a floor by applying a horizontal force of 8.0 N. The kinetic frictional force acting on the box has magnitude 5.0 N. As the box is dragged through 0.70 m along the way, what are (a) the work done by the collie's applied force and (b) the increase in thermal energy of the bed and floor?

••44   A horizontal force of magnitude 35.0 N pushes a block of mass 4.00 kg across a floor where the coefficient of kinetic friction is 0.600. (a) How much work is done by that applied force on the block–floor system when the block slides through a displacement of 3.00 m across the floor? (b) During that displacement, the thermal energy of the block increases by 40.0 J. What is the increase in thermal energy of the floor? (c) What is the increase in the kinetic energy of the block?

••45   SSM   A rope is used to pull a 3.57 kg block at constant speed 4.06 m along a horizontal floor. The force on the block from the rope is 7.68 N and directed 15.0° above the horizontal. What are (a) the work done by the rope's force, (b) the increase in thermal energy of the block–floor system, and (c) the coefficient of kinetic friction between the block and floor?

sec. 8-8   Conservation of Energy

•46   An outfielder throws a baseball with an initial speed of 81.8 mi/h. Just before an infielder catches the ball at the same level, the ball's speed is 110 ft/s. In foot-pounds, by how much is the mechanical energy of the ball–Earth system reduced because of air drag? (The weight of a baseball is 9.0 oz.)

•47 A 75 g Frisbee is thrown from a point 1.1 m above the ground with a speed of 12 m/s. When it has reached a height of 2.1 m, its speed is 10.5 m/s. What was the reduction in E_{mec} of the Frisbee–Earth system because of air drag?

•48 In Fig. 8-49, a block slides down an incline. As it moves from point A to point B, which are 5.0 m apart, force $\vec{F}$ acts on the block, with magnitude 2.0 N and directed down the incline. The magnitude of the frictional force acting on the block is 10 N. If the kinetic energy of the block increases by 35 J between A and B, how much work is done on the block by the gravitational force as the block moves from A to B?

Fig. 8-49 Problems 48 and 71.

•49 SSM ILW A 25 kg bear slides, from rest, 12 m down a lodgepole pine tree, moving with a speed of 5.6 m/s just before hitting the ground. (a) What change occurs in the gravitational potential energy of the bear–Earth system during the slide? (b) What is the kinetic energy of the bear just before hitting the ground? (c) What is the average frictional force that acts on the sliding bear?

•50 A 60 kg skier leaves the end of a ski-jump ramp with a velocity of 24 m/s directed 25° above the horizontal. Suppose that as a result of air drag the skier returns to the ground with a speed of 22 m/s, landing 14 m vertically below the end of the ramp. From the launch to the return to the ground, by how much is the mechanical energy of the skier–Earth system reduced because of air drag?

•51 During a rockslide, a 520 kg rock slides from rest down a hillside that is 500 m long and 300 m high. The coefficient of kinetic friction between the rock and the hill surface is 0.25. (a) If the gravitational potential energy U of the rock–Earth system is zero at the bottom of the hill, what is the value of U just before the slide? (b) How much energy is transferred to thermal energy during the slide? (c) What is the kinetic energy of the rock as it reaches the bottom of the hill? (d) What is its speed then?

••52 A large fake cookie sliding on a horizontal surface is attached to one end of a horizontal spring with spring constant $k = 400$ N/m; the other end of the spring is fixed in place. The cookie has a kinetic energy of 20.0 J as it passes through the spring's equilibrium position. As the cookie slides, a frictional force of magnitude 10.0 N acts on it. (a) How far will the cookie slide from the equilibrium position before coming momentarily to rest? (b) What will be the kinetic energy of the cookie as it slides back through the equilibrium position?

••53 GO In Fig. 8-50, a 3.5 kg block is accelerated from rest by a compressed spring of spring constant 640 N/m. The block leaves

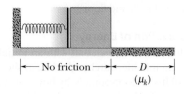

Fig. 8-50 Problem 53.

the spring at the spring's relaxed length and then travels over a horizontal floor with a coefficient of kinetic friction $\mu_k = 0.25$. The frictional force stops the block in distance $D = 7.8$ m. What are (a) the increase in the thermal energy of the block–floor system, (b) the maximum kinetic energy of the block, and (c) the original compression distance of the spring?

••54 A child whose weight is 267 N slides down a 6.1 m playground slide that makes an angle of 20° with the horizontal. The coefficient of kinetic friction between slide and child is 0.10. (a) How much energy is transferred to thermal energy? (b) If she starts at the top with a speed of 0.457 m/s, what is her speed at the bottom?

••55 ILW In Fig. 8-51, a block of mass $m = 2.5$ kg slides head on into a spring of spring constant $k = 320$ N/m. When the block stops, it has compressed the spring by 7.5 cm. The coefficient of kinetic friction between block and floor is 0.25. While the block is in contact with the spring and being brought to rest, what are (a) the work done by the spring force and (b) the increase in thermal energy of the block–floor system? (c) What is the block's speed just as it reaches the spring?

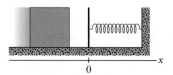

Fig. 8-51 Problem 55.

••56 You push a 2.0 kg block against a horizontal spring, compressing the spring by 15 cm. Then you release the block, and the spring sends it sliding across a tabletop. It stops 75 cm from where you released it. The spring constant is 200 N/m. What is the block–table coefficient of kinetic friction?

••57 GO In Fig. 8-52, a block slides along a track from one level to a higher level after passing through an intermediate valley. The track is frictionless until the block reaches the higher level. There a frictional force stops the block in a distance d. The block's initial speed v_0 is 6.0 m/s, the height difference h is 1.1 m, and μ_k is 0.60. Find d.

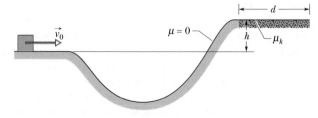

Fig. 8-52 Problem 57.

••58 A cookie jar is moving up a 40° incline. At a point 55 cm from the bottom of the incline (measured along the incline), the jar has a speed of 1.4 m/s. The coefficient of kinetic friction between jar and incline is 0.15. (a) How much farther up the incline will the jar move? (b) How fast will it be going when it has slid back to the bottom of the incline? (c) Do the answers to (a) and (b) increase, decrease, or remain the same if we decrease the coefficient of kinetic friction (but do not change the given speed or location)?

••59 A stone with a weight of 5.29 N is launched vertically from ground level with an initial speed of 20.0 m/s, and the air drag on it is 0.265 N throughout the flight. What are (a) the maximum height reached by the stone and (b) its speed just before it hits the ground?

••60 A 4.0 kg bundle starts up a 30° incline with 128 J of kinetic energy. How far will it slide up the incline if the coefficient of kinetic friction between bundle and incline is 0.30?

••61 [illustration] When a click beetle is upside down on its back, it jumps upward by suddenly arching its back, transferring energy stored in a muscle to mechanical energy. This launching mechanism produces an audible click, giving the beetle its name. Videotape of a certain click-beetle jump shows that a beetle of mass $m = 4.0 \times 10^{-6}$ kg moved directly upward by 0.77 mm during the launch and then to a maximum height of $h = 0.30$ m. During the launch, what are the average magnitudes of (a) the external force on the beetle's back from the floor and (b) the acceleration of the beetle in terms of g?

•••62 [GO] In Fig. 8-53, a block slides along a path that is without friction until the block reaches the section of length $L = 0.75$ m, which begins at height $h = 2.0$ m on a ramp of angle $\theta = 30°$. In that section, the coefficient of kinetic friction is 0.40. The block passes through point A with a speed of 8.0 m/s. If the block can reach point B (where the friction ends), what is its speed there, and if it cannot, what is its greatest height above A?

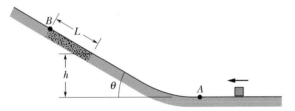

Fig. 8-53 Problem 62.

•••63 The cable of the 1800 kg elevator cab in Fig. 8-54 snaps when the cab is at rest at the first floor, where the cab bottom is a distance $d = 3.7$ m above a spring of spring constant $k = 0.15$ MN/m. A safety device clamps the cab against guide rails so that a constant frictional force of 4.4 kN opposes the cab's motion. (a) Find the speed of the cab just before it hits the spring. (b) Find the maximum distance x that the spring is compressed (the frictional force still acts during this compression). (c) Find the distance that the cab will bounce back up the shaft. (d) Using conservation of energy, find the approximate total distance that the cab will move before coming to rest. (Assume that the frictional force on the cab is negligible when the cab is stationary.)

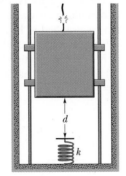

Fig. 8-54
Problem 63.

•••64 In Fig. 8-55, a block is released from rest at height $d = 40$ cm and slides down a frictionless ramp and onto a first plateau, which has length d and where the coefficient of kinetic friction is 0.50. If the block is still moving, it then slides down a second frictionless ramp through height $d/2$ and onto a lower plateau, which has length $d/2$ and where the coefficient of kinetic friction is again 0.50. If the block is still moving, it then slides up a frictionless ramp until it (momentarily) stops. Where does the block stop? If its final

stop is on a plateau, state which one and give the distance L from the left edge of that plateau. If the block reaches the ramp, give the height H above the lower plateau where it momentarily stops.

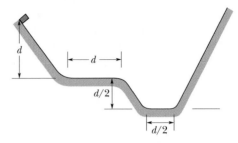

Fig. 8-55 Problem 64.

•••65 A particle can slide along a track with elevated ends and a flat central part, as shown in Fig. 8-56. The flat part has length $L = 40$ cm. The curved portions of the track are frictionless, but for the flat part the coefficient of kinetic friction is $\mu_k = 0.20$. The particle is released from rest at point A, which is at height $h = L/2$. How far from the left edge of the flat part does the particle finally stop?

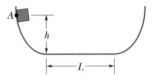

Fig. 8-56 Problem 65.

Additional Problems

66 A 3.2 kg sloth hangs 3.0 m above the ground. (a) What is the gravitational potential energy of the sloth–Earth system if we take the reference point $y = 0$ to be at the ground? If the sloth drops to the ground and air drag on it is assumed to be negligible, what are the (b) kinetic energy and (c) speed of the sloth just before it reaches the ground?

67 **SSM** A spring ($k = 200$ N/m) is fixed at the top of a frictionless plane inclined at angle $\theta = 40°$ (Fig. 8-57). A 1.0 kg block is projected up the plane, from an initial position that is distance $d = 0.60$ m from the end of the relaxed spring, with an initial kinetic energy of 16 J. (a) What is the kinetic energy of the block at the instant it has compressed the spring 0.20 m? (b) With what kinetic energy must the block be projected up the plane if it is to stop momentarily when it has compressed the spring by 0.40 m?

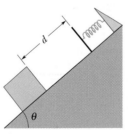

Fig. 8-57 Problem 67.

68 From the edge of a cliff, a 0.55 kg projectile is launched with an initial kinetic energy of 1550 J. The projectile's maximum upward displacement from the launch point is +140 m. What are the

(a) horizontal and (b) vertical components of its launch velocity? (c) At the instant the vertical component of its velocity is 65 m/s, what is its vertical displacement from the launch point?

69 SSM In Fig. 8-58, the pulley has negligible mass, and both it and the inclined plane are frictionless. Block *A* has a mass of 1.0 kg, block *B* has a mass of 2.0 kg, and angle θ is 30°. If the blocks are released from rest with the connecting cord taut, what is their total kinetic energy when block *B* has fallen 25 cm?

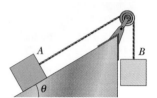

Fig. 8-58 Problem 69.

70 In Fig. 8-36, the string is $L = 120$ cm long, has a ball attached to one end, and is fixed at its other end. A fixed peg is at point *P*. Released from rest, the ball swings down until the string catches on the peg; then the ball swings up, around the peg. If the ball is to swing completely around the peg, what value must distance *d* exceed? (*Hint:* The ball must still be moving at the top of its swing. Do you see why?)

71 SSM In Fig. 8-49, a block is sent sliding down a frictionless ramp. Its speeds at points *A* and *B* are 2.00 m/s and 2.60 m/s, respectively. Next, it is again sent sliding down the ramp, but this time its speed at point *A* is 4.00 m/s. What then is its speed at point *B*?

72 Two snowy peaks are at heights $H = 850$ m and $h = 750$ m above the valley between them. A ski run extends between the peaks, with a total length of 3.2 km and an average slope of $\theta = 30°$ (Fig. 8-59). (a) A skier starts from rest at the top of the higher peak. At what speed will he arrive at the top of the lower peak if he coasts without using ski poles? Ignore friction. (b) Approximately what coefficient of kinetic friction between snow and skis would make him stop just at the top of the lower peak?

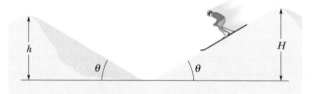

Fig. 8-59 Problem 72.

73 SSM The temperature of a plastic cube is monitored while the cube is pushed 3.0 m across a floor at constant speed by a horizontal force of 15 N. The thermal energy of the cube increases by 20 J. What is the increase in the thermal energy of the floor along which the cube slides?

74 A skier weighing 600 N goes over a frictionless circular hill of radius $R = 20$ m (Fig. 8-60). Assume

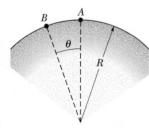

Fig. 8-60 Problem 74.

that the effects of air resistance on the skier are negligible. As she comes up the hill, her speed is 8.0 m/s at point *B*, at angle $\theta = 20°$. (a) What is her speed at the hilltop (point *A*) if she coasts without using her poles? (b) What minimum speed can she have at *B* and still coast to the hilltop? (c) Do the answers to these two questions increase, decrease, or remain the same if the skier weighs 700 N instead of 600 N?

75 SSM To form a pendulum, a 0.092 kg ball is attached to one end of a rod of length 0.62 m and negligible mass, and the other end of the rod is mounted on a pivot. The rod is rotated until it is straight up, and then it is released from rest so that it swings down around the pivot. When the ball reaches its lowest point, what are (a) its speed and (b) the tension in the rod? Next, the rod is rotated until it is horizontal, and then it is again released from rest. (c) At what angle from the vertical does the tension in the rod equal the weight of the ball? (d) If the mass of the ball is increased, does the answer to (c) increase, decrease, or remain the same?

76 We move a particle along an *x* axis, first outward from $x = 1.0$ m to $x = 4.0$ m and then back to $x = 1.0$ m, while an external force acts on it. That force is directed along the *x* axis, and its *x* component can have different values for the outward trip and for the return trip. Here are the values (in newtons) for four situations, where *x* is in meters:

Outward	Inward
(a) +3.0	−3.0
(b) +5.0	+5.0
(c) +2.0x	−2.0x
(d) +3.0x²	+3.0x²

Find the net work done on the particle by the external force *for the round trip* for each of the four situations. (e) For which, if any, is the external force conservative?

77 SSM A conservative force $F(x)$ acts on a 2.0 kg particle that moves along an *x* axis. The potential energy $U(x)$ associated with $F(x)$ is graphed in Fig. 8-61. When the particle is at $x = 2.0$ m, its velocity is −1.5 m/s. What are the (a) magnitude and (b) direction of $F(x)$ at this position? Between what positions on the (c) left and (d) right does the particle move? (e) What is the particle's speed at $x = 7.0$ m?

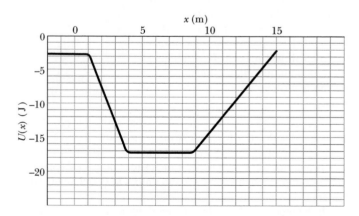

Fig. 8-61 Problem 77.

78 At a certain factory, 300 kg crates are dropped vertically from a packing machine onto a conveyor belt moving at 1.20 m/s (Fig. 8-62). (A motor maintains the belt's constant speed.) The coefficient of kinetic friction between the belt and each crate is 0.400. After a short time, slipping between the belt and the crate ceases, and the crate then moves along with the belt. For the period of time during which the crate is being brought to rest relative to the belt, calculate, for a coordinate system at rest in the factory, (a) the kinetic energy supplied to the crate, (b) the magnitude of the kinetic frictional force acting on the crate, and (c) the energy supplied by the motor. (d) Explain why answers (a) and (c) differ.

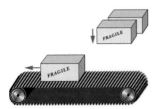

Fig. 8-62 Problem 78.

79 SSM A 1500 kg car begins sliding down a 5.0° inclined road with a speed of 30 km/h. The engine is turned off, and the only forces acting on the car are a net frictional force from the road and the gravitational force. After the car has traveled 50 m along the road, its speed is 40 km/h. (a) How much is the mechanical energy of the car reduced because of the net frictional force? (b) What is the magnitude of that net frictional force?

80 In Fig. 8-63, a 1400 kg block of granite is pulled up an incline at a constant speed of 1.34 m/s by a cable and winch. The indicated distances are $d_1 = 40$ m and $d_2 = 30$ m. The coefficient of kinetic friction between the block and the incline is 0.40. What is the power due to the force applied to the block by the cable?

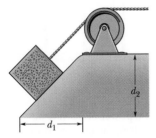

Fig. 8-63 Problem 80.

81 A particle can move along only an x axis, where conservative forces act on it (Fig. 8-64 and the following table). The particle is released at $x = 5.00$ m with a kinetic energy of $K = 14.0$ J and a potential energy of $U = 0$. If its motion is in the negative direction of the x axis, what are its (a) K and (b) U at $x = 2.00$ m and its (c) K and (d) U at $x = 0$? If its motion is in the positive direction of the x axis, what are its (e) K and (f) U at $x = 11.0$ m, its (g) K and (h) U at $x = 12.0$ m, and its (i) K and (j) U at $x = 13.0$ m? (k) Plot $U(x)$ versus x for the range $x = 0$ to $x = 13.0$ m.

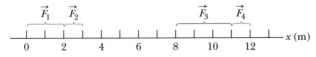

Fig. 8-64 Problems 81 and 82.

Next, the particle is released from rest at $x = 0$. What are (l) its kinetic energy at $x = 5.0$ m and (m) the maximum positive position x_{max} it reaches? (n) What does the particle do after it reaches x_{max}?

Range	Force
0 to 2.00 m	$\vec{F}_1 = +(3.00 \text{ N})\hat{i}$
2.00 m to 3.00 m	$\vec{F}_2 = +(5.00 \text{ N})\hat{i}$
3.00 m to 8.00 m	$F = 0$
8.00 m to 11.0 m	$\vec{F}_3 = -(4.00 \text{ N})\hat{i}$
11.0 m to 12.0 m	$\vec{F}_4 = -(1.00 \text{ N})\hat{i}$
12.0 m to 15.0 m	$F = 0$

82 For the arrangement of forces in Problem 81, a 2.00 kg particle is released at $x = 5.00$ m with an initial velocity of 3.45 m/s in the negative direction of the x axis. (a) If the particle can reach $x = 0$ m, what is its speed there, and if it cannot, what is its turning point? Suppose, instead, the particle is headed in the positive x direction when it is released at $x = 5.00$ m at speed 3.45 m/s. (b) If the particle can reach $x = 13.0$ m, what is its speed there, and if it cannot, what is its turning point?

83 SSM A 15 kg block is accelerated at 2.0 m/s² along a horizontal frictionless surface, with the speed increasing from 10 m/s to 30 m/s. What are (a) the change in the block's mechanical energy and (b) the average rate at which energy is transferred to the block? What is the instantaneous rate of that transfer when the block's speed is (c) 10 m/s and (d) 30 m/s?

84 A certain spring is found *not* to conform to Hooke's law. The force (in newtons) it exerts when stretched a distance x (in meters) is found to have magnitude $52.8x + 38.4x^2$ in the direction opposing the stretch. (a) Compute the work required to stretch the spring from $x = 0.500$ m to $x = 1.00$ m. (b) With one end of the spring fixed, a particle of mass 2.17 kg is attached to the other end of the spring when it is stretched by an amount $x = 1.00$ m. If the particle is then released from rest, what is its speed at the instant the stretch in the spring is $x = 0.500$ m? (c) Is the force exerted by the spring conservative or nonconservative? Explain.

85 SSM Each second, 1200 m³ of water passes over a waterfall 100 m high. Three-fourths of the kinetic energy gained by the water in falling is transferred to electrical energy by a hydroelectric generator. At what rate does the generator produce electrical energy? (The mass of 1 m³ of water is 1000 kg.)

86 GO In Fig. 8-65, a small block is sent through point A with a speed of 7.0 m/s. Its path is without friction until it reaches the section of length $L = 12$ m, where the coefficient of kinetic friction is 0.70. The indicated heights are $h_1 = 6.0$ m and $h_2 = 2.0$ m. What are the speeds of the block at (a) point B and (b) point C? (c) Does

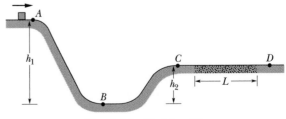

Fig. 8-65 Problem 86.

the block reach point D? If so, what is its speed there; if not, how far through the section of friction does it travel?

87 SSM A massless rigid rod of length L has a ball of mass m attached to one end (Fig. 8-66). The other end is pivoted in such a way that the ball will move in a vertical circle. First, assume that there is no friction at the pivot. The system is launched downward from the horizontal position A with initial speed v_0. The ball just barely reaches point D and then stops. (a) Derive an expression for v_0 in terms of $L, m,$ and g. (b) What is the tension in the rod when the ball passes through B? (c) A little grit is placed on the pivot to increase the friction there. Then the ball just barely reaches C when launched from A with the same speed as before. What is the decrease in the mechanical energy during this motion? (d) What is the decrease in the mechanical energy by the time the ball finally comes to rest at B after several oscillations?

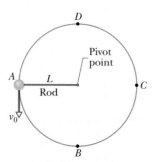

Fig. 8-66 Problem 87.

88 A 1.50 kg water balloon is shot straight up with an initial speed of 3.00 m/s. (a) What is the kinetic energy of the balloon just as it is launched? (b) How much work does the gravitational force do on the balloon during the balloon's full ascent? (c) What is the change in the gravitational potential energy of the balloon–Earth system during the full ascent? (d) If the gravitational potential energy is taken to be zero at the launch point, what is its value when the balloon reaches its maximum height? (e) If, instead, the gravitational potential energy is taken to be zero at the maximum height, what is its value at the launch point? (f) What is the maximum height?

89 A 2.50 kg beverage can is thrown directly downward from a height of 4.00 m, with an initial speed of 3.00 m/s. The air drag on the can is negligible. What is the kinetic energy of the can (a) as it reaches the ground at the end of its fall and (b) when it is halfway to the ground? What are (c) the kinetic energy of the can and (d) the gravitational potential energy of the can–Earth system 0.200 s before the can reaches the ground? For the latter, take the reference point $y = 0$ to be at the ground.

90 A constant horizontal force moves a 50 kg trunk 6.0 m up a 30° incline at constant speed. The coefficient of kinetic friction between the trunk and the incline is 0.20. What are (a) the work done by the applied force and (b) the increase in the thermal energy of the trunk and incline?

91 Two blocks, of masses $M = 2.0$ kg and $2M$, are connected to a spring of spring constant $k = 200$ N/m that has one end fixed, as shown in Fig. 8-67. The horizontal surface and the pulley are frictionless, and the pulley has negligible mass. The blocks are released from rest with the spring relaxed. (a) What is the combined kinetic energy of the two blocks when the hanging block has fallen 0.090 m? (b) What is the kinetic energy of the hanging block when it has

fallen that 0.090 m? (c) What maximum distance does the hanging block fall before momentarily stopping?

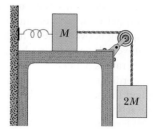

Fig. 8-67 Problem 91.

92 A volcanic ash flow is moving across horizontal ground when it encounters a 10° upslope. The front of the flow then travels 920 m up the slope before stopping. Assume that the gases entrapped in the flow lift the flow and thus make the frictional force from the ground negligible; assume also that the mechanical energy of the front of the flow is conserved. What was the initial speed of the front of the flow?

93 A playground slide is in the form of an arc of a circle that has a radius of 12 m. The maximum height of the slide is $h = 4.0$ m, and the ground is tangent to the circle (Fig. 8-68). A 25 kg child starts from rest at the top of the slide and has a speed of 6.2 m/s at the bottom. (a) What is the length of the slide? (b) What average frictional force acts on the child over this distance? If, instead of the ground, a vertical line through the *top of the slide* is tangent to the circle, what are (c) the length of the slide and (d) the average frictional force on the child?

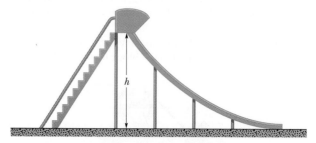

Fig. 8-68 Problem 93.

94 The luxury liner *Queen Elizabeth 2* has a diesel-electric power plant with a maximum power of 92 MW at a cruising speed of 32.5 knots. What forward force is exerted on the ship at this speed? (1 knot = 1.852 km/h.)

95 A factory worker accidentally releases a 180 kg crate that was being held at rest at the top of a ramp that is 3.7 m long and inclined at 39° to the horizontal. The coefficient of kinetic friction between the crate and the ramp, and between the crate and the horizontal factory floor, is 0.28. (a) How fast is the crate moving as it reaches the bottom of the ramp? (b) How far will it subsequently slide across the floor? (Assume that the crate's kinetic energy does not change as it moves from the ramp onto the floor.) (c) Do the answers to (a) and (b) increase, decrease, or remain the same if we halve the mass of the crate?

96 If a 70 kg baseball player steals home by sliding into the plate with an initial speed of 10 m/s just as he hits the ground, (a) what is the decrease in the player's kinetic energy and (b) what is the increase in the thermal energy of his body and the ground along which he slides?

97   A 0.50 kg banana is thrown directly upward with an initial speed of 4.00 m/s and reaches a maximum height of 0.80 m. What change does air drag cause in the mechanical energy of the banana–Earth system during the ascent?

98   A metal tool is sharpened by being held against the rim of a wheel on a grinding machine by a force of 180 N. The frictional forces between the rim and the tool grind off small pieces of the tool. The wheel has a radius of 20.0 cm and rotates at 2.50 rev/s. The coefficient of kinetic friction between the wheel and the tool is 0.320. At what rate is energy being transferred from the motor driving the wheel to the thermal energy of the wheel and tool and to the kinetic energy of the material thrown from the tool?

99   A swimmer moves through the water at an average speed of 0.22 m/s. The average drag force is 110 N. What average power is required of the swimmer?

100   An automobile with passengers has weight 16 400 N and is moving at 113 km/h when the driver brakes, sliding to a stop. The frictional force on the wheels from the road has a magnitude of 8230 N. Find the stopping distance.

101   A 0.63 kg ball thrown directly upward with an initial speed of 14 m/s reaches a maximum height of 8.1 m. What is the change in the mechanical energy of the ball–Earth system during the ascent of the ball to that maximum height?

102   The summit of Mount Everest is 8850 m above sea level. (a) How much energy would a 90 kg climber expend against the gravitational force on him in climbing to the summit from sea level? (b) How many candy bars, at 1.25 MJ per bar, would supply an energy equivalent to this? Your answer should suggest that work done against the gravitational force is a very small part of the energy expended in climbing a mountain.

103   A sprinter who weighs 670 N runs the first 7.0 m of a race in 1.6 s, starting from rest and accelerating uniformly. What are the sprinter's (a) speed and (b) kinetic energy at the end of the 1.6 s? (c) What average power does the sprinter generate during the 1.6 s interval?

104   A 20 kg object is acted on by a conservative force given by $F = -3.0x - 5.0x^2$, with F in newtons and x in meters. Take the potential energy associated with the force to be zero when the object is at $x = 0$. (a) What is the potential energy of the system associated with the force when the object is at $x = 2.0$ m? (b) If the object has a velocity of 4.0 m/s in the negative direction of the x axis when it is at $x = 5.0$ m, what is its speed when it passes through the origin? (c) What are the answers to (a) and (b) if the potential energy of the system is taken to be -8.0 J when the object is at $x = 0$?

105   A machine pulls a 40 kg trunk 2.0 m up a 40° ramp at constant velocity, with the machine's force on the trunk directed parallel to the ramp. The coefficient of kinetic friction between the trunk and the ramp is 0.40. What are (a) the work done on the trunk by the machine's force and (b) the increase in thermal energy of the trunk and the ramp?

106   The spring in the muzzle of a child's spring gun has a spring constant of 700 N/m. To shoot a ball from the gun, first the spring is compressed and then the ball is placed on it. The gun's trigger then releases the spring, which pushes the ball through the muzzle. The ball leaves the spring just as it leaves the outer end of the muzzle. When the gun is inclined upward by 30° to the horizontal, a 57 g ball is shot to a maximum height of 1.83 m above the gun's muzzle.

Assume air drag on the ball is negligible. (a) At what speed does the spring launch the ball? (b) Assuming that friction on the ball within the gun can be neglected, find the spring's initial compression distance.

107   The only force acting on a particle is conservative force $\vec{F}$. If the particle is at point A, the potential energy of the system associated with $\vec{F}$ and the particle is 40 J. If the particle moves from point A to point B, the work done on the particle by $\vec{F}$ is +25 J. What is the potential energy of the system with the particle at B?

108   In 1981, Daniel Goodwin climbed 443 m up the *exterior* of the Sears Building in Chicago using suction cups and metal clips. (a) Approximate his mass and then compute how much energy he had to transfer from biomechanical (internal) energy to the gravitational potential energy of the Earth–Goodwin system to lift himself to that height. (b) How much energy would he have had to transfer if he had, instead, taken the stairs inside the building (to the same height)?

109   A 60.0 kg circus performer slides 4.00 m down a pole to the circus floor, starting from rest. What is the kinetic energy of the performer as she reaches the floor if the frictional force on her from the pole (a) is negligible (she will be hurt) and (b) has a magnitude of 500 N?

110   A 5.0 kg block is projected at 5.0 m/s up a plane that is inclined at 30° with the horizontal. How far up along the plane does the block go (a) if the plane is frictionless and (b) if the coefficient of kinetic friction between the block and the plane is 0.40? (c) In the latter case, what is the increase in thermal energy of block and plane during the block's ascent? (d) If the block then slides back down against the frictional force, what is the block's speed when it reaches the original projection point?

111   A 9.40 kg projectile is fired vertically upward. Air drag decreases the mechanical energy of the projectile–Earth system by 68.0 kJ during the projectile's ascent. How much higher would the projectile have gone were air drag negligible?

112   A 70.0 kg man jumping from a window lands in an elevated fire rescue net 11.0 m below the window. He momentarily stops when he has stretched the net by 1.50 m. Assuming that mechanical energy is conserved during this process and that the net functions like an ideal spring, find the elastic potential energy of the net when it is stretched by 1.50 m.

113   A 30 g bullet moving a horizontal velocity of 500 m/s comes to a stop 12 cm within a solid wall. (a) What is the change in the bullet's mechanical energy? (b) What is the magnitude of the average force from the wall stopping it?

114   A 1500 kg car starts from rest on a horizontal road and gains a speed of 72 km/h in 30 s. (a) What is its kinetic energy at the end of the 30 s? (b) What is the average power required of the car during the 30 s interval? (c) What is the instantaneous power at the end of the 30 s interval, assuming that the acceleration is constant?

115   A 1.50 kg snowball is shot upward at an angle of 34.0° to the horizontal with an initial speed of 20.0 m/s. (a) What is its initial kinetic energy? (b) By how much does the gravitational potential energy of the snowball–Earth system change as the snowball moves from the launch point to the point of maximum height? (c) What is that maximum height?

116   A 68 kg sky diver falls at a constant terminal speed of 59 m/s. (a) At what rate is the gravitational potential energy of the

Earth–sky diver system being reduced? (b) At what rate is the system's mechanical energy being reduced?

117 A 20 kg block on a horizontal surface is attached to a horizontal spring of spring constant $k = 4.0$ kN/m. The block is pulled to the right so that the spring is stretched 10 cm beyond its relaxed length, and the block is then released from rest. The frictional force between the sliding block and the surface has a magnitude of 80 N. (a) What is the kinetic energy of the block when it has moved 2.0 cm from its point of release? (b) What is the kinetic energy of the block when it first slides back through the point at which the spring is relaxed? (c) What is the maximum kinetic energy attained by the block as it slides from its point of release to the point at which the spring is relaxed?

118 Resistance to the motion of an automobile consists of road friction, which is almost independent of speed, and air drag, which is proportional to speed-squared. For a certain car with a weight of 12 000 N, the total resistant force F is given by $F = 300 + 1.8v^2$, with F in newtons and v in meters per second. Calculate the power (in horsepower) required to accelerate the car at 0.92 m/s² when the speed is 80 km/h.

119 SSM A 50 g ball is thrown from a window with an initial velocity of 8.0 m/s at an angle of 30° above the horizontal. Using energy methods, determine (a) the kinetic energy of the ball at the top of its flight and (b) its speed when it is 3.0 m below the window. Does the answer to (b) depend on either (c) the mass of the ball or (d) the initial angle?

120 A spring with a spring constant of 3200 N/m is initially stretched until the elastic potential energy of the spring is 1.44 J. ($U = 0$ for the relaxed spring.) What is ΔU if the initial stretch is changed to (a) a stretch of 2.0 cm, (b) a compression of 2.0 cm, and (c) a compression of 4.0 cm?

121 A locomotive with a power capability of 1.5 MW can accelerate a train from a speed of 10 m/s to 25 m/s in 6.0 min. (a) Calculate the mass of the train. Find (b) the speed of the train and (c) the force accelerating the train as functions of time (in seconds) during the 6.0 min interval. (d) Find the distance moved by the train during the interval.

122 SSM A 0.42 kg shuffleboard disk is initially at rest when a player uses a cue to increase its speed to 4.2 m/s at constant acceleration. The acceleration takes place over a 2.0 m distance, at the end of which the cue loses contact with the disk. Then the disk slides an additional 12 m before stopping. Assume that the shuffleboard court is level and that the force of friction on the disk is constant. What is the increase in the thermal energy of the disk–court system (a) for that additional 12 m and (b) for the entire 14 m distance? (c) How much work is done on the disk by the cue?

CENTER OF MASS AND LINEAR MOMENTUM

9-1 WHAT IS PHYSICS?

Every mechanical engineer hired as an expert witness to reconstruct a traffic accident uses physics. Every trainer who coaches a ballerina on how to leap uses physics. Indeed, analyzing complicated motion of any sort requires simplification via an understanding of physics. In this chapter we discuss how the complicated motion of a system of objects, such as a car or a ballerina, can be simplified if we determine a special point of the system—the *center of mass* of that system.

Here is a quick example. If you toss a ball into the air without much spin on the ball (Fig. 9-1*a*), its motion is simple—it follows a parabolic path, as we discussed in Chapter 4, and the ball can be treated as a particle. If, instead, you flip a baseball bat into the air (Fig. 9-1*b*), its motion is more complicated. Because every part of the bat moves differently, along paths of many different shapes, you cannot represent the bat as a particle. Instead, it is a system of particles each of which follows its own path through the air. However, the bat has one special point—the center of mass—that *does* move in a simple parabolic path. The other parts of the bat move around the center of mass. (To locate the center of mass, balance the bat on an outstretched finger; the point is above your finger, on the bat's central axis.)

You cannot make a career of flipping baseball bats into the air, but you can make a career of advising long-jumpers or dancers on how to leap properly into the air while either moving their arms and legs or rotating their torso. Your starting point would be the person's center of mass because of its simple motion.

9-2 The Center of Mass

We define the **center of mass** (com) of a system of particles (such as a person) in order to predict the possible motion of the system.

> The center of mass of a system of particles is the point that moves as though (1) all of the system's mass were concentrated there and (2) all external forces were applied there.

In this section we discuss how to determine where the center of mass of a system of particles is located. We start with a system of only a few particles, and then we consider a system of a great many particles (a solid body, such as a baseball bat). Later in the chapter, we discuss how the center of mass of a system moves when external forces act on the system.

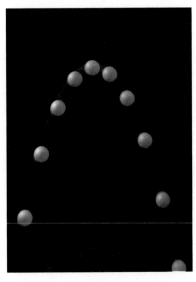

(*a*)

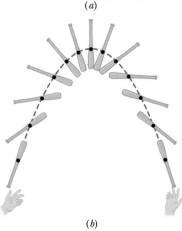

(*b*)

Fig. 9-1 (*a*) A ball tossed into the air follows a parabolic path. (*b*) The center of mass (black dot) of a baseball bat flipped into the air follows a parabolic path, but all other points of the bat follow more complicated curved paths. (*a: Richard Megna/Fundamental Photographs*)

Systems of Particles

Figure 9-2a shows two particles of masses m_1 and m_2 separated by distance d. We have arbitrarily chosen the origin of an x axis to coincide with the particle of mass m_1. We *define* the position of the center of mass (com) of this two-particle system to be

$$x_{com} = \frac{m_2}{m_1 + m_2} d. \tag{9-1}$$

Suppose, as an example, that $m_2 = 0$. Then there is only one particle, of mass m_1, and the center of mass must lie at the position of that particle; Eq. 9-1 dutifully reduces to $x_{com} = 0$. If $m_1 = 0$, there is again only one particle (of mass m_2), and we have, as we expect, $x_{com} = d$. If $m_1 = m_2$, the center of mass should be halfway between the two particles; Eq. 9-1 reduces to $x_{com} = \frac{1}{2}d$, again as we expect. Finally, Eq. 9-1 tells us that if neither m_1 nor m_2 is zero, x_{com} can have only values that lie between zero and d; that is, the center of mass must lie somewhere between the two particles.

Figure 9-2b shows a more generalized situation, in which the coordinate system has been shifted leftward. The position of the center of mass is now defined as

$$x_{com} = \frac{m_1 x_1 + m_2 x_2}{m_1 + m_2}. \tag{9-2}$$

Note that if we put $x_1 = 0$, then x_2 becomes d and Eq. 9-2 reduces to Eq. 9-1, as it must. Note also that in spite of the shift of the coordinate system, the center of mass is still the same distance from each particle.

We can rewrite Eq. 9-2 as

$$x_{com} = \frac{m_1 x_1 + m_2 x_2}{M}, \tag{9-3}$$

in which M is the total mass of the system. (Here, $M = m_1 + m_2$.) We can extend this equation to a more general situation in which n particles are strung out along the x axis. Then the total mass is $M = m_1 + m_2 + \cdots + m_n$, and the location of the center of mass is

$$x_{com} = \frac{m_1 x_1 + m_2 x_2 + m_3 x_3 + \cdots + m_n x_n}{M}$$

$$= \frac{1}{M} \sum_{i=1}^{n} m_i x_i. \tag{9-4}$$

The subscript i is an index that takes on all integer values from 1 to n.

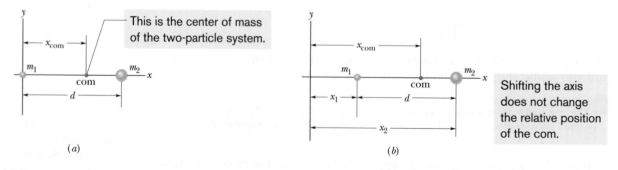

This is the center of mass of the two-particle system.

Shifting the axis does not change the relative position of the com.

(a)   (b)

Fig. 9-2   (a) Two particles of masses m_1 and m_2 are separated by distance d. The dot labeled com shows the position of the center of mass, calculated from Eq. 9-1. (b) The same as (a) except that the origin is located farther from the particles. The position of the center of mass is calculated from Eq. 9-2. The location of the center of mass with respect to the particles is the same in both cases.

If the particles are distributed in three dimensions, the center of mass must be identified by three coordinates. By extension of Eq. 9-4, they are

$$x_{\text{com}} = \frac{1}{M} \sum_{i=1}^{n} m_i x_i, \qquad y_{\text{com}} = \frac{1}{M} \sum_{i=1}^{n} m_i y_i, \qquad z_{\text{com}} = \frac{1}{M} \sum_{i=1}^{n} m_i z_i. \quad (9\text{-}5)$$

We can also define the center of mass with the language of vectors. First recall that the position of a particle at coordinates x_i, y_i, and z_i is given by a position vector:

$$\vec{r}_i = x_i \hat{\imath} + y_i \hat{\jmath} + z_i \hat{k}. \quad (9\text{-}6)$$

Here the index identifies the particle, and $\hat{\imath}$, $\hat{\jmath}$, and $\hat{k}$ are unit vectors pointing, respectively, in the positive direction of the x, y, and z axes. Similarly, the position of the center of mass of a system of particles is given by a position vector:

$$\vec{r}_{\text{com}} = x_{\text{com}} \hat{\imath} + y_{\text{com}} \hat{\jmath} + z_{\text{com}} \hat{k}. \quad (9\text{-}7)$$

The three scalar equations of Eq. 9-5 can now be replaced by a single vector equation,

$$\vec{r}_{\text{com}} = \frac{1}{M} \sum_{i=1}^{n} m_i \vec{r}_i, \quad (9\text{-}8)$$

where again M is the total mass of the system. You can check that this equation is correct by substituting Eqs. 9-6 and 9-7 into it, and then separating out the x, y, and z components. The scalar relations of Eq. 9-5 result.

Solid Bodies

An ordinary object, such as a baseball bat, contains so many particles (atoms) that we can best treat it as a continuous distribution of matter. The "particles" then become differential mass elements dm, the sums of Eq. 9-5 become integrals, and the coordinates of the center of mass are defined as

$$x_{\text{com}} = \frac{1}{M} \int x \, dm, \qquad y_{\text{com}} = \frac{1}{M} \int y \, dm, \qquad z_{\text{com}} = \frac{1}{M} \int z \, dm, \quad (9\text{-}9)$$

where M is now the mass of the object.

Evaluating these integrals for most common objects (such as a television set or a moose) would be difficult, so here we consider only *uniform* objects. Such objects have uniform *density,* or mass per unit volume; that is, the density ρ (Greek letter rho) is the same for any given element of an object as for the whole object. From Eq. 1-8, we can write

$$\rho = \frac{dm}{dV} = \frac{M}{V}, \quad (9\text{-}10)$$

where dV is the volume occupied by a mass element dm, and V is the total volume of the object. Substituting $dm = (M/V) \, dV$ from Eq. 9-10 into Eq. 9-9 gives

$$x_{\text{com}} = \frac{1}{V} \int x \, dV, \qquad y_{\text{com}} = \frac{1}{V} \int y \, dV, \qquad z_{\text{com}} = \frac{1}{V} \int z \, dV. \quad (9\text{-}11)$$

You can bypass one or more of these integrals if an object has a point, a line, or a plane of symmetry. The center of mass of such an object then lies at that point, on that line, or in that plane. For example, the center of mass of a uniform sphere (which has a point of symmetry) is at the center of the sphere (which is the point of symmetry). The center of mass of a uniform cone (whose axis is a line of symmetry) lies on the axis of the cone. The center of mass of a banana (which has a plane of symmetry that splits it into two equal parts) lies somewhere in the plane of simmetry.

The center of mass of an object need not lie within the object. There is no dough at the com of a doughnut, and no iron at the com of a horseshoe.

Sample Problem

com of plate with missing piece

Figure 9-3a shows a uniform metal plate P of radius $2R$ from which a disk of radius R has been stamped out (removed) in an assembly line. The disk is shown in Fig. 9-3b. Using the xy coordinate system shown, locate the center of mass com_P of the remaining plate.

KEY IDEAS

(1) Let us roughly locate the center of plate P by using symmetry. We note that the plate is symmetric about the x axis (we get the portion below that axis by rotating the upper portion about the axis). Thus, com_P must be on the x axis. The plate (with the disk removed) is not symmetric about the y axis. However, because there is somewhat more mass on the right of the y axis, com_P must be somewhat to the right of that axis. Thus, the location of com_P should be roughly as indicated in Fig. 9-3a. Our job here is to find the actual value of that location.

(2) Plate P is an extended solid body, so in principle we can use Eqs. 9-11 to find the actual coordinates of the center of mass of plate P. Here we are simply looking for the xy coordinates of the center of mass because the plate is thin and uniform. If it had any appreciable thickness, we would just say that the center of mass is midway across the thickness. Still, even neglecting the width, using Eqs. 9-11 would be challenging because we would need a function for the shape of the plate with its hole, and then we would need to integrate the function in two dimensions.

(3) Here is a much easier way: In working with centers of mass, we can assume that the mass of a uniform object (as we have here) is concentrated in a particle at the object's center of mass. Thus we can treat the object as a particle and avoid any two-dimensional integration.

Calculations: First, put the stamped-out disk (call it disk S) back into place (Fig. 9-3c) to form the original composite plate (call it plate C). Because of its circular symmetry, the center of mass com_S for disk S is at the center of S, at $x = -R$ (as shown). Similarly, the center of mass com_C for composite plate C is at the center of C, at the origin (as shown). We then have the following:

Plate	Center of Mass	Location of com	Mass
P	com_P	$x_P = ?$	m_P
S	com_S	$x_S = -R$	m_S
C	com_C	$x_C = 0$	$m_C = m_S + m_P$

Assume that mass m_S of disk S is concentrated in a particle at $x_S = -R$, and mass m_P is concentrated in a particle at x_P (Fig. 9-3d). Next treat these two particles as a two-particle system, using Eq. 9-2 to find their center of mass x_{S+P}. We get

$$x_{S+P} = \frac{m_S x_S + m_P x_P}{m_S + m_P}. \tag{9-12}$$

Next note that the combination of disk S and plate P is composite plate C. Thus, the position x_{S+P} of com_{S+P} must coincide with the position x_C of com_C, which is at the origin; so $x_{S+P} = x_C = 0$. Substituting this into Eq. 9-12 and solving for x_P, we get

$$x_P = -x_S \frac{m_S}{m_P}. \tag{9-13}$$

We can relate these masses to the face areas of S and P by noting that

$$\text{mass} = \text{density} \times \text{volume}$$
$$= \text{density} \times \text{thickness} \times \text{area}.$$

Then $\dfrac{m_S}{m_P} = \dfrac{\text{density}_S}{\text{density}_P} \times \dfrac{\text{thickness}_S}{\text{thickness}_P} \times \dfrac{\text{area}_S}{\text{area}_P}.$

Because the plate is uniform, the densities and thicknesses are equal; we are left with

$$\frac{m_S}{m_P} = \frac{\text{area}_S}{\text{area}_P} = \frac{\text{area}_S}{\text{area}_C - \text{area}_S}$$

$$= \frac{\pi R^2}{\pi (2R)^2 - \pi R^2} = \frac{1}{3}.$$

Substituting this and $x_S = -R$ into Eq. 9-13, we have

$$x_P = \tfrac{1}{3}R. \qquad \text{(Answer)}$$

 Additional examples, video, and practice available at *WileyPLUS*

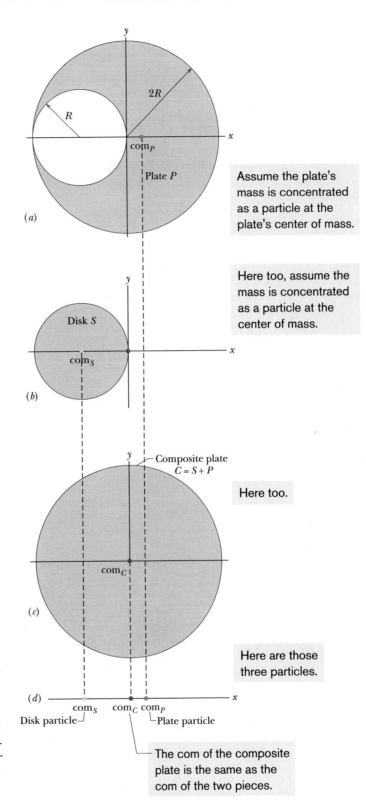

Fig. 9-3 (*a*) Plate *P* is a metal plate of radius 2*R*, with a circular hole of radius *R*. The center of mass of *P* is at point com$_P$. (*b*) Disk *S*. (*c*) Disk *S* has been put back into place to form a composite plate *C*. The center of mass com$_S$ of disk *S* and the center of mass com$_C$ of plate *C* are shown. (*d*) The center of mass com$_{S+P}$ of the combination of *S* and *P* coincides with com$_C$, which is at $x = 0$.

Sample Problem

com of three particles

Three particles of masses $m_1 = 1.2$ kg, $m_2 = 2.5$ kg, and $m_3 = 3.4$ kg form an equilateral triangle of edge length $a = 140$ cm. Where is the center of mass of this system?

KEY IDEA

We are dealing with particles instead of an extended solid body, so we can use Eq. 9-5 to locate their center of mass. The particles are in the plane of the equilateral triangle, so we need only the first two equations.

Calculations: We can simplify the calculations by choosing the x and y axes so that one of the particles is located at the

origin and the x axis coincides with one of the triangle's sides (Fig. 9-4). The three particles then have the following coordinates:

Particle	Mass (kg)	x (cm)	y (cm)
1	1.2	0	0
2	2.5	140	0
3	3.4	70	120

The total mass M of the system is 7.1 kg.

From Eq. 9-5, the coordinates of the center of mass are

$$x_{com} = \frac{1}{M}\sum_{i=1}^{3} m_i x_i = \frac{m_1 x_1 + m_2 x_2 + m_3 x_3}{M}$$

$$= \frac{(1.2\text{ kg})(0) + (2.5\text{ kg})(140\text{ cm}) + (3.4\text{ kg})(70\text{ cm})}{7.1\text{ kg}}$$

$$= 83\text{ cm} \qquad \text{(Answer)}$$

and $$y_{com} = \frac{1}{M}\sum_{i=1}^{3} m_i y_i = \frac{m_1 y_1 + m_2 y_2 + m_3 y_3}{M}$$

$$= \frac{(1.2\text{ kg})(0) + (2.5\text{ kg})(0) + (3.4\text{ kg})(120\text{ cm})}{7.1\text{ kg}}$$

$$= 58\text{ cm}. \qquad \text{(Answer)}$$

In Fig. 9-4, the center of mass is located by the position vector $\vec{r}_{com}$, which has components x_{com} and y_{com}.

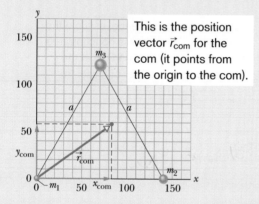

This is the position vector $\vec{r}_{com}$ for the com (it points from the origin to the com).

Fig. 9-4 Three particles form an equilateral triangle of edge length a. The center of mass is located by the position vector $\vec{r}_{com}$.

WILEY **PLUS** Additional examples, video, and practice available at *WileyPLUS*

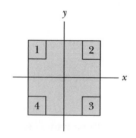

✓CHECKPOINT 1

The figure shows a uniform square plate from which four identical squares at the corners will be removed. (a) Where is the center of mass of the plate originally? Where is it after the removal of (b) square 1; (c) squares 1 and 2; (d) squares 1 and 3; (e) squares 1, 2, and 3; (f) all four squares? Answer in terms of quadrants, axes, or points (without calculation, of course).

9-3 Newton's Second Law for a System of Particles

Now that we know how to locate the center of mass of a system of particles, we discuss how external forces can move a center of mass. Let us start with a simple system of two billiard balls.

If you roll a cue ball at a second billiard ball that is at rest, you expect that the two-ball system will continue to have some forward motion after impact. You would be surprised, for example, if both balls came back toward you or if both moved to the right or to the left.

What continues to move forward, its steady motion completely unaffected by the collision, is the center of mass of the two-ball system. If you focus on this point—which is always halfway between these bodies because they have identical masses—you can easily convince yourself by trial at a billiard table that this is so. No matter whether the collision is glancing, head-on, or somewhere in between, the center of mass continues to move forward, as if the collision had never occurred. Let us look into this center-of-mass motion in more detail.

To do so, we replace the pair of billiard balls with an assemblage of n particles of (possibly) different masses. We are interested not in the individual motions of these particles but *only* in the motion of the center of mass of the assemblage. Although the center of mass is just a point, it moves like a particle whose mass is equal to the total mass of the system; we can assign a position, a velocity, and an acceleration to it. We state (and shall prove next) that the vector equation that governs the motion of the center of mass of such a system of particles is

$$\vec{F}_{net} = M\vec{a}_{com} \quad \text{(system of particles).} \quad (9\text{-}14)$$

This equation is Newton's second law for the motion of the center of mass of a system of particles. Note that its form is the same as the form of the equation ($\vec{F}_{net} = m\vec{a}$) for the motion of a single particle. However, the three quantities that appear in Eq. 9-14 must be evaluated with some care:

1. $\vec{F}_{net}$ is the net force of *all external forces* that act on the system. Forces on one part of the system from another part of the system (*internal forces*) are not included in Eq. 9-14.

2. M is the *total mass* of the system. We assume that no mass enters or leaves the system as it moves, so that M remains constant. The system is said to be **closed.**

3. $\vec{a}_{com}$ is the acceleration of the *center of mass* of the system. Equation 9-14 gives no information about the acceleration of any other point of the system.

Equation 9-14 is equivalent to three equations involving the components of $\vec{F}_{net}$ and $\vec{a}_{com}$ along the three coordinate axes. These equations are

$$F_{net,x} = Ma_{com,x} \qquad F_{net,y} = Ma_{com,y} \qquad F_{net,z} = Ma_{com,z}. \qquad (9\text{-}15)$$

Now we can go back and examine the behavior of the billiard balls. Once the cue ball has begun to roll, no net external force acts on the (two-ball) system. Thus, because $\vec{F}_{net} = 0$, Eq. 9-14 tells us that $\vec{a}_{com} = 0$ also. Because acceleration is the rate of change of velocity, we conclude that the velocity of the center of mass of the system of two balls does not change. When the two balls collide, the forces that come into play are *internal* forces, on one ball from the other. Such forces do not contribute to the net force $\vec{F}_{net}$, which remains zero. Thus, the center of mass of the system, which was moving forward before the collision, must continue to move forward after the collision, with the same speed and in the same direction.

Equation 9-14 applies not only to a system of particles but also to a solid body, such as the bat of Fig. 9-1b. In that case, M in Eq. 9-14 is the mass of the bat and $\vec{F}_{net}$ is the gravitational force on the bat. Equation 9-14 then tells us that $\vec{a}_{com} = \vec{g}$. In other words, the center of mass of the bat moves as if the bat were a single particle of mass M, with force $\vec{F}_g$ acting on it.

Figure 9-5 shows another interesting case. Suppose that at a fireworks display, a rocket is launched on a parabolic path. At a certain point, it explodes into fragments. If the explosion had not occurred, the rocket would have continued along the trajectory shown in the figure. The forces of the explosion are *internal* to the system (at first the system is just the rocket, and later it is its fragments); that is, they are forces on parts of the system from other parts. If we ignore air drag, the net *external* force $\vec{F}_{net}$ acting on the system is the gravitational force on the system, regardless of whether the rocket explodes. Thus, from Eq. 9-14, the acceleration $\vec{a}_{com}$ of the center of mass of the fragments (while they are in flight) remains equal to $\vec{g}$. This means that the center of mass of the fragments follows the same parabolic trajectory that the rocket would have followed had it not exploded.

The internal forces of the explosion cannot change the path of the com.

Fig. 9-5 A fireworks rocket explodes in flight. In the absence of air drag, the center of mass of the fragments would continue to follow the original parabolic path, until fragments began to hit the ground.

Path of head

Path of center of mass

Fig. 9-6 A grand jeté. (Adapted from *The Physics of Dance,* by Kenneth Laws, Schirmer Books, 1984.)

When a ballet dancer leaps across the stage in a grand jeté, she raises her arms and stretches her legs out horizontally as soon as her feet leave the stage (Fig. 9-6). These actions shift her center of mass upward through her body. Although the shifting center of mass faithfully follows a parabolic path across the stage, its movement relative to the body decreases the height that is attained by her head and torso, relative to that of a normal jump. The result is that the head and torso follow a nearly horizontal path, giving an illusion that the dancer is floating.

Proof of Equation 9-14

Now let us prove this important equation. From Eq. 9-8 we have, for a system of n particles,

$$M\vec{r}_{com} = m_1\vec{r}_1 + m_2\vec{r}_2 + m_3\vec{r}_3 + \cdots + m_n\vec{r}_n, \tag{9-16}$$

in which M is the system's total mass and $\vec{r}_{com}$ is the vector locating the position of the system's center of mass.

Differentiating Eq. 9-16 with respect to time gives

$$M\vec{v}_{com} = m_1\vec{v}_1 + m_2\vec{v}_2 + m_3\vec{v}_3 + \cdots + m_n\vec{v}_n. \tag{9-17}$$

Here $\vec{v}_i \ (= d\vec{r}_i/dt)$ is the velocity of the ith particle, and $\vec{v}_{com} \ (= d\vec{r}_{com}/dt)$ is the velocity of the center of mass.

Differentiating Eq. 9-17 with respect to time leads to

$$M\vec{a}_{com} = m_1\vec{a}_1 + m_2\vec{a}_2 + m_3\vec{a}_3 + \cdots + m_n\vec{a}_n. \tag{9-18}$$

Here $\vec{a}_i \ (= d\vec{v}_i/dt)$ is the acceleration of the ith particle, and $\vec{a}_{com} \ (= d\vec{v}_{com}/dt)$ is the acceleration of the center of mass. Although the center of mass is just a geometrical point, it has a position, a velocity, and an acceleration, as if it were a particle.

From Newton's second law, $m_i\vec{a}_i$ is equal to the resultant force $\vec{F}_i$ that acts on the ith particle. Thus, we can rewrite Eq. 9-18 as

$$M\vec{a}_{com} = \vec{F}_1 + \vec{F}_2 + \vec{F}_3 + \cdots + \vec{F}_n. \tag{9-19}$$

Among the forces that contribute to the right side of Eq. 9-19 will be forces that the particles of the system exert on each other (internal forces) and forces exerted on the particles from outside the system (external forces). By Newton's third law, the internal forces form third-law force pairs and cancel out in the sum that appears on the right side of Eq. 9-19. What remains is the vector sum of all the *external* forces that act on the system. Equation 9-19 then reduces to Eq. 9-14, the relation that we set out to prove.

✓CHECKPOINT 2

Two skaters on frictionless ice hold opposite ends of a pole of negligible mass. An axis runs along it, with the origin at the center of mass of the two-skater system. One skater, Fred, weighs twice as much as the other skater, Ethel. Where do the skaters meet if (a) Fred pulls hand over hand along the pole so as to draw himself to Ethel, (b) Ethel pulls hand over hand to draw herself to Fred, and (c) both skaters pull hand over hand?

Sample Problem

Motion of the com of three particles

The three particles in Fig. 9-7a are initially at rest. Each experiences an *external* force due to bodies outside the three-particle system. The directions are indicated, and the magnitudes are $F_1 = 6.0$ N, $F_2 = 12$ N, and $F_3 = 14$ N. What is the acceleration of the center of mass of the system, and in what direction does it move?

KEY IDEAS

The position of the center of mass is marked by a dot in the figure. We can treat the center of mass as if it were a real particle, with a mass equal to the system's total mass $M = 16$ kg.

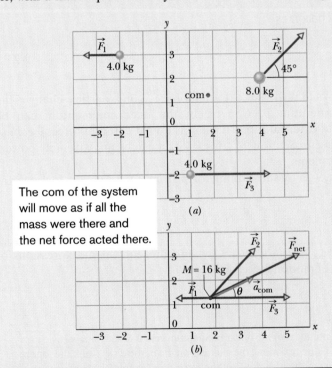

The com of the system will move as if all the mass were there and the net force acted there.

Fig. 9-7   (a) Three particles, initially at rest in the positions shown, are acted on by the external forces shown. The center of mass (com) of the system is marked. (b) The forces are now transferred to the center of mass of the system, which behaves like a particle with a mass M equal to the total mass of the system. The net external force $\vec{F}_{net}$ and the acceleration $\vec{a}_{com}$ of the center of mass are shown.

We can also treat the three external forces as if they act at the center of mass (Fig. 9-7b).

Calculations: We can now apply Newton's second law ($\vec{F}_{net} = m\vec{a}$) to the center of mass, writing

$$\vec{F}_{net} = M\vec{a}_{com} \tag{9-20}$$

or

$$\vec{F}_1 + \vec{F}_2 + \vec{F}_3 = M\vec{a}_{com}$$

so

$$\vec{a}_{com} = \frac{\vec{F}_1 + \vec{F}_2 + \vec{F}_3}{M}. \tag{9-21}$$

Equation 9-20 tells us that the acceleration $\vec{a}_{com}$ of the center of mass is in the same direction as the net external force $\vec{F}_{net}$ on the system (Fig. 9-7b). Because the particles are initially at rest, the center of mass must also be at rest. As the center of mass then begins to accelerate, it must move off in the common direction of $\vec{a}_{com}$ and $\vec{F}_{net}$.

We can evaluate the right side of Eq. 9-21 directly on a vector-capable calculator, or we can rewrite Eq. 9-21 in component form, find the components of $\vec{a}_{com}$, and then find $\vec{a}_{com}$. Along the x axis, we have

$$a_{com, x} = \frac{F_{1x} + F_{2x} + F_{3x}}{M}$$

$$= \frac{-6.0 \text{ N} + (12 \text{ N}) \cos 45° + 14 \text{ N}}{16 \text{ kg}} = 1.03 \text{ m/s}^2.$$

Along the y axis, we have

$$a_{com, y} = \frac{F_{1y} + F_{2y} + F_{3y}}{M}$$

$$= \frac{0 + (12 \text{ N}) \sin 45° + 0}{16 \text{ kg}} = 0.530 \text{ m/s}^2.$$

From these components, we find that $\vec{a}_{com}$ has the magnitude

$$a_{com} = \sqrt{(a_{com, x})^2 + (a_{com, y})^2}$$

$$= 1.16 \text{ m/s}^2 \approx 1.2 \text{ m/s}^2 \qquad \text{(Answer)}$$

and the angle (from the positive direction of the x axis)

$$\theta = \tan^{-1} \frac{a_{com, y}}{a_{com, x}} = 27°. \qquad \text{(Answer)}$$

 Additional examples, video, and practice available at *WileyPLUS*

9-4 Linear Momentum

In this section, we discuss only a single particle instead of a system of particles, in order to define two important quantities. Then in Section 9-5, we extend those definitions to systems of many particles.

The first definition concerns a familiar word—*momentum*—that has several meanings in everyday language but only a single precise meaning in physics and engineering. The **linear momentum** of a particle is a vector quantity $\vec{p}$ that is defined as

$$\vec{p} = m\vec{v} \qquad \text{(linear momentum of a particle),} \qquad (9\text{-}22)$$

in which m is the mass of the particle and $\vec{v}$ is its velocity. (The adjective *linear* is often dropped, but it serves to distinguish $\vec{p}$ from *angular* momentum, which is introduced in Chapter 11 and which is associated with rotation.) Since m is always a positive scalar quantity, Eq. 9-22 tells us that $\vec{p}$ and $\vec{v}$ have the same direction. From Eq. 9-22, the SI unit for momentum is the kilogram-meter per second (kg · m/s).

Newton expressed his second law of motion in terms of momentum:

> The time rate of change of the momentum of a particle is equal to the net force acting on the particle and is in the direction of that force.

In equation form this becomes

$$\vec{F}_{\text{net}} = \frac{d\vec{p}}{dt}. \qquad (9\text{-}23)$$

In words, Eq. 9-23 says that the net external force $\vec{F}_{\text{net}}$ on a particle changes the particle's linear momentum $\vec{p}$. Conversely, the linear momentum can be changed only by a net external force. If there is no net external force, $\vec{p}$ *cannot* change. As we shall see in Section 9-7, this last fact can be an extremely powerful tool in solving problems.

Manipulating Eq. 9-23 by substituting for $\vec{p}$ from Eq. 9-22 gives, for constant mass m,

$$\vec{F}_{\text{net}} = \frac{d\vec{p}}{dt} = \frac{d}{dt}(m\vec{v}) = m\frac{d\vec{v}}{dt} = m\vec{a}.$$

Thus, the relations $\vec{F}_{\text{net}} = d\vec{p}/dt$ and $\vec{F}_{\text{net}} = m\vec{a}$ are equivalent expressions of Newton's second law of motion for a particle.

✔ CHECKPOINT 3

The figure gives the magnitude p of the linear momentum versus time t for a particle moving along an axis. A force directed along the axis acts on the particle. (a) Rank the four regions indicated according to the magnitude of the force, greatest first. (b) In which region is the particle slowing?

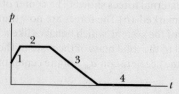

9-5 The Linear Momentum of a System of Particles

Let's extend the definition of linear momentum to a system of particles. Consider a system of n particles, each with its own mass, velocity, and linear momentum. The particles may interact with each other, and external forces may act on them. The system as a whole has a total linear momentum $\vec{P}$, which is defined to be the vector sum of the individual particles' linear momenta. Thus,

$$\vec{P} = \vec{p}_1 + \vec{p}_2 + \vec{p}_3 + \cdots + \vec{p}_n$$
$$= m_1\vec{v}_1 + m_2\vec{v}_2 + m_3\vec{v}_3 + \cdots + m_n\vec{v}_n. \qquad (9\text{-}24)$$

If we compare this equation with Eq. 9-17, we see that

$$\vec{P} = M\vec{v}_{com} \qquad \text{(linear momentum, system of particles)}, \qquad (9\text{-}25)$$

which is another way to define the linear momentum of a system of particles:

> The linear momentum of a system of particles is equal to the product of the total mass M of the system and the velocity of the center of mass.

If we take the time derivative of Eq. 9-25, we find

$$\frac{d\vec{P}}{dt} = M \frac{d\vec{v}_{com}}{dt} = M\vec{a}_{com}. \qquad (9\text{-}26)$$

Comparing Eqs. 9-14 and 9-26 allows us to write Newton's second law for a system of particles in the equivalent form

$$\vec{F}_{net} = \frac{d\vec{P}}{dt} \qquad \text{(system of particles)}, \qquad (9\text{-}27)$$

where $\vec{F}_{net}$ is the net external force acting on the system. This equation is the generalization of the single-particle equation $\vec{F}_{net} = d\vec{p}/dt$ to a system of many particles. In words, the equation says that the net external force $\vec{F}_{net}$ on a system of particles changes the linear momentum $\vec{P}$ of the system. Conversely, the linear momentum can be changed only by a net external force. If there is no net external force, $\vec{P}$ *cannot* change.

9-6 Collision and Impulse

The momentum $\vec{p}$ of any particle-like body cannot change unless a net external force changes it. For example, we could push on the body to change its momentum. More dramatically, we could arrange for the body to collide with a baseball bat. In such a *collision* (or *crash*), the external force on the body is brief, has large magnitude, and suddenly changes the body's momentum. Collisions occur commonly in our world, but before we get to them, we need to consider a simple collision in which a moving particle-like body (a *projectile*) collides with some other body (a *target*).

Single Collision

Let the projectile be a ball and the target be a bat. The collision is brief, and the ball experiences a force that is great enough to slow, stop, or even reverse its motion. Figure 9-8 depicts the collision at one instant. The ball experiences a force $\vec{F}(t)$ that

The collision of a ball with a bat collapses part of the ball. *(Photo by Harold E. Edgerton. ©The Harold and Esther Edgerton Family Trust, courtesy of Palm Press, Inc.)*

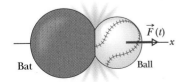

Fig. 9-8 Force $\vec{F}(t)$ acts on a ball as the ball and a bat collide.

The impulse in the collision is equal to the area under the curve.

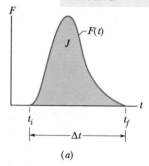

(a)

The average force gives the same area under the curve.

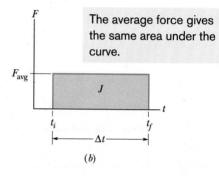

(b)

Fig. 9-9 *(a)* The curve shows the magnitude of the time-varying force $F(t)$ that acts on the ball in the collision of Fig. 9-8. The area under the curve is equal to the magnitude of the impulse $\vec{J}$ on the ball in the collision. *(b)* The height of the rectangle represents the average force F_{avg} acting on the ball over the time interval Δt. The area within the rectangle is equal to the area under the curve in *(a)* and thus is also equal to the magnitude of the impulse $\vec{J}$ in the collision.

varies during the collision and changes the linear momentum $\vec{p}$ of the ball. That change is related to the force by Newton's second law written in the form $\vec{F} = d\vec{p}/dt$. Thus, in time interval dt, the change in the ball's momentum is

$$d\vec{p} = \vec{F}(t)\,dt. \tag{9-28}$$

We can find the net change in the ball's momentum due to the collision if we integrate both sides of Eq. 9-28 from a time t_i just before the collision to a time t_f just after the collision:

$$\int_{t_i}^{t_f} d\vec{p} = \int_{t_i}^{t_f} \vec{F}(t)\,dt. \tag{9-29}$$

The left side of this equation gives us the change in momentum: $\vec{p}_f - \vec{p}_i = \Delta\vec{p}$. The right side, which is a measure of both the magnitude and the duration of the collision force, is called the **impulse** $\vec{J}$ of the collision:

$$\vec{J} = \int_{t_i}^{t_f} \vec{F}(t)\,dt \qquad \text{(impulse defined).} \tag{9-30}$$

Thus, the change in an object's momentum is equal to the impulse on the object:

$$\Delta\vec{p} = \vec{J} \qquad \text{(linear momentum–impulse theorem).} \tag{9-31}$$

This expression can also be written in the vector form

$$\vec{p}_f - \vec{p}_i = \vec{J} \tag{9-32}$$

and in such component forms as

$$\Delta p_x = J_x \tag{9-33}$$

and

$$p_{fx} - p_{ix} = \int_{t_i}^{t_f} F_x\,dt. \tag{9-34}$$

If we have a function for $\vec{F}(t)$, we can evaluate $\vec{J}$ (and thus the change in momentum) by integrating the function. If we have a plot of $\vec{F}$ versus time t, we can evaluate $\vec{J}$ by finding the area between the curve and the t axis, such as in Fig. 9-9a. In many situations we do not know how the force varies with time but we do know the average magnitude F_{avg} of the force and the duration Δt ($= t_f - t_i$) of the collision. Then we can write the magnitude of the impulse as

$$J = F_{avg}\,\Delta t. \tag{9-35}$$

The average force is plotted versus time as in Fig. 9-9b. The area under that curve is equal to the area under the curve for the actual force $F(t)$ in Fig. 9-9a because both areas are equal to impulse magnitude J.

Instead of the ball, we could have focused on the bat in Fig. 9-8. At any instant, Newton's third law tells us that the force on the bat has the same magnitude but the opposite direction as the force on the ball. From Eq. 9-30, this means that the impulse on the bat has the same magnitude but the opposite direction as the impulse on the ball.

✓**CHECKPOINT 4**

A paratrooper whose chute fails to open lands in snow; he is hurt slightly. Had he landed on bare ground, the stopping time would have been 10 times shorter and the collision lethal. Does the presence of the snow increase, decrease, or leave unchanged the values of (a) the paratrooper's change in momentum, (b) the impulse stopping the paratrooper, and (c) the force stopping the paratrooper?

Series of Collisions

Now let's consider the force on a body when it undergoes a series of identical, re-peated collisions. For example, as a prank, we might adjust one of those machines that fire tennis balls to fire them at a rapid rate directly at a wall. Each collision would produce a force on the wall, but that is not the force we are seeking. We want the average force F_{avg} on the wall during the bombardment—that is, the av-erage force during a large number of collisions.

In Fig. 9-10, a steady stream of projectile bodies, with identical mass m and linear momenta $m\vec{v}$, moves along an x axis and collides with a target body that is fixed in place. Let n be the number of projectiles that collide in a time interval Δt. Because the motion is along only the x axis, we can use the components of the momenta along that axis. Thus, each projectile has initial momentum mv and undergoes a change Δp in linear momentum because of the collision. The total change in linear momentum for n projectiles during interval Δt is $n\,\Delta p$. The resulting impulse $\vec{J}$ on the target during Δt is along the x axis and has the same magnitude of $n\,\Delta p$ but is in the opposite direction. We can write this relation in component form as

$$J = -n\,\Delta p, \qquad (9\text{-}36)$$

where the minus sign indicates that J and Δp have opposite directions.

By rearranging Eq. 9-35 and substituting Eq. 9-36, we find the average force F_{avg} acting on the target during the collisions:

$$F_{avg} = \frac{J}{\Delta t} = -\frac{n}{\Delta t}\,\Delta p = -\frac{n}{\Delta t}\,m\,\Delta v. \qquad (9\text{-}37)$$

This equation gives us F_{avg} in terms of $n/\Delta t$, the rate at which the projectiles collide with the target, and Δv, the change in the velocity of those projectiles.

If the projectiles stop upon impact, then in Eq. 9-37 we can substitute, for Δv,

$$\Delta v = v_f - v_i = 0 - v = -v, \qquad (9\text{-}38)$$

where $v_i\,(= v)$ and $v_f\,(= 0)$ are the velocities before and after the collision, respectively. If, instead, the projectiles bounce (rebound) directly backward from the target with no change in speed, then $v_f = -v$ and we can substitute

$$\Delta v = v_f - v_i = -v - v = -2v. \qquad (9\text{-}39)$$

In time interval Δt, an amount of mass $\Delta m = nm$ collides with the target. With this result, we can rewrite Eq. 9-37 as

$$F_{avg} = -\frac{\Delta m}{\Delta t}\,\Delta v. \qquad (9\text{-}40)$$

This equation gives the average force F_{avg} in terms of $\Delta m/\Delta t$, the rate at which mass collides with the target. Here again we can substitute for Δv from Eq. 9-38 or 9-39 depending on what the projectiles do.

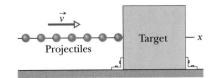

Fig. 9-10 A steady stream of projectiles, with identical linear momenta, collides with a target, which is fixed in place. The average force F_{avg} on the target is to the right and has a magnitude that depends on the rate at which the projectiles collide with the target or, equivalently, the rate at which mass col-lides with the target.

CHECKPOINT 5

The figure shows an overhead view of a ball bounc-ing from a vertical wall without any change in its speed. Consider the change $\Delta\vec{p}$ in the ball's linear momentum. (a) Is Δp_x positive, negative, or zero? (b) Is Δp_y positive, negative, or zero? (c) What is the direction of $\Delta\vec{p}$?

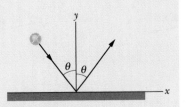

Sample Problem

Two-dimensional impulse, race car–wall collision

Race car–wall collision. Figure 9-11a is an overhead view of the path taken by a race car driver as his car collides with the racetrack wall. Just before the collision, he is traveling at speed $v_i = 70$ m/s along a straight line at 30° from the wall. Just after the collision, he is traveling at speed $v_f = 50$ m/s along a straight line at 10° from the wall. His mass m is 80 kg.

(a) What is the impulse $\vec{J}$ on the driver due to the collision?

KEY IDEAS

We can treat the driver as a particle-like body and thus apply the physics of this section. However, we cannot calculate $\vec{J}$ directly from Eq. 9-30 because we do not know anything about the force $\vec{F}(t)$ on the driver during the collision. That is, we do not have a function of $\vec{F}(t)$ or a plot for it and thus cannot integrate to find $\vec{J}$. However, we *can* find $\vec{J}$ from the change in the driver's linear momentum $\vec{p}$ via Eq. 9-32 ($\vec{J} = \vec{p}_f - \vec{p}_i$).

Calculations: Figure 9-11b shows the driver's momentum $\vec{p}_i$ before the collision (at angle 30° from the positive x direction) and his momentum $\vec{p}_f$ after the collision (at angle −10°). From Eqs. 9-32 and 9-22 ($\vec{p} = m\vec{v}$), we can write

$$\vec{J} = \vec{p}_f - \vec{p}_i = m\vec{v}_f - m\vec{v}_i = m(\vec{v}_f - \vec{v}_i). \quad (9\text{-}41)$$

We could evaluate the right side of this equation directly on a vector-capable calculator because we know m is 80 kg, $\vec{v}_f$ is 50 m/s at −10°, and $\vec{v}_i$ is 70 m/s at 30°. Instead, here we evaluate Eq. 9-41 in component form.

x component: Along the x axis we have

$$J_x = m(v_{fx} - v_{ix})$$
$$= (80 \text{ kg})[(50 \text{ m/s})\cos(-10°) - (70 \text{ m/s})\cos 30°]$$
$$= -910 \text{ kg} \cdot \text{m/s}.$$

y component: Along the y axis,

$$J_y = m(v_{fy} - v_{iy})$$
$$= (80 \text{ kg})[(50 \text{ m/s})\sin(-10°) - (70 \text{ m/s})\sin 30°]$$
$$= -3495 \text{ kg} \cdot \text{m/s} \approx -3500 \text{ kg} \cdot \text{m/s}.$$

Impulse: The impulse is then

$$\vec{J} = (-910\hat{i} - 3500\hat{j}) \text{ kg} \cdot \text{m/s}, \quad \text{(Answer)}$$

which means the impulse magnitude is

$$J = \sqrt{J_x^2 + J_y^2} = 3616 \text{ kg} \cdot \text{m/s} \approx 3600 \text{ kg} \cdot \text{m/s}.$$

The angle of $\vec{J}$ is given by

$$\theta = \tan^{-1}\frac{J_y}{J_x}, \quad \text{(Answer)}$$

which a calculator evaluates as 75.4°. Recall that the physically correct result of an inverse tangent might be the displayed answer plus 180°. We can tell which is correct here by drawing the components of $\vec{J}$ (Fig. 9-11c). We find that θ is actually 75.4° + 180° = 255.4°, which we can write as

$$\theta = -105°. \quad \text{(Answer)}$$

(b) The collision lasts for 14 ms. What is the magnitude of the average force on the driver during the collision?

KEY IDEA

From Eq. 9-35 ($J = F_{avg} \Delta t$), the magnitude F_{avg} of the average force is the ratio of the impulse magnitude J to the duration Δt of the collision.

Calculations: We have

$$F_{avg} = \frac{J}{\Delta t} = \frac{3616 \text{ kg} \cdot \text{m/s}}{0.014 \text{ s}}$$
$$= 2.583 \times 10^5 \text{ N} \approx 2.6 \times 10^5 \text{ N}. \quad \text{(Answer)}$$

Using $F = ma$ with $m = 80$ kg, you can show that the magnitude of the driver's average acceleration during the collision is about 3.22×10^3 m/s² = 329g, which is fatal.

Surviving: Mechanical engineers attempt to reduce the chances of a fatality by designing and building racetrack walls with more "give," so that a collision lasts longer. For example, if the collision here lasted 10 times longer and the other data remained the same, the magnitudes of the average force and average acceleration would be 10 times less and probably survivable.

Fig. 9-11 (a) Overhead view of the path taken by a race car and its driver as the car slams into the racetrack wall. (b) The initial momentum $\vec{p}_i$ and final momentum $\vec{p}_f$ of the driver. (c) The impulse $\vec{J}$ on the driver during the collision.

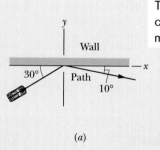

The collision changes the momentum.

The impulse on the car is equal to the change in the momentum.

(a)   *(b)*   *(c)*

WILEY **PLUS**   Additional examples, video, and practice available at *WileyPLUS*

9-7 Conservation of Linear Momentum

Suppose that the net external force $\vec{F}_{net}$ (and thus the net impulse $\vec{J}$) acting on a system of particles is zero (the system is isolated) and that no particles leave or enter the system (the system is closed). Putting $\vec{F}_{net} = 0$ in Eq. 9-27 then yields $d\vec{P}/dt = 0$, or

$$\vec{P} = \text{constant} \qquad \text{(closed, isolated system)}. \qquad (9\text{-}42)$$

In words,

> If no net external force acts on a system of particles, the total linear momentum $\vec{P}$ of the system cannot change.

This result is called the **law of conservation of linear momentum.** It can also be written as

$$\vec{P}_i = \vec{P}_f \qquad \text{(closed, isolated system)}. \qquad (9\text{-}43)$$

In words, this equation says that, for a closed, isolated system,

$$\left(\begin{array}{c} \text{total linear momentum} \\ \text{at some initial time } t_i \end{array} \right) = \left(\begin{array}{c} \text{total linear momentum} \\ \text{at some later time } t_f \end{array} \right).$$

Caution: Momentum should not be confused with energy. In the sample problems of this section, momentum is conserved but energy is definitely not.

Equations 9-42 and 9-43 are vector equations and, as such, each is equivalent to three equations corresponding to the conservation of linear momentum in three mutually perpendicular directions as in, say, an *xyz* coordinate system. Depending on the forces acting on a system, linear momentum might be conserved in one or two directions but not in all directions. However,

> If the component of the net *external* force on a closed system is zero along an axis, then the component of the linear momentum of the system along that axis cannot change.

As an example, suppose that you toss a grapefruit across a room. During its flight, the only external force acting on the grapefruit (which we take as the system) is the gravitational force $\vec{F}_g$, which is directed vertically downward. Thus, the vertical component of the linear momentum of the grapefruit changes, but since no horizontal external force acts on the grapefruit, the horizontal component of the linear momentum cannot change.

Note that we focus on the external forces acting on a closed system. Although internal forces can change the linear momentum of portions of the system, they cannot change the total linear momentum of the entire system.

The sample problems in this section involve explosions that are either one-dimensional (meaning that the motions before and after the explosion are along a single axis) or two-dimensional (meaning that they are in a plane containing two axes). In the following sections we consider collisions.

✔ **CHECKPOINT 6**

An initially stationary device lying on a frictionless floor explodes into two pieces, which then slide across the floor. One piece slides in the positive direction of an *x* axis. (a) What is the sum of the momenta of the two pieces after the explosion? (b) Can the second piece move at an angle to the *x* axis? (c) What is the direction of the momentum of the second piece?

One-dimensional explosion, relative velocity, space hauler

One-dimensional explosion: Figure 9-12a shows a space hauler and cargo module, of total mass M, traveling along an x axis in deep space. They have an initial velocity $\vec{v}_i$ of magnitude 2100 km/h relative to the Sun. With a small explosion, the hauler ejects the cargo module, of mass $0.20M$ (Fig. 9-12b). The hauler then travels 500 km/h faster than the module along the x axis; that is, the relative speed v_{rel} between the hauler and the module is 500 km/h. What then is the velocity $\vec{v}_{HS}$ of the hauler relative to the Sun?

KEY IDEA

Because the hauler–module system is closed and isolated, its total linear momentum is conserved; that is,

$$\vec{P}_i = \vec{P}_f, \qquad (9\text{-}44)$$

> The explosive separation can change the momentum of the parts but not the momentum of the system.

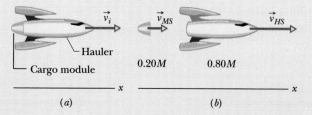

Hauler

Cargo module $0.20M$ $0.80M$

(a)   (b)

Fig. 9-12 (a) A space hauler, with a cargo module, moving at initial velocity $\vec{v}_i$. (b) The hauler has ejected the cargo module. Now the velocities relative to the Sun are $\vec{v}_{MS}$ for the module and $\vec{v}_{HS}$ for the hauler.

where the subscripts i and f refer to values before and after the ejection, respectively.

Calculations: Because the motion is along a single axis, we can write momenta and velocities in terms of their x components, using a sign to indicate direction. Before the ejection, we have

$$P_i = Mv_i. \qquad (9\text{-}45)$$

Let v_{MS} be the velocity of the ejected module relative to the Sun. The total linear momentum of the system after the ejection is then

$$P_f = (0.20M)v_{MS} + (0.80M)v_{HS}, \qquad (9\text{-}46)$$

where the first term on the right is the linear momentum of the module and the second term is that of the hauler.

We do not know the velocity v_{MS} of the module relative to the Sun, but we can relate it to the known velocities with

$$\begin{pmatrix} \text{velocity of} \\ \text{hauler relative} \\ \text{to Sun} \end{pmatrix} = \begin{pmatrix} \text{velocity of} \\ \text{hauler relative} \\ \text{to module} \end{pmatrix} + \begin{pmatrix} \text{velocity of} \\ \text{module relative} \\ \text{to Sun} \end{pmatrix}.$$

In symbols, this gives us

$$v_{HS} = v_{\text{rel}} + v_{MS} \qquad (9\text{-}47)$$

or

$$v_{MS} = v_{HS} - v_{\text{rel}}.$$

Substituting this expression for v_{MS} into Eq. 9-46, and then substituting Eqs. 9-45 and 9-46 into Eq. 9-44, we find

$$Mv_i = 0.20M(v_{HS} - v_{\text{rel}}) + 0.80Mv_{HS},$$

which gives us

$$v_{HS} = v_i + 0.20v_{\text{rel}},$$

or

$$v_{HS} = 2100 \text{ km/h} + (0.20)(500 \text{ km/h})$$

$$= 2200 \text{ km/h}. \qquad \text{(Answer)}$$

Two-dimensional explosion, momentum, coconut

Two-dimensional explosion: A firecracker placed inside a coconut of mass M, initially at rest on a frictionless floor, blows the coconut into three pieces that slide across the floor. An overhead view is shown in Fig. 9-13a. Piece C, with mass $0.30M$, has final speed $v_{fC} = 5.0$ m/s.

(a) What is the speed of piece B, with mass $0.20M$?

KEY IDEA

First we need to see whether linear momentum is conserved. We note that (1) the coconut and its pieces form a closed system, (2) the explosion forces are internal to that

system, and (3) no net external force acts on the system. Therefore, the linear momentum of the system is conserved.

Calculations: To get started, we superimpose an xy coordinate system as shown in Fig. 9-13b, with the negative direction of the x axis coinciding with the direction of $\vec{v}_{fA}$. The x axis is at 80° with the direction of $\vec{v}_{fC}$ and 50° with the direction of $\vec{v}_{fB}$.

Linear momentum is conserved separately along each axis. Let's use the y axis and write

$$P_{iy} = P_{fy}, \qquad (9\text{-}48)$$

where subscript i refers to the initial value (before the explosion), and subscript y refers to the y component of $\vec{P}_i$ or $\vec{P}_f$.

The component P_{iy} of the initial linear momentum is zero, because the coconut is initially at rest. To get an expression for P_{fy}, we find the y component of the final linear momentum of each piece, using the y-component version of Eq. 9-22 ($p_y = mv_y$):

$$p_{fA,y} = 0,$$
$$p_{fB,y} = -0.20Mv_{fB,y} = -0.20Mv_{fB} \sin 50°,$$
$$p_{fC,y} = 0.30Mv_{fC,y} = 0.30Mv_{fC} \sin 80°.$$

(Note that $p_{fA,y} = 0$ because of our choice of axes.) Equation 9-48 can now be written as

$$P_{iy} = P_{fy} = p_{fA,y} + p_{fB,y} + p_{fC,y}.$$

The explosive separation can change the momentum of the parts but not the momentum of the system.

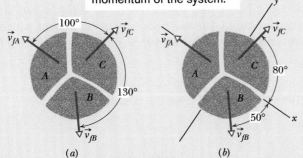

Fig. 9-13 Three pieces of an exploded coconut move off in three directions along a frictionless floor. (*a*) An overhead view of the event. (*b*) The same with a two-dimensional axis system imposed.

Then, with $v_{fC} = 5.0$ m/s, we have

$$0 = 0 - 0.20Mv_{fB} \sin 50° + (0.30M)(5.0 \text{ m/s}) \sin 80°,$$

from which we find

$$v_{fB} = 9.64 \text{ m/s} \approx 9.6 \text{ m/s}. \qquad \text{(Answer)}$$

(b) What is the speed of piece A?

Calculations: Because linear momentum is also conserved along the x axis, we have

$$P_{ix} = P_{fx}, \qquad (9\text{-}49)$$

where $P_{ix} = 0$ because the coconut is initially at rest. To get P_{fx}, we find the x components of the final momenta, using the fact that piece A must have a mass of $0.50M$ ($= M - 0.20M - 0.30M$):

$$p_{fA,x} = -0.50Mv_{fA},$$
$$p_{fB,x} = 0.20Mv_{fB,x} = 0.20Mv_{fB} \cos 50°,$$
$$p_{fC,x} = 0.30Mv_{fC,x} = 0.30Mv_{fC} \cos 80°.$$

Equation 9-49 can now be written as

$$P_{ix} = P_{fx} = p_{fA,x} + p_{fB,x} + p_{fC,x}.$$

Then, with $v_{fC} = 5.0$ m/s and $v_{fB} = 9.64$ m/s, we have

$$0 = -0.50Mv_{fA} + 0.20M(9.64 \text{ m/s}) \cos 50°$$
$$+ 0.30M(5.0 \text{ m/s}) \cos 80°,$$

from which we find

$$v_{fA} = 3.0 \text{ m/s}. \qquad \text{(Answer)}$$

9-8 Momentum and Kinetic Energy in Collisions

In Section 9-6, we considered the collision of two particle-like bodies but focused on only one of the bodies at a time. For the next several sections we switch our focus to the system itself, with the assumption that the system is closed and isolated. In Section 9-7, we discussed a rule about such a system: The total linear momentum $\vec{P}$ of the system cannot change because there is no net external force to change it. This is a very powerful rule because it can allow us to determine the results of a collision *without* knowing the details of the collision (such as how much damage is done).

We shall also be interested in the total kinetic energy of a system of two colliding bodies. If that total happens to be unchanged by the collision, then the kinetic energy of the system is *conserved* (it is the same before and after the collision). Such a collision is called an **elastic collision.** In everyday collisions of common bodies, such as two cars or a ball and a bat, some energy is always transferred from kinetic energy to other forms of energy, such as thermal energy or energy of sound. Thus, the kinetic energy of the system is *not* conserved. Such a collision is called an **inelastic collision.**

However, in some situations, we can *approximate* a collision of common bodies as elastic. Suppose that you drop a Superball onto a hard floor. If the collision

Here is the generic setup for an inelastic collision.

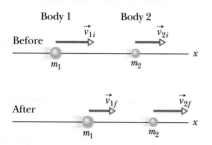

Fig. 9-14 Bodies 1 and 2 move along an x axis, before and after they have an inelastic collision.

between the ball and floor (or Earth) were elastic, the ball would lose no kinetic energy because of the collision and would rebound to its original height. However, the actual rebound height is somewhat short, showing that at least some kinetic energy is lost in the collision and thus that the collision is somewhat inelastic. Still, we might choose to neglect that small loss of kinetic energy to approximate the collision as elastic.

The inelastic collision of two bodies always involves a loss in the kinetic energy of the system. The greatest loss occurs if the bodies stick together, in which case the collision is called a **completely inelastic collision.** The collision of a baseball and a bat is inelastic. However, the collision of a wet putty ball and a bat is completely inelastic because the putty sticks to the bat.

9-9 Inelastic Collisions in One Dimension

One-Dimensional Inelastic Collision

Figure 9-14 shows two bodies just before and just after they have a one-dimensional collision. The velocities before the collision (subscript i) and after the collision (subscript f) are indicated. The two bodies form our system, which is closed and isolated. We can write the law of conservation of linear momentum for this two-body system as

$$\begin{pmatrix} \text{total momentum } \vec{P}_i \\ \text{before the collision} \end{pmatrix} = \begin{pmatrix} \text{total momentum } \vec{P}_f \\ \text{after the collision} \end{pmatrix},$$

which we can symbolize as

$$\vec{p}_{1i} + \vec{p}_{2i} = \vec{p}_{1f} + \vec{p}_{2f} \qquad \text{(conservation of linear momentum)}. \qquad (9\text{-}50)$$

Because the motion is one-dimensional, we can drop the overhead arrows for vectors and use only components along the axis, indicating direction with a sign. Thus, from $p = mv$, we can rewrite Eq. 9-50 as

$$m_1 v_{1i} + m_2 v_{2i} = m_1 v_{1f} + m_2 v_{2f}. \qquad (9\text{-}51)$$

If we know values for, say, the masses, the initial velocities, and one of the final velocities, we can find the other final velocity with Eq. 9-51.

One-Dimensional Completely Inelastic Collision

Figure 9-15 shows two bodies before and after they have a completely inelastic collision (meaning they stick together). The body with mass m_2 happens to be initially at rest ($v_{2i} = 0$). We can refer to that body as the *target* and to the incoming body as the *projectile.* After the collision, the stuck-together bodies move with velocity V. For this situation, we can rewrite Eq. 9-51 as

$$m_1 v_{1i} = (m_1 + m_2)V \qquad (9\text{-}52)$$

In a completely inelastic collision, the bodies stick together.

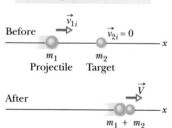

Fig. 9-15 A completely inelastic collision between two bodies. Before the collision, the body with mass m_2 is at rest and the body with mass m_1 moves directly toward it. After the collision, the stuck-together bodies move with the same velocity $\vec{V}$.

or

$$V = \frac{m_1}{m_1 + m_2} v_{1i}. \qquad (9\text{-}53)$$

If we know values for, say, the masses and the initial velocity v_{1i} of the projectile, we can find the final velocity V with Eq. 9-53. Note that V must be less than v_{1i} because the mass ratio $m_1/(m_1 + m_2)$ must be less than unity.

Velocity of the Center of Mass

In a closed, isolated system, the velocity $\vec{v}_{\text{com}}$ of the center of mass of the system cannot be changed by a collision because, with the system isolated, there is no net

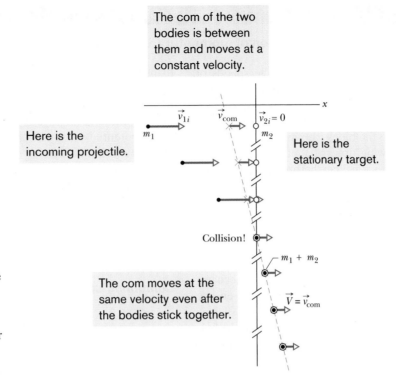

The com of the two bodies is between them and moves at a constant velocity.

Here is the incoming projectile.

Here is the stationary target.

The com moves at the same velocity even after the bodies stick together.

Collision!

Fig. 9-16   Some freeze-frames of the two-body system in Fig. 9-15, which undergoes a completely inelastic collision. The system's center of mass is shown in each freeze-frame. The velocity $\vec{v}_{\text{com}}$ of the center of mass is unaffected by the collision. Because the bodies stick together after the collision, their common velocity $\vec{V}$ must be equal to $\vec{v}_{\text{com}}$.

external force to change it. To get an expression for $\vec{v}_{\text{com}}$, let us return to the two-body system and one-dimensional collision of Fig. 9-14. From Eq. 9-25 ($\vec{P} = M\vec{v}_{\text{com}}$), we can relate $\vec{v}_{\text{com}}$ to the total linear momentum $\vec{P}$ of that two-body system by writing

$$\vec{P} = M\vec{v}_{\text{com}} = (m_1 + m_2)\vec{v}_{\text{com}}. \qquad (9\text{-}54)$$

The total linear momentum $\vec{P}$ is conserved during the collision; so it is given by either side of Eq. 9-50. Let us use the left side to write

$$\vec{P} = \vec{p}_{1i} + \vec{p}_{2i}. \qquad (9\text{-}55)$$

Substituting this expression for $\vec{P}$ in Eq. 9-54 and solving for $\vec{v}_{\text{com}}$ give us

$$\vec{v}_{\text{com}} = \frac{\vec{P}}{m_1 + m_2} = \frac{\vec{p}_{1i} + \vec{p}_{2i}}{m_1 + m_2}. \qquad (9\text{-}56)$$

The right side of this equation is a constant, and $\vec{v}_{\text{com}}$ has that same constant value before and after the collision.

For example, Fig. 9-16 shows, in a series of freeze-frames, the motion of the center of mass for the completely inelastic collision of Fig. 9-15. Body 2 is the target, and its initial linear momentum in Eq. 9-56 is $\vec{p}_{2i} = m_2\vec{v}_{2i} = 0$. Body 1 is the projectile, and its initial linear momentum in Eq. 9-56 is $\vec{p}_{1i} = m_1\vec{v}_{1i}$. Note that as the series of freeze-frames progresses to and then beyond the collision, the center of mass moves at a constant velocity to the right. After the collision, the common final speed V of the bodies is equal to $\vec{v}_{\text{com}}$ because then the center of mass travels with the stuck-together bodies.

✓ CHECKPOINT 7

Body 1 and body 2 are in a completely inelastic one-dimensional collision. What is their final momentum if their initial momenta are, respectively, (a) 10 kg·m/s and 0; (b) 10 kg·m/s and 4 kg·m/s; (c) 10 kg·m/s and −4 kg·m/s?

Conservation of momentum, ballistic pendulum

The *ballistic pendulum* was used to measure the speeds of bullets before electronic timing devices were developed. The version shown in Fig. 9-17 consists of a large block of wood of mass $M = 5.4$ kg, hanging from two long cords. A bullet of mass $m = 9.5$ g is fired into the block, coming quickly to rest. The *block + bullet* then swing upward, their center of mass rising a vertical distance $h = 6.3$ cm before the pendulum comes momentarily to rest at the end of its arc. What is the speed of the bullet just prior to the collision?

We can see that the bullet's speed v must determine the rise height h. However, we cannot use the conservation of mechanical energy to relate these two quantities because surely energy is transferred from mechanical energy to other forms (such as thermal energy and energy to break apart the wood) as the bullet penetrates the block. Nevertheless, we can split this complicated motion into two steps that we can separately analyze: (1) the bullet–block collision and (2) the bullet–block rise, during which mechanical energy *is* conserved.

Reasoning step 1: Because the collision within the bullet–block system is so brief, we can make two important assumptions: (1) During the collision, the gravitational force on the block and the force on the block from the cords are still balanced. Thus, during the collision, the net external impulse on the bullet–block system is zero. Therefore, the system is isolated and its total linear momentum is conserved:

$$\begin{pmatrix} \text{total momentum} \\ \text{before the collision} \end{pmatrix} = \begin{pmatrix} \text{total momentum} \\ \text{after the collision} \end{pmatrix}. \quad (9\text{-}57)$$

(2) The collision is one-dimensional in the sense that the direction of the bullet and block *just after the collision* is in the bullet's original direction of motion.

Because the collision is one-dimensional, the block is initially at rest, and the bullet sticks in the block, we use Eq. 9-53 to express the conservation of linear momentum. Replacing the symbols there with the corresponding symbols here, we have

$$V = \frac{m}{m + M} v. \quad (9\text{-}58)$$

Reasoning step 2: As the bullet and block now swing up together, the mechanical energy of the bullet– block–Earth system is conserved:

$$\begin{pmatrix} \text{mechanical energy} \\ \text{at bottom} \end{pmatrix} = \begin{pmatrix} \text{mechanical energy} \\ \text{at top} \end{pmatrix}. \quad (9\text{-}59)$$

(This mechanical energy is not changed by the force of the cords on the block, because that force is always directed perpendicular to the block's direction of travel.) Let's take the block's initial level as our reference level of zero gravitational potential energy. Then conservation of mechanical energy means that the system's kinetic energy at the start of the swing must equal its gravitational potential energy at the highest point of the swing. Because the speed of the bullet and block at the start of the swing is the speed V immediately after the collision, we may write this conservation as

$$\tfrac{1}{2}(m + M)V^2 = (m + M)gh. \quad (9\text{-}60)$$

Combining steps: Substituting for V from Eq. 9-58 leads to

$$v = \frac{m + M}{m} \sqrt{2gh} \quad (9\text{-}61)$$

$$= \left(\frac{0.0095 \text{ kg} + 5.4 \text{ kg}}{0.0095 \text{ kg}} \right) \sqrt{(2)(9.8 \text{ m/s}^2)(0.063 \text{ m})}$$

$$= 630 \text{ m/s}. \quad \text{(Answer)}$$

The ballistic pendulum is a kind of "transformer," exchanging the high speed of a light object (the bullet) for the low—and thus more easily measurable—speed of a massive object (the block).

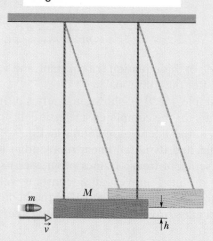

There are two events here. The bullet collides with the block. Then the bullet–block system swings upward by height h.

Fig. 9-17   A ballistic pendulum, used to measure the speeds of bullets.

 Additional examples, video, and practice available at *WileyPLUS*

9-10 Elastic Collisions in One Dimension

As we discussed in Section 9-8, everyday collisions are inelastic but we can approximate some of them as being elastic; that is, we can approximate that the total kinetic energy of the colliding bodies is conserved and is not transferred to other forms of energy:

$$\left(\begin{array}{c}\text{total kinetic energy} \\ \text{before the collision}\end{array}\right) = \left(\begin{array}{c}\text{total kinetic energy} \\ \text{after the collision}\end{array}\right). \quad (9\text{-}62)$$

This does not mean that the kinetic energy of each colliding body cannot change. Rather, it means this:

> In an elastic collision, the kinetic energy of each colliding body may change, but the total kinetic energy of the system does not change.

For example, the collision of a cue ball with an object ball in a game of pool can be approximated as being an elastic collision. If the collision is head-on (the cue ball heads directly toward the object ball), the kinetic energy of the cue ball can be transferred almost entirely to the object ball. (Still, the fact that the collision makes a sound means that at least a little of the kinetic energy is transferred to the energy of the sound.)

Here is the generic setup for an elastic collision with a stationary target.

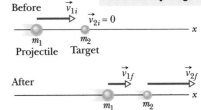

Fig. 9-18 Body 1 moves along an x axis before having an elastic collision with body 2, which is initially at rest. Both bodies move along that axis after the collision.

Stationary Target

Figure 9-18 shows two bodies before and after they have a one-dimensional collision, like a head-on collision between pool balls. A projectile body of mass m_1 and initial velocity v_{1i} moves toward a target body of mass m_2 that is initially at rest ($v_{2i} = 0$). Let's assume that this two-body system is closed and isolated. Then the net linear momentum of the system is conserved, and from Eq. 9-51 we can write that conservation as

$$m_1 v_{1i} = m_1 v_{1f} + m_2 v_{2f} \quad \text{(linear momentum)}. \quad (9\text{-}63)$$

If the collision is also elastic, then the total kinetic energy is conserved and we can write that conservation as

$$\tfrac{1}{2} m_1 v_{1i}^2 = \tfrac{1}{2} m_1 v_{1f}^2 + \tfrac{1}{2} m_2 v_{2f}^2 \quad \text{(kinetic energy)}. \quad (9\text{-}64)$$

In each of these equations, the subscript i identifies the initial velocities and the subscript f the final velocities of the bodies. If we know the masses of the bodies and if we also know v_{1i}, the initial velocity of body 1, the only unknown quantities are v_{1f} and v_{2f}, the final velocities of the two bodies. With two equations at our disposal, we should be able to find these two unknowns.

To do so, we rewrite Eq. 9-63 as

$$m_1(v_{1i} - v_{1f}) = m_2 v_{2f} \quad (9\text{-}65)$$

and Eq. 9-64 as*

$$m_1(v_{1i} - v_{1f})(v_{1i} + v_{1f}) = m_2 v_{2f}^2. \quad (9\text{-}66)$$

After dividing Eq. 9-66 by Eq. 9-65 and doing some more algebra, we obtain

$$v_{1f} = \frac{m_1 - m_2}{m_1 + m_2} v_{1i} \quad (9\text{-}67)$$

and

$$v_{2f} = \frac{2m_1}{m_1 + m_2} v_{1i}. \quad (9\text{-}68)$$

*In this step, we use the identity $a^2 - b^2 = (a - b)(a + b)$. It reduces the amount of algebra needed to solve the simultaneous equations Eqs. 9-65 and 9-66.

We note from Eq. 9-68 that v_{2f} is always positive (the initially stationary target body with mass m_2 always moves forward). From Eq. 9-67 we see that v_{1f} may be of either sign (the projectile body with mass m_1 moves forward if $m_1 > m_2$ but rebounds if $m_1 < m_2$).

Let us look at a few special situations.

1. **Equal masses**   If $m_1 = m_2$, Eqs. 9-67 and 9-68 reduce to

$$v_{1f} = 0 \quad \text{and} \quad v_{2f} = v_{1i},$$

which we might call a pool player's result. It predicts that after a head-on collision of bodies with equal masses, body 1 (initially moving) stops dead in its tracks and body 2 (initially at rest) takes off with the initial speed of body 1. In head-on collisions, bodies of equal mass simply exchange velocities. This is true even if body 2 is not initially at rest.

2. **A massive target**   In Fig. 9-18, a massive target means that $m_2 \gg m_1$. For example, we might fire a golf ball at a stationary cannonball. Equations 9-67 and 9-68 then reduce to

$$v_{1f} \approx -v_{1i} \quad \text{and} \quad v_{2f} \approx \left(\frac{2m_1}{m_2}\right)v_{1i}. \tag{9-69}$$

This tells us that body 1 (the golf ball) simply bounces back along its incoming path, its speed essentially unchanged. Initially stationary body 2 (the cannonball) moves forward at a low speed, because the quantity in parentheses in Eq. 9-69 is much less than unity. All this is what we should expect.

3. **A massive projectile**   This is the opposite case; that is, $m_1 \gg m_2$. This time, we fire a cannonball at a stationary golf ball. Equations 9-67 and 9-68 reduce to

$$v_{1f} \approx v_{1i} \quad \text{and} \quad v_{2f} \approx 2v_{1i}. \tag{9-70}$$

Equation 9-70 tells us that body 1 (the cannonball) simply keeps on going, scarcely slowed by the collision. Body 2 (the golf ball) charges ahead at twice the speed of the cannonball.

You may wonder: Why twice the speed? Recall the collision described by Eq. 9-69, in which the velocity of the incident light body (the golf ball) changed from $+v$ to $-v$, a velocity *change* of $2v$. The same *change* in velocity (but now from zero to $2v$) occurs in this example also.

Moving Target

Now that we have examined the elastic collision of a projectile and a stationary target, let us examine the situation in which both bodies are moving before they undergo an elastic collision.

For the situation of Fig. 9-19, the conservation of linear momentum is written as

$$m_1 v_{1i} + m_2 v_{2i} = m_1 v_{1f} + m_2 v_{2f}, \tag{9-71}$$

and the conservation of kinetic energy is written as

$$\tfrac{1}{2}m_1 v_{1i}^2 + \tfrac{1}{2}m_2 v_{2i}^2 = \tfrac{1}{2}m_1 v_{1f}^2 + \tfrac{1}{2}m_2 v_{2f}^2. \tag{9-72}$$

To solve these simultaneous equations for v_{1f} and v_{2f}, we first rewrite Eq. 9-71 as

$$m_1(v_{1i} - v_{1f}) = -m_2(v_{2i} - v_{2f}), \tag{9-73}$$

and Eq. 9-72 as

$$m_1(v_{1i} - v_{1f})(v_{1i} + v_{1f}) = -m_2(v_{2i} - v_{2f})(v_{2i} + v_{2f}). \tag{9-74}$$

After dividing Eq. 9-74 by Eq. 9-73 and doing some more algebra, we obtain

$$v_{1f} = \frac{m_1 - m_2}{m_1 + m_2} v_{1i} + \frac{2m_2}{m_1 + m_2} v_{2i} \tag{9-75}$$

and

$$v_{2f} = \frac{2m_1}{m_1 + m_2} v_{1i} + \frac{m_2 - m_1}{m_1 + m_2} v_{2i}. \tag{9-76}$$

Here is the generic setup for an elastic collision with a moving target.

Fig. 9-19 Two bodies headed for a one-dimensional elastic collision.

Note that the assignment of subscripts 1 and 2 to the bodies is arbitrary. If we exchange those subscripts in Fig. 9-19 and in Eqs. 9-75 and 9-76, we end up with the same set of equations. Note also that if we set $v_{2i} = 0$, body 2 becomes a stationary target as in Fig. 9-18, and Eqs. 9-75 and 9-76 reduce to Eqs. 9-67 and 9-68, respectively.

 CHECKPOINT 8

What is the final linear momentum of the target in Fig. 9-18 if the initial linear momentum of the projectile is 6 kg · m/s and the final linear momentum of the projectile is (a) 2 kg · m/s and (b) −2 kg · m/s? (c) What is the final kinetic energy of the target if the initial and final kinetic energies of the projectile are, respectively, 5 J and 2 J?

Sample Problem

Elastic collision, two pendulums

Two metal spheres, suspended by vertical cords, initially just touch, as shown in Fig. 9-20. Sphere 1, with mass $m_1 = 30$ g, is pulled to the left to height $h_1 = 8.0$ cm, and then released from rest. After swinging down, it undergoes an elastic collision with sphere 2, whose mass $m_2 = 75$ g. What is the velocity v_{1f} of sphere 1 just after the collision?

KEY IDEA

We can split this complicated motion into two steps that we can analyze separately: (1) the descent of sphere 1 (in which mechanical energy is conserved) and (2) the two-sphere collision (in which momentum is also conserved).

Step 1: As sphere 1 swings down, the mechanical energy of the sphere–Earth system is conserved. (The mechanical energy is not changed by the force of the cord on sphere 1 because that force is always directed perpendicular to the sphere's direction of travel.)

Calculation: Let's take the lowest level as our reference level of zero gravitational potential energy. Then the kinetic energy of sphere 1 at the lowest level must equal the gravitational potential energy of the system when sphere 1 is at height h_1. Thus,

$$\tfrac{1}{2}m_1 v_{1i}^2 = m_1 g h_1,$$

which we solve for the speed v_{1i} of sphere 1 just before the collision:

$$v_{1i} = \sqrt{2gh_1} = \sqrt{(2)(9.8 \text{ m/s}^2)(0.080 \text{ m})}$$
$$= 1.252 \text{ m/s}.$$

Step 2: Here we can make two assumptions in addition to the assumption that the collision is elastic. First, we can assume that the collision is one-dimensional because the motions of the spheres are approximately horizontal from just before the collision to just after it. Second, because the collision is so

brief, we can assume that the two-sphere system is closed and isolated. This means that the total linear momentum of the system is conserved.

Calculation: Thus, we can use Eq. 9-67 to find the velocity of sphere 1 just after the collision:

$$v_{1f} = \frac{m_1 - m_2}{m_1 + m_2} v_{1i}$$
$$= \frac{0.030 \text{ kg} - 0.075 \text{ kg}}{0.030 \text{ kg} + 0.075 \text{ kg}} (1.252 \text{ m/s})$$
$$= -0.537 \text{ m/s} \approx -0.54 \text{ m/s}. \qquad \text{(Answer)}$$

The minus sign tells us that sphere 1 moves to the left just after the collision.

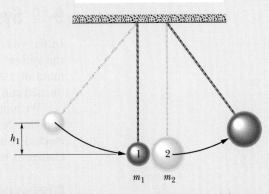

Ball 1 swings down and collides with ball 2, which then swings upward. If the collision is elastic, no mechanical energy is lost.

Fig. 9-20 Two metal spheres suspended by cords just touch when they are at rest. Sphere 1, with mass m_1, is pulled to the left to height h_1 and then released.

 Additional examples, video, and practice available at *WileyPLUS*

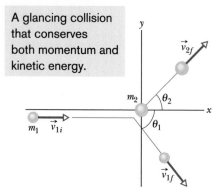

Fig. 9-21 An elastic collision between two bodies in which the collision is not head-on. The body with mass m_2 (the target) is initially at rest.

9-11 Collisions in Two Dimensions

When two bodies collide, the impulse between them determines the directions in which they then travel. In particular, when the collision is not head-on, the bodies do not end up traveling along their initial axis. For such two-dimensional collisions in a closed, isolated system, the total linear momentum must still be conserved:

$$\vec{P}_{1i} + \vec{P}_{2i} = \vec{P}_{1f} + \vec{P}_{2f}. \tag{9-77}$$

If the collision is also elastic (a special case), then the total kinetic energy is also conserved:

$$K_{1i} + K_{2i} = K_{1f} + K_{2f}. \tag{9-78}$$

Equation 9-77 is often more useful for analyzing a two-dimensional collision if we write it in terms of components on an xy coordinate system. For example, Fig. 9-21 shows a *glancing collision* (it is not head-on) between a projectile body and a target body initially at rest. The impulses between the bodies have sent the bodies off at angles θ_1 and θ_2 to the x axis, along which the projectile initially traveled. In this situation we would rewrite Eq. 9-77 for components along the x axis as

$$m_1 v_{1i} = m_1 v_{1f} \cos \theta_1 + m_2 v_{2f} \cos \theta_2, \tag{9-79}$$

and along the y axis as

$$0 = -m_1 v_{1f} \sin \theta_1 + m_2 v_{2f} \sin \theta_2. \tag{9-80}$$

We can also write Eq. 9-78 (for the special case of an elastic collision) in terms of speeds:

$$\tfrac{1}{2} m_1 v_{1i}^2 = \tfrac{1}{2} m_1 v_{1f}^2 + \tfrac{1}{2} m_2 v_{2f}^2 \quad \text{(kinetic energy)}. \tag{9-81}$$

Equations 9-79 to 9-81 contain seven variables: two masses, m_1 and m_2; three speeds, v_{1i}, v_{1f}, and v_{2f}; and two angles, θ_1 and θ_2. If we know any four of these quantities, we can solve the three equations for the remaining three quantities.

**CHECKPOINT 9**

In Fig. 9-21, suppose that the projectile has an initial momentum of 6 kg·m/s, a final x component of momentum of 4 kg·m/s, and a final y component of momentum of −3 kg·m/s. For the target, what then are (a) the final x component of momentum and (b) the final y component of momentum?

9-12 Systems with Varying Mass: A Rocket

In the systems we have dealt with so far, we have assumed that the total mass of the system remains constant. Sometimes, as in a rocket, it does not. Most of the mass of a rocket on its launching pad is fuel, all of which will eventually be burned and ejected from the nozzle of the rocket engine.

We handle the variation of the mass of the rocket as the rocket accelerates by applying Newton's second law, not to the rocket alone but to the rocket and its ejected combustion products taken together. The mass of *this* system does *not* change as the rocket accelerates.

Finding the Acceleration

Assume that we are at rest relative to an inertial reference frame, watching a rocket accelerate through deep space with no gravitational or atmospheric drag forces acting on it. For this one-dimensional motion, let M be the mass of the rocket and v its velocity at an arbitrary time t (see Fig. 9-22a).

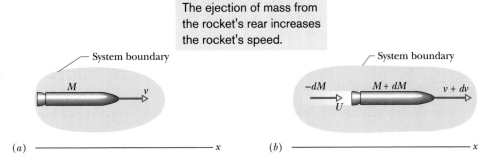

The ejection of mass from the rocket's rear increases the rocket's speed.

Fig. 9-22 (a) An accelerating rocket of mass M at time t, as seen from an inertial reference frame. (b) The same but at time $t + dt$. The exhaust products released during interval dt are shown.

Figure 9-22b shows how things stand a time interval dt later. The rocket now has velocity $v + dv$ and mass $M + dM$, where the change in mass dM is a *negative quantity*. The exhaust products released by the rocket during interval dt have mass $-dM$ and velocity U relative to our inertial reference frame.

Our system consists of the rocket and the exhaust products released during interval dt. The system is closed and isolated, so the linear momentum of the system must be conserved during dt; that is,

$$P_i = P_f, \tag{9-82}$$

where the subscripts i and f indicate the values at the beginning and end of time interval dt. We can rewrite Eq. 9-82 as

$$Mv = -dM\,U + (M + dM)(v + dv), \tag{9-83}$$

where the first term on the right is the linear momentum of the exhaust products released during interval dt and the second term is the linear momentum of the rocket at the end of interval dt.

We can simplify Eq. 9-83 by using the relative speed v_{rel} between the rocket and the exhaust products, which is related to the velocities relative to the frame with

$$\begin{pmatrix} \text{velocity of rocket} \\ \text{relative to frame} \end{pmatrix} = \begin{pmatrix} \text{velocity of rocket} \\ \text{relative to products} \end{pmatrix} + \begin{pmatrix} \text{velocity of products} \\ \text{relative to frame} \end{pmatrix}.$$

In symbols, this means

$$(v + dv) = v_{rel} + U,$$

or

$$U = v + dv - v_{rel}. \tag{9-84}$$

Substituting this result for U into Eq. 9-83 yields, with a little algebra,

$$-dM\,v_{rel} = M\,dv. \tag{9-85}$$

Dividing each side by dt gives us

$$-\frac{dM}{dt}v_{rel} = M\frac{dv}{dt}. \tag{9-86}$$

We replace dM/dt (the rate at which the rocket loses mass) by $-R$, where R is the (positive) mass rate of fuel consumption, and we recognize that dv/dt is the acceleration of the rocket. With these changes, Eq. 9-86 becomes

$$Rv_{rel} = Ma \qquad \text{(first rocket equation).} \tag{9-87}$$

Equation 9-87 holds for the values at any given instant.

Note the left side of Eq. 9-87 has the dimensions of force (kg/s · m/s = kg · m/s² = N) and depends only on design characteristics of the rocket engine—namely, the rate R at which it consumes fuel mass and the speed v_{rel} with which that mass is ejected relative to the rocket. We call this term Rv_{rel} the **thrust** of the rocket engine and represent it with T. Newton's second law emerges clearly if we write Eq. 9-87 as $T = Ma$, in which a is the acceleration of the rocket at the time that its mass is M.

Finding the Velocity

How will the velocity of a rocket change as it consumes its fuel? From Eq. 9-85 we have

$$dv = -v_{rel}\frac{dM}{M}.$$

Integrating leads to

$$\int_{v_i}^{v_f} dv = -v_{rel}\int_{M_i}^{M_f}\frac{dM}{M},$$

in which M_i is the initial mass of the rocket and M_f its final mass. Evaluating the integrals then gives

$$v_f - v_i = v_{rel}\ln\frac{M_i}{M_f} \qquad \text{(second rocket equation)} \qquad (9\text{-}88)$$

for the increase in the speed of the rocket during the change in mass from M_i to M_f. (The symbol "ln" in Eq. 9-88 means the *natural logarithm*.) We see here the advantage of multistage rockets, in which M_f is reduced by discarding successive stages when their fuel is depleted. An ideal rocket would reach its destination with only its payload remaining.

Sample Problem

Rocket engine, thrust, acceleration

A rocket whose initial mass M_i is 850 kg consumes fuel at the rate $R = 2.3$ kg/s. The speed v_{rel} of the exhaust gases relative to the rocket engine is 2800 m/s. What thrust does the rocket engine provide?

KEY IDEA

Thrust T is equal to the product of the fuel consumption rate R and the relative speed v_{rel} at which exhaust gases are expelled, as given by Eq. 9-87.

Calculation: Here we find

$$T = Rv_{rel} = (2.3 \text{ kg/s})(2800 \text{ m/s})$$

$$= 6440 \text{ N} \approx 6400 \text{ N.} \qquad \text{(Answer)}$$

(b) What is the initial acceleration of the rocket?

KEY IDEA

We can relate the thrust T of a rocket to the magnitude a of the resulting acceleration with $T = Ma$, where M is the

rocket's mass. However, M decreases and a increases as fuel is consumed. Because we want the initial value of a here, we must use the intial value M_i of the mass.

Calculation: We find

$$a = \frac{T}{M_i} = \frac{6440 \text{ N}}{850 \text{ kg}} = 7.6 \text{ m/s}^2. \qquad \text{(Answer)}$$

To be launched from Earth's surface, a rocket must have an initial acceleration greater than $g = 9.8$ m/s². That is, it must be greater than the gravitational acceleration at the surface. Put another way, the thrust T of the rocket engine must exceed the initial gravitational force on the rocket, which here has the magnitude $M_i g$, which gives us

$$(850 \text{ kg})(9.8 \text{ m/s}^2) = 8330 \text{ N.}$$

Because the acceleration or thrust requirement is not met (here $T = 6400$ N), our rocket could not be launched from Earth's surface by itself; it would require another, more powerful, rocket.

 Additional examples, video, and practice available at *WileyPLUS*

Center of Mass   The **center of mass** of a system of n particles is defined to be the point whose coordinates are given by

$$x_{com} = \frac{1}{M} \sum_{i=1}^{n} m_i x_i, \quad y_{com} = \frac{1}{M} \sum_{i=1}^{n} m_i y_i, \quad z_{com} = \frac{1}{M} \sum_{i=1}^{n} m_i z_i,$$
(9-5)

or
$$\vec{r}_{com} = \frac{1}{M} \sum_{i=1}^{n} m_i \vec{r}_i,$$
(9-8)

where M is the total mass of the system.

Newton's Second Law for a System of Particles   The motion of the center of mass of any system of particles is governed by **Newton's second law for a system of particles**, which is

$$\vec{F}_{net} = M \vec{a}_{com}.$$
(9-14)

Here $\vec{F}_{net}$ is the net force of all the *external* forces acting on the system, M is the total mass of the system, and $\vec{a}_{com}$ is the acceleration of the system's center of mass.

Linear Momentum and Newton's Second Law   For a single particle, we define a quantity $\vec{p}$ called its **linear momentum** as

$$\vec{p} = m\vec{v},$$
(9-22)

and can write Newton's second law in terms of this momentum:

$$\vec{F}_{net} = \frac{d\vec{p}}{dt}.$$
(9-23)

For a system of particles these relations become

$$\vec{P} = M\vec{v}_{com} \quad \text{and} \quad \vec{F}_{net} = \frac{d\vec{P}}{dt}.$$
(9-25, 9-27)

Collision and Impulse   Applying Newton's second law in momentum form to a particle-like body involved in a collision leads to the **impulse–linear momentum theorem:**

$$\vec{p}_f - \vec{p}_i = \Delta\vec{p} = \vec{J},$$
(9-31, 9-32)

where $\vec{p}_f - \vec{p}_i = \Delta\vec{p}$ is the change in the body's linear momentum, and $\vec{J}$ is the **impulse** due to the force $\vec{F}(t)$ exerted on the body by the other body in the collision:

$$\vec{J} = \int_{t_i}^{t_f} \vec{F}(t) \, dt.$$
(9-30)

If F_{avg} is the average magnitude of $\vec{F}(t)$ during the collision and Δt is the duration of the collision, then for one-dimensional motion

$$J = F_{avg} \Delta t.$$
(9-35)

When a steady stream of bodies, each with mass m and speed v, collides with a body whose position is fixed, the average force on the fixed body is

$$F_{avg} = -\frac{n}{\Delta t} \Delta p = -\frac{n}{\Delta t} m \, \Delta v,$$
(9-37)

where $n/\Delta t$ is the rate at which the bodies collide with the fixed body, and Δv is the change in velocity of each colliding body. This average force can also be written as

$$F_{avg} = -\frac{\Delta m}{\Delta t} \Delta v,$$
(9-40)

where $\Delta m/\Delta t$ is the rate at which mass collides with the fixed body. In Eqs. 9-37 and 9-40, $\Delta v = -v$ if the bodies stop upon impact and $\Delta v = -2v$ if they bounce directly backward with no change in their speed.

Conservation of Linear Momentum   If a system is isolated so that no net *external* force acts on it, the linear momentum $\vec{P}$ of the system remains constant:

$$\vec{P} = \text{constant} \quad \text{(closed, isolated system)}.$$
(9-42)

This can also be written as

$$\vec{P}_i = \vec{P}_f \quad \text{(closed, isolated system)},$$
(9-43)

where the subscripts refer to the values of $\vec{P}$ at some initial time and at a later time. Equations 9-42 and 9-43 are equivalent statements of the **law of conservation of linear momentum.**

Inelastic Collision in One Dimension   In an *inelastic collision* of two bodies, the kinetic energy of the two-body system is not conserved. If the system is closed and isolated, the total linear momentum of the system *must* be conserved, which we can write in vector form as

$$\vec{p}_{1i} + \vec{p}_{2i} = \vec{p}_{1f} + \vec{p}_{2f},$$
(9-50)

where subscripts i and f refer to values just before and just after the collision, respectively.

If the motion of the bodies is along a single axis, the collision is one-dimensional and we can write Eq. 9-50 in terms of velocity components along that axis:

$$m_1 v_{1i} + m_2 v_{2i} = m_1 v_{1f} + m_2 v_{2f}.$$
(9-51)

If the bodies stick together, the collision is a *completely inelastic collision* and the bodies have the same final velocity V (because they *are* stuck together).

Motion of the Center of Mass   The center of mass of a closed, isolated system of two colliding bodies is not affected by a collision. In particular, the velocity $\vec{v}_{com}$ of the center of mass cannot be changed by the collision.

Elastic Collisions in One Dimension   An *elastic collision* is a special type of collision in which the kinetic energy of a system of colliding bodies is conserved. If the system is closed and isolated, its linear momentum is also conserved. For a one-dimensional collision in which body 2 is a target and body 1 is an incoming projectile, conservation of kinetic energy and linear momentum yield the following expressions for the velocities immediately after the collision:

$$v_{1f} = \frac{m_1 - m_2}{m_1 + m_2} v_{1i}$$
(9-67)

and

$$v_{2f} = \frac{2m_1}{m_1 + m_2} v_{1i}.$$
(9-68)

Collisions in Two Dimensions   If two bodies collide and their motion is not along a single axis (the collision is not head-on), the collision is two-dimensional. If the two-body system is closed and isolated, the law of conservation of momentum applies to the collision and can be written as

$$\vec{P}_{1i} + \vec{P}_{2i} = \vec{P}_{1f} + \vec{P}_{2f}. \qquad (9\text{-}77)$$

In component form, the law gives two equations that describe the collision (one equation for each of the two dimensions). If the collision is also elastic (a special case), the conservation of kinetic energy during the collision gives a third equation:

$$K_{1i} + K_{2i} = K_{1f} + K_{2f}. \qquad (9\text{-}78)$$

Variable-Mass Systems In the absence of external forces a rocket accelerates at an instantaneous rate given by

$$Rv_{\text{rel}} = Ma \qquad \text{(first rocket equation),} \qquad (9\text{-}87)$$

in which M is the rocket's instantaneous mass (including unexpended fuel), R is the fuel consumption rate, and v_{rel} is the fuel's exhaust speed relative to the rocket. The term Rv_{rel} is the **thrust** of the rocket engine. For a rocket with constant R and v_{rel}, whose speed changes from v_i to v_f when its mass changes from M_i to M_f,

$$v_f - v_i = v_{\text{rel}} \ln \frac{M_i}{M_f} \qquad \text{(second rocket equation).} \qquad (9\text{-}88)$$

QUESTIONS

1 Figure 9-23 shows an overhead view of three particles on which external forces act. The magnitudes and directions of the forces on two of the particles are indicated. What are the magnitude and direction of the force acting on the third particle if the center of mass of the three-particle system is (a) stationary, (b) moving at a constant velocity rightward, and (c) accelerating rightward?

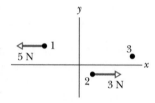

Fig. 9-23 Question 1.

2 Figure 9-24 shows an overhead view of four particles of equal mass sliding over a frictionless surface at constant velocity. The directions of the velocities are indicated; their magnitudes are equal. Consider pairing the particles. Which pairs form a system with a center of mass that (a) is stationary, (b) is stationary and at the origin, and (c) passes through the origin?

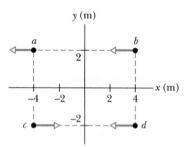

Fig. 9-24 Question 2.

3 Consider a box that explodes into two pieces while moving with a constant positive velocity along an x axis. If one piece, with mass m_1, ends up with positive velocity $\vec{v}_1$, then the second piece, with mass m_2, could end up with (a) a positive velocity $\vec{v}_2$ (Fig. 9-25a), (b) a negative velocity $\vec{v}_2$ (Fig. 9-25b), or (c) zero velocity (Fig. 9-25c). Rank those three possible results for the second piece according to the corresponding magnitude of $\vec{v}_1$, greatest first.

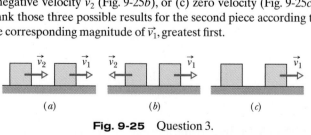

Fig. 9-25 Question 3.

4 Figure 9-26 shows graphs of force magnitude versus time for a body involved in a collision. Rank the graphs according to the magnitude of the impulse on the body, greatest first.

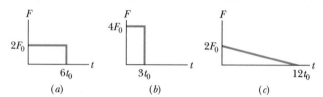

Fig. 9-26 Question 4.

5 The free-body diagrams in Fig. 9-27 give, from overhead views, the horizontal forces acting on three boxes of chocolates as the boxes move over a frictionless confectioner's counter. For each box, is its linear momentum conserved along the x axis and the y axis?

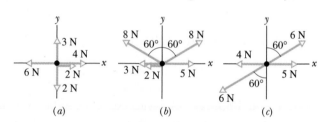

Fig. 9-27 Question 5.

6 Figure 9-28 shows four groups of three or four identical particles that move parallel to either the x axis or the y axis, at identical speeds. Rank the groups according to center-of-mass speed, greatest first.

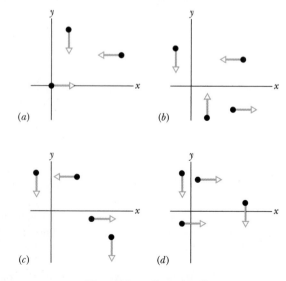

Fig. 9-28 Question 6.

7 A block slides along a frictionless floor and into a stationary second block with the same mass. Figure 9-29 shows four choices for a graph of the kinetic energies K of the blocks. (a) Determine which represent physically impossible situations. Of the others, which best represents (b) an elastic collision and (c) an inelastic collision?

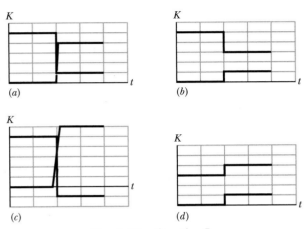

Fig. 9-29   Question 7.

8 Figure 9-30 shows a snapshot of block 1 as it slides along an x axis on a frictionless floor, before it undergoes an elastic collision with stationary block 2. The figure also shows three possible positions of the center of mass (com) of the two-block system at the time of the snapshot. (Point B is halfway between the centers of the two blocks.) Is block 1 stationary, moving forward, or moving backward after the collision if the com is located in the snapshot at (a) A, (b) B, and (c) C?

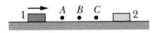

Fig. 9-30   Question 8.

9 Two bodies have undergone an elastic one-dimensional collision along an x axis. Figure 9-31 is a graph of position versus time for those bodies and for their center of mass. (a) Were both bodies initially moving, or was one initially stationary? Which line segment corresponds to the motion of the center of mass (b) before the collision and (c) after the collision? (d) Is the mass of the body that was moving faster before the collision greater than, less than, or equal to that of the other body?

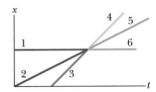

Fig. 9-31   Question 9.

10 Figure 9-32: A block on a horizontal floor is initially either stationary, sliding in the positive direction of an x axis, or sliding in the negative direction of that axis. Then the block explodes into two pieces that slide along the x axis. Assume the block and the two pieces form a closed, isolated system. Six choices for a graph of the momenta of the block and the pieces are given, all versus time t. Determine which choices represent physically impossible situations and explain why.

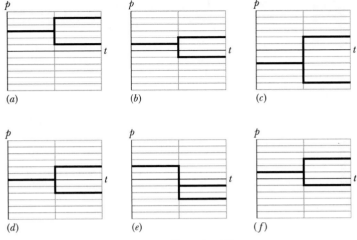

Fig. 9-32   Question 10.

11 Block 1 with mass m_1 slides along an x axis across a frictionless floor and then undergoes an elastic collision with a stationary block 2 with mass m_2. Figure 9-33 shows a plot of position x versus time t of block 1 until the collision occurs at position x_c and time t_c. In which of the lettered regions on the graph will the plot be continued (after the collision) if (a) $m_1 < m_2$ and (b) $m_1 > m_2$? (c) Along which of the numbered dashed lines will the plot be continued if $m_1 = m_2$?

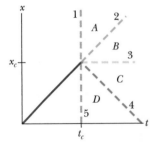

Fig. 9-33   Question 11.

12 Figure 9-34 shows four graphs of position versus time for two bodies and their center of mass. The two bodies form a closed, isolated system and undergo a completely inelastic, one-dimensional collision on an x axis. In graph 1, are (a) the two bodies and (b) the center of mass moving in the positive or negative direction of the x axis? (c) Which graphs correspond to a physically impossible situation? Explain.

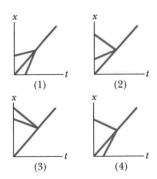

Fig. 9-34   Question 12.

GO Tutoring problem available (at instructor's discretion) in *WileyPLUS* and WebAssign

SSM Worked-out solution available in Student Solutions Manual

• – ••• Number of dots indicates level of problem difficulty

WWW Worked-out solution is at

ILW Interactive solution is at

http://www.wiley.com/college/halliday

Additional information available in *The Flying Circus of Physics* and at flyingcircusofphysics.com

sec. 9-2 The Center of Mass

•1 A 2.00 kg particle has the xy coordinates $(-1.20$ m, 0.500 m), and a 4.00 kg particle has the xy coordinates $(0.600$ m, -0.750 m). Both lie on a horizontal plane. At what (a) x and (b) y coordinates must you place a 3.00 kg particle such that the center of mass of the three-particle system has the coordinates $(-0.500$ m, -0.700 m)?

•2 Figure 9-35 shows a three-particle system, with masses $m_1 = 3.0$ kg, $m_2 = 4.0$ kg, and $m_3 = 8.0$ kg. The scales on the axes are set by $x_s = 2.0$ m and $y_s = 2.0$ m. What are (a) the x coordinate and (b) the y coordinate of the system's center of mass? (c) If m_3 is gradually increased, does the center of mass of the system shift toward or away from that particle, or does it remain stationary?

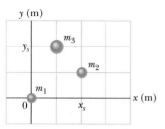

Fig. 9-35 Problem 2.

••3 Figure 9-36 shows a slab with dimensions $d_1 = 11.0$ cm, $d_2 = 2.80$ cm, and $d_3 = 13.0$ cm. Half the slab consists of aluminum (density $= 2.70$ g/cm³) and half consists of iron (density $= 7.85$ g/cm³). What are (a) the x coordinate, (b) the y coordinate, and (c) the z coordinate of the slab's center of mass?

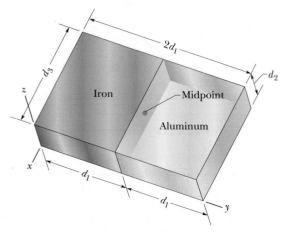

Fig. 9-36 Problem 3.

••4 In Fig. 9-37, three uniform thin rods, each of length $L = 22$ cm, form an inverted U. The vertical rods each have a mass of 14 g; the horizontal rod has a mass of 42 g. What are (a) the x coordinate and (b) the y coordinate of the system's center of mass?

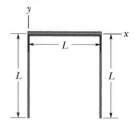

Fig. 9-37 Problem 4.

••5 **GO** What are (a) the x coordinate and (b) the y coordinate of the center of mass for the uniform plate shown in Fig. 9-38 if $L = 5.0$ cm?

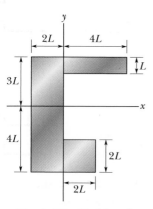

Fig. 9-38 Problem 5.

••6 Figure 9-39 shows a cubical box that has been constructed from uniform metal plate of negligible thickness. The box is open at the top and has edge length $L = 40$ cm. Find (a) the x coordinate, (b) the y coordinate, and (c) the z coordinate of the center of mass of the box.

Fig. 9-39 Problem 6.

•••7 **ILW** In the ammonia (NH₃) molecule of Fig. 9-40, three hydrogen (H) atoms form an equilateral triangle, with the center of the triangle at distance $d = 9.40 \times 10^{-11}$ m from each hydrogen atom. The nitrogen (N) atom is at the apex of a pyramid, with the three hydrogen atoms forming the base. The nitrogen-to-hydrogen atomic mass ratio is 13.9, and the nitrogen-to-hydrogen distance is $L = 10.14 \times 10^{-11}$ m. What are the (a) x and (b) y coordinates of the molecule's center of mass?

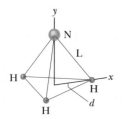

Fig. 9-40 Problem 7.

•••8 A uniform soda can of mass 0.140 kg is 12.0 cm tall and filled with 0.354 kg of soda (Fig. 9-41). Then small holes are drilled in the top and bottom (with negligible loss of metal) to drain the soda. What is the height h of the com of the can and contents (a) initially and (b) after the can loses all the soda? (c) What happens to h as the soda drains out? (d) If x is the height of the remaining soda at any given instant, find x when the com reaches its lowest point.

Fig. 9-41 Problem 8.

sec. 9-3 Newton's Second Law for a System of Particles

•9 ILW A stone is dropped at $t = 0$. A second stone, with twice the mass of the first, is dropped from the same point at $t = 100$ ms. (a) How far below the release point is the center of mass of the two stones at $t = 300$ ms? (Neither stone has yet reached the ground.) (b) How fast is the center of mass of the two-stone system moving at that time?

•10 GO A 1000 kg automobile is at rest at a traffic signal. At the instant the light turns green, the automobile starts to move with a constant acceleration of 4.0 m/s². At the same instant a 2000 kg truck, traveling at a constant speed of 8.0 m/s, overtakes and passes the automobile. (a) How far is the com of the automobile–truck system from the traffic light at $t = 3.0$ s? (b) What is the speed of the com then?

•11 A big olive ($m = 0.50$ kg) lies at the origin of an xy coordinate system, and a big Brazil nut ($M = 1.5$ kg) lies at the point (1.0, 2.0) m. At $t = 0$, a force $\vec{F}_o = (2.0\hat{i} + 3.0\hat{j})$ N begins to act on the olive, and a force $\vec{F}_n = (-3.0\hat{i} - 2.0\hat{j})$ N begins to act on the nut. In unit-vector notation, what is the displacement of the center of mass of the olive–nut system at $t = 4.0$ s, with respect to its position at $t = 0$?

•12 Two skaters, one with mass 65 kg and the other with mass 40 kg, stand on an ice rink holding a pole of length 10 m and negligible mass. Starting from the ends of the pole, the skaters pull themselves along the pole until they meet. How far does the 40 kg skater move?

•••13 SSM A shell is shot with an initial velocity $\vec{v}_0$ of 20 m/s, at an angle of $\theta_0 = 60°$ with the horizontal. At the top of the trajectory, the shell explodes into two fragments of equal mass (Fig. 9-42). One fragment, whose speed immediately after the explosion is zero, falls vertically. How far from the gun does the other fragment land, assuming that the terrain is level and that air drag is negligible?

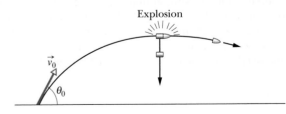

Fig. 9-42 Problem 13.

•••14 In Figure 9-43, two particles are launched from the origin of the coordinate system at time $t = 0$. Particle 1 of mass $m_1 = 5.00$ g is shot directly along the x axis on a frictionless floor, with constant speed 10.0 m/s. Particle 2 of mass $m_2 = 3.00$ g is shot with a velocity of magnitude 20.0 m/s, at an upward angle such that it always stays

directly above particle 1. (a) What is the maximum height H_{max} reached by the com of the two-particle system? In unit-vector notation, what are the (b) velocity and (c) acceleration of the com when the com reaches H_{max}?

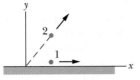

Fig. 9-43 Problem 14.

••15 Figure 9-44 shows an arrangement with an air track, in which a cart is connected by a cord to a hanging block. The cart has mass $m_1 = 0.600$ kg, and its center is initially at xy coordinates $(-0.500$ m, 0 m); the block has mass $m_2 = 0.400$ kg, and its center is initially at xy coordinates $(0, -0.100$ m). The mass of the cord and pulley are negligible. The cart is released from rest, and both cart and block move until the cart hits the pulley. The friction between the cart and the air track and between the pulley and its axle is negligible. (a) In unit-vector notation, what is the acceleration of the center of mass of the cart–block system? (b) What is the velocity of the com as a function of time t? (c) Sketch the path taken by the com. (d) If the path is curved, determine whether it bulges upward to the right or downward to the left, and if it is straight, find the angle between it and the x axis.

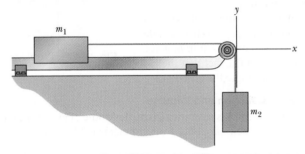

Fig. 9-44 Problem 15.

•••16 Ricardo, of mass 80 kg, and Carmelita, who is lighter, are enjoying Lake Merced at dusk in a 30 kg canoe. When the canoe is at rest in the placid water, they exchange seats, which are 3.0 m apart and symmetrically located with respect to the canoe's center. If the canoe moves 40 cm horizontally relative to a pier post, what is Carmelita's mass?

•••17 In Fig. 9-45a, a 4.5 kg dog stands on an 18 kg flatboat at distance $D = 6.1$ m from the shore. It walks 2.4 m along the boat toward shore and then stops. Assuming no friction between the boat and the water, find how far the dog is then from the shore. (*Hint:* See Fig. 9-45b.)

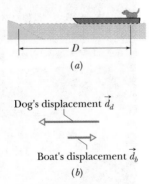

Fig. 9-45 Problem 17.

sec. 9-5 The Linear Momentum of a System of Particles

•18 A 0.70 kg ball moving horizontally at 5.0 m/s strikes a vertical wall and rebounds with speed 2.0 m/s. What is the magnitude of the change in its linear momentum?

•19 ILW A 2100 kg truck traveling north at 41 km/h turns east and accelerates to 51 km/h. (a) What is the change in the truck's kinetic energy? What are the (b) magnitude and (c) direction of the change in its momentum?

••20 GO At time $t = 0$, a ball is struck at ground level and sent over level ground. The momentum p versus t during the flight is given by Fig. 9-46 (p_0 = 6.0 kg·m/s and p_1 = 4.0 kg·m/s). At what initial angle is the ball launched? (*Hint*: find a solution that does not require you to read the time of the low point of the plot.)

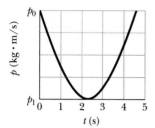

Fig. 9-46 Problem 20.

•21 A 0.30 kg softball has a velocity of 15 m/s at an angle of 35° below the horizontal just before making contact with the bat. What is the magnitude of the change in momentum of the ball while in contact with the bat if the ball leaves with a velocity of (a) 20 m/s, vertically downward, and (b) 20 m/s, horizontally back toward the pitcher?

••22 Figure 9-47 gives an overhead view of the path taken by a 0.165 kg cue ball as it bounces from a rail of a pool table. The ball's initial speed is 2.00 m/s, and the angle θ_1 is 30.0°. The bounce reverses the *y* component of the ball's velocity but does not alter the *x* component. What are (a) angle θ_2 and (b) the change in the ball's linear momentum in unit-vector notation? (The fact that the ball rolls is irrelevant to the problem.)

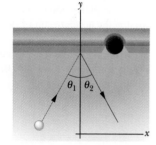

Fig. 9-47 Problem 22.

sec. 9-6 Collision and Impulse

•23 Until his seventies, Henri LaMothe (Fig. 9-48) excited audiences by belly-flopping from a height of 12 m into 30 cm of

Fig. 9-48 Problem 23. Belly-flopping into 30 cm of water. (*George Long/ Sports Illustrated/©Time, Inc.*)

water. Assuming that he stops just as he reaches the bottom of the water and estimating his mass, find the magnitude of the impulse on him from the water.

•24 In February 1955, a paratrooper fell 370 m from an airplane without being able to open his chute but happened to land in snow, suffering only minor injuries. Assume that his speed at impact was 56 m/s (terminal speed), that his mass (including gear) was 85 kg, and that the magnitude of the force on him from the snow was at the survivable limit of 1.2×10^5 N. What are (a) the minimum depth of snow that would have stopped him safely and (b) the magnitude of the impulse on him from the snow?

•25 A 1.2 kg ball drops vertically onto a floor, hitting with a speed of 25 m/s. It rebounds with an initial speed of 10 m/s. (a) What impulse acts on the ball during the contact? (b) If the ball is in contact with the floor for 0.020 s, what is the magnitude of the average force on the floor from the ball?

•26 In a common but dangerous prank, a chair is pulled away as a person is moving downward to sit on it, causing the victim to land hard on the floor. Suppose the victim falls by 0.50 m, the mass that moves downward is 70 kg, and the collision on the floor lasts 0.082 s. What are the magnitudes of the (a) impulse and (b) average force acting on the victim from the floor during the collision?

•27 SSM A force in the negative direction of an *x* axis is applied for 27 ms to a 0.40 kg ball initially moving at 14 m/s in the positive direction of the axis. The force varies in magnitude, and the impulse has magnitude 32.4 N·s. What are the ball's (a) speed and (b) direction of travel just after the force is applied? What are (c) the average magnitude of the force and (d) the direction of the impulse on the ball?

•28 In tae-kwon-do, a hand is slammed down onto a target at a speed of 13 m/s and comes to a stop during the 5.0 ms collision. Assume that during the impact the hand is independent of the arm and has a mass of 0.70 kg. What are the magnitudes of the (a) impulse and (b) average force on the hand from the target?

•29 Suppose a gangster sprays Superman's chest with 3 g bullets at the rate of 100 bullets/min, and the speed of each bullet is 500 m/s. Suppose too that the bullets rebound straight back with no change in speed. What is the magnitude of the average force on Superman's chest?

••30 *Two average forces.* A steady stream of 0.250 kg snowballs is shot perpendicularly into a wall at a speed of 4.00 m/s. Each ball sticks to the wall. Figure 9-49 gives the magnitude *F* of the force on the wall as a function of time *t* for two of the snowball impacts. Impacts occur with a repetition time interval Δt_r = 50.0 ms, last a duration time interval Δt_d = 10 ms, and produce isosceles triangles on the graph, with each impact reaching a force maximum F_{max} = 200 N. During each impact, what are the magnitudes of (a) the impulse and (b) the average force on the wall? (c) During a time in-

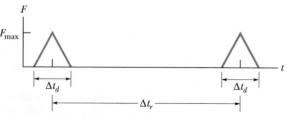

Fig. 9-49 Problem 30.

terval of many impacts, what is the magnitude of the average force on the wall?

••31 ✈ *Jumping up before the elevator hits.* After the cable snaps and the safety system fails, an elevator cab free-falls from a height of 36 m. During the collision at the bottom of the elevator shaft, a 90 kg passenger is stopped in 5.0 ms. (Assume that neither the passenger nor the cab rebounds.) What are the magnitudes of the (a) impulse and (b) average force on the passenger during the collision? If the passenger were to jump upward with a speed of 7.0 m/s relative to the cab floor just before the cab hits the bottom of the shaft, what are the magnitudes of the (c) impulse and (d) average force (assuming the same stopping time)?

••32 A 5.0 kg toy car can move along an x axis; Fig. 9-50 gives F_x of the force acting on the car, which begins at rest at time $t = 0$. The scale on the F_x axis is set by $F_{xs} = 5.0$ N. In unit-vector notation, what is $\vec{p}$ at (a) $t = 4.0$ s and (b) $t = 7.0$ s, and (c) what is $\vec{v}$ at $t = 9.0$ s?

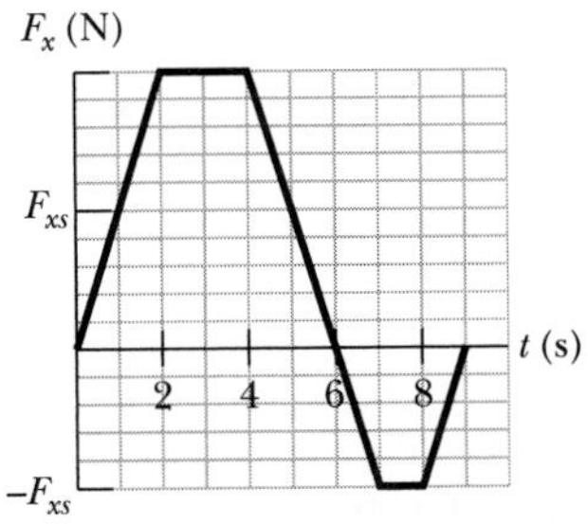

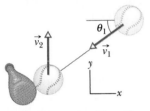

Fig. 9-50   Problem 32.

••33 GO Figure 9-51 shows a 0.300 kg baseball just before and just after it collides with a bat. Just before, the ball has velocity $\vec{v}_1$ of magnitude 12.0 m/s and angle $\theta_1 = 35.0°$. Just after, it is traveling directly upward with velocity $\vec{v}_2$ of magnitude 10.0 m/s. The duration of the collision is 2.00 ms. What are the (a) magnitude and (b) direction (relative to the positive direction of the x axis) of the impulse on the ball from the bat? What are the (c) magnitude and (d) direction of the average force on the ball from the bat?

Fig. 9-51   Problem 33.

••34 ✈ Basilisk lizards can run across the top of a water surface (Fig. 9-52). With each step, a lizard first slaps its foot against the water and then pushes it down into the water rapidly enough to form an air cavity around the top of the foot. To avoid having to pull the foot back up against water drag in order to complete the step, the lizard withdraws the foot before water can flow into the

Fig. 9-52   Problem 34. Lizard running across water. (*Stephen Dalton/Photo Researchers*)

air cavity. If the lizard is not to sink, the average upward impulse on the lizard during this full action of slap, downward push, and withdrawal must match the downward impulse due to the gravitational force. Suppose the mass of a basilisk lizard is 90.0 g, the mass of each foot is 3.00 g, the speed of a foot as it slaps the water is 1.50 m/s, and the time for a single step is 0.600 s. (a) What is the magnitude of the impulse on the lizard during the slap? (Assume this impulse is directly upward.) (b) During the 0.600 s duration of a step, what is the downward impulse on the lizard due to the gravitational force? (c) Which action, the slap or the push, provides the primary support for the lizard, or are they approximately equal in their support?

••35 GO Figure 9-53 shows an approximate plot of force magnitude F versus time t during the collision of a 58 g Superball with a wall. The initial velocity of the ball is 34 m/s perpendicular to the wall; the ball rebounds directly back with approximately the same speed, also perpendicular to the wall. What is F_{max}, the maximum magnitude of the force on the ball from the wall during the collision?

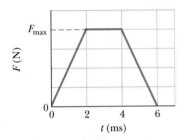

Fig. 9-53   Problem 35.

••36 A 0.25 kg puck is initially stationary on an ice surface with negligible friction. At time $t = 0$, a horizontal force begins to move the puck. The force is given by $\vec{F} = (12.0 - 3.00t^2)\hat{i}$, with $\vec{F}$ in newtons and t in seconds, and it acts until its magnitude is zero. (a) What is the magnitude of the impulse on the puck from the force between $t = 0.500$ s and $t = 1.25$ s? (b) What is the change in momentum of the puck between $t = 0$ and the instant at which $F = 0$?

••37 SSM A soccer player kicks a soccer ball of mass 0.45 kg that is initially at rest. The foot of the player is in contact with the ball for 3.0×10^{-3} s, and the force of the kick is given by

$$F(t) = [(6.0 \times 10^6)t - (2.0 \times 10^9)t^2]\ \text{N}$$

for $0 \le t \le 3.0 \times 10^{-3}$ s, where t is in seconds. Find the magnitudes of (a) the impulse on the ball due to the kick, (b) the average force on the ball from the player's foot during the period of contact, (c) the maximum force on the ball from the player's foot during the period of contact, and (d) the ball's velocity immediately after it loses contact with the player's foot.

••38 In the overhead view of Fig. 9-54, a 300 g ball with a speed v of 6.0 m/s strikes a wall at an angle θ of 30° and then rebounds with the

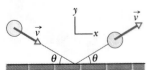

Fig. 9-54   Problem 38.

same speed and angle. It is in contact with the wall for 10 ms. In unit-vector notation, what are (a) the impulse on the ball from the wall and (b) the average force on the wall from the ball?

sec. 9-7   Conservation of Linear Momentum

•39 SSM A 91 kg man lying on a surface of negligible friction shoves a 68 g stone away from himself, giving it a speed of 4.0 m/s. What speed does the man acquire as a result?

•40 A space vehicle is traveling at 4300 km/h relative to Earth when the exhausted rocket motor (mass $4m$) is disengaged and sent backward with a speed of 82 km/h relative to the command module (mass m). What is the speed of the command module relative to Earth just after the separation?

••41 Figure 9-55 shows a two-ended "rocket" that is initially stationary on a frictionless floor, with its center at the origin of an x axis. The rocket consists of a central block C (of mass $M = 6.00$ kg) and blocks L and R (each of mass $m = 2.00$ kg) on the left and right sides. Small explosions can shoot either of the side blocks away from block C and along the x axis. Here is the sequence: (1) At time $t = 0$, block L is shot to the left with a speed of 3.00 m/s *relative* to the velocity that the explosion gives the rest of the rocket. (2) Next, at time $t = 0.80$ s, block R is shot to the right with a speed of 3.00 m/s *relative* to the velocity that block C then has. At $t = 2.80$ s, what are (a) the velocity of block C and (b) the position of its center?

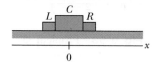

Fig. 9-55   Problem 41.

••42 An object, with mass m and speed v relative to an observer, explodes into two pieces, one three times as massive as the other; the explosion takes place in deep space. The less massive piece stops relative to the observer. How much kinetic energy is added to the system during the explosion, as measured in the observer's reference frame?

••43 〔✈〕 In the Olympiad of 708 B.C., some athletes competing in the standing long jump used handheld weights called *halteres* to lengthen their jumps (Fig. 9-56). The weights were swung up in front just before liftoff and then swung down and thrown backward during the flight. Suppose a modern 78 kg long jumper similarly uses two 5.50 kg halteres, throwing them horizontally to the rear at his maximum height such that their horizontal velocity is zero relative to the ground. Let his liftoff velocity be $\vec{v} = (9.5\hat{i} + 4.0\hat{j})$ m/s

Fig. 9-56   Problem 43. (*Réunion des Musées Nationaux/Art Resource*)

with or without the halteres, and assume that he lands at the liftoff level. What distance would the use of the halteres add to his range?

••44 〔GO〕 In Fig. 9-57, a stationary block explodes into two pieces L and R that slide across a frictionless floor and then into regions with friction, where they stop. Piece L, with a mass of 2.0 kg, encounters a coefficient of kinetic friction $\mu_L = 0.40$ and slides to a stop in distance $d_L = 0.15$ m. Piece R encounters a coefficient of kinetic friction $\mu_R = 0.50$ and slides to a stop in distance $d_R = 0.25$ m. What was the mass of the block?

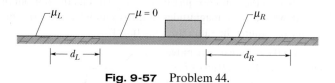

Fig. 9-57   Problem 44.

••45 SSM WWW A 20.0 kg body is moving through space in the positive direction of an x axis with a speed of 200 m/s when, due to an internal explosion, it breaks into three parts. One part, with a mass of 10.0 kg, moves away from the point of explosion with a speed of 100 m/s in the positive y direction. A second part, with a mass of 4.00 kg, moves in the negative x direction with a speed of 500 m/s. (a) In unit-vector notation, what is the velocity of the third part? (b) How much energy is released in the explosion? Ignore effects due to the gravitational force.

••46 A 4.0 kg mess kit sliding on a frictionless surface explodes into two 2.0 kg parts: 3.0 m/s, due north, and 5.0 m/s, 30° north of east. What is the original speed of the mess kit?

••47 A vessel at rest at the origin of an xy coordinate system explodes into three pieces. Just after the explosion, one piece, of mass m, moves with velocity $(-30$ m/s$)\hat{i}$ and a second piece, also of mass m, moves with velocity $(-30$ m/s$)\hat{j}$. The third piece has mass $3m$. Just after the explosion, what are the (a) magnitude and (b) direction of the velocity of the third piece?

•••48 〔GO〕 Particle A and particle B are held together with a compressed spring between them. When they are released, the spring pushes them apart, and they then fly off in opposite directions, free of the spring. The mass of A is 2.00 times the mass of B, and the energy stored in the spring was 60 J. Assume that the spring has negligible mass and that all its stored energy is transferred to the particles. Once that transfer is complete, what are the kinetic energies of (a) particle A and (b) particle B?

sec. 9-9   Inelastic Collisions in One Dimension

•49 A bullet of mass 10 g strikes a ballistic pendulum of mass 2.0 kg. The center of mass of the pendulum rises a vertical distance of 12 cm. Assuming that the bullet remains embedded in the pendulum, calculate the bullet's initial speed.

•50 A 5.20 g bullet moving at 672 m/s strikes a 700 g wooden block at rest on a frictionless surface. The bullet emerges, traveling in the same direction with its speed reduced to 428 m/s. (a) What is the resulting speed of the block? (b) What is the speed of the bullet–block center of mass?

••51 〔GO〕 In Fig. 9-58a, a 3.50 g bullet is fired horizontally at two blocks at rest on a frictionless table. The bullet passes through block 1 (mass 1.20 kg) and embeds itself in block 2 (mass 1.80 kg). The blocks end up with speeds $v_1 = 0.630$ m/s and $v_2 = 1.40$ m/s (Fig. 9-58b). Neglecting the material removed from block 1 by the bullet, find the speed of the bullet as it (a) leaves and (b) enters block 1.

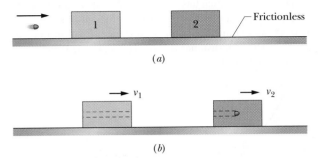

Fig. 9-58   Problem 51.

••**52** GO   In Fig. 9-59, a 10 g bullet moving directly upward at 1000 m/s strikes and passes through the center of mass of a 5.0 kg block initially at rest. The bullet emerges from the block moving directly upward at 400 m/s. To what maximum height does the block then rise above its initial position?

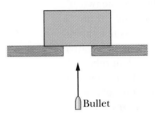

Fig. 9-59   Problem 52.

••**53**   In Anchorage, collisions of a vehicle with a moose are so common that they are referred to with the abbreviation MVC. Suppose a 1000 kg car slides into a stationary 500 kg moose on a very slippery road, with the moose being thrown through the windshield (a common MVC result). (a) What percent of the original kinetic energy is lost in the collision to other forms of energy? A similar danger occurs in Saudi Arabia because of camel–vehicle collisions (CVC). (b) What percent of the original kinetic energy is lost if the car hits a 300 kg camel? (c) Generally, does the percent loss increase or decrease if the animal mass decreases?

••**54**   A completely inelastic collision occurs between two balls of wet putty that move directly toward each other along a vertical axis. Just before the collision, one ball, of mass 3.0 kg, is moving upward at 20 m/s and the other ball, of mass 2.0 kg, is moving downward at 12 m/s. How high do the combined two balls of putty rise above the collision point? (Neglect air drag.)

••**55** ILW   A 5.0 kg block with a speed of 3.0 m/s collides with a 10 kg block that has a speed of 2.0 m/s in the same direction. After the collision, the 10 kg block travels in the original direction with a speed of 2.5 m/s. (a) What is the velocity of the 5.0 kg block immediately after the collision? (b) By how much does the total kinetic energy of the system of two blocks change because of the collision? (c) Suppose, instead, that the 10 kg block ends up with a speed of 4.0 m/s. What then is the change in the total kinetic energy? (d) Account for the result you obtained in (c).

••**56**   In the "before" part of Fig. 9-60, car *A* (mass 1100 kg) is stopped at a traffic light when it is rear-ended by car *B* (mass 1400 kg). Both cars then slide with locked wheels until the frictional force from the slick road (with a low μ_k of 0.13) stops them, at dis-

tances $d_A = 8.2$ m and $d_B = 6.1$ m. What are the speeds of (a) car *A* and (b) car *B* at the start of the sliding, just after the collision? (c) Assuming that linear momentum is conserved during the collision, find the speed of car *B* just before the collision. (d) Explain why this assumption may be invalid.

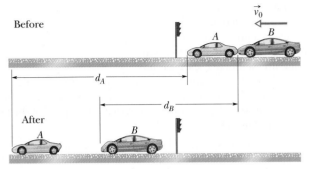

Fig. 9-60   Problem 56.

••**57** GO   In Fig. 9-61, a ball of mass $m = 60$ g is shot with speed $v_i = 22$ m/s into the barrel of a spring gun of mass $M = 240$ g initially at rest on a frictionless surface. The ball sticks in the barrel at the point of maximum compression of the spring. Assume that the increase in thermal energy due to friction between the ball and the barrel is negligible. (a) What is the speed of the spring gun after the ball stops in the barrel? (b) What fraction of the initial kinetic energy of the ball is stored in the spring?

Fig. 9-61   Problem 57.

•••**58**   In Fig. 9-62, block 2 (mass 1.0 kg) is at rest on a frictionless surface and touching the end of an unstretched spring of spring constant 200 N/m. The other end of the spring is fixed to a wall. Block 1 (mass 2.0 kg), traveling at speed $v_1 = 4.0$ m/s, collides with block 2, and the two blocks stick together. When the blocks momentarily stop, by what distance is the spring compressed?

Fig. 9-62   Problem 58.

•••**59** ILW   In Fig. 9-63, block 1 (mass 2.0 kg) is moving rightward at 10 m/s and block 2 (mass 5.0 kg) is moving rightward at 3.0 m/s. The surface is frictionless, and a spring with a spring constant of 1120 N/m is fixed to block 2. When the blocks collide, the compression of the spring is maximum at the instant the blocks have the same velocity. Find the maximum compression.

Fig. 9-63   Problem 59.

sec. 9-10   Elastic Collisions in One Dimension

•**60**   In Fig. 9-64, block *A* (mass 1.6 kg) slides into block *B* (mass 2.4 kg), along a frictionless surface. The directions of three velocities before (*i*) and after (*f*) the collision are indicated; the corresponding speeds are v_{Ai} = 5.5 m/s, v_{Bi} = 2.5 m/s, and v_{Bf} = 4.9 m/s. What are the (a) speed and (b) direction (left or right) of velocity $\vec{v}_{Af}$? (c) Is the collision elastic?

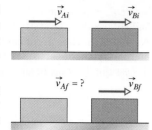

•61 SSM A cart with mass 340 g moving on a frictionless linear air track at an initial speed of 1.2 m/s undergoes an elastic collision with an initially stationary cart of unknown mass. After the collision, the first cart continues in its original direction at 0.66 m/s. (a) What is the mass of the second cart? (b) What is its speed after impact? (c) What is the speed of the two-cart center of mass?

Fig. 9-64 Problem 60.

•62 Two titanium spheres approach each other head-on with the same speed and collide elastically. After the collision, one of the spheres, whose mass is 300 g, remains at rest. (a) What is the mass of the other sphere? (b) What is the speed of the two-sphere center of mass if the initial speed of each sphere is 2.00 m/s?

••63 Block 1 of mass m_1 slides along a frictionless floor and into a one-dimensional elastic collision with stationary block 2 of mass $m_2 = 3m_1$. Prior to the collision, the center of mass of the two-block system had a speed of 3.00 m/s. Afterward, what are the speeds of (a) the center of mass and (b) block 2?

••64 GO A steel ball of mass 0.500 kg is fastened to a cord that is 70.0 cm long and fixed at the far end. The ball is then released when the cord is horizontal (Fig. 9-65). At the bottom of its path, the ball strikes a 2.50 kg steel block initially at rest on a frictionless surface. The collision is elastic. Find (a) the speed of the ball and (b) the speed of the block, both just after the collision.

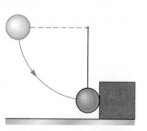

Fig. 9-65 Problem 64.

••65 SSM A body of mass 2.0 kg makes an elastic collision with another body at rest and continues to move in the original direction but with one-fourth of its original speed. (a) What is the mass of the other body? (b) What is the speed of the two-body center of mass if the initial speed of the 2.0 kg body was 4.0 m/s?

••66 Block 1, with mass m_1 and speed 4.0 m/s, slides along an x axis on a frictionless floor and then undergoes a one-dimensional elastic collision with stationary block 2, with mass $m_2 = 0.40m_1$. The two blocks then slide into a region where the coefficient of kinetic friction is 0.50; there they stop. How far into that region do (a) block 1 and (b) block 2 slide?

••67 In Fig. 9-66, particle 1 of mass $m_1 = 0.30$ kg slides rightward along an x axis on a frictionless floor with a speed of 2.0 m/s. When it reaches $x = 0$, it undergoes a one-dimensional elastic collision with stationary particle 2 of mass $m_2 = 0.40$ kg. When particle 2 then reaches a wall at $x_w = 70$ cm, it bounces from the wall with no loss of speed. At what position on the x axis does particle 2 then collide with particle 1?

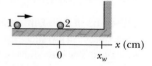

Fig. 9-66 Problem 67.

••68 GO In Fig. 9-67, block 1 of mass m_1 slides from rest along a frictionless ramp from height $h = 2.50$ m and then collides with stationary block 2, which has mass $m_2 = 2.00m_1$. After the collision, block 2 slides into a region where the coefficient of kinetic friction μ_k is 0.500 and comes to a stop in distance d within that region. What is the value of distance d if the collision is (a) elastic and (b) completely inelastic?

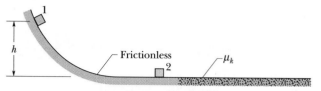

Fig. 9-67 Problem 68.

•••69 A small ball of mass m is aligned above a larger ball of mass $M = 0.63$ kg (with a slight separation, as with the baseball and basketball of Fig. 9-68a), and the two are dropped simultaneously from a height of $h = 1.8$ m. (Assume the radius of each ball is negligible relative to h.) (a) If the larger ball rebounds elastically from the floor and then the small ball rebounds elastically from the larger ball, what value of m results in the larger ball stopping when it collides with the small ball? (b) What height does the small ball then reach (Fig. 9-68b)?

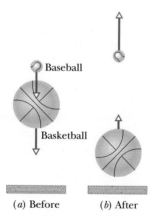

Fig. 9-68 Problem 69.

•••70 In Fig. 9-69, puck 1 of mass $m_1 = 0.20$ kg is sent sliding across a frictionless lab bench, to undergo a one-dimensional elastic collision with stationary puck 2. Puck 2 then slides off the bench and lands a distance d from the base of the bench. Puck 1 rebounds from the collision and slides off the opposite edge of the bench, landing a distance $2d$ from the base of the bench. What is the mass of puck 2? (*Hint:* Be careful with signs.)

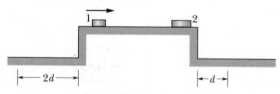

Fig. 9-69 Problem 70.

sec. 9-11 Collisions in Two Dimensions

••71 ILW In Fig. 9-21, projectile particle 1 is an alpha particle and target particle 2 is an oxygen nucleus. The alpha particle is scattered at angle $\theta_1 = 64.0°$ and the oxygen nucleus recoils with speed 1.20×10^5 m/s and at angle $\theta_2 = 51.0°$. In atomic mass units, the mass of the alpha particle is 4.00 u and the mass of the oxygen nucleus is 16.0 u. What are the (a) final and (b) initial speeds of the alpha particle?

••72 Ball B, moving in the positive direction of an x axis at speed v, collides with stationary ball A at the origin. A and B have different masses. After the collision, B moves in the negative direction of the y axis at speed $v/2$. (a) In what direction does A move? (b)

Show that the speed of *A* cannot be determined from the given information.

••73 After a completely inelastic collision, two objects of the same mass and same initial speed move away together at half their initial speed. Find the angle between the initial velocities of the objects.

••74 Two 2.0 kg bodies, *A* and *B*, collide. The velocities before the collision are $\vec{v}_A = (15\hat{i} + 30\hat{j})$ m/s and $\vec{v}_B = (-10\hat{i} + 5.0\hat{j})$ m/s. After the collision, $\vec{v}'_A = (-5.0\hat{i} + 20\hat{j})$ m/s. What are (a) the final velocity of *B* and (b) the change in the total kinetic energy (including sign)?

••75 A projectile proton with a speed of 500 m/s collides elastically with a target proton initially at rest. The two protons then move along perpendicular paths, with the projectile path at 60° from the original direction. After the collision, what are the speeds of (a) the target proton and (b) the projectile proton?

sec. 9-12   Systems with Varying Mass: A Rocket

•76 A 6090 kg space probe moving nose-first toward Jupiter at 105 m/s relative to the Sun fires its rocket engine, ejecting 80.0 kg of exhaust at a speed of 253 m/s relative to the space probe. What is the final velocity of the probe?

•77 SSM In Fig. 9-70, two long barges are moving in the same direction in still water, one with a speed of 10 km/h and the other with a speed of 20 km/h. While they are passing each other, coal is shoveled from the slower to the faster one at a rate of 1000 kg/min. How much additional force must be provided by the driving engines of (a) the faster barge and (b) the slower barge if neither is to change speed? Assume that the shoveling is always perfectly sideways and that the frictional forces between the barges and the water do not depend on the mass of the barges.

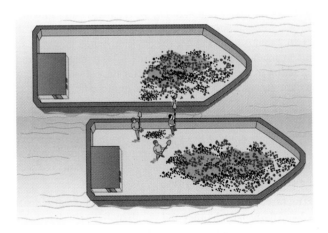

Fig. 9-70   Problem 77.

•78 Consider a rocket that is in deep space and at rest relative to an inertial reference frame. The rocket's engine is to be fired for a certain interval. What must be the rocket's *mass ratio* (ratio of initial to final mass) over that interval if the rocket's original speed relative to the inertial frame is to be equal to (a) the exhaust speed (speed of the exhaust products relative to the rocket) and (b) 2.0 times the exhaust speed?

•79 SSM ILW A rocket that is in deep space and initially at rest relative to an inertial reference frame has a mass of 2.55×10^5 kg,

of which 1.81×10^5 kg is fuel. The rocket engine is then fired for 250 s while fuel is consumed at the rate of 480 kg/s. The speed of the exhaust products relative to the rocket is 3.27 km/s. (a) What is the rocket's thrust? After the 250 s firing, what are (b) the mass and (c) the speed of the rocket?

Additional Problems

80 An object is tracked by a radar station and determined to have a position vector given by $\vec{r} = (3500 - 160t)\hat{i} + 2700\hat{j} + 300\hat{k}$, with $\vec{r}$ in meters and *t* in seconds. The radar station's *x* axis points east, its *y* axis north, and its *z* axis vertically up. If the object is a 250 kg meteorological missile, what are (a) its linear momentum, (b) its direction of motion, and (c) the net force on it?

81 The last stage of a rocket, which is traveling at a speed of 7600 m/s, consists of two parts that are clamped together: a rocket case with a mass of 290.0 kg and a payload capsule with a mass of 150.0 kg. When the clamp is released, a compressed spring causes the two parts to separate with a relative speed of 910.0 m/s. What are the speeds of (a) the rocket case and (b) the payload after they have separated? Assume that all velocities are along the same line. Find the total kinetic energy of the two parts (c) before and (d) after they separate. (e) Account for the difference.

82 *Pancake collapse of a tall building.* In the section of a tall building shown in Fig. 9-71a, the infrastructure of any given floor *K* must support the weight *W* of all higher floors. Normally the infrastructure is constructed with a safety factor *s* so that it can withstand an even greater downward force of *sW*. If, however, the support columns between *K* and *L* suddenly collapse and allow the higher floors to free-fall together onto floor *K* (Fig. 9-71b), the force in the collision can exceed *sW* and, after a brief pause, cause *K* to collapse onto floor *J*, which collapses on floor *I*, and so on until the ground is reached. Assume that the floors are separated by *d* = 4.0 m and have the same mass. Also assume that when the floors above *K* free-fall onto *K*, the collision lasts 1.5 ms. Under these simplified conditions, what value must the safety factor *s* exceed to prevent pancake collapse of the building?

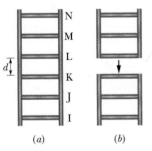

(a)   (b)

Fig. 9-71   Problem 82.

83 *"Relative" is an important word.* In Fig. 9-72, block *L* of mass $m_L = 1.00$ kg and block *R* of mass $m_R = 0.500$ kg are held in place with a compressed spring between them. When the blocks are released, the spring sends them sliding across a frictionless floor. (The spring has negligible mass and falls to the floor after the

Fig. 9-72   Problem 83.

blocks leave it.) (a) If the spring gives block L a release speed of 1.20 m/s *relative* to the floor, how far does block R travel in the next 0.800 s? (b) If, instead, the spring gives block L a release speed of 1.20 m/s *relative* to the velocity that the spring gives block R, how far does block R travel in the next 0.800 s?

84 Figure 9-73 shows an overhead view of two particles sliding at constant velocity over a frictionless surface. The particles have the same mass and the same initial speed $v = 4.00$ m/s, and they collide where their paths intersect. An x axis is arranged to bisect the angle between their incoming paths, such that $\theta = 40.0°$. The region to the right of the collision is divided into four lettered sections by the x axis and four numbered dashed lines. In what region or along what line do the particles travel if the collision is (a) completely inelastic, (b) elastic, and (c) inelastic? What are their final speeds if the collision is (d) completely inelastic and (e) elastic?

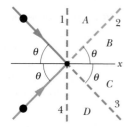

Fig. 9-73 Problem 84.

85 Speed deamplifier. In Fig. 9-74, block 1 of mass m_1 slides along an x axis on a frictionless floor at speed 4.00 m/s. Then it undergoes a one-dimensional elastic collision with stationary block 2 of mass $m_2 = 2.00m_1$. Next, block 2 undergoes a one-dimensional elastic collision with stationary block 3 of mass $m_3 = 2.00m_2$. (a) What then is the speed of block 3? Are (b) the speed, (c) the kinetic energy, and (d) the momentum of block 3 greater than, less than, or the same as the initial values for block 1?

Fig. 9-74 Problem 85.

86 Speed amplifier. In Fig. 9-75, block 1 of mass m_1 slides along an x axis on a frictionless floor with a speed of $v_{1i} = 4.00$ m/s. Then it undergoes a one-dimensional elastic collision with stationary block 2 of mass $m_2 = 0.500m_1$. Next, block 2 undergoes a one-dimensional elastic collision with stationary block 3 of mass $m_3 = 0.500m_2$. (a) What then is the speed of block 3? Are (b) the speed, (c) the kinetic energy, and (d) the momentum of block 3 greater than, less than, or the same as the initial values for block 1?

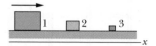

Fig. 9-75 Problem 86.

87 A ball having a mass of 150 g strikes a wall with a speed of 5.2 m/s and rebounds with only 50% of its initial kinetic energy. (a) What is the speed of the ball immediately after rebounding? (b) What is the magnitude of the impulse on the wall from the ball? (c) If the ball is in contact with the wall for 7.6 ms, what is the magnitude of the average force on the ball from the wall during this time interval?

88 A spacecraft is separated into two parts by detonating the explosive bolts that hold them together. The masses of the parts are 1200 kg and 1800 kg; the magnitude of the impulse on each part from the bolts is 300 N·s. With what relative speed do the two parts separate because of the detonation?

89 **SSM** A 1400 kg car moving at 5.3 m/s is initially traveling north along the positive direction of a y axis. After completing a 90° right-hand turn in 4.6 s, the inattentive operator drives into a tree, which stops the car in 350 ms. In unit-vector notation, what is the impulse on the car (a) due to the turn and (b) due to the collision? What is the magnitude of the average force that acts on the car (c) during the turn and (d) during the collision? (e) What is the direction of the average force during the turn?

90 **ILW** A certain radioactive (parent) nucleus transforms to a different (daughter) nucleus by emitting an electron and a neutrino. The parent nucleus was at rest at the origin of an xy coordinate system. The electron moves away from the origin with linear momentum $(-1.2 \times 10^{-22}$ kg·m/s$)\hat{i}$; the neutrino moves away from the origin with linear momentum $(-6.4 \times 10^{-23}$ kg·m/s$)\hat{j}$. What are the (a) magnitude and (b) direction of the linear momentum of the daughter nucleus? (c) If the daughter nucleus has a mass of 5.8×10^{-26} kg, what is its kinetic energy?

91 A 75 kg man rides on a 39 kg cart moving at a velocity of 2.3 m/s. He jumps off with zero horizontal velocity relative to the ground. What is the resulting change in the cart's velocity, including sign?

92 Two blocks of masses 1.0 kg and 3.0 kg are connected by a spring and rest on a frictionless surface. They are given velocities toward each other such that the 1.0 kg block travels initially at 1.7 m/s toward the center of mass, which remains at rest. What is the initial speed of the other block?

93 **SSM** A railroad freight car of mass 3.18×10^4 kg collides with a stationary caboose car. They couple together, and 27.0% of the initial kinetic energy is transferred to thermal energy, sound, vibrations, and so on. Find the mass of the caboose.

94 An old Chrysler with mass 2400 kg is moving along a straight stretch of road at 80 km/h. It is followed by a Ford with mass 1600 kg moving at 60 km/h. How fast is the center of mass of the two cars moving?

95 **SSM** In the arrangement of Fig. 9-21, billiard ball 1 moving at a speed of 2.2 m/s undergoes a glancing collision with identical billiard ball 2 that is at rest. After the collision, ball 2 moves at speed 1.1 m/s, at an angle of $\theta_2 = 60°$. What are (a) the magnitude and (b) the direction of the velocity of ball 1 after the collision? (c) Do the given data suggest the collision is elastic or inelastic?

96 A rocket is moving away from the solar system at a speed of 6.0×10^3 m/s. It fires its engine, which ejects exhaust with a speed of 3.0×10^3 m/s relative to the rocket. The mass of the rocket at this time is 4.0×10^4 kg, and its acceleration is 2.0 m/s². (a) What is the thrust of the engine? (b) At what rate, in kilograms per second, is exhaust ejected during the firing?

97 The three balls in the overhead view of Fig. 9-76 are identical. Balls 2 and 3 touch each other and are aligned perpendicular to the path of ball 1. The velocity of ball 1 has magnitude $v_0 = 10$ m/s and is directed at the contact point of balls 1 and 2. After the collision, what are the (a) speed and (b) direction of the velocity of ball 2, the (c) speed and (d) direction of the velocity of ball 3, and the (e) speed and (f) direction of the velocity of ball 1? (*Hint:* With friction absent, each impulse is directed along the line connecting the centers of the colliding balls, normal to the colliding surfaces.)

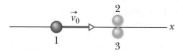

Fig. 9-76 Problem 97.

98 A 0.15 kg ball hits a wall with a velocity of $(5.00 \text{ m/s})\hat{\imath} + (6.50 \text{ m/s})\hat{\jmath} + (4.00 \text{ m/s})\hat{k}$. It rebounds from the wall with a velocity of $(2.00 \text{ m/s})\hat{\imath} + (3.50 \text{ m/s})\hat{\jmath} + (-3.20 \text{ m/s})\hat{k}$. What are (a) the change in the ball's momentum, (b) the impulse on the ball, and (c) the impulse on the wall?

99 In Fig. 9-77, two identical containers of sugar are connected by a cord that passes over a friction-less pulley. The cord and pulley have negligible mass, each container and its sugar together have a mass of 500 g, the centers of the containers are separated by 50 mm, and the containers are held fixed at the same height. What is the horizontal distance between the center of container 1 and the center of mass of the two-container system (a) initially and (b) after 20 g of sugar is transferred from container 1 to container 2? After the transfer and after the containers are released, (c) in what direction and (d) at what acceleration magnitude does the center of mass move?

Fig. 9-77
Problem 99.

100 In a game of pool, the cue ball strikes another ball of the same mass and initially at rest. After the collision, the cue ball moves at 3.50 m/s along a line making an angle of 22.0° with the cue ball's original direction of motion, and the second ball has a speed of 2.00 m/s. Find (a) the angle between the direction of motion of the second ball and the original direction of motion of the cue ball and (b) the original speed of the cue ball. (c) Is kinetic energy (of the centers of mass, don't consider the rotation) conserved?

101 In Fig. 9-78, a 3.2 kg box of running shoes slides on a horizontal frictionless table and collides with a 2.0 kg box of ballet slippers initially at rest on the edge of the table, at height $h = 0.40$ m. The speed of the 3.2 kg box is 3.0 m/s just before the collision. If the two boxes stick together because of packing tape on their sides, what is their kinetic energy just before they strike the floor?

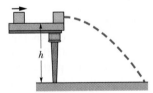

Fig. 9-78 Problem 101.

102 In Fig. 9-79, an 80 kg man is on a ladder hanging from a balloon that has a total mass of 320 kg (including the basket passenger). The balloon is initially stationary relative to the ground. If the man on the ladder begins to climb at 2.5 m/s relative to the ladder, (a) in what direction and (b) at what speed does the balloon move? (c) If the man then stops climbing, what is the speed of the balloon?

103 In Fig. 9-80, block 1 of mass $m_1 = 6.6$ kg is at rest on a long frictionless table that is up against a wall. Block 2 of mass m_2 is placed between block 1 and the wall and sent sliding to the left, toward block 1, with constant speed v_{2i}. Find the value of m_2 for which both blocks move with the same

Fig. 9-79
Problem 102.

velocity after block 2 has collided once with block 1 and once with the wall. Assume all collisions are elastic (the collision with the wall does not change the speed of block 2).

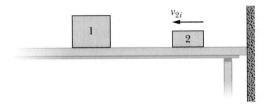

Fig. 9-80 Problem 103.

104 The script for an action movie calls for a small race car (of mass 1500 kg and length 3.0 m) to accelerate along a flattop boat (of mass 4000 kg and length 14 m), from one end of the boat to the other, where the car will then jump the gap between the boat and a somewhat lower dock. You are the technical advisor for the movie. The boat will initially touch the dock, as in Fig. 9-81; the boat can slide through the water without significant resistance; both the car and the boat can be approximated as uniform in their mass distribution. Determine what the width of the gap will be just as the car is about to make the jump.

Fig. 9-81 Problem 104.

105 SSM A 3.0 kg object moving at 8.0 m/s in the positive direction of an x axis has a one-dimensional elastic collision with an object of mass M, initially at rest. After the collision the object of mass M has a velocity of 6.0 m/s in the positive direction of the axis. What is mass M?

106 A 2140 kg railroad flatcar, which can move with negligible friction, is motionless next to a platform. A 242 kg sumo wrestler runs at 5.3 m/s along the platform (parallel to the track) and then jumps onto the flatcar. What is the speed of the flatcar if he then (a) stands on it, (b) runs at 5.3 m/s relative to it in his original direction, and (c) turns and runs at 5.3 m/s relative to the flatcar opposite his original direction?

107 SSM A 6100 kg rocket is set for vertical firing from the ground. If the exhaust speed is 1200 m/s, how much gas must be ejected each second if the thrust (a) is to equal the magnitude of the gravitational force on the rocket and (b) is to give the rocket an initial upward acceleration of 21 m/s²?

108 A 500.0 kg module is attached to a 400.0 kg shuttle craft, which moves at 1000 m/s relative to the stationary main spaceship. Then a small explosion sends the module backward with speed 100.0 m/s relative to the new speed of the shuttle craft. As measured by someone on the main spaceship, by what fraction did the kinetic energy of the module and shuttle craft increase because of the explosion?

109 SSM (a) How far is the center of mass of the Earth–Moon system from the center of Earth? (Appendix C gives the masses of Earth and the Moon and the distance between the two.) (b) What percentage of Earth's radius is that distance?

110 A 140 g ball with speed 7.8 m/s strikes a wall perpendicularly and rebounds in the opposite direction with the same speed. The collision lasts 3.80 ms. What are the magnitudes of the (a) impulse and (b) average force on the wall from the ball?

111 SSM A rocket sled with a mass of 2900 kg moves at 250 m/s on a set of rails. At a certain point, a scoop on the sled dips into a trough of water located between the tracks and scoops water into an empty tank on the sled. By applying the principle of conservation of linear momentum, determine the speed of the sled after 920 kg of water has been scooped up. Ignore any retarding force on the scoop.

112 SSM A pellet gun fires ten 2.0 g pellets per second with a speed of 500 m/s. The pellets are stopped by a rigid wall. What are (a) the magnitude of the momentum of each pellet, (b) the kinetic energy of each pellet, and (c) the magnitude of the average force on the wall from the stream of pellets? (d) If each pellet is in contact with the wall for 0.60 ms, what is the magnitude of the average force on the wall from each pellet during contact? (e) Why is this average force so different from the average force calculated in (c)?

113 A railroad car moves under a grain elevator at a constant speed of 3.20 m/s. Grain drops into the car at the rate of 540 kg/min. What is the magnitude of the force needed to keep the car moving at constant speed if friction is negligible?

114 Figure 9-82 shows a uniform square plate of edge length $6d = 6.0$ m from which a square piece of edge length $2d$ has been removed. What are (a) the x coordinate and (b) the y coordinate of the center of mass of the remaining piece?

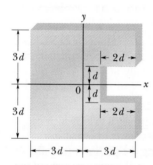

Fig. 9-82 Problem 114.

115 SSM At time $t = 0$, force $\vec{F}_1 = (-4.00\hat{i} + 5.00\hat{j})$ N acts on an initially stationary particle of mass 2.00×10^{-3} kg and force $\vec{F}_2 = (2.00\hat{i} - 4.00\hat{j})$ N acts on an initially stationary particle of mass 4.00×10^{-3} kg. From time $t = 0$ to $t = 2.00$ ms, what are the (a) magnitude and (b) angle (relative to the positive direction of the x axis) of the displacement of the center of mass of the two-particle system? (c) What is the kinetic energy of the center of mass at $t = 2.00$ ms?

116 Two particles P and Q are released from rest 1.0 m apart. P has a mass of 0.10 kg, and Q a mass of 0.30 kg. P and Q attract each other with a constant force of 1.0×10^{-2} N. No external forces act on the

system. (a) What is the speed of the center of mass of P and Q when the separation is 0.50 m? (b) At what distance from P's original position do the particles collide?

117 A collision occurs between a 2.00 kg particle traveling with velocity $\vec{v}_1 = (-4.00 \text{ m/s})\hat{i} + (-5.00 \text{ m/s})\hat{j}$ and a 4.00 kg particle traveling with velocity $\vec{v}_2 = (6.00 \text{ m/s})\hat{i} + (-2.00 \text{ m/s})\hat{j}$. The collision connects the two particles. What then is their velocity in (a) unit-vector notation and as a (b) magnitude and (c) angle?

118 In the two-sphere arrangement of Fig. 9-20, assume that sphere 1 has a mass of 50 g and an initial height of $h_1 = 9.0$ cm, and that sphere 2 has a mass of 85 g. After sphere 1 is released and collides elastically with sphere 2, what height is reached by (a) sphere 1 and (b) sphere 2? After the next (elastic) collision, what height is reached by (c) sphere 1 and (d) sphere 2? (*Hint:* Do not use rounded-off values.)

119 In Fig. 9-83, block 1 slides along an x axis on a frictionless floor with a speed of 0.75 m/s. When it reaches stationary block 2, the two blocks undergo an elastic collision. The following table gives the mass and length of the (uniform) blocks and also the locations of their centers at time $t = 0$. Where is the center of mass of the two-block system located (a) at $t = 0$, (b) when the two blocks first touch, and (c) at $t = 4.0$ s?

Block	Mass (kg)	Length (cm)	Center at $t = 0$
1	0.25	5.0	$x = -1.50$ m
2	0.50	6.0	$x = 0$

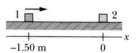

Fig. 9-83 Problem 119.

120 A body is traveling at 2.0 m/s along the positive direction of an x axis; no net force acts on the body. An internal explosion separates the body into two parts, each of 4.0 kg, and increases the total kinetic energy by 16 J. The forward part continues to move in the original direction of motion. What are the speeds of (a) the rear part and (b) the forward part?

121 An electron undergoes a one-dimensional elastic collision with an initially stationary hydrogen atom. What percentage of the electron's initial kinetic energy is transferred to kinetic energy of the hydrogen atom? (The mass of the hydrogen atom is 1840 times the mass of the electron.)

122 A man (weighing 915 N) stands on a long railroad flatcar (weighing 2415 N) as it rolls at 18.2 m/s in the positive direction of an x axis, with negligible friction. Then the man runs along the flatcar in the negative x direction at 4.00 m/s relative to the flatcar. What is the resulting increase in the speed of the flatcar?

ROTATION

10-1 WHAT IS PHYSICS?

As we have discussed, one focus of physics is motion. However, so far we have examined only the motion of **translation**, in which an object moves along a straight or curved line, as in Fig. 10-1a. We now turn to the motion of **rotation**, in which an object turns about an axis, as in Fig. 10-1b.

You see rotation in nearly every machine, you use it every time you open a beverage can with a pull tab, and you pay to experience it every time you go to an amusement park. Rotation is the key to many fun activities, such as hitting a long drive in golf (the ball needs to rotate in order for the air to keep it aloft longer) and throwing a curveball in baseball (the ball needs to rotate in order for the air to push it left or right). Rotation is also the key to more serious matters, such as metal failure in aging airplanes.

We begin our discussion of rotation by defining the variables for the motion, just as we did for translation in Chapter 2. As we shall see, the variables for rotation are analogous to those for one-dimensional motion and, as in Chapter 2, an important special situation is where the acceleration (here the rotational acceleration) is constant. We shall also see that Newton's second law can be written for rotational motion, but we must use a new quantity called *torque* instead of just force. Work and the work–kinetic energy theorem can also be applied to rotational motion, but we must use a new quantity called *rotational inertia* instead of just mass. In short, much of what we have discussed so far can be applied to rotational motion with, perhaps, a few changes.

(a)

10-2 The Rotational Variables

We wish to examine the rotation of a rigid body about a fixed axis. A **rigid body** is a body that can rotate with all its parts locked together and without any change in its shape. A **fixed axis** means that the rotation occurs about an axis that does not move. Thus, we shall not examine an object like the Sun, because the parts of the Sun (a ball of gas) are not locked together. We also shall not examine an object like a bowling ball rolling along a lane, because the ball rotates about a moving axis (the ball's motion is a mixture of rotation and translation).

Fig. 10-1 Figure skater Sasha Cohen in motion of (a) pure translation in a fixed direction and (b) pure rotation about a vertical axis. *(a: Mike Segar/Reuters/Landov LLC; b: Elsa/Getty Images, Inc.)*

(b)

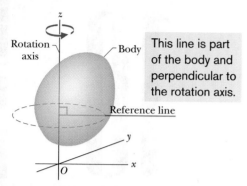

This line is part of the body and perpendicular to the rotation axis.

Fig. 10-2 A rigid body of arbitrary shape in pure rotation about the *z* axis of a coordinate system. The position of the *reference line* with respect to the rigid body is arbitrary, but it is perpendicular to the rotation axis. It is fixed in the body and rotates with the body.

Figure 10-2 shows a rigid body of arbitrary shape in rotation about a fixed axis, called the **axis of rotation** or the **rotation axis**. In pure rotation (*angular motion*), every point of the body moves in a circle whose center lies on the axis of rotation, and every point moves through the same angle during a particular time interval. In pure translation (*linear motion*), every point of the body moves in a straight line, and every point moves through the same *linear distance* during a particular time interval.

We deal now—one at a time—with the angular equivalents of the linear quantities position, displacement, velocity, and acceleration.

Angular Position

Figure 10-2 shows a *reference line*, fixed in the body, perpendicular to the rotation axis and rotating with the body. The **angular position** of this line is the angle of the line relative to a fixed direction, which we take as the **zero angular position.** In Fig. 10-3, the angular position θ is measured relative to the positive direction of the *x* axis. From geometry, we know that θ is given by

$$\theta = \frac{s}{r} \quad \text{(radian measure).} \tag{10-1}$$

Here *s* is the length of a circular arc that extends from the *x* axis (the zero angular position) to the reference line, and *r* is the radius of the circle.

An angle defined in this way is measured in **radians** (rad) rather than in revolutions (rev) or degrees. The radian, being the ratio of two lengths, is a pure number and thus has no dimension. Because the circumference of a circle of radius *r* is $2\pi r$, there are 2π radians in a complete circle:

$$1 \text{ rev} = 360° = \frac{2\pi r}{r} = 2\pi \text{ rad,} \tag{10-2}$$

and thus
$$1 \text{ rad} = 57.3° = 0.159 \text{ rev.} \tag{10-3}$$

We do *not* reset θ to zero with each complete rotation of the reference line about the rotation axis. If the reference line completes two revolutions from the zero angular position, then the angular position θ of the line is $\theta = 4\pi$ rad.

For pure translation along an *x* axis, we can know all there is to know about a moving body if we know $x(t)$, its position as a function of time. Similarly, for pure rotation, we can know all there is to know about a rotating body if we know $\theta(t)$, the angular position of the body's reference line as a function of time.

Angular Displacement

If the body of Fig. 10-3 rotates about the rotation axis as in Fig. 10-4, changing the angular position of the reference line from θ_1 to θ_2, the body undergoes an **angular displacement** $\Delta\theta$ given by

$$\Delta\theta = \theta_2 - \theta_1. \tag{10-4}$$

This definition of angular displacement holds not only for the rigid body as a whole but also for *every particle within that body*.

If a body is in translational motion along an *x* axis, its displacement Δx is either positive or negative, depending on whether the body is moving in the positive or negative direction of the axis. Similarly, the angular displacement $\Delta\theta$ of a rotating body is either positive or negative, according to the following rule:

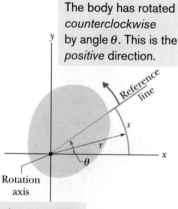

The body has rotated *counterclockwise* by angle θ. This is the *positive* direction.

This dot means that the rotation axis is out toward you.

Fig. 10-3 The rotating rigid body of Fig. 10-2 in cross section, viewed from above. The plane of the cross section is perpendicular to the rotation axis, which now extends out of the page, toward you. In this position of the body, the reference line makes an angle θ with the *x* axis.

 An angular displacement in the counterclockwise direction is positive, and one in the clockwise direction is negative.

The phrase *"clocks are negative"* can help you remember this rule (they certainly are negative when their alarms sound off early in the morning).

**CHECKPOINT 1**

A disk can rotate about its central axis like a merry-go-round. Which of the following pairs of values for its initial and final angular positions, respectively, give a negative angular displacement: (a) −3 rad, +5 rad, (b) −3 rad, −7 rad, (c) 7 rad, −3 rad?

Angular Velocity

Suppose that our rotating body is at angular position θ_1 at time t_1 and at angular position θ_2 at time t_2 as in Fig. 10-4. We define the **average angular velocity** of the body in the time interval Δt from t_1 to t_2 to be

$$\omega_{avg} = \frac{\theta_2 - \theta_1}{t_2 - t_1} = \frac{\Delta\theta}{\Delta t},$$

(10-5)

where $\Delta\theta$ is the angular displacement during Δt (ω is the lowercase omega).

The **(instantaneous) angular velocity** ω, with which we shall be most concerned, is the limit of the ratio in Eq. 10-5 as Δt approaches zero. Thus,

$$\omega = \lim_{\Delta t \to 0} \frac{\Delta\theta}{\Delta t} = \frac{d\theta}{dt}.$$

(10-6)

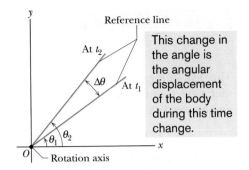

Fig. 10-4 The reference line of the rigid body of Figs. 10-2 and 10-3 is at angular position θ_1 at time t_1 and at angular position θ_2 at a later time t_2. The quantity $\Delta\theta \, (= \theta_2 - \theta_1)$ is the angular displacement that occurs during the interval $\Delta t \, (= t_2 - t_1)$. The body itself is not shown.

If we know $\theta(t)$, we can find the angular velocity ω by differentiation.

Equations 10-5 and 10-6 hold not only for the rotating rigid body as a whole but also for *every particle of that body* because the particles are all locked together. The unit of angular velocity is commonly the radian per second (rad/s) or the revolution per second (rev/s). Another measure of angular velocity was used during at least the first three decades of rock: Music was produced by vinyl (phonograph) records that were played on turntables at "$33\frac{1}{3}$ rpm" or "45 rpm," meaning at $33\frac{1}{3}$ rev/min or 45 rev/min.

If a particle moves in translation along an x axis, its linear velocity v is either positive or negative, depending on its direction along the axis. Similarly, the angular velocity ω of a rotating rigid body is either positive or negative, depending on whether the body is rotating counterclockwise (positive) or clockwise (negative). ("Clocks are negative" still works.) The magnitude of an angular velocity is called the **angular speed,** which is also represented with ω.

Angular Acceleration

If the angular velocity of a rotating body is not constant, then the body has an angular acceleration. Let ω_2 and ω_1 be its angular velocities at times t_2 and t_1, respectively. The **average angular acceleration** of the rotating body in the interval from t_1 to t_2 is defined as

$$\alpha_{avg} = \frac{\omega_2 - \omega_1}{t_2 - t_1} = \frac{\Delta\omega}{\Delta t},$$

(10-7)

in which $\Delta\omega$ is the change in the angular velocity that occurs during the time interval Δt. The **(instantaneous) angular acceleration** α, with which we shall be most concerned, is the limit of this quantity as Δt approaches zero. Thus,

$$\alpha = \lim_{\Delta t \to 0} \frac{\Delta\omega}{\Delta t} = \frac{d\omega}{dt}.$$

(10-8)

Equations 10-7 and 10-8 also hold for *every particle of that body*. The unit of angular acceleration is commonly the radian per second-squared (rad/s²) or the revolution per second-squared (rev/s²).

Sample Problem

Angular velocity derived from angular position

The disk in Fig. 10-5a is rotating about its central axis like a merry-go-round. The angular position $\theta(t)$ of a reference line on the disk is given by

$$\theta = -1.00 - 0.600t + 0.250t^2, \qquad (10\text{-}9)$$

with t in seconds, θ in radians, and the zero angular position as indicated in the figure.

(a) Graph the angular position of the disk versus time from $t = -3.0$ s to $t = 5.4$ s. Sketch the disk and its angular position reference line at $t = -2.0$ s, 0 s, and 4.0 s, and when the curve crosses the t axis.

KEY IDEA

The angular position of the disk is the angular position $\theta(t)$ of its reference line, which is given by Eq. 10-9 as a function of time t. So we graph Eq. 10-9; the result is shown in Fig. 10-5b.

Calculations: To sketch the disk and its reference line at a particular time, we need to determine θ for that time. To do so, we substitute the time into Eq. 10-9. For $t = -2.0$ s, we get

$$\theta = -1.00 - (0.600)(-2.0) + (0.250)(-2.0)^2$$

$$= 1.2 \text{ rad} = 1.2 \text{ rad} \frac{360°}{2\pi \text{ rad}} = 69°.$$

This means that at $t = -2.0$ s the reference line on the disk is rotated counterclockwise from the zero position by 1.2 rad = 69° (counterclockwise because θ is positive). Sketch 1 in Fig. 10-5b shows this position of the reference line.

Similarly, for $t = 0$, we find $\theta = -1.00$ rad = $-57°$, which means that the reference line is rotated clockwise from the zero angular position by 1.0 rad, or 57°, as shown in sketch 3. For $t = 4.0$ s, we find $\theta = 0.60$ rad = 34° (sketch 5). Drawing sketches for when the curve crosses the t axis is easy, because

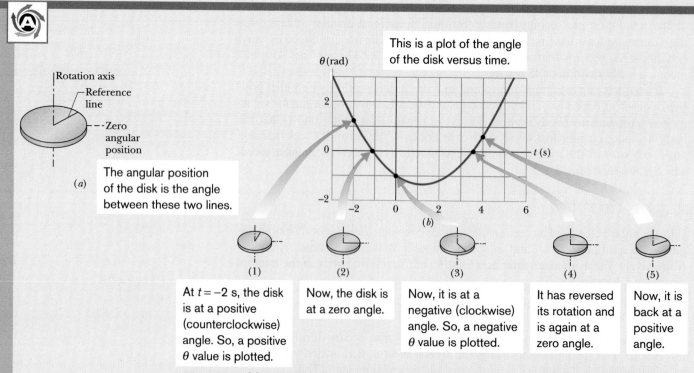

At $t = -2$ s, the disk is at a positive (counterclockwise) angle. So, a positive θ value is plotted.

Now, the disk is at a zero angle.

Now, it is at a negative (clockwise) angle. So, a negative θ value is plotted.

It has reversed its rotation and is again at a zero angle.

Now, it is back at a positive angle.

Fig. 10-5 (a) A rotating disk. (b) A plot of the disk's angular position $\theta(t)$. Five sketches indicate the angular position of the reference line on the disk for five points on the curve. (c) A plot of the disk's angular velocity $\omega(t)$. Positive values of ω correspond to counterclockwise rotation, and negative values to clockwise rotation.

then $\theta = 0$ and the reference line is momentarily aligned with the zero angular position (sketches 2 and 4).

(b) At what time t_{min} does $\theta(t)$ reach the minimum value shown in Fig. 10-5b? What is that minimum value?

KEY IDEA

To find the extreme value (here the minimum) of a function, we take the first derivative of the function and set the result to zero.

Calculations: The first derivative of $\theta(t)$ is

$$\frac{d\theta}{dt} = -0.600 + 0.500t. \qquad (10\text{-}10)$$

Setting this to zero and solving for t give us the time at which $\theta(t)$ is minimum:

$$t_{min} = 1.20 \text{ s.} \qquad \text{(Answer)}$$

To get the minimum value of θ, we next substitute t_{min} into Eq. 10-9, finding

$$\theta = -1.36 \text{ rad} \approx -77.9°. \qquad \text{(Answer)}$$

This *minimum* of $\theta(t)$ (the bottom of the curve in Fig. 10-5b) corresponds to the *maximum clockwise* rotation of the disk from the zero angular position, somewhat more than is shown in sketch 3.

(c) Graph the angular velocity ω of the disk versus time from $t = -3.0$ s to $t = 6.0$ s. Sketch the disk and indicate the direction of turning and the sign of ω at $t = -2.0$ s, 4.0 s, and t_{min}.

KEY IDEA

From Eq. 10-6, the angular velocity ω is equal to $d\theta/dt$ as given in Eq. 10-10. So, we have

$$\omega = -0.600 + 0.500t. \qquad (10\text{-}11)$$

The graph of this function $\omega(t)$ is shown in Fig. 10-5c.

Calculations: To sketch the disk at $t = -2.0$ s, we substitute that value into Eq. 10-11, obtaining

$$\omega = -1.6 \text{ rad/s.} \qquad \text{(Answer)}$$

The minus sign here tells us that at $t = -2.0$ s, the disk is turning clockwise (the left-hand sketch in Fig. 10-5c).

Substituting $t = 4.0$ s into Eq. 10-11 gives us

$$\omega = 1.4 \text{ rad/s.} \qquad \text{(Answer)}$$

The implied plus sign tells us that now the disk is turning counterclockwise (the right-hand sketch in Fig. 10-5c).

For t_{min}, we already know that $d\theta/dt = 0$. So, we must also have $\omega = 0$. That is, the disk momentarily stops when the reference line reaches the minimum value of θ in Fig. 10-5b, as suggested by the center sketch in Fig. 10-5c. On the graph, this momentary stop is the zero point where the plot changes from the negative clockwise motion to the positive counterclockwise motion.

(d) Use the results in parts (a) through (c) to describe the motion of the disk from $t = -3.0$ s to $t = 6.0$ s.

Description: When we first observe the disk at $t = -3.0$ s, it has a positive angular position and is turning clockwise but slowing. It stops at angular position $\theta = -1.36$ rad and then begins to turn counterclockwise, with its angular position eventually becoming positive again.

This is a plot of the angular velocity of the disk versus time.

ω (rad/s)

negative ω zero ω positive ω

(c)

The angular velocity is initially negative and slowing, then momentarily zero during reversal, and then positive and increasing.

 Additional examples, video, and practice available at WileyPLUS

Sample Problem

Angular velocity derived from angular acceleration

A child's top is spun with angular acceleration

$$\alpha = 5t^3 - 4t,$$

with t in seconds and α in radians per second-squared. At $t = 0$, the top has angular velocity 5 rad/s, and a reference line on it is at angular position $\theta = 2$ rad.

(a) Obtain an expression for the angular velocity $\omega(t)$ of the top. That is, find an expression that explicitly indicates how the angular velocity depends on time. (We can tell that there *is* such a dependence because the top is undergoing an angular acceleration, which means that its angular velocity *is* changing.)

KEY IDEA

By definition, $\alpha(t)$ is the derivative of $\omega(t)$ with respect to time. Thus, we can find $\omega(t)$ by integrating $\alpha(t)$ with respect to time.

Calculations: Equation 10-8 tells us

$$d\omega = \alpha \, dt,$$

so

$$\int d\omega = \int \alpha \, dt.$$

From this we find

$$\omega = \int (5t^3 - 4t) \, dt = \tfrac{5}{4}t^4 - \tfrac{4}{2}t^2 + C.$$

To evaluate the constant of integration C, we note that $\omega = 5$ rad/s at $t = 0$. Substituting these values in our expression for ω yields

$$5 \text{ rad/s} = 0 - 0 + C,$$

so $C = 5$ rad/s. Then

$$\omega = \tfrac{5}{4}t^4 - 2t^2 + 5. \qquad \text{(Answer)}$$

(b) Obtain an expression for the angular position $\theta(t)$ of the top.

KEY IDEA

By definition, $\omega(t)$ is the derivative of $\theta(t)$ with respect to time. Therefore, we can find $\theta(t)$ by integrating $\omega(t)$ with respect to time.

Calculations: Since Eq. 10-6 tells us that

$$d\theta = \omega \, dt,$$

we can write

$$\theta = \int \omega \, dt = \int (\tfrac{5}{4}t^4 - 2t^2 + 5) \, dt$$
$$= \tfrac{1}{4}t^5 - \tfrac{2}{3}t^3 + 5t + C'$$
$$= \tfrac{1}{4}t^5 - \tfrac{2}{3}t^3 + 5t + 2, \qquad \text{(Answer)}$$

where C' has been evaluated by noting that $\theta = 2$ rad at $t = 0$.

 Additional examples, video, and practice available at *WileyPLUS*

10-3 Are Angular Quantities Vectors?

We can describe the position, velocity, and acceleration of a single particle by means of vectors. If the particle is confined to a straight line, however, we do not really need vector notation. Such a particle has only two directions available to it, and we can indicate these directions with plus and minus signs.

In the same way, a rigid body rotating about a fixed axis can rotate only clockwise or counterclockwise as seen along the axis, and again we can select between the two directions by means of plus and minus signs. The question arises: "Can we treat the angular displacement, velocity, and acceleration of a rotating body as vectors?" The answer is a qualified "yes" (see the caution below, in connection with angular displacements).

Consider the angular velocity. Figure 10-6a shows a vinyl record rotating on a turntable. The record has a constant angular speed ω ($= 33\tfrac{1}{3}$ rev/min) in the clockwise direction. We can represent its angular velocity as a vector $\vec{\omega}$ pointing along the axis of rotation, as in Fig. 10-6b. Here's how: We choose the length of this vector according to some convenient scale, for example, with 1 cm corresponding to 10 rev/min. Then we establish a direction for the vector $\vec{\omega}$ by using a

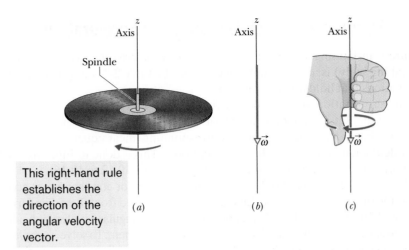

This right-hand rule establishes the direction of the angular velocity vector.

Fig. 10-6 (*a*) A record rotating about a vertical axis that coincides with the axis of the spindle. (*b*) The angular velocity of the rotating record can be represented by the vector $\vec{\omega}$, lying along the axis and pointing down, as shown. (*c*) We establish the direction of the angular velocity vector as downward by using a right-hand rule. When the fingers of the right hand curl around the record and point the way it is moving, the extended thumb points in the direction of $\vec{\omega}$.

right-hand rule, as Fig. 10-6*c* shows: Curl your right hand about the rotating record, your fingers pointing *in the direction of rotation*. Your extended thumb will then point in the direction of the angular velocity vector. If the record were to rotate in the opposite sense, the right-hand rule would tell you that the angular velocity vector then points in the opposite direction.

It is not easy to get used to representing angular quantities as vectors. We instinctively expect that something should be moving *along* the direction of a vector. That is not the case here. Instead, something (the rigid body) is rotating *around* the direction of the vector. In the world of pure rotation, a vector defines an axis of rotation, not a direction in which something moves. Nonetheless, the vector also defines the motion. Furthermore, it obeys all the rules for vector manipulation discussed in Chapter 3. The angular acceleration $\vec{\alpha}$ is another vector, and it too obeys those rules.

In this chapter we consider only rotations that are about a fixed axis. For such situations, we need not consider vectors—we can represent angular velocity with ω and angular acceleration with α, and we can indicate direction with an implied plus sign for counterclockwise or an explicit minus sign for clockwise.

Now for the caution: Angular *displacements* (unless they are very small) *cannot* be treated as vectors. Why not? We can certainly give them both magnitude and direction, as we did for the angular velocity vector in Fig. 10-6. However, to be represented as a vector, a quantity must *also* obey the rules of vector addition, one of which says that if you add two vectors, the order in which you add them does not matter. Angular displacements fail this test.

Figure 10-7 gives an example. An initially horizontal book is given two 90° angular displacements, first in the order of Fig. 10-7*a* and then in the order of Fig. 10-7*b*. Although the two angular displacements are identical, their order is not, and the book ends up with different orientations. Here's another example. Hold your right arm downward, palm toward your thigh. Keeping your wrist rigid, (1) lift the arm forward until it is horizontal, (2) move it horizontally until it points toward the right, and (3) then bring it down to your side. Your palm faces forward. If you start over, but reverse the steps, which way does your palm end up facing? From either example, we must conclude that the addition of two angular displacements depends on their order and they cannot be vectors.

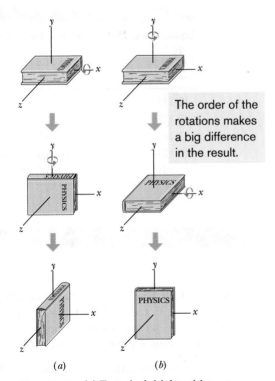

The order of the rotations makes a big difference in the result.

Fig. 10-7 (*a*) From its initial position, at the top, the book is given two successive 90° rotations, first about the (horizontal) *x* axis and then about the (vertical) *y* axis. (*b*) The book is given the same rotations, but in the reverse order.

10-4 **Rotation with Constant Angular Acceleration**

In pure translation, motion with a *constant linear acceleration* (for example, that of a falling body) is an important special case. In Table 2-1, we displayed a series of equations that hold for such motion.

In pure rotation, the case of *constant angular acceleration* is also important, and a parallel set of equations holds for this case also. We shall not derive them here, but simply write them from the corresponding linear equations, substituting equivalent angular quantities for the linear ones. This is done in Table 10-1, which lists both sets of equations (Eqs. 2-11 and 2-15 to 2-18; 10-12 to 10-16).

Recall that Eqs. 2-11 and 2-15 are basic equations for constant linear acceleration—the other equations in the Linear list can be derived from them. Similarly, Eqs. 10-12 and 10-13 are the basic equations for constant angular acceleration, and the other equations in the Angular list can be derived from them. To solve a simple problem involving constant angular acceleration, you can usually use an equation from the Angular list (*if* you have the list). Choose an equation for which the only unknown variable will be the variable requested in the problem. A better plan is to remember only Eqs. 10-12 and 10-13, and then solve them as simultaneous equations whenever needed.

CHECKPOINT 2

In four situations, a rotating body has angular position $\theta(t)$ given by (a) $\theta = 3t - 4$, (b) $\theta = -5t^3 + 4t^2 + 6$, (c) $\theta = 2/t^2 - 4/t$, and (d) $\theta = 5t^2 - 3$. To which situations do the angular equations of Table 10-1 apply?

Table 10-1

Equations of Motion for Constant Linear Acceleration and for Constant Angular Acceleration

Equation Number	Linear Equation	Missing Variable		Angular Equation	Equation Number
(2-11)	$v = v_0 + at$	$x - x_0$	$\theta - \theta_0$	$\omega = \omega_0 + \alpha t$	(10-12)
(2-15)	$x - x_0 = v_0 t + \frac{1}{2}at^2$	v	ω	$\theta - \theta_0 = \omega_0 t + \frac{1}{2}\alpha t^2$	(10-13)
(2-16)	$v^2 = v_0^2 + 2a(x - x_0)$	t	t	$\omega^2 = \omega_0^2 + 2\alpha(\theta - \theta_0)$	(10-14)
(2-17)	$x - x_0 = \frac{1}{2}(v_0 + v)t$	a	α	$\theta - \theta_0 = \frac{1}{2}(\omega_0 + \omega)t$	(10-15)
(2-18)	$x - x_0 = vt - \frac{1}{2}at^2$	v_0	ω_0	$\theta - \theta_0 = \omega t - \frac{1}{2}\alpha t^2$	(10-16)

Sample Problem

Constant angular acceleration, grindstone

A grindstone (Fig. 10-8) rotates at constant angular acceleration $\alpha = 0.35$ rad/s^2. At time $t = 0$, it has an angular velocity of $\omega_0 = -4.6$ rad/s and a reference line on it is horizontal, at the angular position $\theta_0 = 0$.

(a) At what time after $t = 0$ is the reference line at the angular position $\theta = 5.0$ rev?

KEY IDEA

The angular acceleration is constant, so we can use the rota-

tion equations of Table 10-1. We choose Eq. 10-13,

$$\theta - \theta_0 = \omega_0 t + \frac{1}{2}\alpha t^2,$$

because the only unknown variable it contains is the desired time t.

Calculations: Substituting known values and setting $\theta_0 = 0$ and $\theta = 5.0$ rev $= 10\pi$ rad give us

$$10\pi \text{ rad} = (-4.6 \text{ rad/s})t + \frac{1}{2}(0.35 \text{ rad/s}^2)t^2.$$

(We converted 5.0 rev to 10π rad to keep the units consistent.) Solving this quadratic equation for t, we find

$$t = 32 \text{ s.} \qquad \text{(Answer)}$$

Now notice something a bit strange. We first see the wheel when it is rotating in the negative diretion and through the $\theta = 0$ orientation. Yet, we just found out that 32 s later it is at the positive orientation of $\theta = 5.0$ rev. What happened in that time interval so that it could be at a positive orientation?

(b) Describe the grindstone's rotation between $t = 0$ and $t = 32$ s.

Description: The wheel is initially rotating in the negative (clockwise) direction with angular velocity $\omega_0 = -4.6$ rad/s, but its angular acceleration α is positive. This initial opposition of the signs of angular velocity and angular acceleration means that the wheel slows in its rotation in the negative direction, stops, and then reverses to rotate in the positive direction. After the reference line comes back through its initial orientation of $\theta = 0$, the wheel turns an additional 5.0 rev by time $t = 32$ s.

(c) At what time t does the grindstone momentarily stop?

Calculation: We again go to the table of equations for constant angular acceleration, and again we need an equation

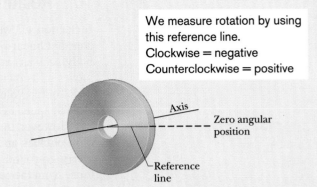

We measure rotation by using this reference line.
Clockwise = negative
Counterclockwise = positive

Fig. 10-8 A grindstone. At $t = 0$ the reference line (which we imagine to be marked on the stone) is horizontal.

that contains only the desired unknown variable t. However, now the equation must also contain the variable ω, so that we can set it to 0 and then solve for the corresponding time t. We choose Eq. 10-12, which yields

$$t = \frac{\omega - \omega_0}{\alpha} = \frac{0 - (-4.6 \text{ rad/s})}{0.35 \text{ rad/s}^2} = 13 \text{ s.} \qquad \text{(Answer)}$$

Sample Problem

Constant angular acceleration, riding a Rotor

While you are operating a Rotor (a large, vertical, rotating cylinder found in amusement parks), you spot a passenger in acute distress and decrease the angular velocity of the cylinder from 3.40 rad/s to 2.00 rad/s in 20.0 rev, at constant angular acceleration. (The passenger is obviously more of a "translation person" than a "rotation person.")

(a) What is the constant angular acceleration during this decrease in angular speed?

KEY IDEA

Because the cylinder's angular acceleration is constant, we can relate it to the angular velocity and angular displacement via the basic equations for constant angular acceleration (Eqs. 10-12 and 10-13).

Calculations: The initial angular velocity is $\omega_0 = 3.40$ rad/s, the angular displacement is $\theta - \theta_0 = 20.0$ rev, and the angular velocity at the end of that displacement is $\omega = 2.00$ rad/s. But we do not know the angular acceleration α and time t, which are in both basic equations.

To eliminate the unknown t, we use Eq. 10-12 to write

$$t = \frac{\omega - \omega_0}{\alpha},$$

which we then substitute into Eq. 10-13 to write

$$\theta - \theta_0 = \omega_0 \left(\frac{\omega - \omega_0}{\alpha} \right) + \tfrac{1}{2} \alpha \left(\frac{\omega - \omega_0}{\alpha} \right)^2.$$

Solving for α, substituting known data, and converting 20 rev to 125.7 rad, we find

$$\alpha = \frac{\omega^2 - \omega_0^2}{2(\theta - \theta_0)} = \frac{(2.00 \text{ rad/s})^2 - (3.40 \text{ rad/s})^2}{2(125.7 \text{ rad})}$$

$$= -0.0301 \text{ rad/s}^2. \qquad \text{(Answer)}$$

(b) How much time did the speed decrease take?

Calculation: Now that we know α, we can use Eq. 10-12 to solve for t:

$$t = \frac{\omega - \omega_0}{\alpha} = \frac{2.00 \text{ rad/s} - 3.40 \text{ rad/s}}{-0.0301 \text{ rad/s}^2}$$

$$= 46.5 \text{ s.} \qquad \text{(Answer)}$$

 Additional examples, video, and practice available at *WileyPLUS*

10-5 Relating the Linear and Angular Variables

In Section 4-7, we discussed uniform circular motion, in which a particle travels at constant linear speed v along a circle and around an axis of rotation. When a rigid body, such as a merry-go-round, rotates around an axis, each particle in the body moves in its own circle around that axis. Since the body is rigid, all the particles make one revolution in the same amount of time; that is, they all have the same angular speed ω.

However, the farther a particle is from the axis, the greater the circumference of its circle is, and so the faster its linear speed v must be. You can notice this on a merry-go-round. You turn with the same angular speed ω regardless of your distance from the center, but your linear speed v increases noticeably if you move to the outside edge of the merry-go-round.

We often need to relate the linear variables s, v, and a for a particular point in a rotating body to the angular variables θ, ω, and α for that body. The two sets of variables are related by r, the *perpendicular distance* of the point from the rotation axis. This perpendicular distance is the distance between the point and the rotation axis, measured along a perpendicular to the axis. It is also the radius r of the circle traveled by the point around the axis of rotation.

The Position

If a reference line on a rigid body rotates through an angle θ, a point within the body at a position r from the rotation axis moves a distance s along a circular arc, where s is given by Eq. 10-1:

$$s = \theta r \qquad \text{(radian measure).} \qquad (10\text{-}17)$$

This is the first of our linear–angular relations. *Caution:* The angle θ here must be measured in radians because Eq. 10-17 is itself the definition of angular measure in radians.

The Speed

Differentiating Eq. 10-17 with respect to time—with r held constant—leads to

$$\frac{ds}{dt} = \frac{d\theta}{dt}\, r.$$

However, ds/dt is the linear speed (the magnitude of the linear velocity) of the point in question, and $d\theta/dt$ is the angular speed ω of the rotating body. So

$$v = \omega r \qquad \text{(radian measure).} \qquad (10\text{-}18)$$

Caution: The angular speed ω must be expressed in radian measure.

Equation 10-18 tells us that since all points within the rigid body have the same angular speed ω, points with greater radius r have greater linear speed v. Figure 10-9a reminds us that the linear velocity is always tangent to the circular path of the point in question.

If the angular speed ω of the rigid body is constant, then Eq. 10-18 tells us that the linear speed v of any point within it is also constant. Thus, each point within the body undergoes uniform circular motion. The period of revolution T for the motion of each point and for the rigid body itself is given by Eq. 4-35:

$$T = \frac{2\pi r}{v}. \qquad (10\text{-}19)$$

This equation tells us that the time for one revolution is the distance $2\pi r$ traveled in one revolution divided by the speed at which that distance is traveled.

Fig. 10-9 The rotating rigid body of Fig. 10-2, shown in cross section viewed from above. Every point of the body (such as P) moves in a circle around the rotation axis. (a) The linear velocity $\vec{v}$ of every point is tangent to the circle in which the point moves. (b) The linear acceleration $\vec{a}$ of the point has (in general) two components: tangential a_t and radial a_r.

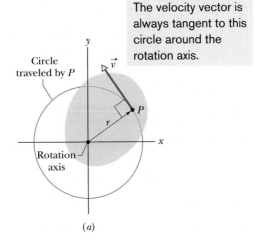

The velocity vector is always tangent to this circle around the rotation axis.

(a)

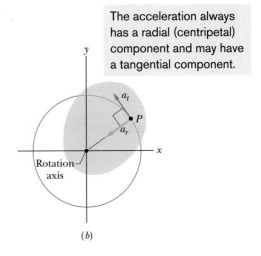

The acceleration always has a radial (centripetal) component and may have a tangential component.

(b)

Substituting for v from Eq. 10-18 and canceling r, we find also that

$$T = \frac{2\pi}{\omega} \quad \text{(radian measure)}. \tag{10-20}$$

This equivalent equation says that the time for one revolution is the angular distance 2π rad traveled in one revolution divided by the angular speed (or rate) at which that angle is traveled.

The Acceleration

Differentiating Eq. 10-18 with respect to time—again with r held constant—leads to

$$\frac{dv}{dt} = \frac{d\omega}{dt}\,r. \tag{10-21}$$

Here we run up against a complication. In Eq. 10-21, dv/dt represents only the part of the linear acceleration that is responsible for changes in the *magnitude v* of the linear velocity $\vec{v}$. Like $\vec{v}$, that part of the linear acceleration is tangent to the path of the point in question. We call it the *tangential component a_t* of the linear acceleration of the point, and we write

$$a_t = \alpha r \quad \text{(radian measure)}, \tag{10-22}$$

where $\alpha = d\omega/dt$. *Caution:* The angular acceleration α in Eq. 10-22 must be expressed in radian measure.

In addition, as Eq. 4-34 tells us, a particle (or point) moving in a circular path has a *radial component* of linear acceleration, $a_r = v^2/r$ (directed radially inward), that is responsible for changes in the *direction* of the linear velocity $\vec{v}$. By substituting for v from Eq. 10-18, we can write this component as

$$a_r = \frac{v^2}{r} = \omega^2 r \quad \text{(radian measure)}. \tag{10-23}$$

Thus, as Fig. 10-9b shows, the linear acceleration of a point on a rotating rigid body has, in general, two components. The radially inward component a_r (given by Eq. 10-23) is present whenever the angular velocity of the body is not zero. The tangential component a_t (given by Eq. 10-22) is present whenever the angular acceleration is not zero.

Sample Problem

Linear and angular variables, roller coaster speedup

In spite of the extreme care taken in engineering a roller coaster, an unlucky few of the millions of people who ride roller coasters each year end up with a medical condition called *roller-coaster headache*. Symptoms, which might not appear for several days, include vertigo and headache, both severe enough to require medical treatment.

Let's investigate the probable cause by designing the track for our own *induction roller coaster* (which can be accelerated by magnetic forces even on a horizontal track). To create an initial thrill, we want each passenger to leave the loading point with acceleration g along the horizontal track. To increase the thrill, we also want that first section of track to form a circular arc (Fig. 10-10), so that the passenger also experiences a centripetal acceleration. As the passenger accelerates along the arc, the magnitude of this centripetal acceleration increases alarmingly. When the magnitude a of the net acceleration reaches $4g$ at some point P and angle θ_P along the arc, we want the passenger then to move in a straight line, along a tangent to the arc.

(a) What angle θ_P should the arc subtend so that a is $4g$ at point P?

KEY IDEAS

(1) At any given time, the passenger's net acceleration $\vec{a}$ is the vector sum of the tangential acceleration $\vec{a}_t$ along the track and the radial acceleration $\vec{a}_r$ toward the arc's center of curvature (as in Fig. 10-9b). (2) The value of a_r at any given time depends on the angular speed ω according to Eq. 10-23 ($a_r = \omega^2 r$, where r is the radius of the circular arc). (3) An angular acceleration α around the arc is associated with the tangential acceleration a_t along the track according to Eq. 10-22 ($a_t = \alpha r$). (4) Because a_t and r are constant, so is α and thus we can use the constant angular-acceleration equations.

Calculations: Because we are trying to determine a value for angular position θ, let's choose Eq. 10-14 from among the constant angular-acceleration equations:

$$\omega^2 = \omega_0^2 + 2\alpha(\theta - \theta_0). \tag{10-24}$$

For the angular acceleration α, we substitute from Eq. 10-22:

$$\alpha = \frac{a_t}{r}. \tag{10-25}$$

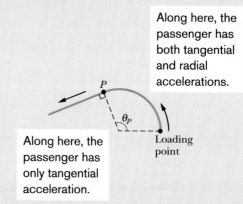

Along here, the passenger has both tangential and radial accelerations.

Along here, the passenger has only tangential acceleration.

Fig. 10-10 An overhead view of a horizontal track for a roller coaster. The track begins as a circular arc at the loading point and then, at point P, continues along a tangent to the arc.

We also substitute $\omega_0 = 0$ and $\theta_0 = 0$, and we find

$$\omega^2 = \frac{2a_t\theta}{r}. \tag{10-26}$$

Substituting this result for ω^2 into

$$a_r = \omega^2 r \tag{10-27}$$

gives a relation between the radial acceleration, the tangential acceleration, and the angular position θ:

$$a_r = 2a_t\theta. \tag{10-28}$$

Because $\vec{a}_t$ and $\vec{a}_r$ are perpendicular vectors, their sum has the magnitude

$$a = \sqrt{a_t^2 + a_r^2}. \tag{10-29}$$

Substituting for a_r from Eq. 10-28 and solving for θ lead to

$$\theta = \frac{1}{2}\sqrt{\frac{a^2}{a_t^2} - 1}. \tag{10-30}$$

When a reaches the design value of $4g$, angle θ is the angle θ_P we want. Substituting $a = 4g$, $\theta = \theta_P$, and $a_t = g$ into Eq. 10-30, we find

$$\theta_P = \frac{1}{2}\sqrt{\frac{(4g)^2}{g^2} - 1} = 1.94 \text{ rad} = 111°. \tag{Answer}$$

(b) What is the magnitude a of the passenger's net acceleration at point P and after point P?

Reasoning: At P, a has the design value of $4g$. Just after P is reached, the passenger moves in a straight line and no longer has centripetal acceleration. Thus, the passenger has only the acceleration magnitude g along the track. Hence,

$$a = 4g \text{ at } P \quad \text{and} \quad a = g \text{ after } P. \qquad \text{(Answer)}$$

Roller-coaster headache can occur when a passenger's head undergoes an abrupt change in acceleration, with the acceleration magnitude large before or after the change. The reason is that the change can cause the brain to move relative to the skull, tearing the veins that bridge the brain and skull. Our design to increase the acceleration from g to $4g$ along the path to P might harm the passenger, but the abrupt change in acceleration as the passenger passes through point P is more likely to cause roller-coaster headache.

 Additional examples, video, and practice available at *WileyPLUS*

10-6 **Kinetic Energy of Rotation**

The rapidly rotating blade of a table saw certainly has kinetic energy due to that rotation. How can we express the energy? We cannot apply the familiar formula $K = \frac{1}{2}mv^2$ to the saw as a whole because that would give us the kinetic energy only of the saw's center of mass, which is zero.

Instead, we shall treat the table saw (and any other rotating rigid body) as a collection of particles with different speeds. We can then add up the kinetic energies of all the particles to find the kinetic energy of the body as a whole. In this way we obtain, for the kinetic energy of a rotating body,

$$K = \frac{1}{2}m_1v_1^2 + \frac{1}{2}m_2v_2^2 + \frac{1}{2}m_3v_3^2 + \cdots$$
$$= \sum \frac{1}{2}m_iv_i^2, \qquad (10\text{-}31)$$

in which m_i is the mass of the ith particle and v_i is its speed. The sum is taken over all the particles in the body.

The problem with Eq. 10-31 is that v_i is not the same for all particles. We solve this problem by substituting for v from Eq. 10-18 ($v = \omega r$), so that we have

$$K = \sum \frac{1}{2}m_i(\omega r_i)^2 = \frac{1}{2}\left(\sum m_i r_i^2\right)\omega^2, \qquad (10\text{-}32)$$

in which ω is the same for all particles.

The quantity in parentheses on the right side of Eq. 10-32 tells us how the mass of the rotating body is distributed about its axis of rotation. We call that quantity the **rotational inertia** (or **moment of inertia**) I of the body with respect to the axis of rotation. It is a constant for a particular rigid body and a particular rotation axis. (That axis must always be specified if the value of I is to be meaningful.)

We may now write

$$I = \sum m_i r_i^2 \quad \text{(rotational inertia)} \qquad (10\text{-}33)$$

and substitute into Eq. 10-32, obtaining

$$K = \frac{1}{2}I\omega^2 \quad \text{(radian measure)} \qquad (10\text{-}34)$$

as the expression we seek. Because we have used the relation $v = \omega r$ in deriving Eq. 10-34, ω must be expressed in radian measure. The SI unit for I is the kilogram–square meter (kg · m²).

Equation 10-34, which gives the kinetic energy of a rigid body in pure rotation, is the angular equivalent of the formula $K = \frac{1}{2}Mv_{com}^2$, which gives the kinetic energy of a rigid body in pure translation. In both formulas there is a factor of $\frac{1}{2}$. Where mass M appears in one equation, I (which involves both mass and its distribution)

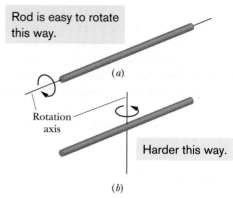

Rod is easy to rotate this way.

(a)

Rotation axis

Harder this way.

(b)

Fig. 10-11 A long rod is much easier to rotate about (*a*) its central (longitudinal) axis than about (*b*) an axis through its center and perpendicular to its length. The reason for the difference is that the mass is distributed closer to the rotation axis in (*a*) than in (*b*).

appears in the other. Finally, each equation contains as a factor the square of a speed—translational or rotational as appropriate. The kinetic energies of translation and of rotation are not different kinds of energy. They are both kinetic energy, expressed in ways that are appropriate to the motion at hand.

We noted previously that the rotational inertia of a rotating body involves not only its mass but also how that mass is distributed. Here is an example that you can literally feel. Rotate a long, fairly heavy rod (a pole, a length of lumber, or something similar), first around its central (longitudinal) axis (Fig. 10-11*a*) and then around an axis perpendicular to the rod and through the center (Fig. 10-11*b*). Both rotations involve the very same mass, but the first rotation is much easier than the second. The reason is that the mass is distributed much closer to the rotation axis in the first rotation. As a result, the rotational inertia of the rod is much smaller in Fig. 10-11*a* than in Fig. 10-11*b*. In general, smaller rotational inertia means easier rotation.

✅ **CHECKPOINT 4**

The figure shows three small spheres that rotate about a vertical axis. The perpendicular distance between the axis and the center of each sphere is given. Rank the three spheres according to their rotational inertia about that axis, greatest first.

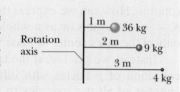

10-7 Calculating the Rotational Inertia

If a rigid body consists of a few particles, we can calculate its rotational inertia about a given rotation axis with Eq. 10-33 ($I = \Sigma\, m_i r_i^2$); that is, we can find the product mr^2 for each particle and then sum the products. (Recall that *r* is the perpendicular distance a particle is from the given rotation axis.)

If a rigid body consists of a great many adjacent particles (it is *continuous*, like a Frisbee), using Eq. 10-33 would require a computer. Thus, instead, we replace the sum in Eq. 10-33 with an integral and define the rotational inertia of the body as

$$I = \int r^2\, dm \qquad \text{(rotational inertia, continuous body).} \qquad (10\text{-}35)$$

Table 10-2 gives the results of such integration for nine common body shapes and the indicated axes of rotation.

Parallel-Axis Theorem

Suppose we want to find the rotational inertia *I* of a body of mass *M* about a given axis. In principle, we can always find *I* with the integration of Eq. 10-35. However, there is a shortcut if we happen to already know the rotational inertia I_{com} of the body about a *parallel* axis that extends through the body's center of mass. Let *h* be the perpendicular distance between the given axis and the axis through the center of mass (remember these two axes must be parallel). Then the rotational inertia *I* about the given axis is

$$I = I_{\text{com}} + Mh^2 \qquad \text{(parallel-axis theorem).} \qquad (10\text{-}36)$$

This equation is known as the **parallel-axis theorem.** We shall now prove it.

Proof of the Parallel-Axis Theorem

Let *O* be the center of mass of the arbitrarily shaped body shown in cross section in Fig. 10-12. Place the origin of the coordinates at *O*. Consider an axis through *O*

Table 10-2

Some Rotational Inertias

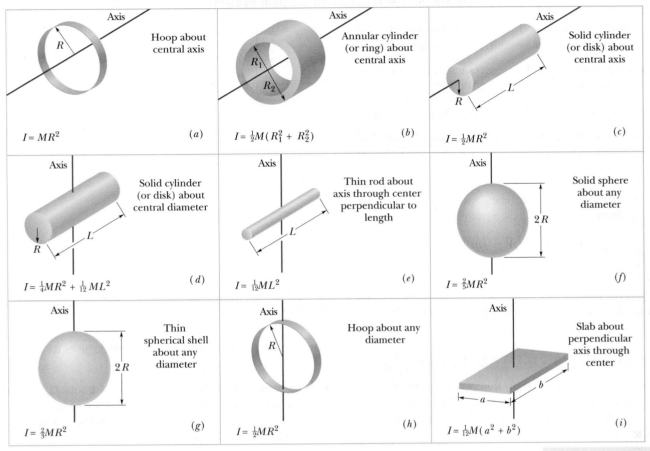

perpendicular to the plane of the figure, and another axis through point P parallel to the first axis. Let the x and y coordinates of P be a and b.

Let dm be a mass element with the general coordinates x and y. The rotational inertia of the body about the axis through P is then, from Eq. 10-35,

$$I = \int r^2 \, dm = \int [(x - a)^2 + (y - b)^2] \, dm,$$

which we can rearrange as

$$I = \int (x^2 + y^2) \, dm - 2a \int x \, dm - 2b \int y \, dm + \int (a^2 + b^2) \, dm. \quad (10\text{-}37)$$

From the definition of the center of mass (Eq. 9-9), the middle two integrals of Eq. 10-37 give the coordinates of the center of mass (multiplied by a constant) and thus must each be zero. Because $x^2 + y^2$ is equal to R^2, where R is the distance from O to dm, the first integral is simply I_{com}, the rotational inertia of the body about an axis through its center of mass. Inspection of Fig. 10-12 shows that the last term in Eq. 10-37 is Mh^2, where M is the body's total mass. Thus, Eq. 10-37 reduces to Eq. 10-36, which is the relation that we set out to prove.

> We need to relate the rotational inertia around the axis at P to that around the axis at the com.

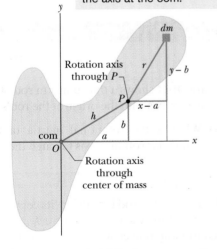

Fig. 10-12 A rigid body in cross section, with its center of mass at O. The parallel-axis theorem (Eq. 10-36) relates the rotational inertia of the body about an axis through O to that about a parallel axis through a point such as P, a distance h from the body's center of mass. Both axes are perpendicular to the plane of the figure.

✓ CHECKPOINT 5

The figure shows a book-like object (one side is longer than the other) and four choices of rotation axes, all perpendicular to the face of the object. Rank the choices according to the rotational inertia of the object about the axis, greatest first.

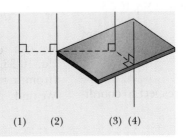

(1)   (2)   (3) (4)

Sample Problem

Rotational inertia of a two-particle system

Figure 10-13*a* shows a rigid body consisting of two particles of mass m connected by a rod of length L and negligible mass.

(a) What is the rotational inertia I_{com} about an axis through the center of mass, perpendicular to the rod as shown?

KEY IDEA

Because we have only two particles with mass, we can find the body's rotational inertia I_{com} by using Eq. 10-33 rather than by integration.

Calculations: For the two particles, each at perpendicular distance $\frac{1}{2}L$ from the rotation axis, we have

$$I = \sum m_i r_i^2 = (m)(\tfrac{1}{2}L)^2 + (m)(\tfrac{1}{2}L)^2$$
$$= \tfrac{1}{2}mL^2. \qquad \text{(Answer)}$$

(b) What is the rotational inertia I of the body about an axis through the left end of the rod and parallel to the first axis (Fig. 10-13*b*)?

KEY IDEAS

This situation is simple enough that we can find I using either of two techniques. The first is similar to the one used in part (a). The other, more powerful one is to apply the parallel-axis theorem.

First technique: We calculate I as in part (a), except here the perpendicular distance r_i is zero for the particle on the left and

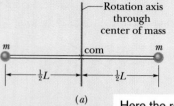

Here the rotation axis is through the com.

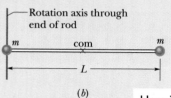

Here it has been shifted from the com without changing the orientation. We can use the parallel-axis theorem.

Fig. 10-13 A rigid body consisting of two particles of mass m joined by a rod of negligible mass.

L for the particle on the right. Now Eq. 10-33 gives us
$$I = m(0)^2 + mL^2 = mL^2. \qquad \text{(Answer)}$$

Second technique: Because we already know I_{com} about an axis through the center of mass and because the axis here is parallel to that "com axis," we can apply the parallel-axis theorem (Eq. 10-36). We find

$$I = I_{com} + Mh^2 = \tfrac{1}{2}mL^2 + (2m)(\tfrac{1}{2}L)^2$$
$$= mL^2. \qquad \text{(Answer)}$$

Sample Problem

Rotational inertia of a uniform rod, integration

Figure 10-14 shows a thin, uniform rod of mass M and length L, on an x axis with the origin at the rod's center.

(a) What is the rotational inertia of the rod about the perpendicular rotation axis through the center?

KEY IDEAS

(1) Because the rod is uniform, its center of mass is at its center. Therefore, we are looking for I_{com}. (2) Because the rod is a continuous object, we must use the integral of Eq. 10-35,

$$I = \int r^2 \, dm, \qquad (10\text{-}38)$$

to find the rotational inertia.

Calculations: We want to integrate with respect to coordi-

nate x (not mass m as indicated in the integral), so we must relate the mass dm of an element of the rod to its length dx along the rod. (Such an element is shown in Fig. 10-14.) Because the rod is uniform, the ratio of mass to length is the same for all the elements and for the rod as a whole. Thus, we can write

$$\frac{\text{element's mass } dm}{\text{element's length } dx} = \frac{\text{rod's mass } M}{\text{rod's length } L}$$

or
$$dm = \frac{M}{L}\, dx.$$

We can now substitute this result for dm and x for r in Eq. 10-38. Then we integrate from end to end of the rod (from $x = -L/2$ to $x = L/2$) to include all the elements. We find

$$I = \int_{x=-L/2}^{x=+L/2} x^2 \left(\frac{M}{L}\right) dx$$

$$= \frac{M}{3L}\left[x^3\right]_{-L/2}^{+L/2} = \frac{M}{3L}\left[\left(\frac{L}{2}\right)^3 - \left(-\frac{L}{2}\right)^3\right]$$

$$= \frac{1}{12}ML^2. \qquad \text{(Answer)}$$

This agrees with the result given in Table 10-2e.

(b) What is the rod's rotational inertia I about a new rotation axis that is perpendicular to the rod and through the left end?

KEY IDEAS

We can find I by shifting the origin of the x axis to the left end of the rod and then integrating from $x = 0$ to $x = L$. However, here we shall use a more powerful (and easier) technique by applying the parallel-axis theorem (Eq. 10-36), in which we shift the rotation axis without changing its orientation.

Calculations: If we place the axis at the rod's end so that it is parallel to the axis through the center of mass, then we can use the parallel-axis theorem (Eq. 10-36). We know from part (a) that I_{com} is $\frac{1}{12}ML^2$. From Fig. 10-14, the perpendicular distance h between the new rotation axis and the center of mass is $\frac{1}{2}L$. Equation 10-36 then gives us

$$I = I_{com} + Mh^2 = \frac{1}{12}ML^2 + (M)(\tfrac{1}{2}L)^2$$

$$= \tfrac{1}{3}ML^2. \qquad \text{(Answer)}$$

Actually, this result holds for any axis through the left or right end that is perpendicular to the rod, whether it is parallel to the axis shown in Fig. 10-14 or not.

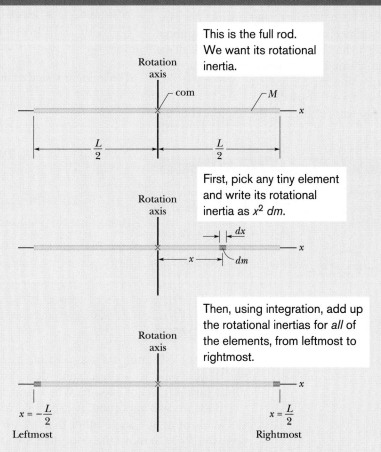

This is the full rod. We want its rotational inertia.

First, pick any tiny element and write its rotational inertia as $x^2\ dm$.

Then, using integration, add up the rotational inertias for *all* of the elements, from leftmost to rightmost.

Fig. 10-14 A uniform rod of length L and mass M. An element of mass dm and length dx is represented.

Additional examples, video, and practice available at *WileyPLUS*

Sample Problem

Rotational kinetic energy, spin test explosion

Large machine components that undergo prolonged, high-speed rotation are first examined for the possibility of failure in a *spin test system*. In this system, a component is *spun up* (brought up to high speed) while inside a cylindrical arrangement of lead bricks and containment liner, all within a steel shell that is closed by a lid clamped into place. If the rotation causes the component to shatter, the soft lead bricks are supposed to catch the pieces for later analysis.

In 1985, Test Devices, Inc. (www.testdevices.com) was spin testing a sample of a solid steel rotor (a disk) of mass $M = 272$ kg and radius $R = 38.0$ cm. When the sample reached an angular speed ω of 14 000 rev/min, the test engineers heard a dull thump from the test system, which was located one floor down and one room over from them. Investigating, they found that lead bricks had been thrown out in the hallway leading to the test room, a door to the room had been hurled into the adjacent parking lot, one lead brick had shot from the test site through the wall of a neighbor's kitchen, the structural beams of the test building had been damaged, the concrete floor beneath the spin chamber had been shoved downward by about 0.5 cm, and the 900 kg lid had been blown upward through the ceiling and had then crashed back onto the test equipment (Fig. 10-15). The exploding pieces had not penetrated the room of the test engineers only by luck.

How much energy was released in the explosion of the rotor?

Fig. 10-15 Some of the destruction caused by the explosion of a rapidly rotating steel disk. (*Courtesy Test Devices, Inc.*)

Calculations: We can find K with Eq. 10-34 ($K = \frac{1}{2}I\omega^2$), but first we need an expression for the rotational inertia I. Because the rotor was a disk that rotated like a merry-go-round, I is given by the expression in Table 10-2c ($I = \frac{1}{2}MR^2$). Thus, we have

$$I = \tfrac{1}{2}MR^2 = \tfrac{1}{2}(272 \text{ kg})(0.38 \text{ m})^2 = 19.64 \text{ kg}\cdot\text{m}^2.$$

The angular speed of the rotor was

$$\omega = (14\,000 \text{ rev/min})(2\pi \text{ rad/rev})\left(\frac{1 \text{ min}}{60 \text{ s}}\right)$$
$$= 1.466 \times 10^3 \text{ rad/s}.$$

Now we can use Eq. 10-34 to write

$$K = \tfrac{1}{2}I\omega^2 = \tfrac{1}{2}(19.64 \text{ kg}\cdot\text{m}^2)(1.466 \times 10^3 \text{ rad/s})^2$$
$$= 2.1 \times 10^7 \text{ J.} \qquad \text{(Answer)}$$

Being near this explosion was quite dangerous.

KEY IDEA

The released energy was equal to the rotational kinetic energy K of the rotor just as it reached the angular speed of 14 000 rev/min.

PLUS Additional examples, video, and practice available at *WileyPLUS*

10-8 Torque

A doorknob is located as far as possible from the door's hinge line for a good reason. If you want to open a heavy door, you must certainly apply a force; that alone, however, is not enough. Where you apply that force and in what direction you push are also important. If you apply your force nearer to the hinge line than the knob, or at any angle other than 90° to the plane of the door, you must use a greater force to move the door than if you apply the force at the knob and perpendicular to the door's plane.

Figure 10-16a shows a cross section of a body that is free to rotate about an axis passing through O and perpendicular to the cross section. A force $\vec{F}$ is applied at point P, whose position relative to O is defined by a position vector $\vec{r}$. The directions of vectors $\vec{F}$ and $\vec{r}$ make an angle ϕ with each other. (For simplicity, we consider only forces that have no component parallel to the rotation axis; thus, $\vec{F}$ is in the plane of the page.)

To determine how $\vec{F}$ results in a rotation of the body around the rotation axis, we resolve $\vec{F}$ into two components (Fig. 10-16b). One component, called the *radial component* F_r, points along $\vec{r}$. This component does not cause rotation, because it acts along a line that extends through O. (If you pull on a door parallel to the plane of the door, you do not rotate the door.) The other component of $\vec{F}$, called the *tangential component* F_t, is perpendicular to $\vec{r}$ and has magnitude $F_t = F \sin \phi$. This component *does* cause rotation. (If you pull on a door perpendicular to its plane, you can rotate the door.)

The ability of $\vec{F}$ to rotate the body depends not only on the magnitude of its tangential component F_t, but also on just how far from O the force is applied. To include both these factors, we define a quantity called **torque** τ as the product of the two factors and write it as

$$\tau = (r)(F \sin \phi). \qquad (10\text{-}39)$$

Two equivalent ways of computing the torque are

$$\tau = (r)(F \sin \phi) = rF_t \qquad (10\text{-}40)$$

and

$$\tau = (r \sin \phi)(F) = r_\perp F, \qquad (10\text{-}41)$$

where $r_\perp$ is the perpendicular distance between the rotation axis at O and an extended line running through the vector $\vec{F}$ (Fig. 10-16c). This extended line is called the **line of action** of $\vec{F}$, and $r_\perp$ is called the **moment arm** of $\vec{F}$. Figure 10-16b shows that we can describe r, the magnitude of $\vec{r}$, as being the moment arm of the force component F_t.

Torque, which comes from the Latin word meaning "to twist," may be loosely identified as the turning or twisting action of the force $\vec{F}$. When you apply a force to an object—such as a screwdriver or torque wrench—with the purpose of turning that object, you are applying a torque. The SI unit of torque is the newton-meter (N·m). *Caution:* The newton-meter is also the unit of work. Torque and work, however, are quite different quantities and must not be confused. Work is often expressed in joules (1 J = 1 N·m), but torque never is.

In the next chapter we shall discuss torque in a general way as being a vector quantity. Here, however, because we consider only rotation around a single axis, we do not need vector notation. Instead, a torque has either a positive or negative value depending on the direction of rotation it would give a body initially at rest: If the body would rotate counterclockwise, the torque is positive. If the object would rotate clockwise, the torque is negative. (The phrase "clocks are negative" from Section 10-2 still works.)

Torques obey the superposition principle that we discussed in Chapter 5 for forces: When several torques act on a body, the **net torque** (or **resultant torque**) is the sum of the individual torques. The symbol for net torque is τ_{net}.

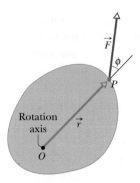

The torque due to this force causes rotation around this axis (which extends out toward you).

(a)

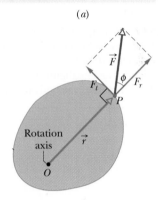

But actually only the *tangential* component of the force causes the rotation.

(b)

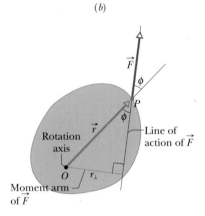

You calculate the same torque by using this moment arm distance and the full force magnitude.

(c)

Fig. 10-16 (a) A force $\vec{F}$ acts on a rigid body, with a rotation axis perpendicular to the page. The torque can be found with (a) angle ϕ, (b) tangential force component F_t, or (c) moment arm $r_\perp$.

✔ **CHECKPOINT 6**

The figure shows an overhead view of a meter stick that can pivot about the dot at the position marked 20 (for 20 cm). All five forces on the stick are horizontal and have the same magnitude. Rank the forces according to the magnitude of the torque they produce, greatest first.

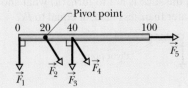

The torque due to the tangential component of the force causes an angular acceleration around the rotation axis.

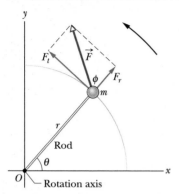

Fig. 10-17   A simple rigid body, free to rotate about an axis through O, consists of a particle of mass m fastened to the end of a rod of length r and negligible mass. An applied force $\vec{F}$ causes the body to rotate.

10-9 Newton's Second Law for Rotation

A torque can cause rotation of a rigid body, as when you use a torque to rotate a door. Here we want to relate the net torque τ_{net} on a rigid body to the angular acceleration α that torque causes about a rotation axis. We do so by analogy with Newton's second law ($F_{net} = ma$) for the acceleration a of a body of mass m due to a net force F_{net} along a coordinate axis. We replace F_{net} with τ_{net}, m with I, and a with α in radian measure, writing

$$\tau_{net} = I\alpha \qquad \text{(Newton's second law for rotation).} \qquad (10\text{-}42)$$

Proof of Equation 10-42

We prove Eq. 10-42 by first considering the simple situation shown in Fig. 10-17. The rigid body there consists of a particle of mass m on one end of a massless rod of length r. The rod can move only by rotating about its other end, around a rotation axis (an axle) that is perpendicular to the plane of the page. Thus, the particle can move only in a circular path that has the rotation axis at its center.

A force $\vec{F}$ acts on the particle. However, because the particle can move only along the circular path, only the tangential component F_t of the force (the component that is tangent to the circular path) can accelerate the particle along the path. We can relate F_t to the particle's tangential acceleration a_t along the path with Newton's second law, writing

$$F_t = ma_t.$$

The torque acting on the particle is, from Eq. 10-40,

$$\tau = F_t r = ma_t r.$$

From Eq. 10-22 ($a_t = \alpha r$) we can write this as

$$\tau = m(\alpha r)r = (mr^2)\alpha. \qquad (10\text{-}43)$$

The quantity in parentheses on the right is the rotational inertia of the particle about the rotation axis (see Eq. 10-33, but here we have only a single particle). Thus, using I for the rotational inertia, Eq. 10-43 reduces to

$$\tau = I\alpha \qquad \text{(radian measure).} \qquad (10\text{-}44)$$

For the situation in which more than one force is applied to the particle, we can generalize Eq. 10-44 as

$$\tau_{net} = I\alpha \qquad \text{(radian measure),} \qquad (10\text{-}45)$$

which we set out to prove. We can extend this equation to any rigid body rotating about a fixed axis, because any such body can always be analyzed as an assembly of single particles.

✔CHECKPOINT 7

The figure shows an overhead view of a meter stick that can pivot about the point indicated, which is to the left of the stick's midpoint. Two horizontal forces, $\vec{F_1}$ and $\vec{F_2}$, are applied to the stick. Only $\vec{F_1}$ is shown. Force $\vec{F_2}$ is perpendicular to the stick and is applied at the right end. If the stick is not to turn, (a) what should be the direction of $\vec{F_2}$, and (b) should F_2 be greater than, less than, or equal to F_1?

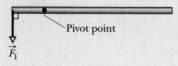

Sample Problem

Newton's 2nd law, rotation, torque, disk

Figure 10-18a shows a uniform disk, with mass $M = 2.5$ kg and radius $R = 20$ cm, mounted on a fixed horizontal axle. A block with mass $m = 1.2$ kg hangs from a massless cord that is wrapped around the rim of the disk. Find the acceleration of the falling block, the angular acceleration of the disk, and the tension in the cord. The cord does not slip, and there is no friction at the axle.

KEY IDEAS

(1) Taking the block as a system, we can relate its acceleration a to the forces acting on it with Newton's second law ($\vec{F}_{net} = m\vec{a}$). (2) Taking the disk as a system, we can relate its angular acceleration α to the torque acting on it with Newton's second law for rotation ($\tau_{net} = I\alpha$). (3) To combine the motions of block and disk, we use the fact that the linear acceleration a of the block and the (tangential) linear acceleration a_t of the disk rim are equal.

Forces on block: The forces are shown in the block's free-body diagram in Fig. 10-18b: The force from the cord is $\vec{T}$, and the gravitational force is $\vec{F}_g$, of magnitude mg. We can now write Newton's second law for components along a vertical y axis ($F_{net,y} = ma_y$) as

$$T - mg = ma. \tag{10-46}$$

However, we cannot solve this equation for a because it also contains the unknown T.

Torque on disk: Previously, when we got stuck on the y axis, we switched to the x axis. Here, we switch to the rotation of the disk. To calculate the torques and the rotational inertia I, we take the rotation axis to be perpendicular to the disk and through its center, at point O in Fig. 10-18c.

The torques are then given by Eq. 10-40 ($\tau = rF_t$). The gravitational force on the disk and the force on the disk from the axle both act at the center of the disk and thus at distance $r = 0$, so their torques are zero. The force $\vec{T}$ on the disk due to the cord acts at distance $r = R$ and is tangent to the rim of the disk. Therefore, its torque is $-RT$, negative because the torque rotates the disk clockwise from rest. From Table 10-2c, the rotational inertia I of the disk is $\frac{1}{2}MR^2$. Thus we can write $\tau_{net} = I\alpha$ as

$$-RT = \frac{1}{2}MR^2\alpha. \tag{10-47}$$

This equation seems useless because it has two unknowns, α and T, neither of which is the desired a. However, mustering physics courage, we can make it useful

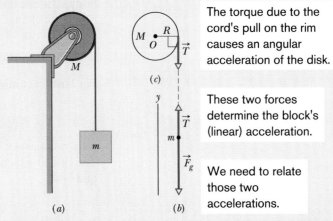

The torque due to the cord's pull on the rim causes an angular acceleration of the disk.

(c)

These two forces determine the block's (linear) acceleration.

We need to relate those two accelerations.

Fig. 10-18 (a) The falling block causes the disk to rotate. (b) A free-body diagram for the block. (c) An incomplete free-body diagram for the disk.

with this fact: Because the cord does not slip, the linear acceleration a of the block and the (tangential) linear acceleration a_t of the rim of the disk are equal. Then, by Eq. 10-22 ($a_t = \alpha r$) we see that here $\alpha = a/R$. Substituting this in Eq. 10-47 yields

$$T = -\frac{1}{2}Ma. \tag{10-48}$$

Combining results: Combining Eqs. 10-46 and 10-48 leads to

$$a = -g\frac{2m}{M + 2m} = -(9.8 \text{ m/s}^2)\frac{(2)(1.2 \text{ kg})}{2.5 \text{ kg} + (2)(1.2 \text{ kg})}$$

$$= -4.8 \text{ m/s}^2. \tag{Answer}$$

We then use Eq. 10-48 to find T:

$$T = -\frac{1}{2}Ma = -\frac{1}{2}(2.5 \text{ kg})(-4.8 \text{ m/s}^2)$$

$$= 6.0 \text{ N}. \tag{Answer}$$

As we should expect, acceleration a of the falling block is less than g, and tension T in the cord (= 6.0 N) is less than the gravitational force on the hanging block (= mg = 11.8 N). We see also that a and T depend on the mass of the disk but not on its radius. As a check, we note that the formulas derived above predict $a = -g$ and $T = 0$ for the case of a massless disk ($M = 0$). This is what we would expect; the block simply falls as a free body. From Eq. 10-22, the angular acceleration of the disk is

$$\alpha = \frac{a}{R} = \frac{-4.8 \text{ m/s}^2}{0.20 \text{ m}} = -24 \text{ rad/s}^2. \tag{Answer}$$

 Additional examples, video, and practice available at *WileyPLUS*

10-10 Work and Rotational Kinetic Energy

As we discussed in Chapter 7, when a force F causes a rigid body of mass m to accelerate along a coordinate axis, the force does work W on the body. Thus, the body's kinetic energy ($K = \frac{1}{2}mv^2$) can change. Suppose it is the only energy of the body that changes. Then we relate the change ΔK in kinetic energy to the work W with the work–kinetic energy theorem (Eq. 7-10), writing

$$\Delta K = K_f - K_i = \tfrac{1}{2}mv_f^2 - \tfrac{1}{2}mv_i^2 = W \quad \text{(work–kinetic energy theorem).} \quad (10\text{-}49)$$

For motion confined to an x axis, we can calculate the work with Eq. 7-32,

$$W = \int_{x_i}^{x_f} F \, dx \quad \text{(work, one-dimensional motion).} \quad (10\text{-}50)$$

This reduces to $W = Fd$ when F is constant and the body's displacement is d. The rate at which the work is done is the power, which we can find with Eqs. 7-43 and 7-48,

$$P = \frac{dW}{dt} = Fv \quad \text{(power, one-dimensional motion).} \quad (10\text{-}51)$$

Now let us consider a rotational situation that is similar. When a torque accelerates a rigid body in rotation about a fixed axis, the torque does work W on the body. Therefore, the body's rotational kinetic energy ($K = \frac{1}{2}I\omega^2$) can change. Suppose that it is the only energy of the body that changes. Then we can still relate the change ΔK in kinetic energy to the work W with the work–kinetic energy theorem, except now the kinetic energy is a rotational kinetic energy:

$$\Delta K = K_f - K_i = \tfrac{1}{2}I\omega_f^2 - \tfrac{1}{2}I\omega_i^2 = W \quad \text{(work–kinetic energy theorem).} \quad (10\text{-}52)$$

Here, I is the rotational inertia of the body about the fixed axis and ω_i and ω_f are the angular speeds of the body before and after the work is done, respectively.

Also, we can calculate the work with a rotational equivalent of Eq. 10-50,

$$W = \int_{\theta_i}^{\theta_f} \tau \, d\theta \quad \text{(work, rotation about fixed axis),} \quad (10\text{-}53)$$

where τ is the torque doing the work W, and θ_i and θ_f are the body's angular positions before and after the work is done, respectively. When τ is constant, Eq. 10-53 reduces to

$$W = \tau(\theta_f - \theta_i) \quad \text{(work, constant torque).} \quad (10\text{-}54)$$

The rate at which the work is done is the power, which we can find with the rotational equivalent of Eq. 10-51,

$$P = \frac{dW}{dt} = \tau\omega \quad \text{(power, rotation about fixed axis).} \quad (10\text{-}55)$$

Table 10-3 summarizes the equations that apply to the rotation of a rigid body about a fixed axis and the corresponding equations for translational motion.

Proof of Eqs. 10-52 through 10-55

Let us again consider the situation of Fig. 10-17, in which force $\vec{F}$ rotates a rigid body consisting of a single particle of mass m fastened to the end of a massless rod. During the rotation, force $\vec{F}$ does work on the body. Let us assume that the

Table 10-3			
Some Corresponding Relations for Translational and Rotational Motion			
Pure Translation (Fixed Direction)		Pure Rotation (Fixed Axis)	
Position	x	Angular position	θ
Velocity	$v = dx/dt$	Angular velocity	$\omega = d\theta/dt$
Acceleration	$a = dv/dt$	Angular acceleration	$\alpha = d\omega/dt$
Mass	m	Rotational inertia	I
Newton's second law	$F_{net} = ma$	Newton's second law	$\tau_{net} = I\alpha$
Work	$W = \int F\,dx$	Work	$W = \int \tau\,d\theta$
Kinetic energy	$K = \frac{1}{2}mv^2$	Kinetic energy	$K = \frac{1}{2}I\omega^2$
Power (constant force)	$P = Fv$	Power (constant torque)	$P = \tau\omega$
Work–kinetic energy theorem	$W = \Delta K$	Work–kinetic energy theorem	$W = \Delta K$

only energy of the body that is changed by $\vec{F}$ is the kinetic energy. Then we can apply the work–kinetic energy theorem of Eq. 10-49:

$$\Delta K = K_f - K_i = W. \tag{10-56}$$

Using $K = \frac{1}{2}mv^2$ and Eq. 10-18 ($v = \omega r$), we can rewrite Eq. 10-56 as

$$\Delta K = \frac{1}{2}mr^2\omega_f^2 - \frac{1}{2}mr^2\omega_i^2 = W. \tag{10-57}$$

From Eq. 10-33, the rotational inertia for this one-particle body is $I = mr^2$. Substituting this into Eq. 10-57 yields

$$\Delta K = \frac{1}{2}I\omega_f^2 - \frac{1}{2}I\omega_i^2 = W,$$

which is Eq. 10-52. We derived it for a rigid body with one particle, but it holds for any rigid body rotated about a fixed axis.

We next relate the work W done on the body in Fig. 10-17 to the torque τ on the body due to force $\vec{F}$. When the particle moves a distance ds along its circular path, only the tangential component F_t of the force accelerates the particle along the path. Therefore, only F_t does work on the particle. We write that work dW as $F_t\,ds$. However, we can replace ds with $r\,d\theta$, where $d\theta$ is the angle through which the particle moves. Thus we have

$$dW = F_t r\,d\theta. \tag{10-58}$$

From Eq. 10-40, we see that the product $F_t r$ is equal to the torque τ, so we can rewrite Eq. 10-58 as

$$dW = \tau\,d\theta. \tag{10-59}$$

The work done during a finite angular displacement from θ_i to θ_f is then

$$W = \int_{\theta_i}^{\theta_f} \tau\,d\theta,$$

which is Eq. 10-53. It holds for any rigid body rotating about a fixed axis. Equation 10-54 comes directly from Eq. 10-53.

We can find the power P for rotational motion from Eq. 10-59:

$$P = \frac{dW}{dt} = \tau\frac{d\theta}{dt} = \tau\omega,$$

which is Eq. 10-55.

Sample Problem

Work, rotational kinetic energy, torque, disk

Let the disk in Fig. 10-18 start from rest at time $t = 0$ and also let the tension in the massless cord be 6.0 N and the angular acceleration of the disk be -24 rad/s^2. What is its rotational kinetic energy K at $t = 2.5$ s?

KEY IDEA

We can find K with Eq. 10-34 ($K = \frac{1}{2}I\omega^2$). We already know that $I = \frac{1}{2}MR^2$, but we do not yet know ω at $t = 2.5$ s. However, because the angular acceleration α has the constant value of -24 rad/s^2, we can apply the equations for constant angular acceleration in Table 10-1.

Calculations: Because we want ω and know α and ω_0 ($= 0$), we use Eq. 10-12:

$$\omega = \omega_0 + \alpha t = 0 + \alpha t = \alpha t.$$

Substituting $\omega = \alpha t$ and $I = \frac{1}{2}MR^2$ into Eq. 10-34, we find

$$K = \frac{1}{2}I\omega^2 = \frac{1}{2}(\frac{1}{2}MR^2)(\alpha t)^2 = \frac{1}{4}M(R\alpha t)^2$$
$$= \frac{1}{4}(2.5 \text{ kg})[(0.20 \text{ m})(-24 \text{ rad/s}^2)(2.5 \text{ s})]^2$$
$$= 90 \text{ J}. \qquad \text{(Answer)}$$

KEY IDEA

We can also get this answer by finding the disk's kinetic energy from the work done on the disk.

Calculations: First, we relate the *change* in the kinetic energy of the disk to the net work W done on the disk, using the work–kinetic energy theorem of Eq. 10-52 ($K_f - K_i = W$). With K substituted for K_f and 0 for K_i, we get

$$K = K_i + W = 0 + W = W. \qquad (10\text{-}60)$$

Next we want to find the work W. We can relate W to the torques acting on the disk with Eq. 10-53 or 10-54. The only torque causing angular acceleration and doing work is the torque due to force $\vec{T}$ on the disk from the cord, which is equal to $-TR$. Because α is constant, this torque also must be constant. Thus, we can use Eq. 10-54 to write

$$W = \tau(\theta_f - \theta_i) = -TR(\theta_f - \theta_i). \qquad (10\text{-}61)$$

Because α is constant, we can use Eq. 10-13 to find $\theta_f - \theta_i$. With $\omega_i = 0$, we have

$$\theta_f - \theta_i = \omega_i t + \tfrac{1}{2}\alpha t^2 = 0 + \tfrac{1}{2}\alpha t^2 = \tfrac{1}{2}\alpha t^2.$$

Now we substitute this into Eq. 10-61 and then substitute the result into Eq. 10-60. Inserting the given values $T = 6.0$ N and $\alpha = -24$ rad/s^2, we have

$$K = W = -TR(\theta_f - \theta_i) = -TR(\tfrac{1}{2}\alpha t^2) = -\tfrac{1}{2}TR\alpha t^2$$
$$= -\tfrac{1}{2}(6.0 \text{ N})(0.20 \text{ m})(-24 \text{ rad/s}^2)(2.5 \text{ s})^2$$
$$= 90 \text{ J}. \qquad \text{(Answer)}$$

 Additional examples, video, and practice available at *WileyPLUS*

REVIEW & SUMMARY

Angular Position To describe the rotation of a rigid body about a fixed axis, called the **rotation axis,** we assume a **reference line** is fixed in the body, perpendicular to that axis and rotating with the body. We measure the **angular position** θ of this line relative to a fixed direction. When θ is measured in **radians,**

$$\theta = \frac{s}{r} \qquad \text{(radian measure)}, \qquad (10\text{-}1)$$

where s is the arc length of a circular path of radius r and angle θ. Radian measure is related to angle measure in revolutions and degrees by

$$1 \text{ rev} = 360° = 2\pi \text{ rad}. \qquad (10\text{-}2)$$

Angular Displacement A body that rotates about a rotation axis, changing its angular position from θ_1 to θ_2, undergoes an **angular displacement**

$$\Delta\theta = \theta_2 - \theta_1, \qquad (10\text{-}4)$$

where $\Delta\theta$ is positive for counterclockwise rotation and negative for clockwise rotation.

Angular Velocity and Speed If a body rotates through an angular displacement $\Delta\theta$ in a time interval Δt, its **average angular velocity** ω_{avg} is

$$\omega_{avg} = \frac{\Delta\theta}{\Delta t}. \qquad (10\text{-}5)$$

The **(instantaneous) angular velocity** ω of the body is

$$\omega = \frac{d\theta}{dt}. \qquad (10\text{-}6)$$

Both ω_{avg} and ω are vectors, with directions given by the **right-hand rule** of Fig. 10-6. They are positive for counterclockwise rotation and negative for clockwise rotation. The magnitude of the body's angular velocity is the **angular speed.**

Angular Acceleration If the angular velocity of a body changes from ω_1 to ω_2 in a time interval $\Delta t = t_2 - t_1$, the **average angular acceleration** α_{avg} of the body is

$$\alpha_{avg} = \frac{\omega_2 - \omega_1}{t_2 - t_1} = \frac{\Delta\omega}{\Delta t}. \tag{10-7}$$

The **(instantaneous) angular acceleration** α of the body is

$$\alpha = \frac{d\omega}{dt}. \tag{10-8}$$

Both α_{avg} and α are vectors.

The Kinematic Equations for Constant Angular Acceleration *Constant angular acceleration* (α = constant) is an important special case of rotational motion. The appropriate kinematic equations, given in Table 10-1, are

$$\omega = \omega_0 + \alpha t, \tag{10-12}$$

$$\theta - \theta_0 = \omega_0 t + \tfrac{1}{2}\alpha t^2, \tag{10-13}$$

$$\omega^2 = \omega_0^2 + 2\alpha(\theta - \theta_0), \tag{10-14}$$

$$\theta - \theta_0 = \tfrac{1}{2}(\omega_0 + \omega)t, \tag{10-15}$$

$$\theta - \theta_0 = \omega t - \tfrac{1}{2}\alpha t^2. \tag{10-16}$$

Linear and Angular Variables Related A point in a rigid rotating body, at a *perpendicular distance r* from the rotation axis, moves in a circle with radius r. If the body rotates through an angle θ, the point moves along an arc with length s given by

$$s = \theta r \quad \text{(radian measure)}, \tag{10-17}$$

where θ is in radians.

The linear velocity $\vec{v}$ of the point is tangent to the circle; the point's linear speed v is given by

$$v = \omega r \quad \text{(radian measure)}, \tag{10-18}$$

where ω is the angular speed (in radians per second) of the body.

The linear acceleration $\vec{a}$ of the point has both *tangential* and *radial* components. The tangential component is

$$a_t = \alpha r \quad \text{(radian measure)}, \tag{10-22}$$

where α is the magnitude of the angular acceleration (in radians per second-squared) of the body. The radial component of $\vec{a}$ is

$$a_r = \frac{v^2}{r} = \omega^2 r \quad \text{(radian measure)}. \tag{10-23}$$

If the point moves in uniform circular motion, the period T of the motion for the point and the body is

$$T = \frac{2\pi r}{v} = \frac{2\pi}{\omega} \quad \text{(radian measure)}. \tag{10-19, 10-20}$$

Rotational Kinetic Energy and Rotational Inertia The kinetic energy K of a rigid body rotating about a fixed axis is given by

$$K = \tfrac{1}{2}I\omega^2 \quad \text{(radian measure)}, \tag{10-34}$$

in which I is the **rotational inertia** of the body, defined as

$$I = \sum m_i r_i^2 \tag{10-33}$$

for a system of discrete particles and defined as

$$I = \int r^2 \, dm \tag{10-35}$$

for a body with continuously distributed mass. The r and r_i in these expressions represent the perpendicular distance from the axis of rotation to each mass element in the body, and the integration is carried out over the entire body so as to include every mass element.

The Parallel-Axis Theorem The *parallel-axis theorem* relates the rotational inertia I of a body about any axis to that of the same body about a parallel axis through the center of mass:

$$I = I_{com} + Mh^2. \tag{10-36}$$

Here h is the perpendicular distance between the two axes, and I_{com} is the rotational inertia of the body about the axis through the com. We can describe h as being the distance the actual rotation axis has been shifted from the rotation axis through the com.

Torque *Torque* is a turning or twisting action on a body about a rotation axis due to a force $\vec{F}$. If $\vec{F}$ is exerted at a point given by the position vector $\vec{r}$ relative to the axis, then the magnitude of the torque is

$$\tau = rF_t = r_\perp F = rF \sin\phi, \tag{10-40, 10-41, 10-39}$$

where F_t is the component of $\vec{F}$ perpendicular to $\vec{r}$ and ϕ is the angle between $\vec{r}$ and $\vec{F}$. The quantity $r_\perp$ is the perpendicular distance between the rotation axis and an extended line running through the $\vec{F}$ vector. This line is called the **line of action** of $\vec{F}$, and $r_\perp$ is called the **moment arm** of $\vec{F}$. Similarly, r is the moment arm of F_t.

The SI unit of torque is the newton-meter (N · m). A torque τ is positive if it tends to rotate a body at rest counterclockwise and negative if it tends to rotate the body clockwise.

Newton's Second Law in Angular Form The rotational analog of Newton's second law is

$$\tau_{net} = I\alpha, \tag{10-45}$$

where τ_{net} is the net torque acting on a particle or rigid body, I is the rotational inertia of the particle or body about the rotation axis, and α is the resulting angular acceleration about that axis.

Work and Rotational Kinetic Energy The equations used for calculating work and power in rotational motion correspond to equations used for translational motion and are

$$W = \int_{\theta_i}^{\theta_f} \tau \, d\theta \tag{10-53}$$

and

$$P = \frac{dW}{dt} = \tau\omega. \tag{10-55}$$

When τ is constant, Eq. 10-53 reduces to

$$W = \tau(\theta_f - \theta_i). \tag{10-54}$$

The form of the work–kinetic energy theorem used for rotating bodies is

$$\Delta K = K_f - K_i = \tfrac{1}{2}I\omega_f^2 - \tfrac{1}{2}I\omega_i^2 = W. \tag{10-52}$$

1 Figure 10-19 is a graph of the angular velocity versus time for a disk rotating like a merry-go-round. For a point on the disk rim, rank the instants a, b, c, and d according to the magnitude of the (a) tangential and (b) radial acceleration, greatest first.

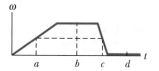

Fig. 10-19 Question 1.

2 Figure 10-20 shows plots of angular position θ versus time t for three cases in which a disk is rotated like a merry-go-round. In each case, the rotation direction changes at a certain angular position θ_{change}. (a) For each case, determine whether θ_{change} is clockwise or counterclockwise from $\theta = 0$, or whether it is at $\theta = 0$. For each case, determine (b) whether ω is zero before, after, or at $t = 0$ and (c) whether α is positive, negative, or zero.

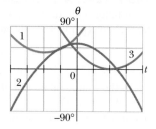

Fig. 10-20 Question 2.

3 A force is applied to the rim of a disk that can rotate like a merry-go-round, so as to change its angular velocity. Its initial and final angular velocities, respectively, for four situations are: (a) −2 rad/s, 5 rad/s; (b) 2 rad/s, 5 rad/s; (c) −2 rad/s, −5 rad/s; and (d) 2 rad/s, −5 rad/s. Rank the situations according to the work done by the torque due to the force, greatest first.

4 Figure 10-21b is a graph of the angular position of the rotating disk of Fig. 10-21a. Is the angular velocity of the disk positive, negative, or zero at (a) $t = 1$ s, (b) $t = 2$ s, and (c) $t = 3$ s? (d) Is the angular acceleration positive or negative?

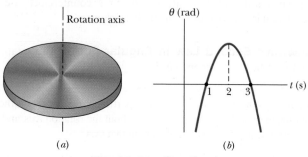

Fig. 10-21 Question 4.

5 In Fig. 10-22, two forces $\vec{F}_1$ and $\vec{F}_2$ act on a disk that turns about its center like a merry-go-round. The forces maintain the indicated angles during the rotation, which is counterclockwise and at a constant rate. However, we are to decrease the angle θ of $\vec{F}_1$ without changing the magnitude of $\vec{F}_1$. (a) To keep the angular speed constant, should we increase, decrease, or maintain the magnitude of

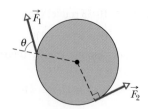

Fig. 10-22 Question 5.

$\vec{F}_2$? Do forces (b) $\vec{F}_1$ and (c) $\vec{F}_2$ tend to rotate the disk clockwise or counterclockwise?

6 In the overhead view of Fig. 10-23, five forces of the same magnitude act on a strange merry-go-round; it is a square that can rotate about point P, at midlength along one of the edges. Rank the forces according to the magnitude of the torque they create about point P, greatest first.

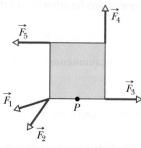

Fig. 10-23 Question 6.

7 Figure 10-24a is an overhead view of a horizontal bar that can pivot; two horizontal forces act on the bar, but it is stationary. If the angle between the bar and $\vec{F}_2$ is now decreased from 90° and the bar is still not to turn, should F_2 be made larger, made smaller, or left the same?

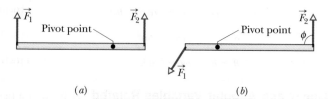

Fig. 10-24 Questions 7 and 8.

8 Figure 10-24b shows an overhead view of a horizontal bar that is rotated about the pivot point by two horizontal forces, $\vec{F}_1$ and $\vec{F}_2$, with $\vec{F}_2$ at angle ϕ to the bar. Rank the following values of ϕ according to the magnitude of the angular acceleration of the bar, greatest first: 90°, 70°, and 110°.

9 Figure 10-25 shows a uniform metal plate that had been square before 25% of it was snipped off. Three lettered points are indicated. Rank them according to the rotational inertia of the plate around a perpendicular axis through them, greatest first.

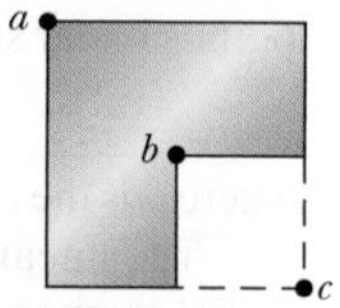

Fig. 10-25 Question 9.

10 Figure 10-26 shows three flat disks (of the same radius) that can rotate about their centers like merry-go-rounds. Each disk consists of the same two materials, one denser than the other (density is mass per unit volume). In disks 1 and 3, the denser material forms the outer half of the disk area. In disk 2, it forms the inner half of the disk area. Forces with identical magnitudes are applied tangentially to the disk, either at the outer edge or at the interface of the two materials, as shown. Rank the disks according to (a) the torque about the disk center, (b) the rotational inertia about the disk center, and (c) the angular acceleration of the disk, greatest first.

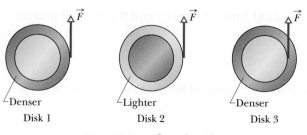

Fig. 10-26 Question 10.

PROBLEMS

GO Tutoring problem available (at instructor's discretion) in *WileyPLUS* and WebAssign

SSM Worked-out solution available in Student Solutions Manual

• – ••• Number of dots indicates level of problem difficulty

WWW Worked-out solution is at

ILW Interactive solution is at   http://www.wiley.com/college/halliday

  Additional information available in *The Flying Circus of Physics* and at flyingcircusofphysics.com

sec. 10-2   The Rotational Variables

•1 A good baseball pitcher can throw a baseball toward home plate at 85 mi/h with a spin of 1800 rev/min. How many revolutions does the baseball make on its way to home plate? For simplicity, assume that the 60 ft path is a straight line.

•2 What is the angular speed of (a) the second hand, (b) the minute hand, and (c) the hour hand of a smoothly running analog watch? Answer in radians per second.

••3  When a slice of buttered toast is accidentally pushed over the edge of a counter, it rotates as it falls. If the distance to the floor is 76 cm and for rotation less than 1 rev, what are the (a) smallest and (b) largest angular speeds that cause the toast to hit and then topple to be butter-side down?

••4 The angular position of a point on a rotating wheel is given by $\theta = 2.0 + 4.0t^2 + 2.0t^3$, where θ is in radians and t is in seconds. At $t = 0$, what are (a) the point's angular position and (b) its angular velocity? (c) What is its angular velocity at $t = 4.0$ s? (d) Calculate its angular acceleration at $t = 2.0$ s. (e) Is its angular acceleration constant?

••5 ILW A diver makes 2.5 revolutions on the way from a 10-m-high platform to the water. Assuming zero initial vertical velocity, find the average angular velocity during the dive.

••6 The angular position of a point on the rim of a rotating wheel is given by $\theta = 4.0t - 3.0t^2 + t^3$, where θ is in radians and t is in seconds. What are the angular velocities at (a) $t = 2.0$ s and (b) $t = 4.0$ s? (c) What is the average angular acceleration for the time interval that begins at $t = 2.0$ s and ends at $t = 4.0$ s? What are the instantaneous angular accelerations at (d) the beginning and (e) the end of this time interval?

•••7 The wheel in Fig. 10-27 has eight equally spaced spokes and a radius of 30 cm. It is mounted on a fixed axle and is spinning at 2.5 rev/s. You want to shoot a 20-cm-long arrow parallel to this axle and through the wheel without hitting any of the spokes. Assume that the arrow and the spokes are very thin.

Fig. 10-27   Problem 7.

(a) What minimum speed must the arrow have? (b) Does it matter where between the axle and rim of the wheel you aim? If so, what is the best location?

•••8 The angular acceleration of a wheel is $\alpha = 6.0t^4 - 4.0t^2$, with α in radians per second-squared and t in seconds. At time $t = 0$, the wheel has an angular velocity of $+2.0$ rad/s and an angular position of $+1.0$ rad. Write expressions for (a) the angular velocity (rad/s) and (b) the angular position (rad) as functions of time (s).

sec. 10-4   Rotation with Constant Angular Acceleration

•9 A drum rotates around its central axis at an angular velocity of 12.60 rad/s. If the drum then slows at a constant rate of 4.20 rad/s², (a) how much time does it take and (b) through what angle does it rotate in coming to rest?

•10 Starting from rest, a disk rotates about its central axis with constant angular acceleration. In 5.0 s, it rotates 25 rad. During that time, what are the magnitudes of (a) the angular acceleration and (b) the average angular velocity? (c) What is the instantaneous angular velocity of the disk at the end of the 5.0 s? (d) With the angular acceleration unchanged, through what additional angle will the disk turn during the next 5.0 s?

•11 A disk, initially rotating at 120 rad/s, is slowed down with a constant angular acceleration of magnitude 4.0 rad/s². (a) How much time does the disk take to stop? (b) Through what angle does the disk rotate during that time?

•12 The angular speed of an automobile engine is increased at a constant rate from 1200 rev/min to 3000 rev/min in 12 s. (a) What is its angular acceleration in revolutions per minute-squared? (b) How many revolutions does the engine make during this 12 s interval?

••13 ILW A flywheel turns through 40 rev as it slows from an angular speed of 1.5 rad/s to a stop. (a) Assuming a constant angular acceleration, find the time for it to come to rest. (b) What is its angular acceleration? (c) How much time is required for it to complete the first 20 of the 40 revolutions?

••14 GO A disk rotates about its central axis starting from rest and accelerates with constant angular acceleration. At one time it is rotating at 10 rev/s; 60 revolutions later, its angular speed is 15 rev/s. Calculate (a) the angular acceleration, (b) the time required to complete the 60 revolutions, (c) the time required to reach the 10 rev/s angular speed, and (d) the number of revolutions from rest until the time the disk reaches the 10 rev/s angular speed.

••15 SSM A wheel has a constant angular acceleration of 3.0 rad/s². During a certain 4.0 s interval, it turns through an angle of 120 rad. Assuming that the wheel started from rest, how long has it been in motion at the start of this 4.0 s interval?

••16 A merry-go-round rotates from rest with an angular acceleration of 1.50 rad/s². How long does it take to rotate through (a) the first 2.00 rev and (b) the next 2.00 rev?

••17 At $t = 0$, a flywheel has an angular velocity of 4.7 rad/s, a constant angular acceleration of -0.25 rad/s², and a reference line at $\theta_0 = 0$. (a) Through what maximum angle θ_{max} will the reference line turn in the positive direction? What are the (b) first and (c) second times the reference line will be at $\theta = \frac{1}{2}\theta_{max}$? At what (d) negative time and (e) positive time will the reference line be at $\theta = 10.5$ rad? (f) Graph θ versus t, and indicate the answers to (a) through (e) on the graph.

sec. 10-5   Relating the Linear and Angular Variables

•18 If an airplane propeller rotates at 2000 rev/min while the airplane flies at a speed of 480 km/h relative to the ground, what is the linear speed of a point on the tip of the propeller, at radius 1.5 m, as seen by (a) the pilot and (b) an observer on the ground? The plane's velocity is parallel to the propeller's axis of rotation.

•19 What are the magnitudes of (a) the angular velocity, (b) the radial acceleration, and (c) the tangential acceleration of a spaceship taking a circular turn of radius 3220 km at a speed of 29 000 km/h?

•20 An object rotates about a fixed axis, and the angular position of a reference line on the object is given by $\theta = 0.40e^{2t}$, where θ is in radians and t is in seconds. Consider a point on the object that is 4.0 cm from the axis of rotation. At $t = 0$, what are the magnitudes of the point's (a) tangential component of acceleration and (b) radial component of acceleration?

•21 ✈ Between 1911 and 1990, the top of the leaning bell tower at Pisa, Italy, moved toward the south at an average rate of 1.2 mm/y. The tower is 55 m tall. In radians per second, what is the average angular speed of the tower's top about its base?

•22 An astronaut is being tested in a centrifuge. The centrifuge has a radius of 10 m and, in starting, rotates according to $\theta = 0.30t^2$, where t is in seconds and θ is in radians. When $t = 5.0$ s, what are the magnitudes of the astronaut's (a) angular velocity, (b) linear velocity, (c) tangential acceleration, and (d) radial acceleration?

•23 SSM WWW A flywheel with a diameter of 1.20 m is rotating at an angular speed of 200 rev/min. (a) What is the angular speed of the flywheel in radians per second? (b) What is the linear speed of a point on the rim of the flywheel? (c) What constant angular acceleration (in revolutions per minute-squared) will increase the wheel's angular speed to 1000 rev/min in 60.0 s? (d) How many revolutions does the wheel make during that 60.0 s?

•24 A vinyl record is played by rotating the record so that an approximately circular groove in the vinyl slides under a stylus. Bumps in the groove run into the stylus, causing it to oscillate. The equipment converts those oscillations to electrical signals and then to sound. Suppose that a record turns at the rate of $33\frac{1}{3}$ rev/min, the groove being played is at a radius of 10.0 cm, and the bumps in the groove are uniformly separated by 1.75 mm. At what rate (hits per second) do the bumps hit the stylus?

••25 SSM (a) What is the angular speed ω about the polar axis of a point on Earth's surface at latitude 40° N? (Earth rotates about that axis.) (b) What is the linear speed v of the point? What are (c) ω and (d) v for a point at the equator?

••26 The flywheel of a steam engine runs with a constant angular velocity of 150 rev/min. When steam is shut off, the friction of the bearings and of the air stops the wheel in 2.2 h. (a) What is the constant angular acceleration, in revolutions per minute-squared, of the wheel during the slowdown? (b) How many revolutions does the wheel make before stopping? (c) At the instant the flywheel is turning at 75 rev/min, what is the tangential component of the linear acceleration of a flywheel particle that is 50 cm from the axis of rotation? (d) What is the magnitude of the net linear acceleration of the particle in (c)?

••27 A record turntable is rotating at $33\frac{1}{3}$ rev/min. A watermelon seed is on the turntable 6.0 cm from the axis of rotation. (a) Calculate the acceleration of the seed, assuming that it does not slip. (b) What is the minimum value of the coefficient of static friction between the seed and the turntable if the seed is not to slip? (c) Suppose that the turntable achieves its angular speed by starting from rest and undergoing a constant angular acceleration for 0.25 s. Calculate the minimum coefficient of static friction required for the seed not to slip during the acceleration period.

••28 In Fig. 10-28, wheel A of radius $r_A = 10$ cm is coupled by belt B to wheel C of radius $r_C = 25$ cm. The angular speed of wheel A is increased from rest at a constant rate

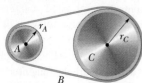

Fig. 10-28 Problem 28.

of 1.6 rad/s². Find the time needed for wheel C to reach an angular speed of 100 rev/min, assuming the belt does not slip. (*Hint:* If the belt does not slip, the linear speeds at the two rims must be equal.)

••29 An early method of measuring the speed of light makes use of a rotating slotted wheel. A beam of light passes through one of the slots at the outside edge of the wheel, as in Fig. 10-29, travels to a distant mirror, and returns to the wheel just in time to pass through the next slot in the wheel. One such slotted wheel has a radius of 5.0 cm and 500 slots around its edge. Measurements taken when the mirror is $L = 500$ m from the wheel indicate a speed of light of 3.0×10^5 km/s. (a) What is the (constant) angular speed of the wheel? (b) What is the linear speed of a point on the edge of the wheel?

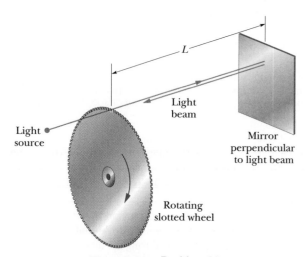

Fig. 10-29 Problem 29.

••30 A gyroscope flywheel of radius 2.83 cm is accelerated from rest at 14.2 rad/s² until its angular speed is 2760 rev/min. (a) What is the tangential acceleration of a point on the rim of the flywheel during this spin-up process? (b) What is the radial acceleration of this point when the flywheel is spinning at full speed? (c) Through what distance does a point on the rim move during the spin-up?

••31 GO A disk, with a radius of 0.25 m, is to be rotated like a merry-go-round through 800 rad, starting from rest, gaining angular speed at the constant rate α_1 through the first 400 rad and then losing angular speed at the constant rate $-\alpha_1$ until it is again at rest. The magnitude of the centripetal acceleration of any portion of the disk is not to exceed 400 m/s². (a) What is the least time required for the rotation? (b) What is the corresponding value of α_1?

•••32 A pulsar is a rapidly rotating neutron star that emits a radio beam the way a lighthouse emits a light beam. We receive a radio pulse for each rotation of the star. The period T of rotation is found by measuring the time between pulses. The pulsar in the Crab nebula has a period of rotation of $T = 0.033$ s that is increasing at the rate of 1.26×10^{-5} s/y. (a) What is the pulsar's angular acceleration α? (b) If α is constant, how many years from now will the pulsar stop rotating? (c) The pulsar originated in a supernova explosion seen in the year 1054. Assuming constant α, find the initial T.

sec. 10-6 Kinetic Energy of Rotation

•33 SSM Calculate the rotational inertia of a wheel that has a kinetic energy of 24 400 J when rotating at 602 rev/min.

•34 Figure 10-30 gives angular speed versus time for a thin rod that rotates around one end. The scale on the ω axis is set by $\omega_s = 6.0$ rad/s. (a) What is the magnitude of the rod's angular acceleration? (b) At $t = 4.0$ s, the rod has a rotational kinetic energy of 1.60 J. What is its kinetic energy at $t = 0$?

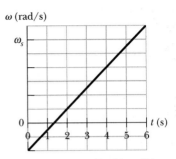

Fig. 10-30 Problem 34.

sec. 10-7 Calculating the Rotational Inertia

•35 SSM Two uniform solid cylinders, each rotating about its central (longitudinal) axis at 235 rad/s, have the same mass of 1.25 kg but differ in radius. What is the rotational kinetic energy of (a) the smaller cylinder, of radius 0.25 m, and (b) the larger cylinder, of radius 0.75 m?

•36 Figure 10-31a shows a disk that can rotate about an axis at a radial distance h from the center of the disk. Figure 10-31b gives the rotational inertia I of the disk about the axis as a function of that distance h, from the center out to the edge of the disk. The scale on the I axis is set by $I_A = 0.050$ kg·m² and $I_B = 0.150$ kg·m². What is the mass of the disk?

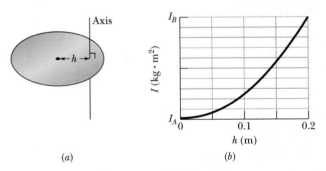

Fig. 10-31 Problem 36.

•37 SSM Calculate the rotational inertia of a meter stick, with mass 0.56 kg, about an axis perpendicular to the stick and located at the 20 cm mark. (Treat the stick as a thin rod.)

•38 Figure 10-32 shows three 0.0100 kg particles that have been glued to a rod of length $L = 6.00$ cm and negligible mass. The assembly can rotate around a perpendicular axis through point O at the left end. If we remove one particle (that is, 33% of the mass), by what percentage does the rotational inertia of the assembly

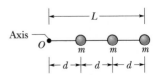

Fig. 10-32 Problems 38 and 62.

around the rotation axis decrease when that removed particle is (a) the innermost one and (b) the outermost one?

••39 Trucks can be run on energy stored in a rotating flywheel, with an electric motor getting the flywheel up to its top speed of 200π rad/s. One such flywheel is a solid, uniform cylinder with a mass of 500 kg and a radius of 1.0 m. (a) What is the kinetic energy of the flywheel after charging? (b) If the truck uses an average power of 8.0 kW, for how many minutes can it operate between chargings?

••40 Figure 10-33 shows an arrangement of 15 identical disks that have been glued together in a rod-like shape of length $L = 1.0000$ m and (total) mass $M = 100.0$ mg. The disk arrangement can rotate about a perpendicular axis through its central disk at point O. (a) What is the rotational inertia of the arrangement about that axis? (b) If we approximated the arrangement as being a uniform rod of mass M and length L, what percentage error would we make in using the formula in Table 10-2e to calculate the rotational inertia?

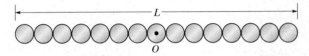

Fig. 10-33 Problem 40.

••41 GO In Fig. 10-34, two particles, each with mass $m = 0.85$ kg, are fastened to each other, and to a rotation axis at O, by two thin rods, each with length $d = 5.6$ cm and mass $M = 1.2$ kg. The combination rotates around the rotation axis with the angular speed $\omega = 0.30$ rad/s. Measured about O, what are the combination's (a) rotational inertia and (b) kinetic energy?

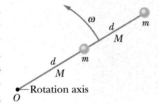

Fig. 10-34 Problem 41.

••42 The masses and coordinates of four particles are as follows: 50 g, $x = 2.0$ cm, $y = 2.0$ cm; 25 g, $x = 0$, $y = 4.0$ cm; 25 g, $x = -3.0$ cm, $y = -3.0$ cm; 30 g, $x = -2.0$ cm, $y = 4.0$ cm. What are the rotational inertias of this collection about the (a) x, (b) y, and (c) z axes? (d) Suppose the answers to (a) and (b) are A and B, respectively. Then what is the answer to (c) in terms of A and B?

••43 SSM WWW The uniform solid block in Fig. 10-35 has mass 0.172 kg and edge lengths $a = 3.5$ cm, $b = 8.4$ cm, and $c = 1.4$ cm. Calculate its rotational inertia about an axis through one corner and perpendicular to the large faces.

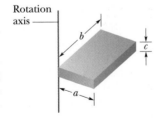

Fig. 10-35 Problem 43.

••44 Four identical particles of mass 0.50 kg each are placed at the vertices of a 2.0 m × 2.0 m square and held there by four massless rods, which form the sides of the square. What is the rotational inertia of this rigid body about an axis that (a) passes through the midpoints of opposite sides and lies in the plane of the square, (b) passes through the midpoint of one of the sides and is perpendicular to the plane of the square, and (c) lies in the plane of the square and passes through two diagonally opposite particles?

sec. 10-8 Torque

•45 SSM ILW The body in Fig. 10-36 is pivoted at O, and two forces act on it as shown. If $r_1 = 1.30$ m, $r_2 = 2.15$ m, $F_1 = 4.20$ N, $F_2 = 4.90$ N, $\theta_1 = 75.0°$, and $\theta_2 = 60.0°$, what is the net torque about the pivot?

Fig. 10-36 Problem 45.

•46 The body in Fig. 10-37 is pivoted at O. Three forces act on it: $F_A = 10$ N at point A, 8.0 m from O; $F_B = 16$ N at B, 4.0 m from O; and $F_C = 19$ N at C, 3.0 m from O. What is the net torque about O?

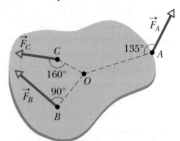

Fig. 10-37 Problem 46.

•47 SSM A small ball of mass 0.75 kg is attached to one end of a 1.25-m-long massless rod, and the other end of the rod is hung from a pivot. When the resulting pendulum is 30° from the vertical, what is the magnitude of the gravitational torque calculated about the pivot?

•48 The length of a bicycle pedal arm is 0.152 m, and a downward force of 111 N is applied to the pedal by the rider. What is the magnitude of the torque about the pedal arm's pivot when the arm is at angle (a) 30°, (b) 90°, and (c) 180° with the vertical?

sec. 10-9 Newton's Second Law for Rotation

•49 SSM ILW During the launch from a board, a diver's angular speed about her center of mass changes from zero to 6.20 rad/s in 220 ms. Her rotational inertia about her center of mass is 12.0 kg·m². During the launch, what are the magnitudes of (a) her average angular acceleration and (b) the average external torque on her from the board?

•50 If a 32.0 N·m torque on a wheel causes angular acceleration 25.0 rad/s², what is the wheel's rotational inertia?

••51 GO In Fig. 10-38, block 1 has mass $m_1 = 460$ g, block 2 has mass $m_2 = 500$ g, and the pulley, which is mounted on a horizontal axle with negligible friction, has radius $R = 5.00$ cm. When released from rest, block 2 falls 75.0 cm in 5.00 s without the cord slipping on the pulley. (a) What is the magnitude of the acceleration of the blocks? What are (b) tension T_2 and (c) tension T_1? (d) What is the magnitude of the pulley's angular acceleration? (e) What is its rotational inertia?

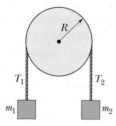

Fig. 10-38
Problems 51 and 83.

••52 GO In Fig. 10-39, a cylinder having a mass of 2.0 kg can rotate about its central axis through point O. Forces are applied as shown: $F_1 = 6.0$ N, $F_2 = 4.0$ N, $F_3 = 2.0$ N, and $F_4 = 5.0$ N. Also, $r = 5.0$ cm and $R = 12$ cm. Find the (a) magnitude and (b) direction of the angular acceleration of the cylinder. (During the rotation, the forces maintain their same angles relative to the cylinder.)

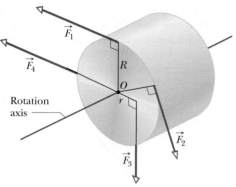

Fig. 10-39 Problem 52.

••53 GO Figure 10-40 shows a uniform disk that can rotate around its center like a merry-go-round. The disk has a radius of 2.00 cm and a mass of 20.0 grams and is initially at rest. Starting at time $t = 0$, two forces are to be applied tangentially to the rim as indicated, so that at time $t = 1.25$ s the disk has an angular velocity of 250 rad/s counterclockwise. Force F_1 has a magnitude of 0.100 N. What is magnitude F_2?

Fig. 10-40 Problem 53.

••54 In a judo foot-sweep move, you sweep your opponent's left foot out from under him while pulling on his gi (uniform) toward that side. As a result, your opponent rotates around his right foot and onto the mat. Figure 10-41 shows a simplified diagram of your opponent as you face him, with his left foot swept out. The rotational axis is through point O. The gravitational force F_g on him effectively acts at his center of mass, which is a horizontal distance $d = 28$ cm from point O. His mass is 70 kg, and his rotational inertia about point O is 65 kg·m². What is the magnitude of his initial angular acceleration about point O if your pull F_a on his gi is (a) negligible and (b) horizontal with a magnitude of 300 N and applied at height $h = 1.4$ m?

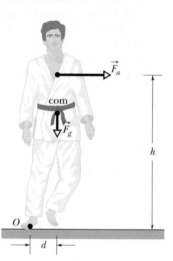

Fig. 10-41 Problem 54.

••55 GO In Fig. 10-42a, an irregularly shaped plastic plate with uniform thickness and density (mass per unit volume) is to be rotated around an axle that is perpendicular to the plate face and through point O. The rotational inertia of the plate about that axle is measured with the following method. A circular disk of mass 0.500 kg and radius 2.00 cm is glued to the plate, with its center aligned with point O (Fig. 10-42b). A string is wrapped around the edge of the disk the way a string is wrapped around a top. Then the string is pulled for 5.00 s. As a result, the disk

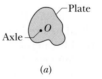

(a)

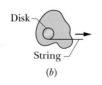

String
(b)

Fig. 10-42
Problem 55.

and plate are rotated by a constant force of 0.400 N that is applied by the string tangentially to the edge of the disk. The resulting angular speed is 114 rad/s. What is the rotational inertia of the plate about the axle?

••56 Figure 10-43 shows particles 1 and 2, each of mass m, attached to the ends of a rigid massless rod of length $L_1 + L_2$, with $L_1 = 20$ cm and $L_2 = 80$ cm. The rod is held horizontally on the fulcrum and then released. What are the magnitudes of the initial accelerations of (a) particle 1 and (b) particle 2?

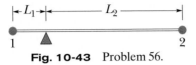

Fig. 10-43   Problem 56.

•••57 A pulley, with a rotational inertia of 1.0×10^{-3} kg·m² about its axle and a radius of 10 cm, is acted on by a force applied tangentially at its rim. The force magnitude varies in time as $F = 0.50t + 0.30t^2$, with F in newtons and t in seconds. The pulley is initially at rest. At $t = 3.0$ s what are its (a) angular acceleration and (b) angular speed?

sec. 10-10   Work and Rotational Kinetic Energy

•58 (a) If $R = 12$ cm, $M = 400$ g, and $m = 50$ g in Fig. 10-18, find the speed of the block after it has descended 50 cm starting from rest. Solve the problem using energy conservation principles. (b) Repeat (a) with $R = 5.0$ cm.

•59 An automobile crankshaft transfers energy from the engine to the axle at the rate of 100 hp ($= 74.6$ kW) when rotating at a speed of 1800 rev/min. What torque (in newton-meters) does the crankshaft deliver?

•60 A thin rod of length 0.75 m and mass 0.42 kg is suspended freely from one end. It is pulled to one side and then allowed to swing like a pendulum, passing through its lowest position with angular speed 4.0 rad/s. Neglecting friction and air resistance, find (a) the rod's kinetic energy at its lowest position and (b) how far above that position the center of mass rises.

•61 A 32.0 kg wheel, essentially a thin hoop with radius 1.20 m, is rotating at 280 rev/min. It must be brought to a stop in 15.0 s. (a) How much work must be done to stop it? (b) What is the required average power?

••62 In Fig. 10-32, three 0.0100 kg particles have been glued to a rod of length $L = 6.00$ cm and negligible mass and can rotate around a perpendicular axis through point O at one end. How much work is required to change the rotational rate (a) from 0 to 20.0 rad/s, (b) from 20.0 rad/s to 40.0 rad/s, and (c) from 40.0 rad/s to 60.0 rad/s? (d) What is the slope of a plot of the assembly's kinetic energy (in joules) versus the square of its rotation rate (in radians-squared per second-squared)?

••63 SSM ILW A meter stick is held vertically with one end on the floor and is then allowed to fall. Find the speed of the other end just before it hits the floor, assuming that the end on the floor does not slip. (*Hint:* Consider the stick to be a thin rod and use the conservation of energy principle.)

••64 A uniform cylinder of radius 10 cm and mass 20 kg is mounted so as to rotate freely about a horizontal axis that is parallel to and 5.0 cm from the central longitudinal axis of the cylinder. (a) What is the rotational inertia of the cylinder about the axis of rotation? (b) If the cylinder is released from rest with its central longitudinal axis at the same height as the axis about which the cylinder rotates, what is the angular speed of the cylinder as it passes through its lowest position?

•••65 A tall, cylindrical chimney falls over when its base is ruptured. Treat the chimney as a thin rod of length 55.0 m. At the instant it makes an angle of 35.0° with the vertical as it falls, what are (a) the radial acceleration of the top, and (b) the tangential acceleration of the top. (*Hint:* Use energy considerations, not a torque.) (c) At what angle θ is the tangential acceleration equal to g?

•••66 A uniform spherical shell of mass $M = 4.5$ kg and radius $R = 8.5$ cm can rotate about a vertical axis on frictionless bearings (Fig. 10-44). A massless cord passes around the equator of the shell, over a pulley of rotational inertia $I = 3.0 \times 10^{-3}$ kg·m² and radius $r = 5.0$ cm, and is attached to a small object of mass $m = 0.60$ kg. There is no friction on the pulley's axle; the cord does not slip on the pulley. What is the speed of the object when it has fallen 82 cm after being released from rest? Use energy considerations.

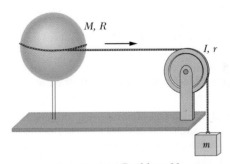

Fig. 10-44   Problem 66.

•••67 GO Figure 10-45 shows a rigid assembly of a thin hoop (of mass m and radius $R = 0.150$ m) and a thin radial rod (of mass m and length $L = 2.00R$). The assembly is upright, but if we give it a slight nudge, it will rotate around a horizontal axis in the plane of the rod and hoop, through the lower end of the rod. Assuming that the energy given to the assembly in such a nudge is negligible, what would be the assembly's angular speed about the rotation axis when it passes through the upside-down (inverted) orientation?

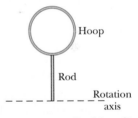

Fig. 10-45   Problem 67.

Additional Problems

68 Two uniform solid spheres have the same mass of 1.65 kg, but one has a radius of 0.226 m and the other has a radius of 0.854 m. Each can rotate about an axis through its center. (a) What is the magnitude τ of the torque required to bring the smaller sphere from rest to an angular speed of 317 rad/s in 15.5 s? (b) What is the magnitude F of the force that must be applied tangentially at the sphere's equator to give that torque? What are the corresponding values of (c) τ and (d) F for the larger sphere?

69 In Fig. 10-46, a small disk of radius $r = 2.00$ cm has been glued to the edge of a larger disk of radius $R = 4.00$ cm so that the disks lie in the same plane. The disks can be rotated around a perpendicular axis through point O at the center of the larger disk. The disks both have a uniform density (mass per unit

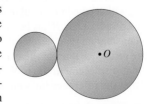

Fig. 10-46   Problem 69.

volume) of 1.40×10^3 kg/m³ and a uniform thickness of 5.00 mm. What is the rotational inertia of the two-disk assembly about the rotation axis through O?

70 A wheel, starting from rest, rotates with a constant angular acceleration of 2.00 rad/s². During a certain 3.00 s interval, it turns through 90.0 rad. (a) What is the angular velocity of the wheel at the start of the 3.00 s interval? (b) How long has the wheel been turning before the start of the 3.00 s interval?

71 SSM In Fig. 10-47, two 6.20 kg blocks are connected by a massless string over a pulley of radius 2.40 cm and rotational inertia 7.40×10^{-4} kg·m². The string does not slip on the pulley; it is not known whether there is friction between the table and the sliding block; the pulley's axis is frictionless. When this system is released from rest, the pulley turns through 0.650 rad in 91.0 ms and the acceleration of the blocks is constant. What are (a) the magnitude of the pulley's angular acceleration, (b) the magnitude of either block's acceleration, (c) string tension T_1, and (d) string tension T_2?

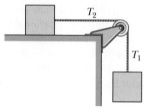

Fig. 10-47 Problem 71.

72 Attached to each end of a thin steel rod of length 1.20 m and mass 6.40 kg is a small ball of mass 1.06 kg. The rod is constrained to rotate in a horizontal plane about a vertical axis through its midpoint. At a certain instant, it is rotating at 39.0 rev/s. Because of friction, it slows to a stop in 32.0 s. Assuming a constant retarding torque due to friction, compute (a) the angular acceleration, (b) the retarding torque, (c) the total energy transferred from mechanical energy to thermal energy by friction, and (d) the number of revolutions rotated during the 32.0 s. (e) Now suppose that the retarding torque is known not to be constant. If any of the quantities (a), (b), (c), and (d) can still be computed without additional information, give its value.

73 A uniform helicopter rotor blade is 7.80 m long, has a mass of 110 kg, and is attached to the rotor axle by a single bolt. (a) What is the magnitude of the force on the bolt from the axle when the rotor is turning at 320 rev/min? (*Hint:* For this calculation the blade can be considered to be a point mass at its center of mass. Why?) (b) Calculate the torque that must be applied to the rotor to bring it to full speed from rest in 6.70 s. Ignore air resistance. (The blade cannot be considered to be a point mass for this calculation. Why not? Assume the mass distribution of a uniform thin rod.) (c) How much work does the torque do on the blade in order for the blade to reach a speed of 320 rev/min?

74 *Racing disks.* Figure 10-48 shows two disks that can rotate about their centers like a merry-go-round. At time $t = 0$, the reference lines of the two disks have the same orientation. Disk A is already rotating, with a constant angular velocity of 9.5 rad/s. Disk B has been stationary but now begins to rotate at a constant angular acceleration of 2.2 rad/s². (a) At what time t will the reference lines of the two disks momentarily have the same angular displacement θ? (b) Will that time t be the first time since $t = 0$ that the reference lines are momentarily aligned?

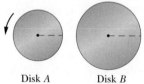

Fig. 10-48 Problem 74.

75 A high-wire walker always attempts to keep his center of mass over the wire (or rope). He normally carries a long, heavy

pole to help: If he leans, say, to his right (his com moves to the right) and is in danger of rotating around the wire, he moves the pole to his left (its com moves to the left) to slow the rotation and allow himself time to adjust his balance. Assume that the walker has a mass of 70.0 kg and a rotational inertia of 15.0 kg·m² about the wire. What is the magnitude of his angular acceleration about the wire if his com is 5.0 cm to the right of the wire and (a) he carries no pole and (b) the 14.0 kg pole he carries has its com 10 cm to the left of the wire?

76 Starting from rest at $t = 0$, a wheel undergoes a constant angular acceleration. When $t = 2.0$ s, the angular velocity of the wheel is 5.0 rad/s. The acceleration continues until $t = 20$ s, when it abruptly ceases. Through what angle does the wheel rotate in the interval $t = 0$ to $t = 40$ s?

77 SSM A record turntable rotating at $33\frac{1}{3}$ rev/min slows down and stops in 30 s after the motor is turned off. (a) Find its (constant) angular acceleration in revolutions per minute-squared. (b) How many revolutions does it make in this time?

78 GO A rigid body is made of three identical thin rods, each with length $L = 0.600$ m, fastened together in the form of a letter **H** (Fig. 10-49). The body is free to rotate about a horizontal axis that runs along the length of one of the legs of the **H**. The body is allowed to fall from rest from a position in which the plane of the **H** is horizontal. What is the angular speed of the body when the plane of the **H** is vertical?

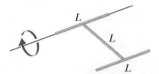

Fig. 10-49 Problem 78.

79 SSM (a) Show that the rotational inertia of a solid cylinder of mass M and radius R about its central axis is equal to the rotational inertia of a thin hoop of mass M and radius $R/\sqrt{2}$ about its central axis. (b) Show that the rotational inertia I of any given body of mass M about any given axis is equal to the rotational inertia of an *equivalent hoop* about that axis, if the hoop has the same mass M and a radius k given by

$$k = \sqrt{\frac{I}{M}}.$$

The radius k of the equivalent hoop is called the *radius of gyration* of the given body.

80 A disk rotates at constant angular acceleration, from angular position $\theta_1 = 10.0$ rad to angular position $\theta_2 = 70.0$ rad in 6.00 s. Its angular velocity at θ_2 is 15.0 rad/s. (a) What was its angular velocity at θ_1? (b) What is the angular acceleration? (c) At what angular position was the disk initially at rest? (d) Graph θ versus time t and angular speed ω versus t for the disk, from the beginning of the motion (let $t = 0$ then).

81 The thin uniform rod in Fig. 10-50 has length 2.0 m and can pivot about a horizontal, frictionless pin through one end. It is released from rest at angle $\theta = 40°$ above the horizontal. Use the principle of conservation of energy to determine the angular speed of the rod as it passes through the horizontal position.

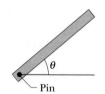

Fig. 10-50 Problem 81.

82 George Washington Gale Ferris, Jr., a civil engineering graduate from Rensselaer Polytechnic Institute, built the original Ferris wheel for the 1893 World's Columbian Exposition in Chicago. The wheel, an astounding engineering construction at the time, carried 36 wooden cars, each holding up to 60 passengers, around a circle 76 m in diameter. The cars were loaded 6 at a time, and once all 36 cars were full, the wheel made a complete rotation at constant angular speed in about 2 min. Estimate the amount of work that was required of the machinery to rotate the passengers alone.

83 In Fig. 10-38, two blocks, of mass $m_1 = 400$ g and $m_2 = 600$ g, are connected by a massless cord that is wrapped around a uniform disk of mass $M = 500$ g and radius $R = 12.0$ cm. The disk can rotate without friction about a fixed horizontal axis through its center; the cord cannot slip on the disk. The system is released from rest. Find (a) the magnitude of the acceleration of the blocks, (b) the tension T_1 in the cord at the left, and (c) the tension T_2 in the cord at the right.

84 At 7:14 A.M. on June 30, 1908, a huge explosion occurred above remote central Siberia, at latitude 61° N and longitude 102° E; the fireball thus created was the brightest flash seen by anyone before nuclear weapons. The *Tunguska Event,* which according to one chance witness "covered an enormous part of the sky," was probably the explosion of a *stony asteroid* about 140 m wide. (a) Considering only Earth's rotation, determine how much later the asteroid would have had to arrive to put the explosion above Helsinki at longitude 25° E. This would have obliterated the city. (b) If the asteroid had, instead, been a *metallic asteroid,* it could have reached Earth's surface. How much later would such an asteroid have had to arrive to put the impact in the Atlantic Ocean at longitude 20° W? (The resulting tsunamis would have wiped out coastal civilization on both sides of the Atlantic.)

85 A golf ball is launched at an angle of 20° to the horizontal, with a speed of 60 m/s and a rotation rate of 90 rad/s. Neglecting air drag, determine the number of revolutions the ball makes by the time it reaches maximum height.

86 Figure 10-51 shows a flat construction of two circular rings that have a common center and are held together by three rods of negligible mass. The construction, which is initially at rest, can rotate around the common center (like a merry-go-round), where another rod of negligible mass lies. The mass, inner radius, and outer radius of the

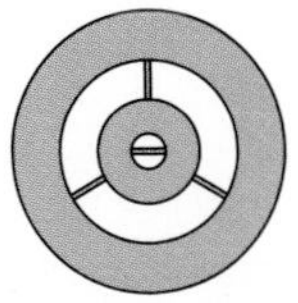

Fig. 10-51
Problem 86.

rings are given in the following table. A tangential force of magnitude 12.0 N is applied to the outer edge of the outer ring for 0.300 s. What is the change in the angular speed of the construction during that time interval?

Ring	Mass (kg)	Inner Radius (m)	Outer Radius (m)
1	0.120	0.0160	0.0450
2	0.240	0.0900	0.1400

87 In Fig. 10-52, a wheel of radius 0.20 m is mounted on a frictionless horizontal axle. A massless cord is wrapped around the wheel and attached to a 2.0 kg box that slides on a frictionless surface inclined at angle $\theta = 20°$ with the horizontal. The box accelerates down the surface at 2.0 m/s². What is the rotational inertia of the wheel about the axle?

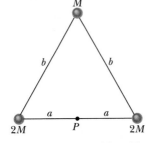

Fig. 10-52 Problem 87.

88 A thin spherical shell has a radius of 1.90 m. An applied torque of 960 N · m gives the shell an angular acceleration of 6.20 rad/s² about an axis through the center of the shell. What are (a) the rotational inertia of the shell about that axis and (b) the mass of the shell?

89 A bicyclist of mass 70 kg puts all his mass on each downward-moving pedal as he pedals up a steep road. Take the diameter of the circle in which the pedals rotate to be 0.40 m, and determine the magnitude of the maximum torque he exerts about the rotation axis of the pedals.

90 The flywheel of an engine is rotating at 25.0 rad/s. When the engine is turned off, the flywheel slows at a constant rate and stops in 20.0 s. Calculate (a) the angular acceleration of the flywheel, (b) the angle through which the flywheel rotates in stopping, and (c) the number of revolutions made by the flywheel in stopping.

91 SSM In Fig. 10-18a, a wheel of radius 0.20 m is mounted on a frictionless horizontal axis. The rotational inertia of the wheel about the axis is 0.40 kg · m². A massless cord wrapped around the wheel's circumference is attached to a 6.0 kg box. The system is released from rest. When the box has a kinetic energy of 6.0 J, what are (a) the wheel's rotational kinetic energy and (b) the distance the box has fallen?

92 Our Sun is 2.3×10^4 ly (light-years) from the center of our Milky Way galaxy and is moving in a circle around that center at a speed of 250 km/s. (a) How long does it take the Sun to make one revolution about the galactic center? (b) How many revolutions has the Sun completed since it was formed about 4.5×10^9 years ago?

93 SSM A wheel of radius 0.20 m is mounted on a frictionless horizontal axis. The rotational inertia of the wheel about the axis is 0.050 kg · m². A massless cord wrapped around the wheel is attached to a 2.0 kg block that slides on a horizontal frictionless surface. If a horizontal

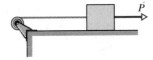

Fig. 10-53 Problem 93.

force of magnitude $P = 3.0$ N is applied to the block as shown in Fig. 10-53, what is the magnitude of the angular acceleration of the wheel? Assume the cord does not slip on the wheel.

94 A car starts from rest and moves around a circular track of radius 30.0 m. Its speed increases at the constant rate of 0.500 m/s². (a) What is the magnitude of its *net* linear acceleration 15.0 s later? (b) What angle does this net acceleration vector make with the car's velocity at this time?

95 The rigid body shown in Fig. 10-54 consists of three particles connected by massless rods. It is to be rotated about an axis perpendicular to its plane through point P. If $M = 0.40$ kg, $a = 30$ cm, and $b = 50$ cm, how much work is required to take the body from rest to an angular speed of 5.0 rad/s?

96 *Beverage engineering.* The pull tab was a major advance in the engineering design of beverage containers. The tab pivots on a central bolt in the can's top. When you pull upward on one end of the tab, the other end presses downward on a portion of the can's top that has been scored. If you pull upward with a 10 N force, approximately what is the magnitude of the force applied to the scored section? (You will need to examine a can with a pull tab.)

Fig. 10-54 Problem 95.

97 Figure 10-55 shows a propeller blade that rotates at 2000 rev/min about a perpendicular axis at point B. Point A is at the outer tip of the blade, at radial distance 1.50 m. (a) What is the difference in the magnitudes a of the centripetal acceleration of point A and of a point at radial distance 0.150 m? (b) Find the slope of a plot of a versus radial distance along the blade.

Fig. 10-55
Problem 97.

98 A yo-yo-shaped device mounted on a horizontal frictionless axis is used to lift a 30 kg box as shown in Fig. 10-56. The outer radius R of the device is 0.50 m, and the radius r of the hub is 0.20 m. When a constant horizontal force $\vec{F}_{app}$ of magnitude 140 N is applied to a rope wrapped around the outside of the device, the box, which is suspended from a rope wrapped around the hub, has an upward acceleration of magnitude 0.80 m/s². What is the rotational inertia of the device about its axis of rotation?

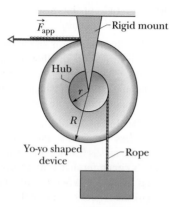

Fig. 10-56 Problem 98.

99 A small ball with mass 1.30 kg is mounted on one end of a rod 0.780 m long and of negligible mass. The system rotates in a horizontal circle about the other end of the rod at 5010 rev/min. (a) Calculate the rotational inertia of the system about the axis of rotation. (b) There is an air drag of 2.30×10^{-2} N on the ball, directed opposite its motion. What torque must be applied to the system to keep it rotating at constant speed?

100 Two thin rods (each of mass 0.20 kg) are joined together to form a rigid body as shown in Fig. 10-57. One of the rods has length $L_1 = 0.40$ m, and the other has length $L_2 = 0.50$ m. What is the rotational inertia of this rigid body about (a) an axis that is perpendicular to the plane of the paper and passes through the center of the shorter rod and (b) an axis that is perpendicular to the plane of the paper and passes through the center of the longer rod?

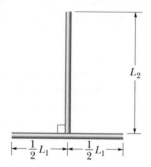

Fig. 10-57 Problem 100.

101 In Fig. 10-58, four pulleys are connected by two belts. Pulley A (radius 15 cm) is the drive pulley, and it rotates at 10 rad/s. Pulley B (radius 10 cm) is connected by belt 1 to pulley A. Pulley B' (radius 5 cm) is concentric with pulley B and is rigidly attached to it. Pulley C (radius 25 cm) is connected by belt 2 to pulley B'. Calculate (a) the linear speed of a point on belt 1, (b) the an-

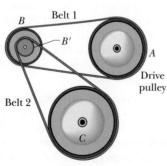

Fig. 10-58 Problem 101.

gular speed of pulley B, (c) the angular speed of pulley B', (d) the linear speed of a point on belt 2, and (e) the angular speed of pulley C. (*Hint:* If the belt between two pulleys does not slip, the linear speeds at the rims of the two pulleys must be equal.)

102 The rigid object shown in Fig. 10-59 consists of three balls and three connecting rods, with $M = 1.6$ kg, $L = 0.60$ m, and $\theta = 30°$. The balls may be treated as particles, and the connecting rods have negligible mass. Determine the rotational kinetic energy of the object if it has an angular speed of 1.2 rad/s about (a) an axis that passes through point P and is perpendicular to the plane of the figure and (b) an axis that passes through point P, is perpendicular to the rod of length $2L$, and lies in the plane of the figure.

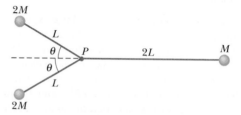

Fig. 10-59 Problem 102.

103 In Fig. 10-60, a thin uniform rod (mass 3.0 kg, length 4.0 m) rotates freely about a horizontal axis A that is perpendicular to the rod and passes through a point at distance $d = 1.0$ m from the end of the rod. The kinetic energy of the rod as it passes through the vertical position is 20 J. (a) What is the rotational inertia of the rod about axis A? (b) What is the (linear) speed of the end B of the rod as the rod passes through the vertical position? (c) At what angle θ will the rod momentarily stop in its upward swing?

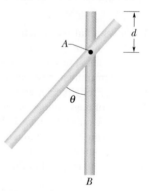

Fig. 10-60 Problem 103.

104 Four particles, each of mass, 0.20 kg, are placed at the vertices of a square with sides of length 0.50 m. The particles are connected by rods of negligible mass. This rigid body can rotate in a vertical plane about a horizontal axis A that passes through one of the particles. The body is released from rest with rod AB horizontal (Fig. 10-61). (a) What is the rotational inertia of the body about axis A? (b) What is the angular speed of the body about axis A when rod AB swings through the vertical position?

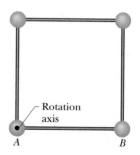

Fig. 10-61 Problem 104.

ROLLING, TORQUE, AND ANGULAR MOMENTUM

11

11-1 | WHAT IS PHYSICS?

As we discussed in Chapter 10, physics includes the study of rotation. Arguably, the most important application of that physics is in the rolling motion of wheels and wheel-like objects. This applied physics has long been used. For example, when the prehistoric people of Easter Island moved their gigantic stone statues from the quarry and across the island, they dragged them over logs acting as rollers. Much later, when settlers moved westward across America in the 1800s, they rolled their possessions first by wagon and then later by train. Today, like it or not, the world is filled with cars, trucks, motorcycles, bicycles, and other rolling vehicles.

The physics and engineering of rolling have been around for so long that you might think no fresh ideas remain to be developed. However, skateboards and in-line skates were invented and engineered fairly recently, to become huge financial successes. Street luge is now catching on, and the self-righting Segway (Fig. 11-1) may change the way people move around in large cities. Applying the physics of rolling can still lead to surprises and rewards. Our starting point in exploring that physics is to simplify rolling motion.

Fig. 11-1 The self-righting Segway Human Transporter. *(Justin Sullivan/Getty Images News and Sport Services)*

11-2 Rolling as Translation and Rotation Combined

Here we consider only objects that *roll smoothly* along a surface; that is, the objects roll without slipping or bouncing on the surface. Figure 11-2 shows how complicated smooth rolling motion can be: Although the center of the object moves in a straight line parallel to the surface, a point on the rim certainly does not. However, we can study this motion by treating it as a combination of translation of the center of mass and rotation of the rest of the object around that center.

Fig. 11-2 A time-exposure photograph of a rolling disk. Small lights have been attached to the disk, one at its center and one at its edge. The latter traces out a curve called a *cycloid*. *(Richard Megna/Fundamental Photographs)*

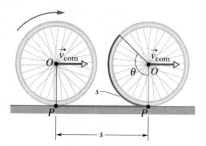

Fig. 11-3 The center of mass O of a rolling wheel moves a distance s at velocity $\vec{v}_{com}$ while the wheel rotates through angle θ. The point P at which the wheel makes contact with the surface over which the wheel rolls also moves a distance s.

To see how we do this, pretend you are standing on a sidewalk watching the bicycle wheel of Fig. 11-3 as it rolls along a street. As shown, you see the center of mass O of the wheel move forward at constant speed v_{com}. The point P on the street where the wheel makes contact with the street surface also moves forward at speed v_{com}, so that P always remains directly below O.

During a time interval t, you see both O and P move forward by a distance s. The bicycle rider sees the wheel rotate through an angle θ about the center of the wheel, with the point of the wheel that was touching the street at the beginning of t moving through arc length s. Equation 10-17 relates the arc length s to the rotation angle θ:

$$s = \theta R, \tag{11-1}$$

where R is the radius of the wheel. The linear speed v_{com} of the center of the wheel (the center of mass of this uniform wheel) is ds/dt. The angular speed ω of the wheel about its center is $d\theta/dt$. Thus, differentiating Eq. 11-1 with respect to time (with R held constant) gives us

$$v_{com} = \omega R \qquad \text{(smooth rolling motion).} \tag{11-2}$$

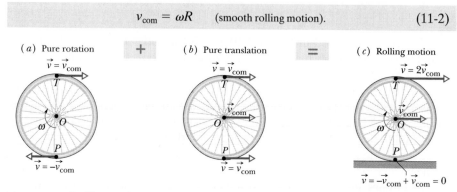

Fig. 11-4 Rolling motion of a wheel as a combination of purely rotational motion and purely translational motion. (a) The purely rotational motion: All points on the wheel move with the same angular speed ω. Points on the outside edge of the wheel all move with the same linear speed $v = v_{com}$. The linear velocities $\vec{v}$ of two such points, at top (T) and bottom (P) of the wheel, are shown. (b) The purely translational motion: All points on the wheel move to the right with the same linear velocity $\vec{v}_{com}$. (c) The rolling motion of the wheel is the combination of (a) and (b).

Figure 11-4 shows that the rolling motion of a wheel is a combination of purely translational and purely rotational motions. Figure 11-4a shows the purely rotational motion (as if the rotation axis through the center were stationary): Every point on the wheel rotates about the center with angular speed ω. (This is the type of motion we considered in Chapter 10.) Every point on the outside edge of the wheel has linear speed v_{com} given by Eq. 11-2. Figure 11-4b shows the purely translational motion (as if the wheel did not rotate at all): Every point on the wheel moves to the right with speed v_{com}.

The combination of Figs. 11-4a and 11-4b yields the actual rolling motion of the wheel, Fig. 11-4c. Note that in this combination of motions, the portion of the wheel at the bottom (at point P) is stationary and the portion at the top (at point T) is moving at speed $2v_{com}$, faster than any other portion of the wheel. These results are demonstrated in Fig. 11-5, which is a time exposure of a rolling bicycle wheel. You can tell that the wheel is moving faster near its top than near its bottom because the spokes are more blurred at the top than at the bottom.

The motion of any round body rolling smoothly over a surface can be separated into purely rotational and purely translational motions, as in Figs. 11-4a and 11-4b.

Fig. 11-5 A photograph of a rolling bicycle wheel. The spokes near the wheel's top are more blurred than those near the bottom because the top ones are moving faster, as Fig. 11-4c shows. (*Courtesy Alice Halliday*)

Rolling as Pure Rotation

Figure 11-6 suggests another way to look at the rolling motion of a wheel—namely, as pure rotation about an axis that always extends through the point where the wheel contacts the street as the wheel moves. We consider the rolling motion to be pure rotation about an axis passing through point P in Fig. 11-4c and perpendicular to the plane of the figure. The vectors in Fig. 11-6 then represent the instantaneous velocities of points on the rolling wheel.

Question: What angular speed about this new axis will a stationary observer assign to a rolling bicycle wheel?

Answer: The same ω that the rider assigns to the wheel as she or he observes it in pure rotation about an axis through its center of mass.

To verify this answer, let us use it to calculate the linear speed of the top of the rolling wheel from the point of view of a stationary observer. If we call the wheel's radius R, the top is a distance $2R$ from the axis through P in Fig. 11-6, so the linear speed at the top should be (using Eq. 11-2)

$$v_{top} = (\omega)(2R) = 2(\omega R) = 2v_{com},$$

in exact agreement with Fig. 11-4c. You can similarly verify the linear speeds shown for the portions of the wheel at points O and P in Fig. 11-4c.

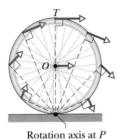

Rotation axis at P

Fig. 11-6 Rolling can be viewed as pure rotation, with angular speed ω, about an axis that always extends through P. The vectors show the instantaneous linear velocities of selected points on the rolling wheel. You can obtain the vectors by combining the translational and rotational motions as in Fig. 11-4.

CHECKPOINT 1

The rear wheel on a clown's bicycle has twice the radius of the front wheel. (a) When the bicycle is moving, is the linear speed at the very top of the rear wheel greater than, less than, or the same as that of the very top of the front wheel? (b) Is the angular speed of the rear wheel greater than, less than, or the same as that of the front wheel?

11-3 The Kinetic Energy of Rolling

Let us now calculate the kinetic energy of the rolling wheel as measured by the stationary observer. If we view the rolling as pure rotation about an axis through P in Fig. 11-6, then from Eq. 10-34 we have

$$K = \tfrac{1}{2}I_P\omega^2, \tag{11-3}$$

in which ω is the angular speed of the wheel and I_P is the rotational inertia of the wheel about the axis through P. From the parallel-axis theorem of Eq. 10-36 ($I = I_{com} + Mh^2$), we have

$$I_P = I_{com} + MR^2, \tag{11-4}$$

in which M is the mass of the wheel, I_{com} is its rotational inertia about an axis through its center of mass, and R (the wheel's radius) is the perpendicular distance h. Substituting Eq. 11-4 into Eq. 11-3, we obtain

$$K = \tfrac{1}{2}I_{com}\omega^2 + \tfrac{1}{2}MR^2\omega^2,$$

and using the relation $v_{com} = \omega R$ (Eq. 11-2) yields

$$K = \tfrac{1}{2}I_{com}\omega^2 + \tfrac{1}{2}Mv_{com}^2. \tag{11-5}$$

We can interpret the term $\tfrac{1}{2}I_{com}\omega^2$ as the kinetic energy associated with the rotation of the wheel about an axis through its center of mass (Fig. 11-4a), and the term $\tfrac{1}{2}Mv_{com}^2$ as the kinetic energy associated with the translational motion of the wheel's center of mass (Fig. 11-4b). Thus, we have the following rule:

 A rolling object has two types of kinetic energy: a rotational kinetic energy $(\tfrac{1}{2}I_{com}\omega^2)$ due to its rotation about its center of mass and a translational kinetic energy $(\tfrac{1}{2}Mv_{com}^2)$ due to translation of its center of mass..

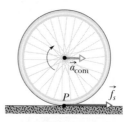

Fig. 11-7 A wheel rolls horizontally without sliding while accelerating with linear acceleration $\vec{a}_{com}$. A static frictional force $\vec{f}_s$ acts on the wheel at P, opposing its tendency to slide.

11-4 The Forces of Rolling

Friction and Rolling

If a wheel rolls at constant speed, as in Fig. 11-3, it has no tendency to slide at the point of contact P, and thus no frictional force acts there. However, if a net force acts on the rolling wheel to speed it up or to slow it, then that net force causes acceleration $\vec{a}_{com}$ of the center of mass along the direction of travel. It also causes the wheel to rotate faster or slower, which means it causes an angular acceleration α. These accelerations tend to make the wheel slide at P. Thus, a frictional force must act on the wheel at P to oppose that tendency.

If the wheel *does not* slide, the force is a *static* frictional force $\vec{f}_s$ and the motion is smooth rolling. We can then relate the magnitudes of the linear acceleration $\vec{a}_{com}$ and the angular acceleration α by differentiating Eq. 11-2 with respect to time (with R held constant). On the left side, dv_{com}/dt is a_{com}, and on the right side $d\omega/dt$ is α. So, for smooth rolling we have

$$a_{com} = \alpha R \qquad \text{(smooth rolling motion).} \qquad (11\text{-}6)$$

If the wheel *does* slide when the net force acts on it, the frictional force that acts at P in Fig. 11-3 is a *kinetic* frictional force $\vec{f}_k$. The motion then is not smooth rolling, and Eq. 11-6 does not apply to the motion. In this chapter we discuss only smooth rolling motion.

Figure 11-7 shows an example in which a wheel is being made to rotate faster while rolling to the right along a flat surface, as on a bicycle at the start of a race. The faster rotation tends to make the bottom of the wheel slide to the left at point P. A frictional force at P, directed to the right, opposes this tendency to slide. If the wheel does not slide, that frictional force is a static frictional force $\vec{f}_s$ (as shown), the motion is smooth rolling, and Eq. 11-6 applies to the motion. (Without friction, bicycle races would be stationary and very boring.)

If the wheel in Fig. 11-7 were made to rotate slower, as on a slowing bicycle, we would change the figure in two ways: The directions of the center-of-mass acceleration $\vec{a}_{com}$ and the frictional force $\vec{f}_s$ at point P would now be to the left.

Rolling Down a Ramp

Figure 11-8 shows a round uniform body of mass M and radius R rolling smoothly down a ramp at angle θ, along an x axis. We want to find an expression for the body's acceleration $a_{com,x}$ down the ramp. We do this by using Newton's second law in both its linear version ($F_{net} = Ma$) and its angular version ($\tau_{net} = I\alpha$).

We start by drawing the forces on the body as shown in Fig. 11-8:

1. The gravitational force $\vec{F}_g$ on the body is directed downward. The tail of the vector is placed at the center of mass of the body. The component along the ramp is $F_g \sin\theta$, which is equal to $Mg \sin\theta$.

2. A normal force $\vec{F}_N$ is perpendicular to the ramp. It acts at the point of contact P, but in Fig. 11-8 the vector has been shifted along its direction until its tail is at the body's center of mass.

3. A static frictional force $\vec{f}_s$ acts at the point of contact P and is directed up the ramp. (Do you see why? If the body were to slide at P, it would slide *down* the ramp. Thus, the frictional force opposing the sliding must be *up* the ramp.)

We can write Newton's second law for components along the x axis in Fig. 11-8 ($F_{net,x} = ma_x$) as

$$f_s - Mg \sin\theta = Ma_{com,x}. \qquad (11\text{-}7)$$

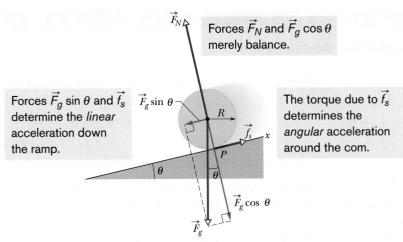

Fig. 11-8 A round uniform body of radius R rolls down a ramp. The forces that act on it are the gravitational force $\vec{F}_g$, a normal force $\vec{F}_N$, and a frictional force $\vec{f}_s$ pointing up the ramp. (For clarity, vector $\vec{F}_N$ has been shifted in the direction it points until its tail is at the center of the body.)

This equation contains two unknowns, f_s and $a_{com,x}$. (We should *not* assume that f_s is at its maximum value $f_{s,max}$. All we know is that the value of f_s is just right for the body to roll smoothly down the ramp, without sliding.)

We now wish to apply Newton's second law in angular form to the body's rotation about its center of mass. First, we shall use Eq. 10-41 ($\tau = r_\perp F$) to write the torques on the body about that point. The frictional force $\vec{f}_s$ has moment arm R and thus produces a torque Rf_s, which is positive because it tends to rotate the body counterclockwise in Fig. 11-8. Forces $\vec{F}_g$ and $\vec{F}_N$ have zero moment arms about the center of mass and thus produce zero torques. So we can write the angular form of Newton's second law ($\tau_{net} = I\alpha$) about an axis through the body's center of mass as

$$Rf_s = I_{com}\alpha. \qquad (11\text{-}8)$$

This equation contains two unknowns, f_s and α.

Because the body is rolling smoothly, we can use Eq. 11-6 ($a_{com} = \alpha R$) to relate the unknowns $a_{com,x}$ and α. But we must be cautious because here $a_{com,x}$ is negative (in the negative direction of the x axis) and α is positive (counterclockwise). Thus we substitute $-a_{com,x}/R$ for α in Eq. 11-8. Then, solving for f_s, we obtain

$$f_s = -I_{com}\frac{a_{com,x}}{R^2}. \qquad (11\text{-}9)$$

Substituting the right side of Eq. 11-9 for f_s in Eq. 11-7, we then find

$$a_{com,x} = -\frac{g \sin \theta}{1 + I_{com}/MR^2}. \qquad (11\text{-}10)$$

We can use this equation to find the linear acceleration $a_{com,x}$ of any body rolling along an incline of angle θ with the horizontal.

✔ CHECKPOINT 2

Disks A and B are identical and roll across a floor with equal speeds. Then disk A rolls up an incline, reaching a maximum height h, and disk B moves up an incline that is identical except that it is frictionless. Is the maximum height reached by disk B greater than, less than, or equal to h?

Ball rolling down a ramp

A uniform ball, of mass $M = 6.00$ kg and radius R, rolls smoothly from rest down a ramp at angle $\theta = 30.0°$ (Fig. 11-8).

(a) The ball descends a vertical height $h = 1.20$ m to reach the bottom of the ramp. What is its speed at the bottom?

KEY IDEAS

The mechanical energy E of the ball–Earth system is conserved as the ball rolls down the ramp. The reason is that the only force doing work on the ball is the gravitational force, a conservative force. The normal force on the ball from the ramp does zero work because it is perpendicular to the ball's path. The frictional force on the ball from the ramp does not transfer any energy to thermal energy because the ball does not slide (it *rolls smoothly*).

Therefore, we can write the conservation of mechanical energy ($E_f = E_i$) as

$$K_f + U_f = K_i + U_i, \qquad (11\text{-}11)$$

where subscripts f and i refer to the final values (at the bottom) and initial values (at rest), respectively. The gravitational potential energy is initially $U_i = Mgh$ (where M is the ball's mass) and finally $U_f = 0$. The kinetic energy is initially $K_i = 0$. For the final kinetic energy K_f, we need an additional idea: Because the ball rolls, the kinetic energy involves both translation *and* rotation, so we include them both by using the right side of Eq. 11-5.

Calculations: Substituting into Eq. 11-11 gives us

$$(\tfrac{1}{2}I_{\text{com}}\omega^2 + \tfrac{1}{2}Mv_{\text{com}}^2) + 0 = 0 + Mgh, \qquad (11\text{-}12)$$

where I_{com} is the ball's rotational inertia about an axis through its center of mass, v_{com} is the requested speed at the bottom, and ω is the angular speed there.

Because the ball rolls smoothly, we can use Eq. 11-2 to substitute v_{com}/R for ω to reduce the unknowns in Eq. 11-12.

Doing so, substituting $\tfrac{2}{5}MR^2$ for I_{com} (from Table 10-2f), and then solving for v_{com} give us

$$v_{\text{com}} = \sqrt{(\tfrac{10}{7})gh} = \sqrt{(\tfrac{10}{7})(9.8 \text{ m/s}^2)(1.20 \text{ m})}$$
$$= 4.10 \text{ m/s}. \qquad \text{(Answer)}$$

Note that the answer does not depend on M or R.

(b) What are the magnitude and direction of the frictional force on the ball as it rolls down the ramp?

KEY IDEA

Because the ball rolls smoothly, Eq. 11-9 gives the frictional force on the ball.

Calculations: Before we can use Eq. 11-9, we need the ball's acceleration $a_{\text{com},x}$ from Eq. 11-10:

$$a_{\text{com},x} = -\frac{g\sin\theta}{1 + I_{\text{com}}/MR^2} = -\frac{g\sin\theta}{1 + \tfrac{2}{5}MR^2/MR^2}$$
$$= -\frac{(9.8 \text{ m/s}^2)\sin 30.0°}{1 + \tfrac{2}{5}} = -3.50 \text{ m/s}^2.$$

Note that we needed neither mass M nor radius R to find $a_{\text{com},x}$. Thus, any size ball with any uniform mass would have this acceleration down a 30.0° ramp, provided the ball rolls smoothly.

We can now solve Eq. 11-9 as

$$f_s = -I_{\text{com}}\frac{a_{\text{com},x}}{R^2} = -\tfrac{2}{5}MR^2\frac{a_{\text{com},x}}{R^2} = -\tfrac{2}{5}Ma_{\text{com},x}$$
$$= -\tfrac{2}{5}(6.00 \text{ kg})(-3.50 \text{ m/s}^2) = 8.40 \text{ N}. \qquad \text{(Answer)}$$

Note that we needed mass M but not radius R. Thus, the frictional force on any 6.00 kg ball rolling smoothly down a 30.0° ramp would be 8.40 N regardless of the ball's radius but would be larger for a larger mass.

 Additional examples, video, and practice available at *WileyPLUS*

11-5 The Yo-Yo

A yo-yo is a physics lab that you can fit in your pocket. If a yo-yo rolls down its string for a distance h, it loses potential energy in amount mgh but gains kinetic energy in both translational ($\tfrac{1}{2}Mv_{\text{com}}^2$) and rotational ($\tfrac{1}{2}I_{\text{com}}\omega^2$) forms. As it climbs back up, it loses kinetic energy and regains potential energy.

In a modern yo-yo, the string is not tied to the axle but is looped around it. When the yo-yo "hits" the bottom of its string, an upward force on the axle from the string stops the descent. The yo-yo then spins, axle inside loop, with only rotational kinetic energy. The yo-yo keeps spinning ("sleeping") until you "wake it" by jerking on the string, causing the string to catch on the axle and the yo-yo to climb back up. The rotational kinetic energy of the yo-yo at the bottom of its string (and thus the sleeping time) can be considerably increased by throwing the yo-yo downward so that it starts down the string with initial speeds v_{com} and ω instead of rolling down from rest.

To find an expression for the linear acceleration a_{com} of a yo-yo rolling down a string, we could use Newton's second law just as we did for the body rolling down a ramp in Fig. 11-8. The analysis is the same except for the following:

1. Instead of rolling down a ramp at angle θ with the horizontal, the yo-yo rolls down a string at angle $\theta = 90°$ with the horizontal.

2. Instead of rolling on its outer surface at radius R, the yo-yo rolls on an axle of radius R_0 (Fig. 11-9a).

3. Instead of being slowed by frictional force $\vec{f_s}$, the yo-yo is slowed by the force $\vec{T}$ on it from the string (Fig. 11-9b).

The analysis would again lead us to Eq. 11-10. Therefore, let us just change the notation in Eq. 11-10 and set $\theta = 90°$ to write the linear acceleration as

$$a_{com} = -\frac{g}{1 + I_{com}/MR_0^2},$$ (11-13)

where I_{com} is the yo-yo's rotational inertia about its center and M is its mass. A yo-yo has the same downward acceleration when it is climbing back up.

11-6 Torque Revisited

In Chapter 10 we defined torque τ for a rigid body that can rotate around a fixed axis, with each particle in the body forced to move in a path that is a circle centered on that axis. We now expand the definition of torque to apply it to an individual particle that moves along any path relative to a fixed *point* (rather than a fixed axis). The path need no longer be a circle, and we must write the torque as a vector $\vec{\tau}$ that may have any direction.

Figure 11-10a shows such a particle at point A in an xy plane. A single force $\vec{F}$ in that plane acts on the particle, and the particle's position relative to the origin O is given by position vector $\vec{r}$. The torque $\vec{\tau}$ acting on the particle relative to the fixed point O is a vector quantity defined as

$$\vec{\tau} = \vec{r} \times \vec{F} \quad \text{(torque defined).}$$ (11-14)

We can evaluate the vector (or cross) product in this definition of $\vec{\tau}$ by using the rules for such products given in Section 3-8. To find the direction of $\vec{\tau}$, we slide

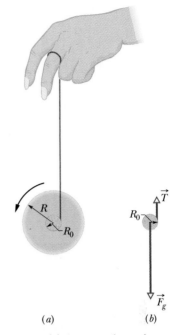

Fig. 11-9 (a) A yo-yo, shown in cross section. The string, of assumed negligible thickness, is wound around an axle of radius R_0. (b) A free-body diagram for the falling yo-yo. Only the axle is shown.

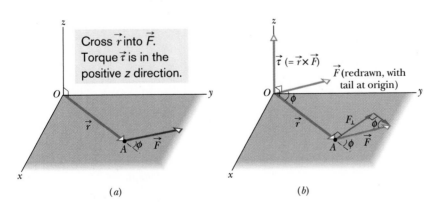
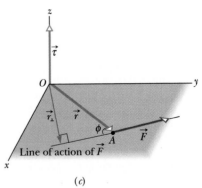

Fig. 11-10 Defining torque. (a) A force $\vec{F}$, lying in an xy plane, acts on a particle at point A. (b) This force produces a torque $\vec{\tau}$ ($= \vec{r} \times \vec{F}$) on the particle with respect to the origin O. By the right-hand rule for vector (cross) products, the torque vector points in the positive direction of z. Its magnitude is given by $rF_\perp$ in (b) and by $r_\perp F$ in (c).

the vector $\vec{F}$ (without changing its direction) until its tail is at the origin O, so that the two vectors in the vector product are tail to tail as in Fig. 11-10b. We then use the right-hand rule for vector products in Fig. 3-19a, sweeping the fingers of the right hand from $\vec{r}$ (the first vector in the product) into $\vec{F}$ (the second vector). The outstretched right thumb then gives the direction of $\vec{\tau}$. In Fig. 11-10b, the direction of $\vec{\tau}$ is in the positive direction of the z axis.

To determine the magnitude of $\vec{\tau}$, we apply the general result of Eq. 3-27 ($c = ab \sin \phi$), finding

$$\tau = rF \sin \phi, \tag{11-15}$$

where ϕ is the smaller angle between the directions of $\vec{r}$ and $\vec{F}$ when the vectors are tail to tail. From Fig. 11-10b, we see that Eq. 11-15 can be rewritten as

$$\tau = rF_\perp, \tag{11-16}$$

Sample Problem

Torque on a particle due to a force

In Fig. 11-11a, three forces, each of magnitude 2.0 N, act on a particle. The particle is in the xz plane at point A given by position vector $\vec{r}$, where $r = 3.0$ m and $\theta = 30°$. Force $\vec{F}_1$ is parallel to the x axis, force $\vec{F}_2$ is parallel to the z axis, and force $\vec{F}_3$ is parallel to the y axis. What is the torque, about the origin O, due to each force?

KEY IDEA

Because the three force vectors do not lie in a plane, we cannot evaluate their torques as in Chapter 10. Instead, we must use

vector (or cross) products, with magnitudes given by Eq. 11-15 ($\tau = rF \sin \phi$) and directions given by the right-hand rule for vector products.

Calculations: Because we want the torques with respect to the origin O, the vector $\vec{r}$ required for each cross product is the given position vector. To determine the angle ϕ between the direction of $\vec{r}$ and the direction of each force, we shift the force vectors of Fig. 11-11a, each in turn, so that their tails are at the origin. Figures 11-11b, c, and d, which are direct views of the xz plane, show the shifted force vectors $\vec{F}_1$,

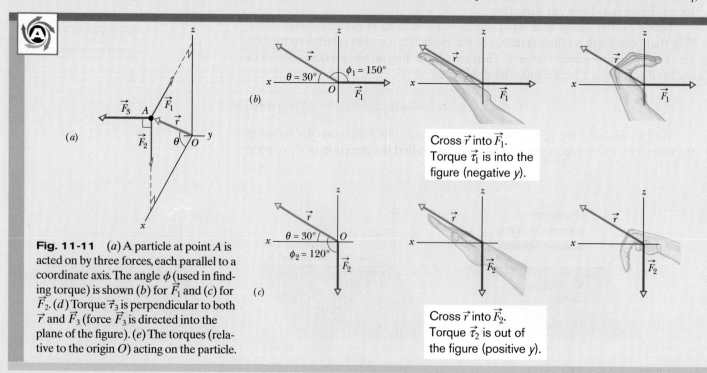

Fig. 11-11 (a) A particle at point A is acted on by three forces, each parallel to a coordinate axis. The angle ϕ (used in finding torque) is shown (b) for $\vec{F}_1$ and (c) for $\vec{F}_2$. (d) Torque $\vec{\tau}_3$ is perpendicular to both $\vec{r}$ and $\vec{F}_3$ (force $\vec{F}_3$ is directed into the plane of the figure). (e) The torques (relative to the origin O) acting on the particle.

Cross $\vec{r}$ into $\vec{F}_1$. Torque $\vec{\tau}_1$ is into the figure (negative y).

Cross $\vec{r}$ into $\vec{F}_2$. Torque $\vec{\tau}_2$ is out of the figure (positive y).

where $F_\perp (= F \sin \phi)$ is the component of $\vec{F}$ perpendicular to $\vec{r}$. From Fig. 11-10c, we see that Eq. 11-15 can also be rewritten as

$$\tau = r_\perp F, \qquad (11\text{-}17)$$

where $r_\perp (= r \sin \phi)$ is the moment arm of $\vec{F}$ (the perpendicular distance between O and the line of action of $\vec{F}$).

✔ CHECKPOINT 3

The position vector $\vec{r}$ of a particle points along the positive direction of a z axis. If the torque on the particle is (a) zero, (b) in the negative direction of x, and (c) in the negative direction of y, in what direction is the force causing the torque?

$\vec{F_2}$, and $\vec{F_3}$, respectively. (Note how much easier the angles between the force vectors and the position vector are to see.) In Fig. 11-11d, the angle between the directions of $\vec{r}$ and $\vec{F_3}$ is 90° and the symbol $\otimes$ means $\vec{F_3}$ is directed into the page. If it were directed out of the page, it would be represented with the symbol $\odot$.

Now, applying Eq. 11-15 for each force, we find the magnitudes of the torques to be

$$\tau_1 = rF_1 \sin \phi_1 = (3.0 \text{ m})(2.0 \text{ N})(\sin 150°) = 3.0 \text{ N} \cdot \text{m},$$
$$\tau_2 = rF_2 \sin \phi_2 = (3.0 \text{ m})(2.0 \text{ N})(\sin 120°) = 5.2 \text{ N} \cdot \text{m},$$

and
$$\tau_3 = rF_3 \sin \phi_3 = (3.0 \text{ m})(2.0 \text{ N})(\sin 90°)$$
$$= 6.0 \text{ N} \cdot \text{m}. \qquad \text{(Answer)}$$

To find the directions of these torques, we use the right-hand rule, placing the fingers of the right hand so as to rotate $\vec{r}$ into $\vec{F}$ through the *smaller* of the two angles between their directions. The thumb points in the direction of the torque. Thus $\vec{\tau_1}$ is directed into the page in Fig. 11-11b; $\vec{\tau_2}$ is directed out of the page in Fig. 11-11c; and $\vec{\tau_3}$ is directed as shown in Fig. 11-11d. All three torque vectors are shown in Fig. 11-11e.

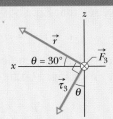

(d)

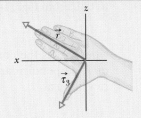

Cross $\vec{r}$ into $\vec{F_3}$.
Torque $\vec{\tau_3}$ is
in the *xz* plane.

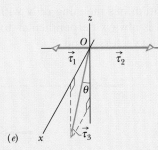

(e)

These are the three torques acting on the particle, each measured about the origin O.

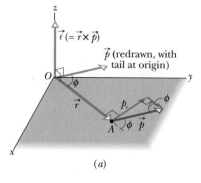

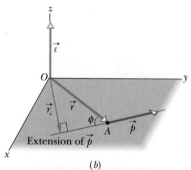

Fig. 11-12 Defining angular momentum. A particle passing through point A has linear momentum $\vec{p}\,(= m\vec{v})$, with the vector $\vec{p}$ lying in an xy plane. The particle has angular momentum $\vec{\ell}\,(= \vec{r} \times \vec{p})$ with respect to the origin O. By the right-hand rule, the angular momentum vector points in the positive direction of z. (a) The magnitude of $\vec{\ell}$ is given by $\ell = rp_\perp = rmv_\perp$. (b) The magnitude of $\vec{\ell}$ is also given by $\ell = r_\perp p = r_\perp mv$.

11-7 Angular Momentum

Recall that the concept of linear momentum $\vec{p}$ and the principle of conservation of linear momentum are extremely powerful tools. They allow us to predict the outcome of, say, a collision of two cars without knowing the details of the collision. Here we begin a discussion of the angular counterpart of $\vec{p}$, winding up in Section 11-11 with the angular counterpart of the conservation principle.

Figure 11-12 shows a particle of mass m with linear momentum $\vec{p}\,(= m\vec{v})$ as it passes through point A in an xy plane. The **angular momentum** $\vec{\ell}$ of this particle with respect to the origin O is a vector quantity defined as

$$\vec{\ell} = \vec{r} \times \vec{p} = m(\vec{r} \times \vec{v}) \quad \text{(angular momentum defined),} \quad (11\text{-}18)$$

where $\vec{r}$ is the position vector of the particle with respect to O. As the particle moves relative to O in the direction of its momentum $\vec{p}\,(= m\vec{v})$, position vector $\vec{r}$ rotates around O. Note carefully that to have angular momentum about O, the particle does *not* itself have to rotate around O. Comparison of Eqs. 11-14 and 11-18 shows that angular momentum bears the same relation to linear momentum that torque does to force. The SI unit of angular momentum is the kilogram-meter-squared per second ($\text{kg} \cdot \text{m}^2/\text{s}$), equivalent to the joule-second ($\text{J} \cdot \text{s}$).

To find the direction of the angular momentum vector $\vec{\ell}$ in Fig. 11-12, we slide the vector $\vec{p}$ until its tail is at the origin O. Then we use the right-hand rule for vector products, sweeping the fingers from $\vec{r}$ into $\vec{p}$. The outstretched thumb then shows that the direction of $\vec{\ell}$ is in the positive direction of the z axis in Fig. 11-12. This positive direction is consistent with the counterclockwise rotation of position vector $\vec{r}$ about the z axis, as the particle moves. (A negative direction of $\vec{\ell}$ would be consistent with a clockwise rotation of $\vec{r}$ about the z axis.)

To find the magnitude of $\vec{\ell}$, we use the general result of Eq. 3-27 to write

$$\ell = rmv \sin \phi, \quad (11\text{-}19)$$

where ϕ is the smaller angle between $\vec{r}$ and $\vec{p}$ when these two vectors are tail to tail. From Fig. 11-12a, we see that Eq. 11-19 can be rewritten as

$$\ell = rp_\perp = rmv_\perp, \quad (11\text{-}20)$$

where $p_\perp$ is the component of $\vec{p}$ perpendicular to $\vec{r}$ and $v_\perp$ is the component of $\vec{v}$ perpendicular to $\vec{r}$. From Fig. 11-12b, we see that Eq. 11-19 can also be rewritten as

$$\ell = r_\perp p = r_\perp mv, \quad (11\text{-}21)$$

where $r_\perp$ is the perpendicular distance between O and the extension of $\vec{p}$.

Note two features here: (1) angular momentum has meaning only with respect to a specified origin and (2) its direction is always perpendicular to the plane formed by the position and linear momentum vectors $\vec{r}$ and $\vec{p}$.

CHECKPOINT 4

In part a of the figure, particles 1 and 2 move around point O in circles with radii 2 m and 4 m. In part b, particles 3 and 4 travel along straight lines at perpendicular distances of 4 m and 2 m from point O. Particle 5 moves directly away from O.

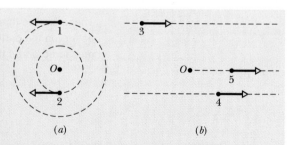

All five particles have the same mass and the same constant speed. (a) Rank the particles according to the magnitudes of their angular momentum about point O, greatest first. (b) Which particles have negative angular momentum about point O?

Angular momentum of a two-particle system

Figure 11-13 shows an overhead view of two particles moving at constant momentum along horizontal paths. Particle 1, with momentum magnitude $p_1 = 5.0$ kg·m/s, has position vector $\vec{r}_1$ and will pass 2.0 m from point O. Particle 2, with momentum magnitude $p_2 = 2.0$ kg·m/s, has position vector $\vec{r}_2$ and will pass 4.0 m from point O. What are the magnitude and direction of the net angular momentum $\vec{L}$ about point O of the two-particle system?

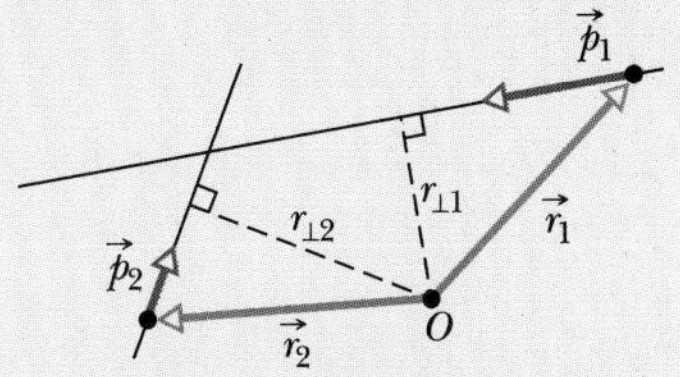

Fig. 11-13   Two particles pass near point O.

KEY IDEA

To find $\vec{L}$, we can first find the individual angular momenta $\vec{\ell}_1$ and $\vec{\ell}_2$ and then add them. To evaluate their magnitudes, we can use any one of Eqs. 11-18 through 11-21. However, Eq. 11-21 is easiest, because we are given the perpendicular distances $r_{\perp 1}$ (= 2.0 m) and $r_{\perp 2}$ (= 4.0 m) and the momentum magnitudes p_1 and p_2.

Calculations: For particle 1, Eq. 11-21 yields

$$\ell_1 = r_{\perp 1}p_1 = (2.0 \text{ m})(5.0 \text{ kg·m/s})$$
$$= 10 \text{ kg·m}^2/\text{s}.$$

To find the direction of vector $\vec{\ell}_1$, we use Eq. 11-18 and the right-hand rule for vector products. For $\vec{r}_1 \times \vec{p}_1$, the vector product is out of the page, perpendicular to the plane of Fig. 11-13. This is the positive direction, consistent with the counterclockwise rotation of the particle's position vector $\vec{r}_1$ around O as particle 1 moves. Thus, the angular momentum vector for particle 1 is

$$\ell_1 = +10 \text{ kg·m}^2/\text{s}.$$

Similarly, the magnitude of $\vec{\ell}_2$ is

$$\ell_2 = r_{\perp 2}p_2 = (4.0 \text{ m})(2.0 \text{ kg·m/s})$$
$$= 8.0 \text{ kg·m}^2/\text{s},$$

and the vector product $\vec{r}_2 \times \vec{p}_2$ is into the page, which is the negative direction, consistent with the clockwise rotation of $\vec{r}_2$ around O as particle 2 moves. Thus, the angular momentum vector for particle 2 is

$$\ell_2 = -8.0 \text{ kg·m}^2/\text{s}.$$

The net angular momentum for the two-particle system is

$$L = \ell_1 + \ell_2 = +10 \text{ kg·m}^2/\text{s} + (-8.0 \text{ kg·m}^2/\text{s})$$
$$= +2.0 \text{ kg·m}^2/\text{s}. \qquad \text{(Answer)}$$

The plus sign means that the system's net angular momentum about point O is out of the page.

 Additional examples, video, and practice available at *WileyPLUS*

11-8 Newton's Second Law in Angular Form

Newton's second law written in the form

$$\vec{F}_{\text{net}} = \frac{d\vec{p}}{dt} \qquad \text{(single particle)} \qquad (11\text{-}22)$$

expresses the close relation between force and linear momentum for a single particle. We have seen enough of the parallelism between linear and angular quantities to be pretty sure that there is also a close relation between torque and angular momentum. Guided by Eq. 11-22, we can even guess that it must be

$$\vec{\tau}_{\text{net}} = \frac{d\vec{\ell}}{dt} \qquad \text{(single particle).} \qquad (11\text{-}23)$$

Equation 11-23 is indeed an angular form of Newton's second law for a single particle:

> The (vector) sum of all the torques acting on a particle is equal to the time rate of change of the angular momentum of that particle.

Equation 11-23 has no meaning unless the torques $\vec{\tau}$ and the angular momentum $\vec{\ell}$ are defined with respect to the same point, usually the origin of the coordinate system being used.

Proof of Equation 11-23

We start with Eq. 11-18, the definition of the angular momentum of a particle:

$$\vec{\ell} = m(\vec{r} \times \vec{v}),$$

where $\vec{r}$ is the position vector of the particle and $\vec{v}$ is the velocity of the particle. Differentiating* each side with respect to time t yields

$$\frac{d\vec{\ell}}{dt} = m\left(\vec{r} \times \frac{d\vec{v}}{dt} + \frac{d\vec{r}}{dt} \times \vec{v}\right). \qquad (11\text{-}24)$$

However, $d\vec{v}/dt$ is the acceleration $\vec{a}$ of the particle, and $d\vec{r}/dt$ is its velocity $\vec{v}$. Thus, we can rewrite Eq. 11-24 as

$$\frac{d\vec{\ell}}{dt} = m(\vec{r} \times \vec{a} + \vec{v} \times \vec{v}).$$

Now $\vec{v} \times \vec{v} = 0$ (the vector product of any vector with itself is zero because the angle between the two vectors is necessarily zero). Thus, the last term of this expression is eliminated and we then have

$$\frac{d\vec{\ell}}{dt} = m(\vec{r} \times \vec{a}) = \vec{r} \times m\vec{a}.$$

We now use Newton's second law ($\vec{F}_{net} = m\vec{a}$) to replace $m\vec{a}$ with its equal, the vector sum of the forces that act on the particle, obtaining

$$\frac{d\vec{\ell}}{dt} = \vec{r} \times \vec{F}_{net} = \sum(\vec{r} \times \vec{F}). \qquad (11\text{-}25)$$

Here the symbol Σ indicates that we must sum the vector products $\vec{r} \times \vec{F}$ for all the forces. However, from Eq. 11-14, we know that each one of those vector products is the torque associated with one of the forces. Therefore, Eq. 11-25 tells us that

$$\vec{\tau}_{net} = \frac{d\vec{\ell}}{dt}.$$

This is Eq. 11-23, the relation that we set out to prove.

✓ CHECKPOINT 5

The figure shows the position vector $\vec{r}$ of a particle at a certain instant, and four choices for the direction of a force that is to accelerate the particle. All four choices lie in the xy plane. (a) Rank the choices according to the magnitude of the time rate of change ($d\vec{\ell}/dt$) they produce in the angular momentum of the particle about point O, greatest first. (b) Which choice results in a negative rate of change about O?

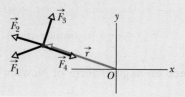

---

*In differentiating a vector product, be sure not to change the order of the two quantities (here $\vec{r}$ and $\vec{v}$) that form that product. (See Eq. 3-28.)

Sample Problem

Torque, time derivative of angular momentum, penguin fall

In Fig. 11-14, a penguin of mass m falls from rest at point A, a horizontal distance D from the origin O of an xyz coordinate system. (The positive direction of the z axis is directly outward from the plane of the figure.)

(a) What is the angular momentum $\vec{\ell}$ of the falling penguin about O?

KEY IDEA

We can treat the penguin as a particle, and thus its angular momentum $\vec{\ell}$ is given by Eq. 11-18 ($\vec{\ell} = \vec{r} \times \vec{p}$), where $\vec{r}$ is the penguin's position vector (extending from O to the penguin) and $\vec{p}$ is the penguin's linear momentum. (The penguin has *angular* momentum about O even though it moves in a straight line, because vector $\vec{r}$ rotates about O as the penguin falls.)

Calculations: To find the magnitude of $\vec{\ell}$, we can use any one of the scalar equations derived from Eq. 11-18— namely, Eqs. 11-19 through 11-21. However, Eq. 11-21 ($\ell = r_{\perp}mv$) is easiest because the perpendicular distance $r_{\perp}$ between O and an extension of vector $\vec{p}$ is the given distance D. The speed of an object that has fallen from rest for a time t is $v = gt$. We can now write Eq. 11-21 in terms of given quantities as

$$\ell = r_{\perp}mv = Dmgt. \qquad \text{(Answer)}$$

To find the direction of $\vec{\ell}$, we use the right-hand rule for the vector product $\vec{r} \times \vec{p}$ in Eq. 11-18. Mentally shift $\vec{p}$ until its tail is at the origin, and then use the fingers of your right hand to rotate $\vec{r}$ into $\vec{p}$ through the smaller angle between the two vectors. Your outstretched thumb then points into the plane of the figure, indicating that the product $\vec{r} \times \vec{p}$ and thus also $\vec{\ell}$ are directed into that plane, in the negative direction of the z axis. We represent $\vec{\ell}$ with an encircled cross $\otimes$ at O. The vector $\vec{\ell}$ changes with time in magnitude only; its direction remains unchanged.

(b) About the origin O, what is the torque $\vec{\tau}$ on the penguin due to the gravitational force $\vec{F}_g$?

KEY IDEAS

(1) The torque is given by Eq. 11-14 ($\vec{\tau} = \vec{r} \times \vec{F}$), where now the force is $\vec{F}_g$. (2) Force $\vec{F}_g$ causes a torque on the penguin, even though the penguin moves in a straight line, because $\vec{r}$ rotates about O as the penguin moves.

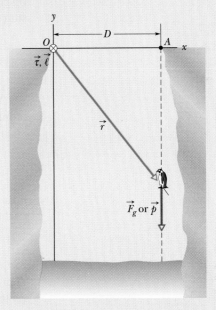

Fig. 11-14 A penguin falls vertically from point A. The torque $\vec{\tau}$ and the angular momentum $\vec{\ell}$ of the falling penguin with respect to the origin O are directed into the plane of the figure at O.

Calculations: To find the magnitude of $\vec{\tau}$, we can use any one of the scalar equations derived from Eq. 11-14— namely, Eqs. 11-15 through 11-17. However, Eq. 11-17 ($\tau = r_{\perp}F$) is easiest because the perpendicular distance $r_{\perp}$ between O and the line of action of $\vec{F}_g$ is the given distance D. So, substituting D and using mg for the magnitude of $\vec{F}_g$, we can write Eq. 11-17 as

$$\tau = DF_g = Dmg. \qquad \text{(Answer)}$$

Using the right-hand rule for the vector product $\vec{r} \times \vec{F}$ in Eq. 11-14, we find that the direction of $\vec{\tau}$ is the negative direction of the z axis, the same as $\vec{\ell}$.

The results we obtained in parts (a) and (b) must be consistent with Newton's second law in the angular form of Eq. 11-23 ($\vec{\tau}_{net} = d\vec{\ell}/dt$). To check the magnitudes we got, we write Eq. 11-23 in component form for the z axis and then substitute our result $\ell = Dmgt$. We find

$$\tau = \frac{d\ell}{dt} = \frac{d(Dmgt)}{dt} = Dmg,$$

which is the magnitude we found for $\vec{\tau}$. To check the directions, we note that Eq. 11-23 tells us that $\vec{\tau}$ and $d\vec{\ell}/dt$ must have the same direction. So $\vec{\tau}$ and $\vec{\ell}$ must also have the same direction, which is what we found.

 Additional examples, video, and practice available at *WileyPLUS*

11-9 The Angular Momentum of a System of Particles

Now we turn our attention to the angular momentum of a system of particles with respect to an origin. The total angular momentum $\vec{L}$ of the system is the (vector) sum of the angular momenta $\vec{\ell}$ of the individual particles (here with label i):

$$\vec{L} = \vec{\ell}_1 + \vec{\ell}_2 + \vec{\ell}_3 + \cdots + \vec{\ell}_n = \sum_{i=1}^{n} \vec{\ell}_i. \tag{11-26}$$

With time, the angular momenta of individual particles may change because of interactions between the particles or with the outside. We can find the resulting change in $\vec{L}$ by taking the time derivative of Eq. 11-26. Thus,

$$\frac{d\vec{L}}{dt} = \sum_{i=1}^{n} \frac{d\vec{\ell}_i}{dt}. \tag{11-27}$$

From Eq. 11-23, we see that $d\vec{\ell}_i/dt$ is equal to the net torque $\vec{\tau}_{net,i}$ on the ith particle. We can rewrite Eq. 11-27 as

$$\frac{d\vec{L}}{dt} = \sum_{i=1}^{n} \vec{\tau}_{net,i}. \tag{11-28}$$

That is, the rate of change of the system's angular momentum $\vec{L}$ is equal to the vector sum of the torques on its individual particles. Those torques include *internal torques* (due to forces between the particles) and *external torques* (due to forces on the particles from bodies external to the system). However, the forces between the particles always come in third-law force pairs so their torques sum to zero. Thus, the only torques that can change the total angular momentum $\vec{L}$ of the system are the external torques acting on the system.

Let $\vec{\tau}_{net}$ represent the net external torque, the vector sum of all external torques on all particles in the system. Then we can write Eq. 11-28 as

$$\vec{\tau}_{net} = \frac{d\vec{L}}{dt} \quad \text{(system of particles),} \tag{11-29}$$

which is Newton's second law in angular form. It says:

> The net external torque $\vec{\tau}_{net}$ acting on a system of particles is equal to the time rate of change of the system's total angular momentum $\vec{L}$.

Equation 11-29 is analogous to $\vec{F}_{net} = d\vec{P}/dt$ (Eq. 9-27) but requires extra caution: Torques and the system's angular momentum must be measured relative to the same origin. If the center of mass of the system is not accelerating relative to an inertial frame, that origin can be any point. However, if it *is* accelerating, then it *must* be the origin. For example, consider a wheel as the system of particles. If it is rotating about an axis that is fixed relative to the ground, then the origin for applying Eq. 11-29 can be any point that is stationary relative to the ground. However, if it is rotating about an axis that is accelerating (such as when it rolls down a ramp), then the origin can be only at its center of mass.

11-10 The Angular Momentum of a Rigid Body Rotating About a Fixed Axis

We next evaluate the angular momentum of a system of particles that form a rigid body that rotates about a fixed axis. Figure 11-15a shows such a body. The fixed axis of rotation is a z axis, and the body rotates about it with constant angular speed ω. We wish to find the angular momentum of the body about that axis.

We can find the angular momentum by summing the z components of the angular momenta of the mass elements in the body. In Fig. 11-15a, a typical mass element, of mass Δm_i, moves around the z axis in a circular path. The position of the mass element is located relative to the origin O by position vector $\vec{r}_i$. The radius of the mass element's circular path is $r_{\perp i}$, the perpendicular distance between the element and the z axis.

The magnitude of the angular momentum $\vec{\ell}_i$ of this mass element, with respect to O, is given by Eq. 11-19:

$$\ell_i = (r_i)(p_i)(\sin 90°) = (r_i)(\Delta m_i\, v_i),$$

where p_i and v_i are the linear momentum and linear speed of the mass element, and 90° is the angle between $\vec{r}_i$ and $\vec{p}_i$. The angular momentum vector $\vec{\ell}_i$ for the mass element in Fig. 11-15a is shown in Fig. 11-15b; its direction must be perpendicular to those of $\vec{r}_i$ and $\vec{p}_i$.

We are interested in the component of $\vec{\ell}_i$ that is parallel to the rotation axis, here the z axis. That z component is

$$\ell_{iz} = \ell_i \sin\theta = (r_i \sin\theta)(\Delta m_i\, v_i) = r_{\perp i}\,\Delta m_i\, v_i.$$

The z component of the angular momentum for the rotating rigid body as a whole is found by adding up the contributions of all the mass elements that make up the body. Thus, because $v = \omega r_\perp$, we may write

$$L_z = \sum_{i=1}^{n} \ell_{iz} = \sum_{i=1}^{n} \Delta m_i\, v_i r_{\perp i} = \sum_{i=1}^{n} \Delta m_i(\omega r_{\perp i})r_{\perp i}$$

$$= \omega\left(\sum_{i=1}^{n} \Delta m_i\, r_{\perp i}^2\right). \qquad (11\text{-}30)$$

We can remove ω from the summation here because it has the same value for all points of the rotating rigid body.

The quantity $\sum \Delta m_i\, r_{\perp i}^2$ in Eq. 11-30 is the rotational inertia I of the body about the fixed axis (see Eq. 10-33). Thus Eq. 11-30 reduces to

$$L = I\omega \qquad \text{(rigid body, fixed axis).} \qquad (11\text{-}31)$$

We have dropped the subscript z, but you must remember that the angular momentum defined by Eq. 11-31 is the angular momentum about the rotation axis. Also, I in that equation is the rotational inertia about that same axis.

Table 11-1, which supplements Table 10-3, extends our list of corresponding linear and angular relations.

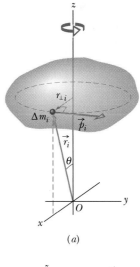

(a)

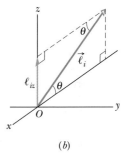

(b)

Fig. 11-15 (a) A rigid body rotates about a z axis with angular speed ω. A mass element of mass Δm_i within the body moves about the z axis in a circle with radius $r_{\perp i}$. The mass element has linear momentum $\vec{p}_i$, and it is located relative to the origin O by position vector $\vec{r}_i$. Here the mass element is shown when $r_{\perp i}$ is parallel to the x axis. (b) The angular momentum $\vec{\ell}_i$, with respect to O, of the mass element in (a). The z component ℓ_{iz} is also shown.

Table 11-1

More Corresponding Variables and Relations for Translational and Rotational Motion[a]

Translational		Rotational	
Force	$\vec{F}$	Torque	$\vec{\tau}\ (= \vec{r} \times \vec{F})$
Linear momentum	$\vec{p}$	Angular momentum	$\vec{\ell}\ (= \vec{r} \times \vec{p})$
Linear momentum[b]	$\vec{P}\ (= \Sigma \vec{p}_i)$	Angular momentum[b]	$\vec{L}\ (= \Sigma \vec{\ell}_i)$
Linear momentum[b]	$\vec{P} = M\vec{v}_{\text{com}}$	Angular momentum[c]	$L = I\omega$
Newton's second law[b]	$\vec{F}_{\text{net}} = \dfrac{d\vec{P}}{dt}$	Newton's second law[b]	$\vec{\tau}_{\text{net}} = \dfrac{d\vec{L}}{dt}$
Conservation law[d]	$\vec{P} = $ a constant	Conservation law[d]	$\vec{L} = $ a constant

[a]See also Table 10-3.
[b]For systems of particles, including rigid bodies.
[c]For a rigid body about a fixed axis, with L being the component along that axis.
[d]For a closed, isolated system.

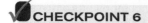

CHECKPOINT 6

In the figure, a disk, a hoop, and a solid sphere are made to spin about fixed central axes (like a top) by means of strings wrapped around them, with the strings producing the same constant tangential force $\vec{F}$ on all three objects. The three objects have the same mass and radius, and they are initially stationary. Rank the objects according to (a) their angular momentum about their central axes and (b) their angular speed, greatest first, when the strings have been pulled for a certain time t.

11-11 Conservation of Angular Momentum

So far we have discussed two powerful conservation laws, the conservation of energy and the conservation of linear momentum. Now we meet a third law of this type, involving the conservation of angular momentum. We start from Eq. 11-29 ($\vec{\tau}_{net} = d\vec{L}/dt$), which is Newton's second law in angular form. If no net external torque acts on the system, this equation becomes $d\vec{L}/dt = 0$, or

$$\vec{L} = \text{a constant} \qquad \text{(isolated system).} \qquad (11\text{-}32)$$

This result, called the **law of conservation of angular momentum,** can also be written as

$$\left(\begin{array}{c} \text{net angular momentum} \\ \text{at some initial time } t_i \end{array} \right) = \left(\begin{array}{c} \text{net angular momentum} \\ \text{at some later time } t_f \end{array} \right),$$

or $\qquad\qquad\qquad \vec{L}_i = \vec{L}_f \qquad \text{(isolated system).} \qquad (11\text{-}33)$

Equations 11-32 and 11-33 tell us:

> If the net external torque acting on a system is zero, the angular momentum $\vec{L}$ of the system remains constant, no matter what changes take place within the system.

Equations 11-32 and 11-33 are vector equations; as such, they are equivalent to three component equations corresponding to the conservation of angular momentum in three mutually perpendicular directions. Depending on the torques acting on a system, the angular momentum of the system might be conserved in only one or two directions but not in all directions:

> If the component of the net *external* torque on a system along a certain axis is zero, then the component of the angular momentum of the system along that axis cannot change, no matter what changes take place within the system.

We can apply this law to the isolated body in Fig. 11-15, which rotates around the z axis. Suppose that the initially rigid body somehow redistributes its mass relative to that rotation axis, changing its rotational inertia about that axis. Equations 11-32 and 11-33 state that the angular momentum of the body cannot change. Substituting Eq. 11-31 (for the angular momentum along the rotational axis) into Eq. 11-33, we write this conservation law as

$$I_i \omega_i = I_f \omega_f. \qquad (11\text{-}34)$$

Here the subscripts refer to the values of the rotational inertia I and angular speed ω before and after the redistribution of mass.

Like the other two conservation laws that we have discussed, Eqs. 11-32 and 11-33 hold beyond the limitations of Newtonian mechanics. They hold for particles whose speeds approach that of light (where the theory of special relativity reigns), and they remain true in the world of subatomic particles (where quantum physics reigns). No exceptions to the law of conservation of angular momentum have ever been found.

We now discuss four examples involving this law.

1. ***The spinning volunteer***   Figure 11-16 shows a student seated on a stool that can rotate freely about a vertical axis. The student, who has been set into rotation at a modest initial angular speed ω_i, holds two dumbbells in his outstretched hands. His angular momentum vector $\vec{L}$ lies along the vertical rotation axis, pointing upward.

   The instructor now asks the student to pull in his arms; this action reduces his rotational inertia from its initial value I_i to a smaller value I_f because he moves mass closer to the rotation axis. His rate of rotation increases markedly, from ω_i to ω_f. The student can then slow down by extending his arms once more, moving the dumbbells outward.

   No net external torque acts on the system consisting of the student, stool, and dumbbells. Thus, the angular momentum of that system about the rotation axis must remain constant, no matter how the student maneuvers the dumbbells. In Fig. 11-16a, the student's angular speed ω_i is relatively low and his rotational inertia I_i is relatively high. According to Eq. 11-34, his angular speed in Fig. 11-16b must be greater to compensate for the decreased I_f.

2. ***The springboard diver***   Figure 11-17 shows a diver doing a forward one-and-a-half-somersault dive. As you should expect, her center of mass follows a parabolic path. She leaves the springboard with a definite angular momentum $\vec{L}$ about an axis through her center of mass, represented by a vector pointing into the plane of Fig. 11-17, perpendicular to the page. When she is in the air, no net external torque acts on her about her center of mass, so her angular momentum about her center of mass cannot change. By pulling her arms and legs into the closed *tuck position,* she can considerably reduce her rotational inertia about the same axis and thus, according to Eq. 11-34, considerably increase her angular speed. Pulling out of the tuck position (into the *open layout position*) at the end of the dive increases her rotational inertia and thus slows her rotation rate so she can enter the water with little splash. Even in a more complicated dive involving both twisting and somersaulting, the angular momentum of the diver must be conserved, in both magnitude *and* direction, throughout the dive.

3. ***Long jump***   When an athlete takes off from the ground in a running long jump, the forces on the launching foot give the athlete an angular momentum with a forward rotation around a horizontal axis. Such rotation would not allow

Rotation axis
(a)

(b)

Fig. 11-16   (a) The student has a relatively large rotational inertia about the rotation axis and a relatively small angular speed. (b) By decreasing his rotational inertia, the student automatically increases his angular speed. The angular momentum $\vec{L}$ of the rotating system remains unchanged.

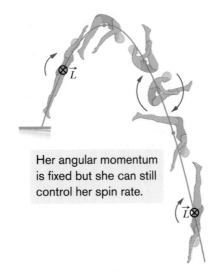

Fig. 11-17   The diver's angular momentum $\vec{L}$ is constant throughout the dive, being represented by the tail $\otimes$ of an arrow that is perpendicular to the plane of the figure. Note also that her center of mass (see the dots) follows a parabolic path.

Her angular momentum is fixed but she can still control her spin rate.

Fig. 11-18 Windmill motion of the arms during a long jump helps maintain body orientation for a proper landing.

(a)

(b)

Fig. 11-19 (a) Initial phase of a tour jeté: large rotational inertia and small angular speed. (b) Later phase: smaller rotational inertia and larger angular speed.

the jumper to land properly: In the landing, the legs should be together and extended forward at an angle so that the heels mark the sand at the greatest distance. Once airborne, the angular momentum cannot change (it is conserved) because no external torque acts to change it. However, the jumper can shift most of the angular momentum to the arms by rotating them in windmill fashion (Fig. 11-18). Then the body remains upright and in the proper orientation for landing.

4. **Tour jeté** In a tour jeté, a ballet performer leaps with a small twisting motion on the floor with one foot while holding the other leg perpendicular to the body (Fig. 11-19a). The angular speed is so small that it may not be perceptible to the audience. As the performer ascends, the outstretched leg is brought down and the other leg is brought up, with both ending up at angle θ to the body (Fig. 11-19b). The motion is graceful, but it also serves to increase the rotation because bringing in the initially outstretched leg decreases the performer's rotational inertia. Since no external torque acts on the airborne performer, the angular momentum cannot change. Thus, with a decrease in rotational inertia, the angular speed must increase. When the jump is well executed, the performer seems to suddenly begin to spin and rotates 180° before the initial leg orientations are reversed in preparation for the landing. Once a leg is again outstretched, the rotation seems to vanish.

CHECKPOINT 7

A rhinoceros beetle rides the rim of a small disk that rotates like a merry-go-round. If the beetle crawls toward the center of the disk, do the following (each relative to the central axis) increase, decrease, or remain the same for the beetle–disk system: (a) rotational inertia, (b) angular momentum, and (c) angular speed?

Sample Problem

Conservation of angular momentum, rotating wheel demo

Figure 11-20a shows a student, again sitting on a stool that can rotate freely about a vertical axis. The student, initially at rest, is holding a bicycle wheel whose rim is loaded with lead and whose rotational inertia I_{wh} about its central axis is 1.2 kg · m². (The rim contains lead in order to make the value of I_{wh} substantial.) The wheel is rotating at an angular speed ω_{wh} of 3.9 rev/s; as seen from overhead, the rotation is counterclockwise. The axis of the wheel is vertical, and the angular momentum $\vec{L}_{wh}$ of the wheel points vertically upward. The student now inverts the wheel (Fig. 11-20b) so

that, as seen from overhead, it is rotating clockwise. Its angular momentum is now $-\vec{L}_{wh}$. The inversion results in the student, the stool, and the wheel's center rotating together as a composite rigid body about the stool's rotation axis, with rotational inertia $I_b = 6.8$ kg · m². (The fact that the wheel is also rotating about its center does not affect the mass distribution of this composite body; thus, I_b has the same value whether or not the wheel rotates.) With what angular speed ω_b and in what direction does the composite body rotate after the inversion of the wheel?

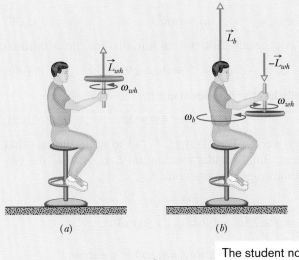

The student now has angular momentum, and the net of these two vectors equals the initial vector.

Fig. 11-20 (a) A student holds a bicycle wheel rotating around a vertical axis. (b) The student inverts the wheel, setting himself into rotation. (c) The net angular momentum of the system must remain the same in spite of the inversion.

KEY IDEAS

1. The angular speed ω_b we seek is related to the final angular momentum $\vec{L}_b$ of the composite body about the stool's rotation axis by Eq. 11-31 ($L = I\omega$).

2. The initial angular speed ω_{wh} of the wheel is related to the angular momentum $\vec{L}_{wh}$ of the wheel's rotation about its center by the same equation.

3. The vector addition of $\vec{L}_b$ and $\vec{L}_{wh}$ gives the total angular momentum $\vec{L}_{tot}$ of the system of the student, stool, and wheel.

4. As the wheel is inverted, no net *external* torque acts on that system to change $\vec{L}_{tot}$ about any vertical axis. (Torques due to forces between the student and the wheel as the student inverts the wheel are *internal* to the system.) So, the system's total angular momentum is conserved about any vertical axis.

Calculations: The conservation of $\vec{L}_{tot}$ is represented with vectors in Fig. 11-20c. We can also write this conservation in terms of components along a vertical axis as

$$L_{b,f} + L_{wh,f} = L_{b,i} + L_{wh,i}, \tag{11-35}$$

where i and f refer to the initial state (before inversion of the wheel) and the final state (after inversion). Because inversion of the wheel inverted the angular momentum vector of the wheel's rotation, we substitute $-L_{wh,i}$ for $L_{wh,f}$. Then, if we set $L_{b,i} = 0$ (because the student, the stool, and the wheel's center were initially at rest), Eq. 11-35 yields

$$L_{b,f} = 2L_{wh,i}.$$

Using Eq. 11-31, we next substitute $I_b\omega_b$ for $L_{b,f}$ and $I_{wh}\omega_{wh}$ for $L_{wh,i}$ and solve for ω_b, finding

$$\omega_b = \frac{2I_{wh}}{I_b}\,\omega_{wh}$$

$$= \frac{(2)(1.2\ \text{kg}\cdot\text{m}^2)(3.9\ \text{rev/s})}{6.8\ \text{kg}\cdot\text{m}^2} = 1.4\ \text{rev/s}. \quad \text{(Answer)}$$

This positive result tells us that the student rotates counterclockwise about the stool axis as seen from overhead. If the student wishes to stop rotating, he has only to invert the wheel once more.

Sample Problem

Conservation of angular momentum, cockroach on disk

In Fig. 11-21, a cockroach with mass m rides on a disk of mass $6.00m$ and radius R. The disk rotates like a merry-go-round around its central axis at angular speed $\omega_i = 1.50$ rad/s. The cockroach is initially at radius $r = 0.800R$, but then it crawls out to the rim of the disk. Treat the cockroach as a particle. What then is the angular speed?

KEY IDEAS

(1) The cockroach's crawl changes the mass distribution (and thus the rotational inertia) of the cockroach–disk system. (2) The angular momentum of the system does not change because there is no external torque to change it. (The forces

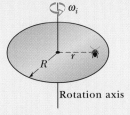

Fig. 11-21 A cockroach rides at radius r on a disk rotating like a merry-go-round.

and torques due to the cockroach's crawl are internal to the system.) (3) The magnitude of the angular momentum of a rigid body or a particle is given by Eq. 11-31 ($L = I\omega$).

Calculations: We want to find the final angular speed. Our key is to equate the final angular momentum L_f to the initial angular momentum L_i, because both involve angular speed. They also involve rotational inertia I. So, let's start by finding the rotational inertia of the system of cockroach and disk before and after the crawl.

The rotational inertia of a disk rotating about its central axis is given by Table 10-2c as $\frac{1}{2}MR^2$. Substituting $6.00m$ for the mass M, our disk here has rotational inertia

$$I_d = 3.00mR^2. \qquad (11\text{-}36)$$

(We don't have values for m and R, but we shall continue with physics courage.)

From Eq. 10-33, we know that the rotational inertia of the cockroach (a particle) is equal to mr^2. Substituting the cockroach's initial radius ($r = 0.800R$) and final radius ($r = R$), we find that its initial rotational inertia about the rotation axis is

$$I_{ci} = 0.64mR^2 \qquad (11\text{-}37)$$

and its final rotational inertia about the rotation axis is

$$I_{cf} = mR^2. \qquad (11\text{-}38)$$

So, the cockroach–disk system initially has the rotational inertia

$$I_i = I_d + I_{ci} = 3.64mR^2, \qquad (11\text{-}39)$$

and finally has the rotational inertia

$$I_f = I_d + I_{cf} = 4.00mR^2. \qquad (11\text{-}40)$$

Next, we use Eq. 11-31 ($L = I\omega$) to write the fact that the system's final angular momentum L_f is equal to the system's initial angular momentum L_i:

$$I_f\omega_f = I_i\omega_i$$

or $\quad 4.00mR^2\omega_f = 3.64mR^2(1.50 \text{ rad/s}).$

After canceling the unknowns m and R, we come to

$$\omega_f = 1.37 \text{ rad/s}. \qquad \text{(Answer)}$$

Note that the angular speed decreased because part of the mass moved outward from the rotation axis, thus increasing the rotational inertia of the system.

 Additional examples, video, and practice available at *WileyPLUS*

11-12 Precession of a Gyroscope

A simple gyroscope consists of a wheel fixed to a shaft and free to spin about the axis of the shaft. If one end of the shaft of a *nonspinning* gyroscope is placed on a support as in Fig. 11-22a and the gyroscope is released, the gyroscope falls by rotating downward about the tip of the support. Since the fall involves rotation, it is governed by Newton's second law in angular form, which is given by Eq. 11-29:

$$\vec{\tau} = \frac{d\vec{L}}{dt}. \qquad (11\text{-}41)$$

This equation tells us that the torque causing the downward rotation (the fall) changes the angular momentum $\vec{L}$ of the gyroscope from its initial value of zero. The torque $\vec{\tau}$ is due to the gravitational force $M\vec{g}$ acting at the gyroscope's center of mass, which we take to be at the center of the wheel. The moment arm relative to the support tip, located at O in Fig. 11-22a, is $\vec{r}$. The magnitude of $\vec{\tau}$ is

$$\tau = Mgr \sin 90° = Mgr \qquad (11\text{-}42)$$

(because the angle between $M\vec{g}$ and $\vec{r}$ is 90°), and its direction is as shown in Fig. 11-22a.

A rapidly spinning gyroscope behaves differently. Assume it is released with the shaft angled slightly upward. It first rotates slightly downward but then, while it is still spinning about its shaft, it begins to rotate horizontally about a vertical axis through support point O in a motion called **precession.**

Why does the spinning gyroscope stay aloft instead of falling over like the nonspinning gyroscope? The clue is that when the spinning gyroscope is released, the torque due to $M\vec{g}$ must change not an initial angular momentum of zero but rather some already existing nonzero angular momentum due to the spin.

To see how this nonzero initial angular momentum leads to precession, we first consider the angular momentum $\vec{L}$ of the gyroscope due to its spin. To simplify the situation, we assume the spin rate is so rapid that the angular momentum due to precession is negligible relative to $\vec{L}$. We also assume the shaft is horizontal when precession begins, as in Fig. 11-22b. The magnitude of $\vec{L}$ is given by Eq. 11-31:

$$L = I\omega, \tag{11-43}$$

where I is the rotational moment of the gyroscope about its shaft and ω is the angular speed at which the wheel spins about the shaft. The vector $\vec{L}$ points along the shaft, as in Fig. 11-22b. Since $\vec{L}$ is parallel to $\vec{r}$, torque $\vec{\tau}$ must be perpendicular to $\vec{L}$.

According to Eq. 11-41, torque $\vec{\tau}$ causes an incremental change $d\vec{L}$ in the angular momentum of the gyroscope in an incremental time interval dt; that is,

$$d\vec{L} = \vec{\tau}\, dt. \tag{11-44}$$

However, for a *rapidly spinning* gyroscope, the magnitude of $\vec{L}$ is fixed by Eq. 11-43. Thus the torque can change only the direction of $\vec{L}$, not its magnitude.

From Eq. 11-44 we see that the direction of $d\vec{L}$ is in the direction of $\vec{\tau}$, perpendicular to $\vec{L}$. The only way that $\vec{L}$ can be changed in the direction of $\vec{\tau}$ without the magnitude L being changed is for $\vec{L}$ to rotate around the z axis as shown in Fig. 11-22c. $\vec{L}$ maintains its magnitude, the head of the $\vec{L}$ vector follows a circular path, and $\vec{\tau}$ is always tangent to that path. Since $\vec{L}$ must always point along the shaft, the shaft must rotate about the z axis in the direction of $\vec{\tau}$. Thus we have precession. Because the spinning gyroscope must obey Newton's law in angular form in response to any change in its initial angular momentum, it must precess instead of merely toppling over.

We can find the **precession rate** Ω by first using Eqs. 11-44 and 11-42 to get the magnitude of $d\vec{L}$:

$$dL = \tau\, dt = Mgr\, dt. \tag{11-45}$$

As $\vec{L}$ changes by an incremental amount in an incremental time interval dt, the shaft and $\vec{L}$ precess around the z axis through incremental angle $d\phi$. (In Fig. 11-22c, angle $d\phi$ is exaggerated for clarity.) With the aid of Eqs. 11-43 and 11-45, we find that $d\phi$ is given by

$$d\phi = \frac{dL}{L} = \frac{Mgr\, dt}{I\omega}.$$

Dividing this expression by dt and setting the rate $\Omega = d\phi/dt$, we obtain

$$\Omega = \frac{Mgr}{I\omega} \qquad \text{(precession rate).} \tag{11-46}$$

This result is valid under the assumption that the spin rate ω is rapid. Note that Ω decreases as ω is increased. Note also that there would be no precession if the gravitational force $M\vec{g}$ did not act on the gyroscope, but because I is a function of M, mass cancels from Eq. 11-46; thus Ω is independent of the mass.

Equation 11-46 also applies if the shaft of a spinning gyroscope is at an angle to the horizontal. It holds as well for a spinning top, which is essentially a spinning gyroscope at an angle to the horizontal.

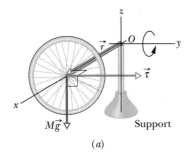

(a)

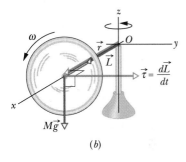

(b)

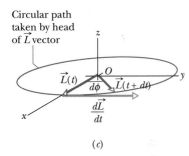

(c)

Fig. 11-22 (a) A nonspinning gyroscope falls by rotating in an xz plane because of torque $\vec{\tau}$. (b) A rapidly spinning gyroscope, with angular momentum $\vec{L}$, precesses around the z axis. Its precessional motion is in the xy plane. (c) The change $d\vec{L}/dt$ in angular momentum leads to a rotation of $\vec{L}$ about O.

Rolling Bodies   For a wheel of radius R rolling smoothly,

$$v_{\text{com}} = \omega R, \tag{11-2}$$

where v_{com} is the linear speed of the wheel's center of mass and ω is the angular speed of the wheel about its center. The wheel may also be viewed as rotating instantaneously about the point P of the "road" that is in contact with the wheel. The angular speed of the

wheel about this point is the same as the angular speed of the wheel about its center. The rolling wheel has kinetic energy

$$K = \tfrac{1}{2}I_{\text{com}}\omega^2 + \tfrac{1}{2}Mv_{\text{com}}^2, \tag{11-5}$$

where I_{com} is the rotational moment of the wheel about its center of mass and M is the mass of the wheel. If the wheel is being accelerated but is still rolling smoothly, the acceleration of the center of mass $\vec{a}_{com}$ is related to the angular acceleration α about the center with

$$a_{com} = \alpha R. \quad (11\text{-}6)$$

If the wheel rolls smoothly down a ramp of angle θ, its acceleration along an x axis extending up the ramp is

$$a_{com,x} = -\frac{g \sin \theta}{1 + I_{com}/MR^2}. \quad (11\text{-}10)$$

Torque as a Vector In three dimensions, *torque* $\vec{\tau}$ is a vector quantity defined relative to a fixed point (usually an origin); it is

$$\vec{\tau} = \vec{r} \times \vec{F}, \quad (11\text{-}14)$$

where $\vec{F}$ is a force applied to a particle and $\vec{r}$ is a position vector locating the particle relative to the fixed point. The magnitude of $\vec{\tau}$ is given by

$$\tau = rF \sin \phi = rF_{\perp} = r_{\perp}F, \quad (11\text{-}15, 11\text{-}16, 11\text{-}17)$$

where ϕ is the angle between $\vec{F}$ and $\vec{r}$, $F_{\perp}$ is the component of $\vec{F}$ perpendicular to $\vec{r}$, and $r_{\perp}$ is the moment arm of $\vec{F}$. The direction of $\vec{\tau}$ is given by the right-hand rule.

Angular Momentum of a Particle The *angular momentum* $\vec{\ell}$ of a particle with linear momentum $\vec{p}$, mass m, and linear velocity $\vec{v}$ is a vector quantity defined relative to a fixed point (usually an origin) as

$$\vec{\ell} = \vec{r} \times \vec{p} = m(\vec{r} \times \vec{v}). \quad (11\text{-}18)$$

The magnitude of $\vec{\ell}$ is given by

$$\ell = rmv \sin \phi \quad (11\text{-}19)$$
$$= rp_{\perp} = rmv_{\perp} \quad (11\text{-}20)$$
$$= r_{\perp}p = r_{\perp}mv, \quad (11\text{-}21)$$

where ϕ is the angle between $\vec{r}$ and $\vec{p}$, $p_{\perp}$ and $v_{\perp}$ are the components of $\vec{p}$ and $\vec{v}$ perpendicular to $\vec{r}$, and $r_{\perp}$ is the perpendicular distance between the fixed point and the extension of $\vec{p}$. The direction of $\vec{\ell}$ is given by the right-hand rule for cross products.

Newton's Second Law in Angular Form Newton's second law for a particle can be written in angular form as

$$\vec{\tau}_{net} = \frac{d\vec{\ell}}{dt}, \quad (11\text{-}23)$$

where $\vec{\tau}_{net}$ is the net torque acting on the particle and $\vec{\ell}$ is the angular momentum of the particle.

Angular Momentum of a System of Particles The angular momentum $\vec{L}$ of a system of particles is the vector sum of the angular momenta of the individual particles:

$$\vec{L} = \vec{\ell}_1 + \vec{\ell}_2 + \cdots + \vec{\ell}_n = \sum_{i=1}^{n} \vec{\ell}_i. \quad (11\text{-}26)$$

The time rate of change of this angular momentum is equal to the net external torque on the system (the vector sum of the torques due to interactions of the particles of the system with particles external to the system):

$$\vec{\tau}_{net} = \frac{d\vec{L}}{dt} \quad \text{(system of particles).} \quad (11\text{-}29)$$

Angular Momentum of a Rigid Body For a rigid body rotating about a fixed axis, the component of its angular momentum parallel to the rotation axis is

$$L = I\omega \quad \text{(rigid body, fixed axis).} \quad (11\text{-}31)$$

Conservation of Angular Momentum The angular momentum $\vec{L}$ of a system remains constant if the net external torque acting on the system is zero:

$$\vec{L} = \text{a constant} \quad \text{(isolated system)} \quad (11\text{-}32)$$

or
$$\vec{L}_i = \vec{L}_f \quad \text{(isolated system).} \quad (11\text{-}33)$$

This is the **law of conservation of angular momentum.**

Precession of a Gyroscope A spinning gyroscope can precess about a vertical axis through its support at the rate

$$\Omega = \frac{Mgr}{I\omega}, \quad (11\text{-}46)$$

where M is the gyroscope's mass, r is the moment arm, I is the rotational inertia, and ω is the spin rate.

QUESTIONS

1 Figure 11-23 shows three particles of the same mass and the same constant speed moving as indicated by the velocity vectors. Points a, b, c, and d form a square, with point e at the center. Rank the points according to the magnitude of the net angular momentum of the three-particle system when measured about the points, greatest first.

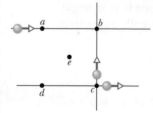

Fig. 11-23 Question 1.

2 Figure 11-24 shows two particles A and B at xyz coordinates $(1 \text{ m}, 1 \text{ m}, 0)$ and $(1 \text{ m}, 0, 1 \text{ m})$. Acting on each particle are three numbered forces, all of the same magnitude and each directed parallel to an axis. (a) Which of the forces produce a torque about the origin that is directed parallel to y? (b) Rank the forces according to the magnitudes of the torques they produce on the particles about the origin, greatest first.

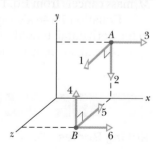

Fig. 11-24 Question 2.

3 What happens to the initially stationary yo-yo in Fig. 11-25 if you pull it via its string with (a) force $\vec{F}_2$ (the line of action passes

through the point of contact on the
table, as indicated), (b) force $\vec{F}_1$ (the
line of action passes above the point
of contact), and (c) force $\vec{F}_3$ (the line
of action passes to the right of the
point of contact)?

4 The position vector $\vec{r}$ of a particle
relative to a certain point has a magnitude of 3 m, and the force $\vec{F}$ on the
particle has a magnitude of 4 N. What
is the angle between the directions of
$\vec{r}$ and $\vec{F}$ if the magnitude of the associated torque equals (a) zero and (b)
12 N·m?

5 In Fig. 11-26, three forces of the
same magnitude are applied to a par-
ticle at the origin ($\vec{F}_1$ acts directly into
the plane of the figure). Rank the
forces according to the magnitudes of
the torques they create about (a)
point P_1, (b) point P_2, and (c) point
P_3, greatest first.

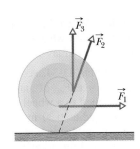

Fig. 11-25 Question 3.

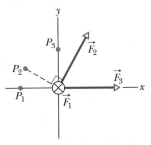

Fig. 11-26 Question 5.

6 The angular momenta $\ell(t)$ of a particle in four situations are
(1) $\ell = 3t + 4$; (2) $\ell = -6t^2$; (3) $\ell = 2$; (4) $\ell = 4/t$. In which situation is the net torque on the particle (a) zero, (b) positive and constant, (c) negative and increasing in magnitude ($t > 0$), and (d)
negative and decreasing in magnitude ($t > 0$)?

7 A rhinoceros beetle rides the rim of a horizontal disk rotating
counterclockwise like a merry-go-round. If the beetle then walks
along the rim in the direction of the rotation, will the magnitudes
of the following quantities (each measured about the rotation axis)
increase, decrease, or remain the same (the disk is still rotating in
the counterclockwise direction): (a) the angular momentum of the
beetle–disk system, (b) the angular momentum and angular velocity of the beetle, and (c) the angular momentum and angular velocity

of the disk? (d) What are your answers if the beetle walks in the direction opposite the rotation?

8 Figure 11-27 shows an overhead
view of a rectangular slab that can
spin like a merry-go-round about its
center at O. Also shown are seven
paths along which wads of bubble
gum can be thrown (all with the
same speed and mass) to stick onto
the stationary slab. (a) Rank the paths according to the angular
speed that the slab (and gum) will have after the gum sticks, greatest first. (b) For which paths will the angular momentum of the slab
(and gum) about O be negative from the view of Fig. 11-27?

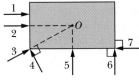

Fig. 11-27 Question 8.

9 Figure 11-28 gives the angular momentum magnitude L of a wheel versus
time t. Rank the four lettered time intervals according to the magnitude of the
torque acting on the wheel, greatest first.

10 Figure 11-29 shows a particle
moving at constant velocity $\vec{v}$ and five
points with their xy coordinates. Rank
the points according to the magnitude of the angular momentum
of the particle measured about them, greatest first.

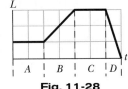

Fig. 11-28
Question 9.

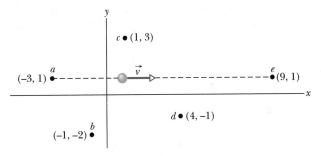

Fig. 11-29 Question 10.

sec. 11-2   Rolling as Translation and Rotation Combined

•1 A car travels at 80 km/h on a level road in the positive direction
of an x axis. Each tire has a diameter of 66 cm. Relative to a woman
riding in the car and in unit-vector notation, what are the velocity $\vec{v}$
at the (a) center, (b) top, and (c) bottom of the tire and the magnitude a of the acceleration at the (d) center, (e) top, and (f) bottom
of each tire? Relative to a hitchhiker sitting next to the road and in
unit-vector notation, what are the velocity $\vec{v}$ at the (g) center,
(h) top, and (i) bottom of the tire and the magnitude a of the
acceleration at the (j) center, (k) top, and (l) bottom of each tire?

•2 An automobile traveling at 80.0 km/h has tires of 75.0 cm diameter. (a) What is the angular speed of the tires about their axles?
(b) If the car is brought to a stop uniformly in 30.0 complete turns

of the tires (without skidding), what is the magnitude of the angular acceleration of the wheels? (c) How far does the car move during the braking?

sec. 11-4   The Forces of Rolling

•3 SSM A 140 kg hoop rolls along a horizontal floor so that the
hoop's center of mass has a speed of 0.150 m/s. How much work
must be done on the hoop to stop it?

•4 A uniform solid sphere rolls down an incline. (a) What must be
the incline angle if the linear acceleration of the center of the
sphere is to have a magnitude of 0.10g? (b) If a frictionless block
were to slide down the incline at that angle, would its acceleration
magnitude be more than, less than, or equal to 0.10g? Why?

•5 **ILW** A 1000 kg car has four 10 kg wheels. When the car is moving, what fraction of its total kinetic energy is due to rotation of the wheels about their axles? Assume that the wheels have the same rotational inertia as uniform disks of the same mass and size. Why do you not need to know the radius of the wheels?

••6 Figure 11-30 gives the speed v versus time t for a 0.500 kg object of radius 6.00 cm that rolls smoothly down a 30° ramp. The scale on the velocity axis is set by $v_s = 4.0$ m/s. What is the rotational inertia of the object?

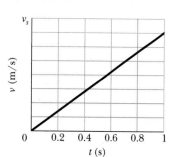

Fig. 11-30 Problem 6.

••7 **ILW** In Fig. 11-31, a solid cylinder of radius 10 cm and mass 12 kg starts from rest and rolls without slipping a distance $L = 6.0$ m down a roof that is inclined at the angle $\theta = 30°$. (a) What is the angular speed of the cylinder about its center as it leaves the roof? (b) The roof's edge is at height $H = 5.0$ m. How far horizontally from the roof's edge does the cylinder hit the level ground?

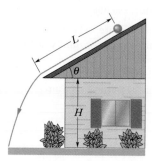

Fig. 11-31 Problem 7.

••8 Figure 11-32 shows the potential energy $U(x)$ of a solid ball that can roll along an x axis. The scale on the U axis is set by $U_s = 100$ J. The ball is uniform, rolls smoothly, and has a mass of 0.400 kg. It is released at $x = 7.0$ m headed in the negative direction of the x axis with a mechanical energy of 75 J. (a) If the ball can reach $x = 0$ m, what is its speed there, and if it cannot, what is its turning point? Suppose, instead, it is headed in the positive direction of the x axis when it is released at $x = 7.0$ m with 75 J. (b) If the ball can reach $x = 13$ m, what is its speed there, and if it cannot, what is its turning point?

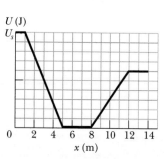

Fig. 11-32 Problem 8.

••9 **GO** In Fig. 11-33, a solid ball rolls smoothly from rest (starting at height $H = 6.0$ m) until it leaves the horizontal section at the end of the track, at height $h = 2.0$ m. How far horizontally from point A does the ball hit the floor?

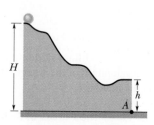

Fig. 11-33 Problem 9.

••10 A hollow sphere of radius 0.15 m, with rotational inertia $I = 0.040$ kg·m² about a line through its center of mass, rolls without slipping up a surface inclined at 30° to the horizontal. At a certain initial position, the sphere's total kinetic energy is 20 J. (a) How much of this initial kinetic energy is rotational? (b) What is the speed of the center of mass of the sphere at the initial

position? When the sphere has moved 1.0 m up the incline from its initial position, what are (c) its total kinetic energy and (d) the speed of its center of mass?

••11 In Fig. 11-34, a constant horizontal force $\vec{F}_{app}$ of magnitude 10 N is applied to a wheel of mass 10 kg and radius 0.30 m. The wheel rolls smoothly on the horizontal surface, and the acceleration of its center of mass has magnitude 0.60 m/s². (a) In unit-vector notation, what is the frictional force on the wheel? (b) What is the rotational inertia of the wheel about the rotation axis through its center of mass?

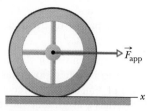

Fig. 11-34 Problem 11.

••12 **GO** In Fig. 11-35, a solid brass ball of mass 0.280 g will roll smoothly along a loop-the-loop track when released from rest along the straight section. The circular loop has radius $R = 14.0$ cm, and the ball has radius $r \ll R$. (a) What is h if the ball is on the verge of leaving the track when it reaches the top of the loop? If the ball is released at height $h = 6.00R$, what are the (b) magnitude and (c) direction of the horizontal force component acting on the ball at point Q?

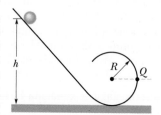

Fig. 11-35 Problem 12.

•••13 *Nonuniform ball.* In Fig. 11-36, a ball of mass M and radius R rolls smoothly from rest down a ramp and onto a circular loop of radius 0.48 m. The initial height of the ball is $h = 0.36$ m. At the loop bottom, the magnitude of the nor-

Fig. 11-36 Problem 13.

mal force on the ball is $2.00Mg$. The ball consists of an outer spherical shell (of a certain uniform density) that is glued to a central sphere (of a different uniform density). The rotational inertia of the ball can be expressed in the general form $I = \beta MR^2$, but β is not 0.4 as it is for a ball of uniform density. Determine β.

•••14 **GO** In Fig. 11-37, a small, solid, uniform ball is to be shot from point P so that it rolls smoothly along a horizontal path, up along a ramp, and onto a plateau. Then it leaves the plateau horizontally to land on a game board, at a horizontal distance d from the right edge of the plateau. The vertical heights are $h_1 = 5.00$ cm and $h_2 = 1.60$ cm. With what speed must the ball be shot at point P for it to land at $d = 6.00$ cm?

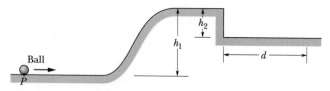

Fig. 11-37 Problem 14.

•••15 ✈ A bowler throws a bowling ball of radius $R = 11$ cm along a lane. The ball (Fig. 11-38) slides on the lane with initial speed $v_{com,0} = 8.5$ m/s and initial angular speed $\omega_0 = 0$. The coefficient of kinetic friction between the ball and the lane is 0.21. The kinetic frictional force $\vec{f}_k$ acting on the ball causes a linear acceleration of the ball while producing a torque that causes an angular acceleration of the ball. When speed v_{com} has decreased enough and angular speed ω has increased enough, the ball stops sliding and then rolls smoothly. (a) What then is v_{com} in terms of ω? During the sliding, what are the ball's (b) linear acceleration and (c) angular acceleration? (d) How long does the ball slide? (e) How far does the ball slide? (f) What is the linear speed of the ball when smooth rolling begins?

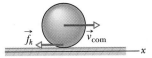

Fig. 11-38 Problem 15.

•••**16** *Nonuniform cylindrical object.* In Fig. 11-39, a cylindrical object of mass M and radius R rolls smoothly from rest down a ramp and onto a horizontal section. From there it rolls off the ramp and onto the floor, landing a horizontal distance $d = 0.506$ m from the end of the ramp. The initial height of the object is $H = 0.90$ m; the end of the ramp is at height $h = 0.10$ m. The object consists of an outer cylindrical shell (of a certain uniform density) that is glued to a central cylinder (of a different uniform density). The rotational inertia of the object can be expressed in the general form $I = \beta MR^2$, but β is not 0.5 as it is for a cylinder of uniform density. Determine β.

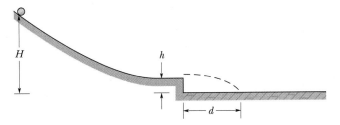

Fig. 11-39 Problem 16.

sec. 11-5 The Yo-Yo

•**17** SSM A yo-yo has a rotational inertia of 950 g·cm² and a mass of 120 g. Its axle radius is 3.2 mm, and its string is 120 cm long. The yo-yo rolls from rest down to the end of the string. (a) What is the magnitude of its linear acceleration? (b) How long does it take to reach the end of the string? As it reaches the end of the string, what are its (c) linear speed, (d) translational kinetic energy, (e) rotational kinetic energy, and (f) angular speed?

•**18** In 1980, over San Francisco Bay, a large yo-yo was released from a crane. The 116 kg yo-yo consisted of two uniform disks of radius 32 cm connected by an axle of radius 3.2 cm. What was the magnitude of the acceleration of the yo-yo during (a) its fall and (b) its rise? (c) What was the tension in the cord on which it rolled? (d) Was that tension near the cord's limit of 52 kN? Suppose you build a scaled-up version of the yo-yo (same shape and materials but larger). (e) Will the magnitude of your yo-yo's acceleration as it falls be greater than, less than, or the same as that of the San Francisco yo-yo? (f) How about the tension in the cord?

sec. 11-6 Torque Revisited

•**19** In unit-vector notation, what is the net torque about the origin on a flea located at coordinates $(0, -4.0$ m, 5.0 m$)$ when forces $\vec{F}_1 = (3.0$ N$)\hat{k}$ and $\vec{F}_2 = (-2.0$ N$)\hat{j}$ act on the flea?

•**20** A plum is located at coordinates $(-2.0$ m, $0, 4.0$ m$)$. In unit-vector notation, what is the torque about the origin on the plum if that torque is due to a force $\vec{F}$ whose only component is (a) $F_x = 6.0$ N, (b) $F_x = -6.0$ N, (c) $F_z = 6.0$ N, and (d) $F_z = -6.0$ N?

•**21** In unit-vector notation, what is the torque about the origin on a particle located at coordinates $(0, -4.0$ m, 3.0 m$)$ if that torque is due to (a) force $\vec{F}_1$ with components $F_{1x} = 2.0$ N, $F_{1y} = F_{1z} = 0$, and (b) force $\vec{F}_2$ with components $F_{2x} = 0, F_{2y} = 2.0$ N, $F_{2z} = 4.0$ N?

••**22** A particle moves through an xyz coordinate system while a force acts on the particle. When the particle has the position vector $\vec{r} = (2.00$ m$)\hat{i} - (3.00$ m$)\hat{j} + (2.00$ m$)\hat{k}$, the force is given by $\vec{F} = F_x\hat{i} + (7.00$ N$)\hat{j} - (6.00$ N$)\hat{k}$ and the corresponding torque about the origin is $\vec{\tau} = (4.00$ N·m$)\hat{i} + (2.00$ N·m$)\hat{j} - (1.00$ N·m$)\hat{k}$. Determine F_x.

••**23** Force $\vec{F} = (2.0$ N$)\hat{i} - (3.0$ N$)\hat{k}$ acts on a pebble with position vector $\vec{r} = (0.50$ m$)\hat{j} - (2.0$ m$)\hat{k}$ relative to the origin. In unit-vector notation, what is the resulting torque on the pebble about (a) the origin and (b) the point $(2.0$ m, $0, -3.0$ m$)$?

••**24** In unit-vector notation, what is the torque about the origin on a jar of jalapeño peppers located at coordinates $(3.0$ m, -2.0 m, 4.0 m$)$ due to (a) force $\vec{F}_1 = (3.0$ N$)\hat{i} - (4.0$ N$)\hat{j} + (5.0$ N$)\hat{k}$, (b) force $\vec{F}_2 = (-3.0$ N$)\hat{i} - (4.0$ N$)\hat{j} - (5.0$ N$)\hat{k}$, and (c) the vector sum of $\vec{F}_1$ and $\vec{F}_2$? (d) Repeat part (c) for the torque about the point with coordinates $(3.0$ m, 2.0 m, 4.0 m$)$.

••**25** SSM Force $\vec{F} = (-8.0$ N$)\hat{i} + (6.0$ N$)\hat{j}$ acts on a particle with position vector $\vec{r} = (3.0$ m$)\hat{i} + (4.0$ m$)\hat{j}$. What are (a) the torque on the particle about the origin, in unit-vector notation, and (b) the angle between the directions of $\vec{r}$ and $\vec{F}$?

sec. 11-7 Angular Momentum

•**26** At the instant of Fig. 11-40, a 2.0 kg particle P has a position vector $\vec{r}$ of magnitude 3.0 m and angle $\theta_1 = 45°$ and a velocity vector $\vec{v}$ of magnitude 4.0 m/s and angle $\theta_2 = 30°$. Force $\vec{F}$, of magnitude 2.0 N and angle $\theta_3 = 30°$, acts on P. All three vectors lie in the xy plane. About the origin, what are the (a) magnitude and (b) direction of the angular momentum of P and the (c) magnitude and (d) direction of the torque acting on P?

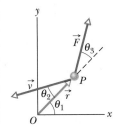

Fig. 11-40
Problem 26.

•**27** SSM At one instant, force $\vec{F} = 4.0\hat{j}$ N acts on a 0.25 kg object that has position vector $\vec{r} = (2.0\hat{i} - 2.0\hat{k})$ m and velocity vector $\vec{v} = (-5.0\hat{i} + 5.0\hat{k})$ m/s. About the origin and in unit-vector notation, what are (a) the object's angular momentum and (b) the torque acting on the object?

•**28** A 2.0 kg particle-like object moves in a plane with velocity components $v_x = 30$ m/s and $v_y = 60$ m/s as it passes through the point with (x, y) coordinates of $(3.0, -4.0)$ m. Just then, in unit-vector notation, what is its angular momentum relative to (a) the origin and (b) the point located at $(-2.0, -2.0)$ m?

•**29** ILW In the instant of Fig. 11-41, two particles move in an xy plane. Particle P_1 has mass 6.5 kg and speed $v_1 = 2.2$ m/s, and it is at distance $d_1 = 1.5$ m from point O. Particle P_2 has mass 3.1 kg and speed

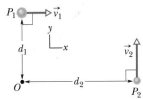

Fig. 11-41 Problem 29.

$v_2 = 3.6$ m/s, and it is at distance $d_2 = 2.8$ m from point O. What are the (a) magnitude and (b) direction of the net angular momentum of the two particles about O?

••30 At the instant the displacement of a 2.00 kg object relative to the origin is $\vec{d} = (2.00 \text{ m})\hat{i} + (4.00 \text{ m})\hat{j} - (3.00 \text{ m})\hat{k}$, its velocity is $\vec{v} = -(6.00 \text{ m/s})\hat{i} + (3.00 \text{ m/s})\hat{j} + (3.00 \text{ m/s})\hat{k}$ and it is subject to a force $\vec{F} = (6.00 \text{ N})\hat{i} - (8.00 \text{ N})\hat{j} + (4.00 \text{ N})\hat{k}$. Find (a) the acceleration of the object, (b) the angular momentum of the object about the origin, (c) the torque about the origin acting on the object, and (d) the angle between the velocity of the object and the force acting on the object.

••31 In Fig. 11-42, a 0.400 kg ball is shot directly upward at initial speed 40.0 m/s. What is its angular momentum about P, 2.00 m horizontally from the launch point, when the ball is (a) at maximum height and (b) halfway back to the ground? What is the torque on the ball about P due to the gravitational force when the ball is (c) at maximum height and (d) halfway back to the ground?

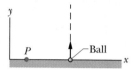

Fig. 11-42 Problem 31.

sec. 11-8 Newton's Second Law in Angular Form

•32 A particle is acted on by two torques about the origin: $\vec{\tau}_1$ has a magnitude of 2.0 N · m and is directed in the positive direction of the x axis, and $\vec{\tau}_2$ has a magnitude of 4.0 N·m and is directed in the negative direction of the y axis. In unit-vector notation, find $d\vec{\ell}/dt$, where $\vec{\ell}$ is the angular momentum of the particle about the origin.

•33 SSM ILW WWW At time $t = 0$, a 3.0 kg particle with velocity $\vec{v} = (5.0 \text{ m/s})\hat{i} - (6.0 \text{ m/s})\hat{j}$ is at $x = 3.0$ m, $y = 8.0$ m. It is pulled by a 7.0 N force in the negative x direction. About the origin, what are (a) the particle's angular momentum, (b) the torque acting on the particle, and (c) the rate at which the angular momentum is changing?

•34 A particle is to move in an xy plane, clockwise around the origin as seen from the positive side of the z axis. In unit-vector notation, what torque acts on the particle if the magnitude of its angular momentum about the origin is (a) 4.0 kg·m²/s, (b) $4.0t^2$ kg·m²/s, (c) $4.0\sqrt{t}$ kg·m²/s, and (d) $4.0/t^2$ kg·m²/s?

••35 At time t, the vector $\vec{r} = 4.0t^2\hat{i} - (2.0t + 6.0t^2)\hat{j}$ gives the position of a 3.0 kg particle relative to the origin of an xy coordinate system ($\vec{r}$ is in meters and t is in seconds). (a) Find an expression for the torque acting on the particle relative to the origin. (b) Is the magnitude of the particle's angular momentum relative to the origin increasing, decreasing, or unchanging?

sec. 11-10 The Angular Momentum of a Rigid Body Rotating About a Fixed Axis

•36 Figure 11-43 shows three rotating, uniform disks that are coupled by belts. One belt runs around the rims of disks A and C. Another belt runs around a central hub on disk A and the rim of disk B. The belts move smoothly without slippage on the rims and hub. Disk A has radius R; its hub has radius $0.5000R$; disk B has radius $0.2500R$; and disk C has radius $2.000R$. Disks B and C have the

same density (mass per unit volume) and thickness. What is the ratio of the magnitude of the angular momentum of disk C to that of disk B?

•37 GO In Fig. 11-44, three particles of mass $m = 23$ g are fastened to three rods of length $d = 12$ cm and negligible mass. The rigid assembly rotates around point O at the angular speed $\omega = 0.85$ rad/s. About O, what are (a) the rotational inertia of the assembly, (b) the magnitude of the angular momentum of the middle particle, and (c) the magnitude of the angular momentum of the asssembly?

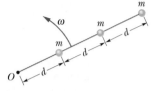

Fig. 11-44 Problem 37.

•38 A sanding disk with rotational inertia 1.2×10^{-3} kg·m² is attached to an electric drill whose motor delivers a torque of magnitude 16 N·m about the central axis of the disk. About that axis and with the torque applied for 33 ms, what is the magnitude of the (a) angular momentum and (b) angular velocity of the disk?

•39 SSM The angular momentum of a flywheel having a rotational inertia of 0.140 kg · m² about its central axis decreases from 3.00 to 0.800 kg · m²/s in 1.50 s. (a) What is the magnitude of the average torque acting on the flywheel about its central axis during this period? (b) Assuming a constant angular acceleration, through what angle does the flywheel turn? (c) How much work is done on the wheel? (d) What is the average power of the flywheel?

••40 A disk with a rotational inertia of 7.00 kg·m² rotates like a merry-go-round while undergoing a variable torque given by $\tau = (5.00 + 2.00t)$ N·m. At time $t = 1.00$ s, its angular momentum is 5.00 kg·m²/s. What is its angular momentum at $t = 3.00$ s?

••41 GO Figure 11-45 shows a rigid structure consisting of a circular hoop of radius R and mass m, and a square made of four thin bars, each of length R and mass m. The rigid structure rotates at a constant speed about a vertical axis, with a period of rotation of 2.5 s. Assuming $R = 0.50$ m and $m = 2.0$ kg, calculate (a) the structure's rotational inertia about the axis of rotation and (b) its angular momentum about that axis.

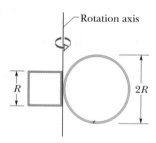

Fig. 11-45 Problem 41.

••42 Figure 11-46 gives the torque τ that acts on an initially stationary disk that can rotate about its center like a merry-go-round.

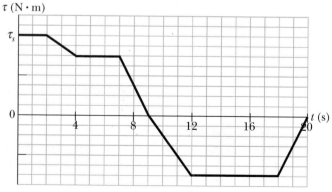

Fig. 11-46 Problem 42.

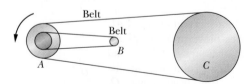

Fig. 11-43 Problem 36.

The scale on the τ axis is set by $\tau_s = 4.0$ N·m. What is the angular momentum of the disk about the rotation axis at times (a) $t = 7.0$ s and (b) $t = 20$ s?

sec. 11-11   Conservation of Angular Momentum

•43 In Fig. 11-47, two skaters, each of mass 50 kg, approach each other along parallel paths separated by 3.0 m. They have opposite velocities of 1.4 m/s each. One skater carries one end of a long pole of negligible mass, and the other skater grabs the other end as she passes. The skaters then

Fig. 11-47   Problem 43.

rotate around the center of the pole. Assume that the friction between skates and ice is negligible. What are (a) the radius of the circle, (b) the angular speed of the skaters, and (c) the kinetic energy of the two-skater system? Next, the skaters pull along the pole until they are separated by 1.0 m. What then are (d) their angular speed and (e) the kinetic energy of the system? (f) What provided the energy for the increased kinetic energy?

•44 A Texas cockroach of mass 0.17 kg runs counterclockwise around the rim of a lazy Susan (a circular disk mounted on a vertical axle) that has radius 15 cm, rotational inertia 5.0×10^{-3} kg·m², and frictionless bearings. The cockroach's speed (relative to the ground) is 2.0 m/s, and the lazy Susan turns clockwise with angular speed $\omega_0 = 2.8$ rad/s. The cockroach finds a bread crumb on the rim and, of course, stops. (a) What is the angular speed of the lazy Susan after the cockroach stops? (b) Is mechanical energy conserved as it stops?

•45 SSM WWW A man stands on a platform that is rotating (without friction) with an angular speed of 1.2 rev/s; his arms are outstretched and he holds a brick in each hand. The rotational inertia of the system consisting of the man, bricks, and platform about the central vertical axis of the platform is 6.0 kg·m². If by moving the bricks the man decreases the rotational inertia of the system to 2.0 kg·m², what are (a) the resulting angular speed of the platform and (b) the ratio of the new kinetic energy of the system to the original kinetic energy? (c) What source provided the added kinetic energy?

•46 The rotational inertia of a collapsing spinning star drops to $\frac{1}{3}$ its initial value. What is the ratio of the new rotational kinetic energy to the initial rotational kinetic energy?

•47 SSM A track is mounted on a large wheel that is free to turn with negligible friction about a vertical axis (Fig. 11-48). A toy train of mass m is placed on the track and, with the system initially at rest, the train's electrical power is turned on. The train reaches speed 0.15 m/s with respect to the track. What is the angular speed of the wheel if its mass is $1.1m$ and its radius is 0.43 m? (Treat the wheel as a hoop, and neglect the mass of the spokes and hub.)

Fig. 11-48   Problem 47.

•48 A Texas cockroach first rides at the center of a circular disk that rotates freely like a merry-go-round without external torques. The cockroach then walks out to

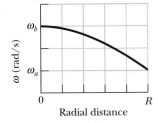

Fig. 11-49   Problem 48.

the edge of the disk, at radius R. Figure 11-49 gives the angular speed ω of the cockroach–disk system during the walk. The scale on the ω axis is set by $\omega_a = 5.0$ rad/s and $\omega_b = 6.0$ rad/s. When the cockroach is on the edge at radius R, what is the ratio of the bug's rotational inertia to that of the disk, both calculated about the rotation axis?

•49 Two disks are mounted (like a merry-go-round) on low-friction bearings on the same axle and can be brought together so that they couple and rotate as one unit. The first disk, with rotational inertia 3.30 kg·m² about its central axis, is set spinning counterclockwise at 450 rev/min. The second disk, with rotational inertia 6.60 kg·m² about its central axis, is set spinning counterclockwise at 900 rev/min. They then couple together. (a) What is their angular speed after coupling? If instead the second disk is set spinning clockwise at 900 rev/min, what are their (b) angular speed and (c) direction of rotation after they couple together?

•50 The rotor of an electric motor has rotational inertia $I_m = 2.0 \times 10^{-3}$ kg·m² about its central axis. The motor is used to change the orientation of the space probe in which it is mounted. The motor axis is mounted along the central axis of the probe; the probe has rotational inertia $I_p = 12$ kg·m² about this axis. Calculate the number of revolutions of the rotor required to turn the probe through 30° about its central axis.

•51 SSM ILW A wheel is rotating freely at angular speed 800 rev/min on a shaft whose rotational inertia is negligible. A second wheel, initially at rest and with twice the rotational inertia of the first, is suddenly coupled to the same shaft. (a) What is the angular speed of the resultant combination of the shaft and two wheels? (b) What fraction of the original rotational kinetic energy is lost?

••52 GO A cockroach of mass m lies on the rim of a uniform disk of mass $4.00m$ that can rotate freely about its center like a merry-go-round. Initially the cockroach and disk rotate together with an angular velocity of 0.260 rad/s. Then the cockroach walks halfway to the center of the disk. (a) What then is the angular velocity of the cockroach–disk system? (b) What is the ratio K/K_0 of the new kinetic energy of the system to its initial kinetic energy? (c) What accounts for the change in the kinetic energy?

••53 GO A uniform thin rod of length 0.500 m and mass 4.00 kg can rotate in a horizontal plane about a vertical axis through its center. The rod is at rest when a 3.00 g bullet traveling in the rotation plane is fired into one end of the rod. As viewed from above, the bullet's path makes angle $\theta = 60.0°$ with the rod (Fig. 11-50). If the bullet lodges in the rod and the angular velocity of the rod is 10 rad/s immediately after the collision, what is the bullet's speed just before impact?

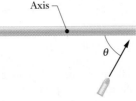

Fig. 11-50   Problem 53.

••54 GO Figure 11-51 shows an overhead view of a ring that can rotate about its center like a merry-go-round. Its outer radius R_2 is 0.800 m, its inner radius R_1 is $R_2/2.00$, its mass M is 8.00 kg, and the mass of the crossbars at its center is negligible. It initially rotates at an angular speed of 8.00 rad/s with a cat of

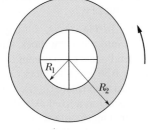

Fig. 11-51   Problem 54.

mass $m = M/4.00$ on its outer edge, at radius R_2. By how much does the cat increase the kinetic energy of the cat–ring system if the cat crawls to the inner edge, at radius R_1?

••55 A horizontal vinyl record of mass 0.10 kg and radius 0.10 m rotates freely about a vertical axis through its center with an angular speed of 4.7 rad/s. The rotational inertia of the record about its axis of rotation is 5.0×10^{-4} kg·m². A wad of wet putty of mass 0.020 kg drops vertically onto the record from above and sticks to the edge of the record. What is the angular speed of the record immediately after the putty sticks to it?

••56 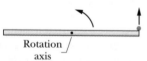 In a long jump, an athlete leaves the ground with an initial angular momentum that tends to rotate her body forward, threatening to ruin her landing. To counter this tendency, she rotates her outstretched arms to "take up" the angular momentum (Fig. 11-18). In 0.700 s, one arm sweeps through 0.500 rev and the other arm sweeps through 1.000 rev. Treat each arm as a thin rod of mass 4.0 kg and length 0.60 m, rotating around one end. In the athlete's reference frame, what is the magnitude of the total angular momentum of the arms around the common rotation axis through the shoulders?

••57 A uniform disk of mass $10m$ and radius $3.0r$ can rotate freely about its fixed center like a merry-go-round. A smaller uniform disk of mass m and radius r lies on top of the larger disk, concentric with it. Initially the two disks rotate together with an angular velocity of 20 rad/s. Then a slight disturbance causes the smaller disk to slide outward across the larger disk, until the outer edge of the smaller disk catches on the outer edge of the larger disk. Afterward, the two disks again rotate together (without further sliding). (a) What then is their angular velocity about the center of the larger disk? (b) What is the ratio K/K_0 of the new kinetic energy of the two-disk system to the system's initial kinetic energy?

••58 A horizontal platform in the shape of a circular disk rotates on a frictionless bearing about a vertical axle through the center of the disk. The platform has a mass of 150 kg, a radius of 2.0 m, and a rotational inertia of 300 kg·m² about the axis of rotation. A 60 kg student walks slowly from the rim of the platform toward the center. If the angular speed of the system is 1.5 rad/s when the student starts at the rim, what is the angular speed when she is 0.50 m from the center?

••59 Figure 11-52 is an overhead view of a thin uniform rod of length 0.800 m and mass M rotating horizontally at angular speed 20.0 rad/s about an axis through its center. A particle of mass $M/3.00$ initially attached to one end is ejected from the rod and travels along a path that is perpendicular to the rod at the instant of ejection. If the particle's speed v_p is 6.00 m/s greater than the speed of the rod end just after ejection, what is the value of v_p?

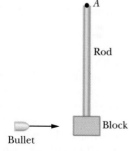

Fig. 11-52 Problem 59.

••60 In Fig. 11-53, a 1.0 g bullet is fired into a 0.50 kg block attached to the end of a 0.60 m nonuniform rod of mass 0.50 kg. The block–rod–bullet system then rotates in the plane of the figure, about a fixed axis at A. The rotational inertia of the rod alone about that axis at A is 0.060 kg·m². Treat the block as a particle. (a) What then is

Fig. 11-53 Problem 60.

the rotational inertia of the block–rod–bullet system about point A? (b) If the angular speed of the system about A just after impact is 4.5 rad/s, what is the bullet's speed just before impact?

••61 The uniform rod (length 0.60 m, mass 1.0 kg) in Fig. 11-54 rotates in the plane of the figure about an axis through one end, with a rotational inertia of 0.12 kg·m². As the rod swings through its lowest position, it collides with a 0.20 kg putty wad that sticks to the end of the rod. If the rod's angular speed just before collision is 2.4 rad/s, what is the angular speed of the rod–putty system immediately after collision?

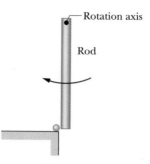

Fig. 11-54 Problem 61.

•••62 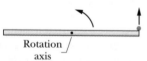 During a jump to his partner, an aerialist is to make a quadruple somersault lasting a time $t = 1.87$ s. For the first and last quarter-revolution, he is in the extended orientation shown in Fig. 11-55, with rotational inertia $I_1 = 19.9$ kg·m² around his center of mass (the dot). During the rest of the flight he is in a tight tuck, with rotational inertia $I_2 = 3.93$ kg·m². What must be his angular speed ω_2 around his center of mass during the tuck?

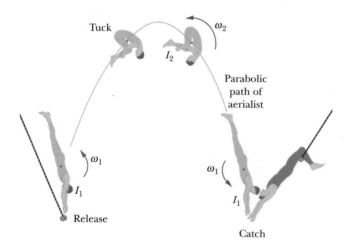

Fig. 11-55 Problem 62.

•••63 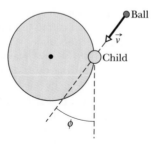 GO In Fig. 11-56, a 30 kg child stands on the edge of a stationary merry-go-round of radius 2.0 m. The rotational inertia of the merry-go-round about its rotation axis is 150 kg·m². The child catches a ball of mass 1.0 kg thrown by a friend. Just before the ball is caught, it has a horizontal velocity $\vec{v}$ of magnitude 12 m/s, at angle $\phi = 37°$ with a line tangent to the outer edge of the merry-go-round, as shown. What is the angular speed of the merry-go-round just after the ball is caught?

Fig. 11-56 Problem 63.

•••64 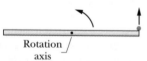 A ballerina begins a tour jeté (Fig. 11-19a) with angular speed ω_i and a rotational inertia consisting of two parts: $I_{leg} = 1.44$ kg·m² for her leg extended outward at angle $\theta = 90.0°$ to her body and $I_{trunk} = 0.660$ kg·m² for the rest of her body (primarily her trunk). Near her maximum height she holds both legs at angle $\theta = 30.0°$ to her body and has angular speed ω_f (Fig. 11-19b). Assuming that I_{trunk} has not changed, what is the ratio ω_f/ω_i?

•••65 SSM WWW Two 2.00 kg balls are attached to the ends of a thin rod of length 50.0 cm and negligible mass. The rod is free to rotate in a vertical plane without friction about a horizontal axis through its center. With the rod initially horizontal (Fig. 11-57), a 50.0 g wad of wet putty drops onto one of the balls, hitting it with a speed of 3.00 m/s and then sticking to it. (a) What is the angular speed of the system just after the putty wad hits? (b) What is the ratio of the kinetic energy of the system after the collision to that of the putty wad just before? (c) Through what angle will the system rotate before it momentarily stops?

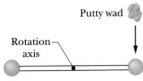

Fig. 11-57   Problem 65.

•••66 In Fig. 11-58, a small 50 g block slides down a frictionless surface through height $h = 20$ cm and then sticks to a uniform rod of mass 100 g and length 40 cm. The rod pivots about point O through angle θ before momentarily stopping. Find θ.

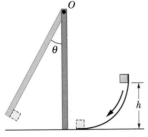

Fig. 11-58   Problem 66.

•••67 Figure 11-59 is an overhead view of a thin uniform rod of length 0.600 m and mass M rotating horizontally at 80.0 rad/s counterclockwise about an axis through its center. A particle of mass $M/3.00$ and traveling horizontally at speed 40.0 m/s hits the rod and sticks. The particle's path is perpendicular to the rod at the instant of the hit, at a distance d from the rod's center. (a) At what value of d are rod and particle stationary after the hit? (b) In which direction do rod and particle rotate if d is greater than this value?

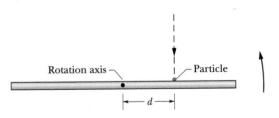

Fig. 11-59   Problem 67.

sec. 11-12   Precession of a Gyroscope
••68 A top spins at 30 rev/s about an axis that makes an angle of 30° with the vertical. The mass of the top is 0.50 kg, its rotational inertia about its central axis is 5.0×10^{-4} kg·m², and its center of mass is 4.0 cm from the pivot point. If the spin is clockwise from an overhead view, what are the (a) precession rate and (b) direction of the precession as viewed from overhead?

••69 A certain gyroscope consists of a uniform disk with a 50 cm radius mounted at the center of an axle that is 11 cm long and of negligible mass. The axle is horizontal and supported at one end. If the disk is spinning around the axle at 1000 rev/min, what is the precession rate?

Additional Problems

70 A uniform solid ball rolls smoothly along a floor, then up a ramp inclined at 15.0°. It momentarily stops when it has rolled 1.50 m along the ramp. What was its initial speed?

71 SSM In Fig. 11-60, a constant horizontal force $\vec{F}_{app}$ of magnitude 12 N is applied to a uniform solid cylinder by fishing line wrapped around the cylinder. The mass of the cylinder is 10 kg, its radius is 0.10 m, and the cylinder rolls smoothly on the horizontal surface. (a) What is the magnitude of the acceleration of the center of mass of the cylinder? (b) What is the magnitude of the angular acceleration of the cylinder about the center of mass? (c) In unit-vector notation, what is the frictional force acting on the cylinder?

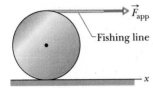

Fig. 11-60   Problem 71.

72 A thin-walled pipe rolls along the floor. What is the ratio of its translational kinetic energy to its rotational kinetic energy about the central axis parallel to its length?

73 SSM A 3.0 kg toy car moves along an x axis with a velocity given by $\vec{v} = -2.0t^3\hat{i}$ m/s, with t in seconds. For $t > 0$, what are (a) the angular momentum $\vec{L}$ of the car and (b) the torque $\vec{\tau}$ on the car, both calculated about the origin? What are (c) $\vec{L}$ and (d) $\vec{\tau}$ about the point (2.0 m, 5.0 m, 0)? What are (e) $\vec{L}$ and (f) $\vec{\tau}$ about the point (2.0 m, −5.0 m, 0)?

74 A wheel rotates clockwise about its central axis with an angular momentum of 600 kg·m²/s. At time $t = 0$, a torque of magnitude 50 N·m is applied to the wheel to reverse the rotation. At what time t is the angular speed zero?

75 SSM In a playground, there is a small merry-go-round of radius 1.20 m and mass 180 kg. Its radius of gyration (see Problem 79 of Chapter 10) is 91.0 cm. A child of mass 44.0 kg runs at a speed of 3.00 m/s along a path that is tangent to the rim of the initially stationary merry-go-round and then jumps on. Neglect friction between the bearings and the shaft of the merry-go-round. Calculate (a) the rotational inertia of the merry-go-round about its axis of rotation, (b) the magnitude of the angular momentum of the running child about the axis of rotation of the merry-go-round, and (c) the angular speed of the merry-go-round and child after the child has jumped onto the merry-go-round.

76 A uniform block of granite in the shape of a book has face dimensions of 20 cm and 15 cm and a thickness of 1.2 cm. The density (mass per unit volume) of granite is 2.64 g/cm³. The block rotates around an axis that is perpendicular to its face and halfway between its center and a corner. Its angular momentum about that axis is 0.104 kg·m²/s. What is its rotational kinetic energy about that axis?

77 SSM Two particles, each of mass 2.90×10^{-4} kg and speed 5.46 m/s, travel in opposite directions along parallel lines separated by 4.20 cm. (a) What is the magnitude L of the angular momentum of the two-particle system around a point midway between the two lines? (b) Does the value of L change if the point about which it is calculated is not midway between the lines? If the direction of travel for one of the particles is reversed, what would be (c) the answer to part (a) and (d) the answer to part (b)?

78 A wheel of radius 0.250 m, which is moving initially at 43.0 m/s, rolls to a stop in 225 m. Calculate the magnitudes of (a) its lin-

ear acceleration and (b) its angular acceleration. (c) The wheel's rotational inertia is 0.155 kg·m² about its central axis. Calculate the magnitude of the torque about the central axis due to friction on the wheel.

79 Wheels A and B in Fig. 11-61 are connected by a belt that does not slip. The radius of B is 3.00 times the radius of A. What would be the ratio of the rotational inertias I_A/I_B if the two wheels had (a) the same angular momentum about their central axes and (b) the same rotational kinetic energy?

Fig. 11-61 Problem 79.

80 A 2.50 kg particle that is moving horizontally over a floor with velocity $(-3.00 \text{ m/s})\hat{j}$ undergoes a completely inelastic collision with a 4.00 kg particle that is moving horizontally over the floor with velocity $(4.50 \text{ m/s})\hat{i}$. The collision occurs at xy coordinates $(-0.500 \text{ m}, -0.100 \text{ m})$. After the collision and in unit-vector notation, what is the angular momentum of the stuck-together particles with respect to the origin?

81 **SSM** A uniform wheel of mass 10.0 kg and radius 0.400 m is mounted rigidly on a massless axle through its center (Fig. 11-62). The radius of the axle is 0.200 m, and the rotational inertia of the wheel–axle combination about its central axis is 0.600 kg·m². The wheel is initially at rest at the top of a surface that is inclined at angle $\theta = 30.0°$ with the horizontal; the axle rests on the surface while the wheel extends into a groove in the surface without touching the surface. Once released, the axle rolls down along the surface smoothly and without slipping. When the wheel–axle combination has moved down the surface by 2.00 m, what are (a) its rotational kinetic energy and (b) its translational kinetic energy?

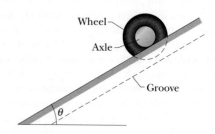

Fig. 11-62 Problem 81.

82 A uniform rod rotates in a horizontal plane about a vertical axis through one end. The rod is 6.00 m long, weighs 10.0 N, and rotates at 240 rev/min. Calculate (a) its rotational inertia about the axis of rotation and (b) the magnitude of its angular momentum about that axis.

83 A solid sphere of weight 36.0 N rolls up an incline at an angle of 30.0°. At the bottom of the incline the center of mass of the sphere has a translational speed of 4.90 m/s. (a) What is the kinetic energy of the sphere at the bottom of the incline? (b) How far does the sphere travel up along the incline? (c) Does the answer to (b) depend on the sphere's mass?

84 Suppose that the yo-yo in Problem 17, instead of rolling from rest, is thrown so that its initial speed down the string is 1.3 m/s. (a) How long does the yo-yo take to reach the end of the string? As it reaches the end of the string, what are its (b) total ki-

netic energy, (c) linear speed, (d) translational kinetic energy, (e) angular speed, and (f) rotational kinetic energy?

85 A girl of mass M stands on the rim of a frictionless merry-go-round of radius R and rotational inertia I that is not moving. She throws a rock of mass m horizontally in a direction that is tangent to the outer edge of the merry-go-round. The speed of the rock, relative to the ground, is v. Afterward, what are (a) the angular speed of the merry-go-round and (b) the linear speed of the girl?

86 At time $t = 0$, a 2.0 kg particle has the position vector $\vec{r} = (4.0 \text{ m})\hat{i} - (2.0 \text{ m})\hat{j}$ relative to the origin. Its velocity is given by $\vec{v} = (-6.0t^2 \text{ m/s})\hat{i}$ for $t \geq 0$ in seconds. About the origin, what are (a) the particle's angular momentum $\vec{L}$ and (b) the torque $\vec{\tau}$ acting on the particle, both in unit-vector notation and for $t > 0$? About the point $(-2.0 \text{ m}, -3.0 \text{ m}, 0)$, what are (c) $\vec{L}$ and (d) $\vec{\tau}$ for $t > 0$?

87 If Earth's polar ice caps fully melted and the water returned to the oceans, the oceans would be deeper by about 30 m. What effect would this have on Earth's rotation? Make an estimate of the resulting change in the length of the day.

88 A 1200 kg airplane is flying in a straight line at 80 m/s, 1.3 km above the ground. What is the magnitude of its angular momentum with respect to a point on the ground directly under the path of the plane?

89 With axle and spokes of negligible mass and a thin rim, a certain bicycle wheel has a radius of 0.350 m and weighs 37.0 N; it can turn on its axle with negligible friction. A man holds the wheel above his head with the axle vertical while he stands on a turntable that is free to rotate without friction; the wheel rotates clockwise, as seen from above, with an angular speed of 57.7 rad/s, and the turntable is initially at rest. The rotational inertia of *wheel* + *man* + *turntable* about the common axis of rotation is 2.10 kg·m². The man's free hand suddenly stops the rotation of the wheel (relative to the turntable). Determine the resulting (a) angular speed and (b) direction of rotation of the system.

90 For an 84 kg person standing at the equator, what is the magnitude of the angular momentum about Earth's center due to Earth's rotation?

91 A small solid sphere with radius 0.25 cm and mass 0.56 g rolls without slipping on the inside of a large fixed hemisphere with radius 15 cm and a vertical axis of symmetry. The sphere starts at the top from rest. (a) What is its kinetic energy at the bottom? (b) What fraction of its kinetic energy at the bottom is associated with rotation about an axis through its com? (c) What is the magnitude of the normal force on the hemisphere from the sphere when the sphere reaches the bottom?

92 An automobile has a total mass of 1700 kg. It accelerates from rest to 40 km/h in 10 s. Assume each wheel is a uniform 32 kg disk. Find, for the end of the 10 s interval, (a) the rotational kinetic energy of each wheel about its axle, (b) the total kinetic energy of each wheel, and (c) the total kinetic energy of the automobile.

93 A body of radius R and mass m is rolling smoothly with speed v on a horizontal surface. It then rolls up a hill to a maximum height h. (a) If $h = 3v^2/4g$, what is the body's rotational inertia about the rotational axis through its center of mass? (b) What might the body be?

The International System of Units (SI)*

TABLE 1

The SI Base Units

Quantity	Name	Symbol	Definition
length	meter	m	"... the length of the path traveled by light in vacuum in 1/299,792,458 of a second." (1983)
mass	kilogram	kg	"... this prototype [a certain platinum–iridium cylinder] shall henceforth be considered to be the unit of mass." (1889)
time	second	s	"... the duration of 9,192,631,770 periods of the radiation corresponding to the transition between the two hyperfine levels of the ground state of the cesium-133 atom." (1967)
electric current	ampere	A	"... that constant current which, if maintained in two straight parallel conductors of infinite length, of negligible circular cross section, and placed 1 meter apart in vacuum, would produce between these conductors a force equal to 2×10^{-7} newton per meter of length." (1946)
thermodynamic temperature	kelvin	K	"... the fraction 1/273.16 of the thermodynamic temperature of the triple point of water." (1967)
amount of substance	mole	mol	"... the amount of substance of a system which contains as many elementary entities as there are atoms in 0.012 kilogram of carbon-12." (1971)
luminous intensity	candela	cd	"... the luminous intensity, in a given direction, of a source that emits monochromatic radiation of frequency 540×10^{12} hertz and that has a radiant intensity in that direction of 1/683 watt per steradian." (1979)

*Adapted from "The International System of Units (SI)," National Bureau of Standards Special Publication 330, 1972 edition. The definitions above were adopted by the General Conference of Weights and Measures, an international body, on the dates shown. In this book we do not use the candela.

TABLE 2

Some SI Derived Units

Quantity	Name of Unit	Symbol	
area	square meter	m^2	
volume	cubic meter	m^3	
frequency	hertz	Hz	s^{-1}
mass density (density)	kilogram per cubic meter	kg/m^3	
speed, velocity	meter per second	m/s	
angular velocity	radian per second	rad/s	
acceleration	meter per second per second	m/s^2	
angular acceleration	radian per second per second	rad/s^2	
force	newton	N	$kg \cdot m/s^2$
pressure	pascal	Pa	N/m^2
work, energy, quantity of heat	joule	J	$N \cdot m$
power	watt	W	J/s
quantity of electric charge	coulomb	C	$A \cdot s$
potential difference, electromotive force	volt	V	W/A
electric field strength	volt per meter (or newton per coulomb)	V/m	N/C
electric resistance	ohm	Ω	V/A
capacitance	farad	F	$A \cdot s/V$
magnetic flux	weber	Wb	$V \cdot s$
inductance	henry	H	$V \cdot s/A$
magnetic flux density	tesla	T	Wb/m^2
magnetic field strength	ampere per meter	A/m	
entropy	joule per kelvin	J/K	
specific heat	joule per kilogram kelvin	$J/(kg \cdot K)$	
thermal conductivity	watt per meter kelvin	$W/(m \cdot K)$	
radiant intensity	watt per steradian	W/sr	

TABLE 3

The SI Supplementary Units

Quantity	Name of Unit	Symbol
plane angle	radian	rad
solid angle	steradian	sr

Some Fundamental Constants of Physics*

Constant	Symbol	Computational Value	Best (1998) Value Value[a]	Best (1998) Value Uncertainty[b]
Speed of light in a vacuum	c	3.00×10^8 m/s	2.997 924 58	exact
Elementary charge	e	1.60×10^{-19} C	1.602 176 487	0.025
Gravitational constant	G	6.67×10^{-11} m³/s²·kg	6.674 28	100
Universal gas constant	R	8.31 J/mol·K	8.314 472	1.7
Avogadro constant	N_A	6.02×10^{23} mol⁻¹	6.022 141 79	0.050
Boltzmann constant	k	1.38×10^{-23} J/K	1.380 650 4	1.7
Stefan–Boltzmann constant	σ	5.67×10^{-8} W/m²·K⁴	5.670 400	7.0
Molar volume of ideal gas at STP[d]	V_m	2.27×10^{-2} m³/mol	2.271 098 1	1.7
Permittivity constant	ϵ_0	8.85×10^{-12} F/m	8.854 187 817 62	exact
Permeability constant	μ_0	1.26×10^{-6} H/m	1.256 637 061 43	exact
Planck constant	h	6.63×10^{-34} J·s	6.626 068 96	0.050
Electron mass[c]	m_e	9.11×10^{-31} kg	9.109 382 15	0.050
		5.49×10^{-4} u	5.485 799 094 3	4.2×10^{-4}
Proton mass[c]	m_p	1.67×10^{-27} kg	1.672 621 637	0.050
		1.0073 u	1.007 276 466 77	1.0×10^{-4}
Ratio of proton mass to electron mass	m_p/m_e	1840	1836.152 672 47	4.3×10^{-4}
Electron charge-to-mass ratio	e/m_e	1.76×10^{11} C/kg	1.758 820 150	0.025
Neutron mass[c]	m_n	1.68×10^{-27} kg	1.674 927 211	0.050
		1.0087 u	1.008 664 915 97	4.3×10^{-4}
Hydrogen atom mass[c]	m_{1_H}	1.0078 u	1.007 825 031 6	0.0005
Deuterium atom mass[c]	m_{2_H}	2.0136 u	2.013 553 212 724	3.9×10^{-5}
Helium atom mass[c]	$m_{4_{He}}$	4.0026 u	4.002 603 2	0.067
Muon mass	m_μ	1.88×10^{-28} kg	1.883 531 30	0.056
Electron magnetic moment	μ_e	9.28×10^{-24} J/T	9.284 763 77	0.025
Proton magnetic moment	μ_p	1.41×10^{-26} J/T	1.410 606 662	0.026
Bohr magneton	μ_B	9.27×10^{-24} J/T	9.274 009 15	0.025
Nuclear magneton	μ_N	5.05×10^{-27} J/T	5.050 783 24	0.025
Bohr radius	a	5.29×10^{-11} m	5.291 772 085 9	6.8×10^{-4}
Rydberg constant	R	1.10×10^7 m⁻¹	1.097 373 156 852 7	6.6×10^{-6}
Electron Compton wavelength	λ_C	2.43×10^{-12} m	2.426 310 217 5	0.0014

[a]Values given in this column should be given the same unit and power of 10 as the computational value.

[b]Parts per million.

[c]Masses given in u are in unified atomic mass units, where 1 u = 1.660 538 782 $\times 10^{-27}$ kg.

[d]STP means standard temperature and pressure: 0°C and 1.0 atm (0.1 MPa).

*The values in this table were selected from the 1998 CODATA recommended values (www.physics.nist.gov).

Some Astronomical Data

Some Distances from Earth

To the Moon*	3.82×10^8 m	To the center of our galaxy	2.2×10^{20} m
To the Sun*	1.50×10^{11} m	To the Andromeda Galaxy	2.1×10^{22} m
To the nearest star (Proxima Centauri)	4.04×10^{16} m	To the edge of the observable universe	$\sim 10^{26}$ m

*Mean distance.

The Sun, Earth, and the Moon

Property	Unit	Sun	Earth	Moon
Mass	kg	1.99×10^{30}	5.98×10^{24}	7.36×10^{22}
Mean radius	m	6.96×10^8	6.37×10^6	1.74×10^6
Mean density	kg/m^3	1410	5520	3340
Free-fall acceleration at the surface	m/s^2	274	9.81	1.67
Escape velocity	km/s	618	11.2	2.38
Period of rotation[a]	—	37 d at poles[b] 26 d at equator[b]	23 h 56 min	27.3 d
Radiation power[c]	W	3.90×10^{26}		

[a]Measured with respect to the distant stars.
[b]The Sun, a ball of gas, does not rotate as a rigid body.
[c]Just outside Earth's atmosphere solar energy is received, assuming normal incidence, at the rate of 1340 W/m^2.

Some Properties of the Planets

	Mercury	Venus	Earth	Mars	Jupiter	Saturn	Uranus	Neptune	Pluto
Mean distance from Sun, 10^6 km	57.9	108	150	228	778	1430	2870	4500	5900
Period of revolution, y	0.241	0.615	1.00	1.88	11.9	29.5	84.0	165	248
Period of rotation,[a] d	58.7	-243^b	0.997	1.03	0.409	0.426	-0.451^b	0.658	6.39
Orbital speed, km/s	47.9	35.0	29.8	24.1	13.1	9.64	6.81	5.43	4.74
Inclination of axis to orbit	<28°	≈3°	23.4°	25.0°	3.08°	26.7°	97.9°	29.6°	57.5°
Inclination of orbit to Earth's orbit	7.00°	3.39°		1.85°	1.30°	2.49°	0.77°	1.77°	17.2°
Eccentricity of orbit	0.206	0.0068	0.0167	0.0934	0.0485	0.0556	0.0472	0.0086	0.250
Equatorial diameter, km	4880	12 100	12 800	6790	143 000	120 000	51 800	49 500	2300
Mass (Earth = 1)	0.0558	0.815	1.000	0.107	318	95.1	14.5	17.2	0.002
Density (water = 1)	5.60	5.20	5.52	3.95	1.31	0.704	1.21	1.67	2.03
Surface value of g,[c] m/s^2	3.78	8.60	9.78	3.72	22.9	9.05	7.77	11.0	0.5
Escape velocity,[c] km/s	4.3	10.3	11.2	5.0	59.5	35.6	21.2	23.6	1.3
Known satellites	0	0	1	2	63 + ring	60 + rings	27 + rings	13 + rings	3

[a]Measured with respect to the distant stars.
[b]Venus and Uranus rotate opposite their orbital motion.
[c]Gravitational acceleration measured at the planet's equator.

Conversion Factors

Conversion factors may be read directly from these tables. For example, 1 degree = 2.778×10^{-3} revolutions, so $16.7° = 16.7 \times 2.778 \times 10^{-3}$ rev. The SI units are fully capitalized. Adapted in part from G. Shortley and D. Williams, *Elements of Physics,* 1971, Prentice-Hall, Englewood Cliffs, NJ.

Plane Angle

	°	′	″	RADIAN	rev
1 degree =	1	60	3600	1.745×10^{-2}	2.778×10^{-3}
1 minute =	1.667×10^{-2}	1	60	2.909×10^{-4}	4.630×10^{-5}
1 second =	2.778×10^{-4}	1.667×10^{-2}	1	4.848×10^{-6}	7.716×10^{-7}
1 RADIAN =	57.30	3438	2.063×10^{5}	1	0.1592
1 revolution =	360	2.16×10^{4}	1.296×10^{6}	6.283	1

Solid Angle

1 sphere = 4π steradians = 12.57 steradians

Length

	cm	METER	km	in.	ft	mi
1 centimeter =	1	10^{-2}	10^{-5}	0.3937	3.281×10^{-2}	6.214×10^{-6}
1 METER =	100	1	10^{-3}	39.37	3.281	6.214×10^{-4}
1 kilometer =	10^{5}	1000	1	3.937×10^{4}	3281	0.6214
1 inch =	2.540	2.540×10^{-2}	2.540×10^{-5}	1	8.333×10^{-2}	1.578×10^{-5}
1 foot =	30.48	0.3048	3.048×10^{-4}	12	1	1.894×10^{-4}
1 mile =	1.609×10^{5}	1609	1.609	6.336×10^{4}	5280	1

1 angström = 10^{-10} m
1 nautical mile = 1852 m
  = 1.151 miles = 6076 ft

1 fermi = 10^{-15} m
1 light-year = 9.461×10^{12} km
1 parsec = 3.084×10^{13} km

1 fathom = 6 ft
1 Bohr radius = 5.292×10^{-11} m
1 yard = 3 ft

1 rod = 16.5 ft
1 mil = 10^{-3} in.
1 nm = 10^{-9} m

Area

	METER^2	cm^2	ft^2	in.^{2}
1 SQUARE METER =	1	10^{4}	10.76	1550
1 square centimeter =	10^{-4}	1	1.076×10^{-3}	0.1550
1 square foot =	9.290×10^{-2}	929.0	1	144
1 square inch =	6.452×10^{-4}	6.452	6.944×10^{-3}	1

1 square mile = 2.788×10^{7} ft^2 = 640 acres
1 barn = 10^{-28} m^2

1 acre = 43 560 ft^2
1 hectare = 10^{4} m^2 = 2.471 acres

Volume

	METER^3	cm^3	L	ft^3	in.^{3}
1 CUBIC METER = 1		10^6	1000	35.31	6.102×10^4
1 cubic centimeter = 10^{-6}		1	1.000×10^{-3}	3.531×10^{-5}	6.102×10^{-2}
1 liter = 1.000×10^{-3}		1000	1	3.531×10^{-2}	61.02
1 cubic foot = 2.832×10^{-2}		2.832×10^4	28.32	1	1728
1 cubic inch = 1.639×10^{-5}		16.39	1.639×10^{-2}	5.787×10^{-4}	1

1 U.S. fluid gallon = 4 U.S. fluid quarts = 8 U.S. pints = 128 U.S. fluid ounces = 231 in.^{3}
1 British imperial gallon = 277.4 in.^{3} = 1.201 U.S. fluid gallons

Mass

Quantities in the colored areas are not mass units but are often used as such. For example, when we write 1 kg "=" 2.205 lb, this means that a kilogram is a *mass* that *weighs* 2.205 pounds at a location where g has the standard value of 9.80665 m/s^2.

	g	KILOGRAM	slug	u	oz	lb	ton
1 gram = 1		0.001	6.852×10^{-5}	6.022×10^{23}	3.527×10^{-2}	2.205×10^{-3}	1.102×10^{-6}
1 KILOGRAM = 1000		1	6.852×10^{-2}	6.022×10^{26}	35.27	2.205	1.102×10^{-3}
1 slug = 1.459×10^4		14.59	1	8.786×10^{27}	514.8	32.17	1.609×10^{-2}
1 atomic mass unit = 1.661×10^{-24}		1.661×10^{-27}	1.138×10^{-28}	1	5.857×10^{-26}	3.662×10^{-27}	1.830×10^{-30}
1 ounce = 28.35		2.835×10^{-2}	1.943×10^{-3}	1.718×10^{25}	1	6.250×10^{-2}	3.125×10^{-5}
1 pound = 453.6		0.4536	3.108×10^{-2}	2.732×10^{26}	16	1	0.0005
1 ton = 9.072×10^5		907.2	62.16	5.463×10^{29}	3.2×10^4	2000	1

1 metric ton = 1000 kg

Density

Quantities in the colored areas are weight densities and, as such, are dimensionally different from mass densities. See the note for the mass table.

	slug/ft^3	KILOGRAM/ METER^3	g/cm^3	lb/ft^3	lb/in.^{3}
1 slug per foot^3 = 1		515.4	0.5154	32.17	1.862×10^{-2}
1 KILOGRAM per METER^3 = 1.940×10^{-3}		1	0.001	6.243×10^{-2}	3.613×10^{-5}
1 gram per centimeter^3 = 1.940		1000	1	62.43	3.613×10^{-2}
1 pound per foot^3 = 3.108×10^{-2}		16.02	16.02×10^{-2}	1	5.787×10^{-4}
1 pound per inch^3 = 53.71		2.768×10^4	27.68	1728	1

Time

	y	d	h	min	SECOND
1 year = 1		365.25	8.766×10^3	5.259×10^5	3.156×10^7
1 day = 2.738×10^{-3}		1	24	1440	8.640×10^4
1 hour = 1.141×10^{-4}		4.167×10^{-2}	1	60	3600
1 minute = 1.901×10^{-6}		6.944×10^{-4}	1.667×10^{-2}	1	60
1 SECOND = 3.169×10^{-8}		1.157×10^{-5}	2.778×10^{-4}	1.667×10^{-2}	1

Speed

	ft/s	km/h	METER/SECOND	mi/h	cm/s
1 foot per second = 1		1.097	0.3048	0.6818	30.48
1 kilometer per hour = 0.9113		1	0.2778	0.6214	27.78
1 METER per SECOND = 3.281		3.6	1	2.237	100
1 mile per hour = 1.467		1.609	0.4470	1	44.70
1 centimeter per second = 3.281×10^{-2}		3.6×10^{-2}	0.01	2.237×10^{-2}	1

1 knot = 1 nautical mi/h = 1.688 ft/s   1 mi/min = 88.00 ft/s = 60.00 mi/h

Force

Force units in the colored areas are now little used. To clarify: 1 gram-force (= 1 gf) is the force of gravity that would act on an object whose mass is 1 gram at a location where g has the standard value of 9.80665 m/s^2.

	dyne	NEWTON	lb	pdl	gf	kgf
1 dyne = 1		10^{-5}	2.248×10^{-6}	7.233×10^{-5}	1.020×10^{-3}	1.020×10^{-6}
1 NEWTON = 10^5		1	0.2248	7.233	102.0	0.1020
1 pound = 4.448×10^5		4.448	1	32.17	453.6	0.4536
1 poundal = 1.383×10^4		0.1383	3.108×10^{-2}	1	14.10	1.410×10^2
1 gram-force = 980.7		9.807×10^{-3}	2.205×10^{-3}	7.093×10^{-2}	1	0.001
1 kilogram-force = 9.807×10^5		9.807	2.205	70.93	1000	1

1 ton = 2000 lb

Pressure

	atm	dyne/cm^2	inch of water	cm Hg	PASCAL	lb/in.^{2}	lb/ft^2
1 atmosphere = 1		1.013×10^6	406.8	76	1.013×10^5	14.70	2116
1 dyne per centimeter^2 = 9.869×10^{-7}		1	4.015×10^{-4}	7.501×10^{-5}	0.1	1.405×10^{-5}	2.089×10^{-3}
1 inch of water^a at 4°C = 2.458×10^{-3}		2491	1	0.1868	249.1	3.613×10^{-2}	5.202
1 centimeter of mercury^a at 0°C = 1.316×10^{-2}		1.333×10^4	5.353	1	1333	0.1934	27.85
1 PASCAL = 9.869×10^{-6}		10	4.015×10^{-3}	7.501×10^{-4}	1	1.450×10^{-4}	2.089×10^{-2}
1 pound per inch^2 = 6.805×10^{-2}		6.895×10^4	27.68	5.171	6.895×10^3	1	144
1 pound per foot^2 = 4.725×10^{-4}		478.8	0.1922	3.591×10^{-2}	47.88	6.944×10^{-3}	1

^{a}Where the acceleration of gravity has the standard value of 9.80665 m/s^2.

1 bar = 10^6 dyne/cm^2 = 0.1 MPa   1 millibar = 10^3 dyne/cm^2 = 10^2 Pa   1 torr = 1 mm Hg

Energy, Work, Heat

Quantities in the colored areas are not energy units but are included for convenience. They arise from the relativistic mass−energy equivalence formula $E = mc^2$ and represent the energy released if a kilogram or unified atomic mass unit (u) is completely converted to energy (bottom two rows) or the mass that would be completely converted to one unit of energy (rightmost two columns).

	Btu	erg	ft·lb	hp·h	JOULE	cal	kW·h	eV	MeV	kg	u
1 British thermal unit =	1	1.055×10^{10}	777.9	3.929×10^{-4}	1055	252.0	2.930×10^{-4}	6.585×10^{21}	6.585×10^{15}	1.174×10^{-14}	7.070×10^{12}
1 erg =	9.481×10^{-11}	1	7.376×10^{-8}	3.725×10^{-14}	10^{-7}	2.389×10^{-8}	2.778×10^{-14}	6.242×10^{11}	6.242×10^{5}	1.113×10^{-24}	670.2
1 foot-pound =	1.285×10^{-3}	1.356×10^{7}	1	5.051×10^{-7}	1.356	0.3238	3.766×10^{-7}	8.464×10^{18}	8.464×10^{12}	1.509×10^{-17}	9.037×10^{9}
1 horsepower-hour =	2545	2.685×10^{13}	1.980×10^{6}	1	2.685×10^{6}	6.413×10^{5}	0.7457	1.676×10^{25}	1.676×10^{19}	2.988×10^{-11}	1.799×10^{16}
1 JOULE =	9.481×10^{-4}	10^{7}	0.7376	3.725×10^{-7}	1	0.2389	2.778×10^{-7}	6.242×10^{18}	6.242×10^{12}	1.113×10^{-17}	6.702×10^{9}
1 calorie =	3.968×10^{-3}	4.1868×10^{7}	3.088	1.560×10^{-6}	4.1868	1	1.163×10^{-6}	2.613×10^{19}	2.613×10^{13}	4.660×10^{-17}	2.806×10^{10}
1 kilowatt-hour =	3413	3.600×10^{13}	2.655×10^{6}	1.341	3.600×10^{6}	8.600×10^{5}	1	2.247×10^{25}	2.247×10^{19}	4.007×10^{-11}	2.413×10^{16}
1 electron-volt =	1.519×10^{-22}	1.602×10^{-12}	1.182×10^{-19}	5.967×10^{-26}	1.602×10^{-19}	3.827×10^{-20}	4.450×10^{-26}	1	10^{-6}	1.783×10^{-36}	1.074×10^{-9}
1 million electron-volts =	1.519×10^{-16}	1.602×10^{-6}	1.182×10^{-13}	5.967×10^{-20}	1.602×10^{-13}	3.827×10^{-14}	4.450×10^{-20}	10^{6}	1	1.783×10^{-30}	1.074×10^{-3}
1 kilogram =	8.521×10^{13}	8.987×10^{23}	6.629×10^{16}	3.348×10^{10}	8.987×10^{16}	2.146×10^{16}	2.497×10^{10}	5.610×10^{35}	5.610×10^{29}	1	6.022×10^{26}
1 unified atomic mass unit =	1.415×10^{-13}	1.492×10^{-3}	1.101×10^{-10}	5.559×10^{-17}	1.492×10^{-10}	3.564×10^{-11}	4.146×10^{-17}	9.320×10^{8}	932.0	1.661×10^{-27}	1

Power

	Btu/h	ft·lb/s	hp	cal/s	kW	WATT
1 British thermal unit per hour =	1	0.2161	3.929×10^{-4}	6.998×10^{-2}	2.930×10^{-4}	0.2930
1 foot-pound per second =	4.628	1	1.818×10^{-3}	0.3239	1.356×10^{-3}	1.356
1 horsepower =	2545	550	1	178.1	0.7457	745.7
1 calorie per second =	14.29	3.088	5.615×10^{-3}	1	4.186×10^{-3}	4.186
1 kilowatt =	3413	737.6	1.341	238.9	1	1000
1 WATT =	3.413	0.7376	1.341×10^{-3}	0.2389	0.001	1

Magnetic Field

	gauss	TESLA	milligauss
1 gauss =	1	10^{-4}	1000
1 TESLA =	10^{4}	1	10^{7}
1 milligauss =	0.001	10^{-7}	1

Magnetic Flux

	maxwell	WEBER
1 maxwell =	1	10^{-8}
1 WEBER =	10^{8}	1

1 tesla = 1 weber/meter^2

Mathematical Formulas

Geometry

Circle of radius r: circumference $= 2\pi r$; area $= \pi r^2$.

Sphere of radius r: area $= 4\pi r^2$; volume $= \frac{4}{3}\pi r^3$.

Right circular cylinder of radius r and height h:
area $= 2\pi r^2 + 2\pi rh$; volume $= \pi r^2 h$.

Triangle of base a and altitude h: area $= \frac{1}{2}ah$.

Quadratic Formula

If $ax^2 + bx + c = 0$, then $x = \dfrac{-b \pm \sqrt{b^2 - 4ac}}{2a}$.

Trigonometric Functions of Angle θ

$\sin \theta = \dfrac{y}{r} \quad \cos \theta = \dfrac{x}{r}$

$\tan \theta = \dfrac{y}{x} \quad \cot \theta = \dfrac{x}{y}$

$\sec \theta = \dfrac{r}{x} \quad \csc \theta = \dfrac{r}{y}$

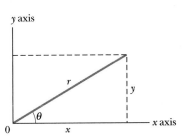

Pythagorean Theorem

In this right triangle,
$$a^2 + b^2 = c^2$$

Triangles

Angles are A, B, C

Opposite sides are a, b, c

Angles $A + B + C = 180°$

$\dfrac{\sin A}{a} = \dfrac{\sin B}{b} = \dfrac{\sin C}{c}$

$c^2 = a^2 + b^2 - 2ab \cos C$

Exterior angle $D = A + C$

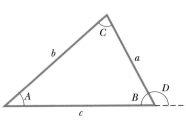

Mathematical Signs and Symbols

$=$ equals

$\approx$ equals approximately

$\sim$ is the order of magnitude of

$\neq$ is not equal to

$\equiv$ is identical to, is defined as

$>$ is greater than ($\gg$ is much greater than)

$<$ is less than ($\ll$ is much less than)

$\geq$ is greater than or equal to (or, is no less than)

$\leq$ is less than or equal to (or, is no more than)

$\pm$ plus or minus

$\propto$ is proportional to

Σ the sum of

x_{avg} the average value of x

Trigonometric Identities

$\sin(90° - \theta) = \cos \theta$

$\cos(90° - \theta) = \sin \theta$

$\sin \theta / \cos \theta = \tan \theta$

$\sin^2 \theta + \cos^2 \theta = 1$

$\sec^2 \theta - \tan^2 \theta = 1$

$\csc^2 \theta - \cot^2 \theta = 1$

$\sin 2\theta = 2 \sin \theta \cos \theta$

$\cos 2\theta = \cos^2 \theta - \sin^2 \theta = 2\cos^2 \theta - 1 = 1 - 2\sin^2 \theta$

$\sin(\alpha \pm \beta) = \sin \alpha \cos \beta \pm \cos \alpha \sin \beta$

$\cos(\alpha \pm \beta) = \cos \alpha \cos \beta \mp \sin \alpha \sin \beta$

$\tan(\alpha \pm \beta) = \dfrac{\tan \alpha \pm \tan \beta}{1 \mp \tan \alpha \tan \beta}$

$\sin \alpha \pm \sin \beta = 2 \sin \frac{1}{2}(\alpha \pm \beta) \cos \frac{1}{2}(\alpha \mp \beta)$

$\cos \alpha + \cos \beta = 2 \cos \frac{1}{2}(\alpha + \beta) \cos \frac{1}{2}(\alpha - \beta)$

$\cos \alpha - \cos \beta = -2 \sin \frac{1}{2}(\alpha + \beta) \sin \frac{1}{2}(\alpha - \beta)$

Binomial Theorem

$$(1 + x)^n = 1 + \frac{nx}{1!} + \frac{n(n-1)x^2}{2!} + \cdots \qquad (x^2 < 1)$$

Exponential Expansion

$$e^x = 1 + x + \frac{x^2}{2!} + \frac{x^3}{3!} + \cdots$$

Logarithmic Expansion

$$\ln(1 + x) = x - \tfrac{1}{2}x^2 + \tfrac{1}{3}x^3 - \cdots \qquad (|x| < 1)$$

Trigonometric Expansions (θ in radians)

$$\sin \theta = \theta - \frac{\theta^3}{3!} + \frac{\theta^5}{5!} - \cdots$$

$$\cos \theta = 1 - \frac{\theta^2}{2!} + \frac{\theta^4}{4!} - \cdots$$

$$\tan \theta = \theta + \frac{\theta^3}{3} + \frac{2\theta^5}{15} + \cdots$$

Cramer's Rule

Two simultaneous equations in unknowns x and y,

$$a_1 x + b_1 y = c_1 \quad \text{and} \quad a_2 x + b_2 y = c_2,$$

have the solutions

$$x = \frac{\begin{vmatrix} c_1 & b_1 \\ c_2 & b_2 \end{vmatrix}}{\begin{vmatrix} a_1 & b_1 \\ a_2 & b_2 \end{vmatrix}} = \frac{c_1 b_2 - c_2 b_1}{a_1 b_2 - a_2 b_1}$$

and

$$y = \frac{\begin{vmatrix} a_1 & c_1 \\ a_2 & c_2 \end{vmatrix}}{\begin{vmatrix} a_1 & b_1 \\ a_2 & b_2 \end{vmatrix}} = \frac{a_1 c_2 - a_2 c_1}{a_1 b_2 - a_2 b_1}.$$

Products of Vectors

Let $\hat{\imath}$, $\hat{\jmath}$, and $\hat{k}$ be unit vectors in the x, y, and z directions. Then

$$\hat{\imath} \cdot \hat{\imath} = \hat{\jmath} \cdot \hat{\jmath} = \hat{k} \cdot \hat{k} = 1, \quad \hat{\imath} \cdot \hat{\jmath} = \hat{\jmath} \cdot \hat{k} = \hat{k} \cdot \hat{\imath} = 0,$$

$$\hat{\imath} \times \hat{\imath} = \hat{\jmath} \times \hat{\jmath} = \hat{k} \times \hat{k} = 0,$$

$$\hat{\imath} \times \hat{\jmath} = \hat{k}, \quad \hat{\jmath} \times \hat{k} = \hat{\imath}, \quad \hat{k} \times \hat{\imath} = \hat{\jmath}$$

Any vector $\vec{a}$ with components a_x, a_y, and a_z along the x, y, and z axes can be written as

$$\vec{a} = a_x \hat{\imath} + a_y \hat{\jmath} + a_z \hat{k}.$$

Let $\vec{a}$, $\vec{b}$, and $\vec{c}$ be arbitrary vectors with magnitudes a, b, and c. Then

$$\vec{a} \times (\vec{b} + \vec{c}) = (\vec{a} \times \vec{b}) + (\vec{a} \times \vec{c})$$

$$(s\vec{a}) \times \vec{b} = \vec{a} \times (s\vec{b}) = s(\vec{a} \times \vec{b}) \qquad (s = \text{a scalar}).$$

Let θ be the smaller of the two angles between $\vec{a}$ and $\vec{b}$. Then

$$\vec{a} \cdot \vec{b} = \vec{b} \cdot \vec{a} = a_x b_x + a_y b_y + a_z b_z = ab \cos \theta$$

$$\vec{a} \times \vec{b} = -\vec{b} \times \vec{a} = \begin{vmatrix} \hat{\imath} & \hat{\jmath} & \hat{k} \\ a_x & a_y & a_z \\ b_x & b_y & b_z \end{vmatrix}$$

$$= \hat{\imath} \begin{vmatrix} a_y & a_z \\ b_y & b_z \end{vmatrix} - \hat{\jmath} \begin{vmatrix} a_x & a_z \\ b_x & b_z \end{vmatrix} + \hat{k} \begin{vmatrix} a_x & a_y \\ b_x & b_y \end{vmatrix}$$

$$= (a_y b_z - b_y a_z)\hat{\imath} + (a_z b_x - b_z a_x)\hat{\jmath}$$
$$+ (a_x b_y - b_x a_y)\hat{k}$$

$$|\vec{a} \times \vec{b}| = ab \sin \theta$$

$$\vec{a} \cdot (\vec{b} \times \vec{c}) = \vec{b} \cdot (\vec{c} \times \vec{a}) = \vec{c} \cdot (\vec{a} \times \vec{b})$$

$$\vec{a} \times (\vec{b} \times \vec{c}) = (\vec{a} \cdot \vec{c})\vec{b} - (\vec{a} \cdot \vec{b})\vec{c}$$

Derivatives and Integrals

In what follows, the letters u and v stand for any functions of x, and a and m are constants. To each of the indefinite integrals should be added an arbitrary constant of integration. The *Handbook of Chemistry and Physics* (CRC Press Inc.) gives a more extensive tabulation.

1. $\dfrac{dx}{dx} = 1$

2. $\dfrac{d}{dx}(au) = a\dfrac{du}{dx}$

3. $\dfrac{d}{dx}(u + v) = \dfrac{du}{dx} + \dfrac{dv}{dx}$

4. $\dfrac{d}{dx}x^m = mx^{m-1}$

5. $\dfrac{d}{dx}\ln x = \dfrac{1}{x}$

6. $\dfrac{d}{dx}(uv) = u\dfrac{dv}{dx} + v\dfrac{du}{dx}$

7. $\dfrac{d}{dx}e^x = e^x$

8. $\dfrac{d}{dx}\sin x = \cos x$

9. $\dfrac{d}{dx}\cos x = -\sin x$

10. $\dfrac{d}{dx}\tan x = \sec^2 x$

11. $\dfrac{d}{dx}\cot x = -\csc^2 x$

12. $\dfrac{d}{dx}\sec x = \tan x \sec x$

13. $\dfrac{d}{dx}\csc x = -\cot x \csc x$

14. $\dfrac{d}{dx}e^u = e^u\dfrac{du}{dx}$

15. $\dfrac{d}{dx}\sin u = \cos u\dfrac{du}{dx}$

16. $\dfrac{d}{dx}\cos u = -\sin u\dfrac{du}{dx}$

1. $\displaystyle\int dx = x$

2. $\displaystyle\int au\, dx = a\int u\, dx$

3. $\displaystyle\int (u + v)\, dx = \int u\, dx + \int v\, dx$

4. $\displaystyle\int x^m\, dx = \dfrac{x^{m+1}}{m + 1}\ (m \neq -1)$

5. $\displaystyle\int \dfrac{dx}{x} = \ln |x|$

6. $\displaystyle\int u\dfrac{dv}{dx}\, dx = uv - \int v\dfrac{du}{dx}\, dx$

7. $\displaystyle\int e^x\, dx = e^x$

8. $\displaystyle\int \sin x\, dx = -\cos x$

9. $\displaystyle\int \cos x\, dx = \sin x$

10. $\displaystyle\int \tan x\, dx = \ln |\sec x|$

11. $\displaystyle\int \sin^2 x\, dx = \tfrac{1}{2}x - \tfrac{1}{4}\sin 2x$

12. $\displaystyle\int e^{-ax}\, dx = -\dfrac{1}{a}e^{-ax}$

13. $\displaystyle\int xe^{-ax}\, dx = -\dfrac{1}{a^2}(ax + 1)\,e^{-ax}$

14. $\displaystyle\int x^2 e^{-ax}\, dx = -\dfrac{1}{a^3}(a^2x^2 + 2ax + 2)e^{-ax}$

15. $\displaystyle\int_0^\infty x^n e^{-ax}\, dx = \dfrac{n!}{a^{n+1}}$

16. $\displaystyle\int_0^\infty x^{2n}e^{-ax^2}\, dx = \dfrac{1\cdot3\cdot5\cdots(2n-1)}{2^{n+1}a^n}\sqrt{\dfrac{\pi}{a}}$

17. $\displaystyle\int \dfrac{dx}{\sqrt{x^2 + a^2}} = \ln(x + \sqrt{x^2 + a^2})$

18. $\displaystyle\int \dfrac{x\, dx}{(x^2 + a^2)^{3/2}} = -\dfrac{1}{(x^2 + a^2)^{1/2}}$

19. $\displaystyle\int \dfrac{dx}{(x^2 + a^2)^{3/2}} = \dfrac{x}{a^2(x^2 + a^2)^{1/2}}$

20. $\displaystyle\int_0^\infty x^{2n+1}e^{-ax^2}\, dx = \dfrac{n!}{2a^{n+1}}\ (a > 0)$

21. $\displaystyle\int \dfrac{x\, dx}{x + d} = x - d\ln(x + d)$

Properties of the Elements

All physical properties are for a pressure of 1 atm unless otherwise specified.

Element	Symbol	Atomic Number Z	Molar Mass, g/mol	Density, g/cm³ at 20°C	Melting Point, °C	Boiling Point, °C	Specific Heat, J/(g·°C) at 25°C
Actinium	Ac	89	(227)	10.06	1323	(3473)	0.092
Aluminum	Al	13	26.9815	2.699	660	2450	0.900
Americium	Am	95	(243)	13.67	1541	—	—
Antimony	Sb	51	121.75	6.691	630.5	1380	0.205
Argon	Ar	18	39.948	1.6626×10^{-3}	−189.4	−185.8	0.523
Arsenic	As	33	74.9216	5.78	817 (28 atm)	613	0.331
Astatine	At	85	(210)	—	(302)	—	—
Barium	Ba	56	137.34	3.594	729	1640	0.205
Berkelium	Bk	97	(247)	14.79	—	—	—
Beryllium	Be	4	9.0122	1.848	1287	2770	1.83
Bismuth	Bi	83	208.980	9.747	271.37	1560	0.122
Bohrium	Bh	107	262.12	—	—	—	—
Boron	B	5	10.811	2.34	2030	—	1.11
Bromine	Br	35	79.909	3.12 (liquid)	−7.2	58	0.293
Cadmium	Cd	48	112.40	8.65	321.03	765	0.226
Calcium	Ca	20	40.08	1.55	838	1440	0.624
Californium	Cf	98	(251)	—	—	—	—
Carbon	C	6	12.01115	2.26	3727	4830	0.691
Cerium	Ce	58	140.12	6.768	804	3470	0.188
Cesium	Cs	55	132.905	1.873	28.40	690	0.243
Chlorine	Cl	17	35.453	3.214×10^{-3} (0°C)	−101	−34.7	0.486
Chromium	Cr	24	51.996	7.19	1857	2665	0.448
Cobalt	Co	27	58.9332	8.85	1495	2900	0.423
Copernicium	Cp	112	(285)	—	—	—	—
Copper	Cu	29	63.54	8.96	1083.40	2595	0.385
Curium	Cm	96	(247)	13.3	—	—	—
Darmstadtium	Ds	110	(271)	—	—	—	—
Dubnium	Db	105	262.114	—	—	—	—
Dysprosium	Dy	66	162.50	8.55	1409	2330	0.172
Einsteinium	Es	99	(254)	—	—	—	—
Erbium	Er	68	167.26	9.15	1522	2630	0.167
Europium	Eu	63	151.96	5.243	817	1490	0.163
Fermium	Fm	100	(237)	—	—	—	—
Fluorine	F	9	18.9984	1.696×10^{-3} (0°C)	−219.6	−188.2	0.753
Francium	Fr	87	(223)	—	(27)	—	—
Gadolinium	Gd	64	157.25	7.90	1312	2730	0.234
Gallium	Ga	31	69.72	5.907	29.75	2237	0.377
Germanium	Ge	32	72.59	5.323	937.25	2830	0.322
Gold	Au	79	196.967	19.32	1064.43	2970	0.131

Element	Symbol	Atomic Number Z	Molar Mass, g/mol	Density, g/cm³ at 20°C	Melting Point, °C	Boiling Point, °C	Specific Heat, J/(g·°C) at 25°C
Hafnium	Hf	72	178.49	13.31	2227	5400	0.144
Hassium	Hs	108	(265)	—	—	—	—
Helium	He	2	4.0026	0.1664×10^{-3}	−269.7	−268.9	5.23
Holmium	Ho	67	164.930	8.79	1470	2330	0.165
Hydrogen	H	1	1.00797	0.08375×10^{-3}	−259.19	−252.7	14.4
Indium	In	49	114.82	7.31	156.634	2000	0.233
Iodine	I	53	126.9044	4.93	113.7	183	0.218
Iridium	Ir	77	192.2	22.5	2447	(5300)	0.130
Iron	Fe	26	55.847	7.874	1536.5	3000	0.447
Krypton	Kr	36	83.80	3.488×10^{-3}	−157.37	−152	0.247
Lanthanum	La	57	138.91	6.189	920	3470	0.195
Lawrencium	Lr	103	(257)	—	—	—	—
Lead	Pb	82	207.19	11.35	327.45	1725	0.129
Lithium	Li	3	6.939	0.534	180.55	1300	3.58
Lutetium	Lu	71	174.97	9.849	1663	1930	0.155
Magnesium	Mg	12	24.312	1.738	650	1107	1.03
Manganese	Mn	25	54.9380	7.44	1244	2150	0.481
Meitnerium	Mt	109	(266)	—	—	—	—
Mendelevium	Md	101	(256)	—	—	—	—
Mercury	Hg	80	200.59	13.55	−38.87	357	0.138
Molybdenum	Mo	42	95.94	10.22	2617	5560	0.251
Neodymium	Nd	60	144.24	7.007	1016	3180	0.188
Neon	Ne	10	20.183	0.8387×10^{-3}	−248.597	−246.0	1.03
Neptunium	Np	93	(237)	20.25	637	—	1.26
Nickel	Ni	28	58.71	8.902	1453	2730	0.444
Niobium	Nb	41	92.906	8.57	2468	4927	0.264
Nitrogen	N	7	14.0067	1.1649×10^{-3}	−210	−195.8	1.03
Nobelium	No	102	(255)	—	—	—	—
Osmium	Os	76	190.2	22.59	3027	5500	0.130
Oxygen	O	8	15.9994	1.3318×10^{-3}	−218.80	−183.0	0.913
Palladium	Pd	46	106.4	12.02	1552	3980	0.243
Phosphorus	P	15	30.9738	1.83	44.25	280	0.741
Platinum	Pt	78	195.09	21.45	1769	4530	0.134
Plutonium	Pu	94	(244)	19.8	640	3235	0.130
Polonium	Po	84	(210)	9.32	254	—	—
Potassium	K	19	39.102	0.862	63.20	760	0.758
Praseodymium	Pr	59	140.907	6.773	931	3020	0.197
Promethium	Pm	61	(145)	7.22	(1027)	—	—
Protactinium	Pa	91	(231)	15.37 (estimated)	(1230)	—	—
Radium	Ra	88	(226)	5.0	700	—	—
Radon	Rn	86	(222)	9.96×10^{-3} (0°C)	(−71)	−61.8	0.092
Rhenium	Re	75	186.2	21.02	3180	5900	0.134
Rhodium	Rh	45	102.905	12.41	1963	4500	0.243
Roentgenium	Rg	111	(280)	—	—	—	—
Rubidium	Rb	37	85.47	1.532	39.49	688	0.364
Ruthenium	Ru	44	101.107	12.37	2250	4900	0.239
Rutherfordium	Rf	104	261.11	—	—	—	—

Element	Symbol	Atomic Number Z	Molar Mass, g/mol	Density, g/cm³ at 20°C	Melting Point, °C	Boiling Point, °C	Specific Heat, J/(g·°C) at 25°C
Samarium	Sm	62	150.35	7.52	1072	1630	0.197
Scandium	Sc	21	44.956	2.99	1539	2730	0.569
Seaborgium	Sg	106	263.118	—	—	—	—
Selenium	Se	34	78.96	4.79	221	685	0.318
Silicon	Si	14	28.086	2.33	1412	2680	0.712
Silver	Ag	47	107.870	10.49	960.8	2210	0.234
Sodium	Na	11	22.9898	0.9712	97.85	892	1.23
Strontium	Sr	38	87.62	2.54	768	1380	0.737
Sulfur	S	16	32.064	2.07	119.0	444.6	0.707
Tantalum	Ta	73	180.948	16.6	3014	5425	0.138
Technetium	Tc	43	(99)	11.46	2200	—	0.209
Tellurium	Te	52	127.60	6.24	449.5	990	0.201
Terbium	Tb	65	158.924	8.229	1357	2530	0.180
Thallium	Tl	81	204.37	11.85	304	1457	0.130
Thorium	Th	90	(232)	11.72	1755	(3850)	0.117
Thulium	Tm	69	168.934	9.32	1545	1720	0.159
Tin	Sn	50	118.69	7.2984	231.868	2270	0.226
Titanium	Ti	22	47.90	4.54	1670	3260	0.523
Tungsten	W	74	183.85	19.3	3380	5930	0.134
Unnamed	Uut	113	(284)	—	—	—	—
Unnamed	Unq	114	(289)	—	—	—	—
Unnamed	Uup	115	(288)	—	—	—	—
Unnamed	Uuh	116	(293)	—	—	—	—
Unnamed	Uus	117	—	—	—	—	—
Unnamed	Uuo	118	(294)	—	—	—	—
Uranium	U	92	(238)	18.95	1132	3818	0.117
Vanadium	V	23	50.942	6.11	1902	3400	0.490
Xenon	Xe	54	131.30	5.495×10^{-3}	−111.79	−108	0.159
Ytterbium	Yb	70	173.04	6.965	824	1530	0.155
Yttrium	Y	39	88.905	4.469	1526	3030	0.297
Zinc	Zn	30	65.37	7.133	419.58	906	0.389
Zirconium	Zr	40	91.22	6.506	1852	3580	0.276

The values in parentheses in the column of molar masses are the mass numbers of the longest-lived isotopes of those elements that are radioactive. Melting points and boiling points in parentheses are uncertain.

The data for gases are valid only when these are in their usual molecular state, such as H_2, He, O_2, Ne, etc. The specific heats of the gases are the values at constant pressure.

Source: Adapted from J. Emsley, *The Elements*, 3rd ed., 1998, Clarendon Press, Oxford. See also www.webelements.com for the latest values and newest elements.

Periodic Table of the Elements

Metals

Metalloids

Nonmetals

Noble gases 0

THE HORIZONTAL PERIODS

Alkali metals
IA

	IA														IIIA	IVA	VA	VIA	VIIA	0
1	1 H	IIA																		2 He
2	3 Li	4 Be						Transition metals							5 B	6 C	7 N	8 O	9 F	10 Ne
3	11 Na	12 Mg	IIIB	IVB	VB	VIB	VIIB	VIIIB			IB	IIB			13 Al	14 Si	15 P	16 S	17 Cl	18 Ar
4	19 K	20 Ca	21 Sc	22 Ti	23 V	24 Cr	25 Mn	26 Fe	27 Co	28 Ni	29 Cu	30 Zn			31 Ga	32 Ge	33 As	34 Se	35 Br	36 Kr
5	37 Rb	38 Sr	39 Y	40 Zr	41 Nb	42 Mo	43 Tc	44 Ru	45 Rh	46 Pd	47 Ag	48 Cd			49 In	50 Sn	51 Sb	52 Te	53 I	54 Xe
6	55 Cs	56 Ba	57-71 *	72 Hf	73 Ta	74 W	75 Re	76 Os	77 Ir	78 Pt	79 Au	80 Hg	81 Tl	82 Pb		83 Bi	84 Po	85 At	86 Rn	
7	87 Fr	88 Ra	89-103 †	104 Rf	105 Db	106 Sg	107 Bh	108 Hs	109 Mt	110 Ds	111 Rg	112 Cp	113	114	115	116	117	118		

Inner transition metals

Lanthanide series *	57 La	58 Ce	59 Pr	60 Nd	61 Pm	62 Sm	63 Eu	64 Gd	65 Tb	66 Dy	67 Ho	68 Er	69 Tm	70 Yb	71 Lu
Actinide series †	89 Ac	90 Th	91 Pa	92 U	93 Np	94 Pu	95 Am	96 Cm	97 Bk	98 Cf	99 Es	100 Fm	101 Md	102 No	103 Lr

Evidence for the discovery of elements 113 through 118 has been reported. See www.webelements.com for the latest information and newest elements.

CHAPTER 1

P **1.** (a) 4.00×10^4 km; (b) 5.10×10^8 km^2; (c) 1.08×10^{12} km^3
3. (a) 10^9 μm; (b) 10^{-4}; (c) 9.1×10^5 μm **5.** (a) 160 rods; (b) 40
chains **7.** 1.1×10^3 acre-feet **9.** 1.9×10^{22} cm^3 **11.** (a) 1.43; (b)
0.864 **13.** (a) 495 s; (b) 141 s; (c) 198 s; (d) -245 s **15.** 1.21×10^{12}
μs **17.** C, D, A, B, E; the important criterion is the consistency of
the daily variation, not its magnitude **19.** 5.2×10^6 m **21.** $9.0 \times$
10^{49} atoms **23.** (a) 1×10^3 kg; (b) 158 kg/s **25.** 1.9×10^5 kg **27.**
(a) 1.18×10^{-29} m^3; (b) 0.282 nm **29.** 1.75×10^3 kg **31.** 1.43
kg/min **33.** (a) 293 U.S. bushels; (b) 3.81×10^3 U.S. bushels **35.**
(a) 22 pecks; (b) 5.5 Imperial bushels; (c) 200 L **37.** 8×10^2 km
39. (a) 18.8 gallons; (b) 22.5 gallons **41.** 0.3 cord **43.** 3.8 mg/s
45. (a) yes; (b) 8.6 universe seconds **47.** 0.12 AU/min **49.** (a)
3.88; (b) 7.65; (c) 156 ken^3; (d) 1.19×10^3 m^3 **51.** (a) 3.9 m, 4.8 m;
(b) 3.9×10^3 mm, 4.8×10^3 mm; (c) 2.2 m^3, 4.2 m^3 **53.** (a) $4.9 \times$
10^{-6} pc; (b) 1.6×10^{-5} ly

CHAPTER 2

CP **1.** b and c **2.** (check the derivative dx/dt) (a) 1 and 4;
(b) 2 and 3 **3.** (a) plus; (b) minus; (c) minus; (d) plus **4.** 1 and 4
($a = d^2x/dt^2$ must be constant) **5.** (a) plus (upward displacement
on y axis); (b) minus (downward displacement on y axis); (c) $a =$
$-g = -9.8$ m/s^2
Q **1.** (a) negative; (b) positive; (c) yes; (d) positive; (e) constant
3. (a) all tie; (b) 4, tie of 1 and 2, then 3 **5.** (a) positive direction;
(b) negative direction; (c) 3 and 5; (d) 2 and 6 tie, then 3 and 5 tie,
then 1 and 4 tie (zero) **7.** (a) D; (b) E **9.** (a) 3, 2, 1; (b) 1, 2, 3; (c)
all tie; (d) 1, 2, 3
P **1.** 13 m **3.** (a) $+40$ km/h; (b) 40 km/h **5.** (a) 0; (b) -2 m; (c) 0;
(d) 12 m; (e) $+12$ m; (f) $+7$ m/s **7.** 60 km **9.** 1.4 m **11.** 128 km/h
13. (a) 73 km/h; (b) 68 km/h; (c) 70 km/h; (d) 0 **15.** (a) -6 m/s; (b)
$-x$ direction; (c) 6 m/s; (d) decreasing; (e) 2 s; (f) no **17.** (a) 28.5
cm/s; (b) 18.0 cm/s; (c) 40.5 cm/s; (d) 28.1 cm/s; (e) 30.3 cm/s **19.**
-20 m/s^2 **21.** (a) 1.10 m/s; (b) 6.11 mm/s^2; (c) 1.47 m/s; (d) 6.11
mm/s^2 **23.** 1.62×10^{15} m/s^2 **25.** (a) 30 s; (b) 300 m **27.** (a) $+1.6$
m/s; (b) $+18$ m/s **29.** (a) 10.6 m; (b) 41.5 s **31.** (a) 3.1×10^6 s; (b)
4.6×10^{13} m **33.** (a) 3.56 m/s^2; (b) 8.43 m/s^2 **35.** 0.90 m/s^2 **37.** (a)
4.0 m/s^2; (b) $+x$ **39.** (a) -2.5 m/s^2; (b) 1; (d) 0; (e) 2 **41.** 40 m **43.**
(a) 0.994 m/s^2 **45.** (a) 31 m/s; (b) 6.4 s **47.** (a) 29.4 m; (b) 2.45 s
49. (a) 5.4 s; (b) 41 m/s **51.** (a) 20 m; (b) 59 m **53.** 4.0 m/s **55.** (a)
857 m/s^2; (b) up **57.** (a) 1.26×10^3 m/s^2; (b) up **59.** (a) 89 cm; (b)
22 cm **61.** 20.4 m **63.** 2.34 m **65.** (a) 2.25 m/s; (b) 3.90 m/s **67.**
0.56 m/s **69.** 100 m **71.** (a) 2.00 m/s; (b) 12 cm; (c) -9.00 cm/s^2; (d)
right; (e) left; (f) 3.46 s **73.** (a) 82 m; (b) 19 m/s **75.** (a) 0.74 s; (b)
6.2 m/s^2 **77.** (a) 3.1 m/s^2; (b) 45 m; (c) 13 s **79.** 17 m/s **81.** $+47$
m/s **83.** (a) 1.23 cm; (b) 4 times; (c) 9 times; (d) 16 times; (e) 25
times **85.** 25 km/h **87.** 1.2 h **89.** $4H$ **91.** (a) 3.2 s; (b) 1.3 s **93.**
(a) 8.85 m/s; (b) 1.00 m **95.** (a) 2.0 m/s^2; (b) 12 m/s; (c) 45 m **97.**
(a) 48.5 m/s; (b) 4.95 s; (c) 34.3 m/s; (d) 3.50 s **99.** 22.0 m/s **101.**
(a) $v = (v_0^2 + 2gh)^{0.5}$; (b) $t = [(v_0^2 + 2gh)^{0.5} - v_0] / g$; (c) same as (a);
(d) $t = [(v_0^2 + 2gh)^{0.5} + v_0] / g$, greater

CHAPTER 3

CP **1.** (a) 7 m ($\vec{a}$ and $\vec{b}$ are in same direction); (b) 1 m ($\vec{a}$ and $\vec{b}$ are in
opposite directions) **2.** c, d, f (components must be head to tail; $\vec{a}$
must extend from tail of one component to head of the other) **3.** (a)
$+, +$; (b) $+, -$; (c) $+, +$ (draw vector from tail of $\vec{d}_1$ to head of $\vec{d}_2$)

4. (a) 90°; (b) 0° (vectors are parallel—same direction); (c) 180° (vectors are antiparallel—opposite directions) **5.** (a) 0° or 180°; (b) 90°
Q **1.** yes, when the vectors are in same direction **3.** Either the se-
quence $\vec{d}_2, \vec{d}_1$ or the sequence $\vec{d}_2, \vec{d}_2, \vec{d}_3$ **5.** all but (e) **7.** (a) yes;
(b) yes; (c) no **9.** (a) $+x$ for (1), $+z$ for (2), $+z$ for (3); (b) $-x$ for (1),
$-z$ for (2), $-z$ for (3)
P **1.** (a) -2.5 m; (b) -6.9 m **3.** (a) 47.2 m; (b) 122° **5.** (a) 156
km; (b) 39.8° west of due north **7.** (a) 6.42 m; (b) no; (c) yes; (d)
yes; (e) a possible answer: $(4.30$ m$)\hat{\text{i}} + (3.70$ m$)\hat{\text{j}} + (3.00$ m$)\hat{\text{k}}$; (f)
7.96 m **9.** (a) $(3.0$ m$)\hat{\text{i}} - (2.0$ m$)\hat{\text{j}} + (5.0$ m$)\hat{\text{k}}$; (b) $(5.0$ m$)\hat{\text{i}} - (4.0$
m$)\hat{\text{j}} - (3.0$ m$)\hat{\text{k}}$; (c) $(-5.0$ m$)\hat{\text{i}} + (4.0$ m$)\hat{\text{j}} + (3.0$ m$)\hat{\text{k}}$ **11.** (a)
$(-9.0$ m$)\hat{\text{i}} + (10$ m$)\hat{\text{j}}$; (b) 13 m; (c) 132° **13.** 4.74 km **15.** (a) 1.59
m; (b) 12.1 m; (c) 12.2 m; (d) 82.5° **17.** (a) 38 m; (b) $-37.5°$; (c) 130
m; (d) 1.2°; (e) 62 m; (f) 130° **19.** 5.39 m at 21.8° left of forward
21. (a) -70.0 cm; (b) 80.0 cm; (c) 141 cm; (d) $-172°$ **23.** 3.2 **25.**
2.6 km **27.** (a) $8\hat{\text{i}} + 16\hat{\text{j}}$; (b) $2\hat{\text{i}} + 4\hat{\text{j}}$ **29.** (a) 7.5 cm; (b) 90°; (c)
8.6 cm; (d) 48° **31.** (a) $a\hat{\text{i}} + a\hat{\text{j}} + a\hat{\text{k}}$; (b) $-a\hat{\text{i}} + a\hat{\text{j}} + a\hat{\text{k}}$; (c) $a\hat{\text{i}} -$
$a\hat{\text{j}} + a\hat{\text{k}}$; (d) $-a\hat{\text{i}} - a\hat{\text{j}} + a\hat{\text{k}}$; (e) 54.7°; (f) $3^{0.5}a$ **33.** (a) 12; (b) $+z$;
(c) 12; (d) $-z$; (e) 12; (f) $+z$ **35.** (a) -18.8 units; (b) 26.9 units, $+z$
direction **37.** (a) -21; (b) -9; (c) $5\hat{\text{i}} - 11\hat{\text{j}} - 9\hat{\text{k}}$ **39.** 70.5° **41.**
22° **43.** (a) 3.00 m; (b) 0; (c) 3.46 m; (d) 2.00 m; (e) -5.00 m; (f)
8.66 m; (g) -6.67; (h) 4.33 **45.** (a) -83.4; (b) $(1.14 \times 10^3)\hat{\text{k}}$; (c)
1.14×10^3, θ not defined, $\phi = 0°$; (d) 90.0°; (e) $-5.14\hat{\text{i}} + 6.13\hat{\text{j}} +$
$3.00\hat{\text{k}}$; (f) 8.54, $\theta = 130°$, $\phi = 69.4°$ **47.** (a) 140°; (b) 90.0°; (c) 99.1°
49. (a) 103 km; (b) 60.9° north of due west **51.** (a) 27.8 m; (b) 13.4
m **53.** (a) 30; (b) 52 **55.** (a) -2.83 m; (b) -2.83 m; (c) 5.00 m; (d)
0; (e) 3.00 m; (f) 5.20 m; (g) 5.17 m; (h) 2.37 m; (i) 5.69 m; (j) 25°
north of due east; (k) 5.69 m; (l) 25° south of due west **57.** 4.1
59. (a) $(9.19$ m$)\hat{\text{i}}' + (7.71$ m$)\hat{\text{j}}'$; (b) $(14.0$ m$)\hat{\text{i}}' + (3.41$ m$)\hat{\text{j}}'$ **61.** (a)
$11\hat{\text{i}} + 5.0\hat{\text{j}} - 7.0\hat{\text{k}}$; (b) 120°; (c) -4.9; (d) 7.3 **63.** (a) 3.0 m^2; (b) 52
m^3; (c) $(11$ m$^2)\hat{\text{i}} + (9.0$ m$^2)\hat{\text{j}} + (3.0$ m$^2)\hat{\text{k}}$ **65.** (a) $(-40\hat{\text{i}} - 20\hat{\text{j}} +$
$25\hat{\text{k}})$ m; (b) 45 m

CHAPTER 4

CP **1.** (draw $\vec{v}$ tangent to path, tail on path) (a) first; (b) third
2. (take second derivative with respect to time) (1) and (3) a_x and
a_y are both constant and thus $\vec{a}$ is constant; (2) and (4) a_y is constant but a_x is not, thus $\vec{a}$ is not **3.** yes **4.** (a) v_x constant; (b) v_y
initially positive, decreases to zero, and then becomes progres-
sively more negative; (c) $a_x = 0$ throughout; (d) $a_y = -g$ throughout **5.** (a) $-(4$ m/s$)\hat{\text{i}}$; (b) $-(8$ m/s$^2)\hat{\text{j}}$
Q **1.** a and c tie, then b **3.** decreases **5.** a, b, c **7.** (a) 0; (b) 350
km/h; (c) 350 km/h; (d) same (nothing changed about the vertical
motion) **9.** (a) all tie; (b) all tie; (c) 3, 2, 1; (d) 3, 2, 1 **11.** 2, then 1
and 4 tie, then 3 **13.** (a) yes; (b) no; (c) yes
P **1.** (a) 6.2 m **3.** $(-2.0$ m$)\hat{\text{i}} + (6.0$ m$)\hat{\text{j}} - (10$ m$)\hat{\text{k}}$ **5.** (a) 7.59
km/h; (b) 22.5° east of due north **7.** $(-0.70$ m/s$)\hat{\text{i}} + (1.4$ m/s$)\hat{\text{j}} -$
$(0.40$ m/s$)\hat{\text{k}}$ **9.** (a) 0.83 cm/s; (b) 0°; (c) 0.11 m/s; (d) $-63°$ **11.** (a)
$(6.00$ m$)\hat{\text{i}} - (106$ m$)\hat{\text{j}}$; (b) $(19.0$ m/s$)\hat{\text{i}} - (224$ m/s$)\hat{\text{j}}$; (c) $(24.0$
m/s$^2)\hat{\text{i}} - (336$ m/s$^2)\hat{\text{j}}$; (d) $-85.2°$ **13.** (a) $(8$ m/s$^2)t\hat{\text{i}} + (1$ m/s$)\hat{\text{k}}$; (b)
$(8$ m/s$^2)\hat{\text{j}}$ **15.** (a) $(-1.50$ m/s$)\hat{\text{j}}$; (b) $(4.50$ m$)\hat{\text{i}} - (2.25$ m$)\hat{\text{j}}$ **17.** (32
m/s$)\hat{\text{i}}$ **19.** (a) $(72.0$ m$)\hat{\text{i}} + (90.7$ m$)\hat{\text{j}}$; (b) 49.5° **21.** (a) 18 cm; (b)
1.9 m **23.** (a) 3.03 s; (b) 758 m; (c) 29.7 m/s **25.** 43.1 m/s (155
km/h) **27.** (a) 10.0 s; (b) 897 m **29.** 78.5° **31.** 3.35 m **33.** (a)
202 m/s; (b) 806 m; (c) 161 m/s; (d) -171 m/s **35.** 4.84 cm **37.** (a)
1.60 m; (b) 6.86 m; (c) 2.86 m **39.** (a) 32.3 m; (b) 21.9 m/s; (c) 40.4°;
(d) below **41.** 55.5° **43.** (a) 11 m; (b) 23 m; (c) 17 m/s; (d) 63°

45. (a) ramp; (b) 5.82 m; (c) 31.0° **47.** (a) yes; (b) 2.56 m **49.** (a) 31°; (b) 63° **51.** (a) 2.3°; (b) 1.4 m; (c) 18° **53.** (a) 75.0 m; (b) 31.9 m/s; (c) 66.9°; (d) 25.5 m **55.** the third **57.** (a) 7.32 m; (b) west; (c) north **59.** (a) 12 s; (b) 4.1 m/s²; (c) down; (d) 4.1 m/s²; (e) up **61.** (a) 1.3×10^5 m/s; (b) 7.9×10^5 m/s²; (c) increase **63.** 2.92 m **65.** $(3.00 \text{ m/s}^2)\hat{i} + (6.00 \text{ m/s}^2)\hat{j}$ **67.** 160 m/s² **69.** (a) 13 m/s²; (b) eastward; (c) 13 m/s²; (d) eastward **71.** 1.67 **73.** (a) $(80 \text{ km/h})\hat{i} - (60 \text{ km/h})\hat{j}$; (b) 0°; (c) answers do not change **75.** 32 m/s **77.** 60° **79.** (a) 38 knots; (b) 1.5° east of due north; (c) 4.2 h; (d) 1.5° west of due south **81.** (a) $(-32 \text{ km/h})\hat{i} - (46 \text{ km/h})\hat{j}$; (b) $[(2.5 \text{ km}) - (32 \text{ km/h})t]\hat{i} + [(4.0 \text{ km}) - (46 \text{ km/h})t]\hat{j}$; (c) 0.084 h; (d) 2×10^2 m **83.** (a) $-30°$; (b) 69 min; (c) 80 min; (d) 80 min; (e) 0°; (f) 60 min **85.** (a) 2.7 km; (b) 76° clockwise **87.** (a) 44 m; (b) 13 m; (c) 8.9 m **89.** (a) 45 m; (b) 22 m/s **91.** (a) 2.6×10^2 m/s; (b) 45 s; (c) increase **93.** (a) 63 km; (b) 18° south of due east; (c) 0.70 km/h; (d) 18° south of due east; (e) 1.6 km/h; (f) 1.2 km/h; (g) 33° north of due east **95.** (a) 1.5; (b) (36 m, 54 m) **97.** (a) 62 ms; (b) 4.8×10^2 m/s **99.** 2.64 m **101.** (a) 2.5 m; (b) 0.82 m; (c) 9.8 m/s²; (d) 9.8 m/s² **103.** (a) 6.79 km/h; (b) 6.96° **105.** (a) 16 m/s; (b) 23°; (c) above; (d) 27 m/s; (e) 57°; (f) below **107.** (a) 4.2 m, 45°; (b) 5.5 m, 68°; (c) 6.0 m, 90°; (d) 4.2 m, 135°; (e) 0.85 m/s, 135°; (f) 0.94 m/s, 90°; (g) 0.94 m/s, 180°; (h) 0.30 m/s², 180°; (i) 0.30 m/s², 270° **109.** (a) 5.4×10^{-13} m; (b) decrease **111.** (a) 0.034 m/s²; (b) 84 min **113.** (a) 8.43 m; (b) $-129°$ **115.** (a) 2.00 ns; (b) 2.00 mm; (c) 1.00×10^7 m/s; (d) 2.00×10^6 m/s **117.** (a) 24 m/s; (b) 65° **119.** 93° from the car's direction of motion

CHAPTER 5

CP 1. c, d, and e ($\vec{F}_1$ and $\vec{F}_2$ must be head to tail, $\vec{F}_{net}$ must be from tail of one of them to head of the other) **2.** (a) and (b) 2 N, leftward (acceleration is zero in each situation) **3.** (a) equal; (b) greater (acceleration is upward, thus net force on body must be upward) **4.** (a) equal; (b) greater; (c) less **5.** (a) increase; (b) yes; (c) same; (d) yes

Q 1. (a) 2, 3, 4; (b) 1, 3, 4; (c) 1, $+y$; 2, $+x$; 3, fourth quadrant; 4, third quadrant **3.** increase **5.** (a) 2 and 4; (b) 2 and 4 **7.** (a) M; (b) M; (c) M; (d) $2M$; (e) $3M$ **9.** (a) 20 kg; (b) 18 kg; (c) 10 kg; (d) all tie; (e) 3, 2, 1 **11.** (a) increases from initial value mg; (b) decreases from mg to zero (after which the block moves up away from the floor)

P 1. 2.9 m/s² **3.** (a) 1.88 N; (b) 0.684 N; (c) $(1.88 \text{ N})\hat{i} + (0.684 \text{ N})\hat{j}$ **5.** (a) $(0.86 \text{ m/s}^2)\hat{i} - (0.16 \text{ m/s}^2)\hat{j}$; (b) 0.88 m/s²; (c) $-11°$ **7.** (a) $(-32.0 \text{ N})\hat{i} - (20.8 \text{ N})\hat{j}$; (b) 38.2 N; (c) $-147°$ **9.** (a) 8.37 N; (b) $-133°$; (c) $-125°$ **11.** 9.0 m/s² **13.** (a) 4.0 kg; (b) 1.0 kg; (c) 4.0 kg; (d) 1.0 kg **15.** (a) 108 N; (b) 108 N; (c) 108 N **17.** (a) 42 N; (b) 72 N; (c) 4.9 m/s² **19.** 1.2×10^5 N **21.** (a) 11.7 N; (b) $-59.0°$ **23.** (a) $(285 \text{ N})\hat{i} + (705 \text{ N})\hat{j}$; (b) $(285 \text{ N})\hat{i} - (115 \text{ N})\hat{j}$; (c) 307 N; (d) $-22.0°$; (e) 3.67 m/s²; (f) $-22.0°$ **25.** (a) 0.022 m/s²; (b) 8.3×10^4 km; (c) 1.9×10^3 m/s **27.** 1.5 mm **29.** (a) 494 N; (b) up; (c) 494 N; (d) down **31.** (a) 1.18 m; (b) 0.674 s; (c) 3.50 m/s **33.** 1.8×10^4 N **35.** (a) 46.7°; (b) 28.0° **37.** (a) 0.62 m/s²; (b) 0.13 m/s²; (c) 2.6 m **39.** (a) 2.2×10^{-3} N; (b) 3.7×10^{-3} N **41.** (a) 1.4 m/s²; (b) 4.1 m/s **43.** (a) 1.23 N; (b) 2.46 N; (c) 3.69 N; (d) 4.92 N; (e) 6.15 N; (f) 0.250 N **45.** (a) 31.3 kN; (b) 24.3 kN **47.** 6.4×10^3 N **49.** (a) 2.18 m/s²; (b) 116 N; (c) 21.0 m/s² **51.** (a) 3.6 m/s²; (b) 17 N **53.** (a) 0.970 m/s²; (b) 11.6 N; (c) 34.9 N **55.** (a) 1.1 N **57.** (a) 0.735 m/s²; (b) down; (c) 20.8 N **59.** (a) 4.9 m/s²; (b) 2.0 m/s²; (c) up; (d) 120 N **61.** $2Ma/(a + g)$ **63.** (a) 8.0 m/s; (b) $+x$ **65.** (a) 0.653 m/s³; (b) 0.896 m/s³; (c) 6.50 s **67.** 81.7 N **69.** 2.4 N **71.** 16 N **73.** (a) 2.6 N; (b) 17° **75.** (a) 0; (b) 0.83 m/s²; (c) 0 **77.** (a) 0.74 m/s²; (b) 7.3 m/s² **79.** (a) 11 N; (b) 2.2 kg; (c) 0; (d) 2.2 kg **81.** 195 N **83.** (a) 4.6 m/s²; (b) 2.6 m/s² **85.** (a) rope breaks; (b) 1.6 m/s² **87.** (a) 65 N; (b) 49 N **89.** (a) 4.6×10^3 N; (b) 5.8×10^3 N **91.** (a) 1.8×10^2 N; (b) 6.4 $\times$

10^2 N **93.** (a) 44 N; (b) 78 N; (c) 54 N; (d) 152 N **95.** (a) 4 kg; (b) 6.5 m/s²; (c) 13 N

CHAPTER 6

CP 1. (a) zero (because there is no attempt at sliding); (b) 5 N; (c) no; (d) yes; (e) 8 N **2.** ($\vec{a}$ is directed toward center of circular path) (a) $\vec{a}$ downward, $\vec{F}_N$ upward; (b) $\vec{a}$ and $\vec{F}_N$ upward

Q 1. (a) decrease; (b) decrease; (c) increase; (d) increase; (e) increase **3.** (a) same; (b) increases; (c) increases; (d) no **5.** (a) upward; (b) horizontal, toward you; (c) no change; (d) increases; (e) increases **7.** At first, $\vec{f}_s$ is directed up the ramp and its magnitude increases from $mg \sin \theta$ until it reaches $f_{s,max}$. Thereafter the force is kinetic friction directed up the ramp, with magnitude f_k (a constant value smaller than $f_{s,max}$). **9.** 4, 3, then 1, 2, and 5 tie **11.** (a) all tie; (b) all tie; (c) 2, 3, 1

P 1. 36 m **3.** (a) 2.0×10^2 N; (b) 1.2×10^2 N **5.** (a) 6.0 N; (b) 3.6 N; (c) 3.1 N **7.** (a) 1.9×10^2 N; (b) 0.56 m/s² **9.** (a) 11 N; (b) 0.14 m/s² **11.** (a) 3.0×10^2 N; (b) 1.3 m/s² **13.** (a) 1.3×10^2 N; (b) no; (c) 1.1×10^2 N; (d) 46 N; (e) 17 N **15.** 2° **17.** (a) $(17 \text{ N})\hat{i}$; (b) $(20 \text{ N})\hat{i}$; (c) $(15 \text{ N})\hat{i}$ **19.** (a) no; (b) $(-12 \text{ N})\hat{i} + (5.0 \text{ N})\hat{j}$ **21.** (a) 19°; (b) 3.3 kN **23.** 0.37 **25.** 1.0×10^2 N **27.** (a) 0; (b) $(-3.9 \text{ m/s}^2)\hat{i}$; (c) $(-1.0 \text{ m/s}^2)\hat{i}$ **29.** (a) 66 N; (b) 2.3 m/s² **31.** (a) 3.5 m/s²; (b) 0.21 N **33.** 9.9 s **35.** 4.9×10^2 N **37.** (a) 3.2×10^2 km/h; (b) 6.5×10^2 km/h; (c) no **39** 2.3 **41.** 0.60 **43.** 21 m **45.** (a) light; (b) 778 N; (c) 223 N; (d) 1.11 kN **47.** (a) 10 s; (b) 4.9×10^2 N; (c) 1.1×10^3 N **49.** 1.37×10^3 N **51.** 2.2 km **53.** 12° **55.** 2.6×10^3 N **57.** 1.81 m/s **59.** (a) 8.74 N; (b) 37.9 N; (c) 6.45 m/s; (d) radially inward **61.** (a) 27 N; (b) 3.0 m/s² **63.** (b) 240 N; (c) 0.60 **65.** (a) 69 km/h; (b) 139 km/h; (c) yes **67.** $g(\sin \theta - 2^{0.5}\mu_k \cos \theta)$ **69.** 3.4 m/s² **71.** (a) 35.3 N; (b) 39.7 N; (c) 320 N **73.** (a) 7.5 m/s²; (b) down; (c) 9.5 m/s²; (d) down **75.** (a) 3.0×10^5 N; (b) 1.2° **77.** 147 m/s **79.** (a) 13 N; (b) 1.6 m/s² **81.** (a) 275 N; (b) 877 N **83.** (a) 84.2 N; (b) 52.8 N; (c) 1.87 m/s² **85.** 3.4% **87.** (a) 3.21×10^3 N; (b) yes **89.** (a) 222 N; (b) 334 N; (c) 311 N; (d) 311 N; (e) c, d **91.** (a) $v_0^2/(4g \sin \theta)$; (b) no **93.** (a) 0.34; (b) 0.24 **95.** (a) $\mu_k mg/(\sin \theta - \mu_k \cos \theta)$; (b) $\theta_0 = \tan^{-1} \mu_s$ **97.** 0.18

CHAPTER 7

CP 1. (a) decrease; (b) same; (c) negative, zero **2.** (a) positive; (b) negative; (c) zero **3.** zero

Q 1. all tie **3.** (a) positive; (b) negative; (c) negative **5.** b (positive work), a (zero work), c (negative work), d (more negative work) **7.** all tie **9.** (a) A; (b) B

P 1. (a) 2.9×10^7 m/s; (b) 2.1×10^{-13} J **3.** (a) 5×10^{14} J; (b) 0.1 megaton TNT; (c) 8 bombs **5.** (a) 2.4 m/s; (b) 4.8 m/s **7.** 0.96 J **9.** 20 J **11.** (a) 62.3°; (b) 118° **13.** (a) 1.7×10^2 N; (b) 3.4×10^2 m; (c) -5.8×10^4 J; (d) 3.4×10^2 N; (e) 1.7×10^2 m; (f) -5.8×10^4 J **15.** (a) 1.50 J; (b) increases **17.** (a) 12 kJ; (b) -11 kJ; (c) 1.1 kJ; (d) 5.4 m/s **19.** 25 J **21.** (a) $-3Mgd/4$; (b) Mgd; (c) $Mgd/4$; (d) $(gd/2)^{0.5}$ **23.** 4.41 J **25.** (a) 25.9 kJ; (b) 2.45 N **27.** (a) 7.2 J; (b) 7.2 J; (c) 0; (d) -25 J **29.** (a) 0.90 J; (b) 2.1 J; (c) 0 **31.** (a) 6.6 m/s; (b) 4.7 m **33.** (a) 0.12 m; (b) 0.36 J; (c) -0.36 J; (d) 0.060 m; (e) 0.090 J **35.** (a) 0; (b) 0 **37.** (a) 42 J; (b) 30 J; (c) 12 J; (d) 6.5 m/s, $+x$ axis; (e) 5.5 m/s, $+x$ axis; (f) 3.5 m/s, $+x$ axis **39.** 4.00 N/m **41.** 5.3×10^2 J **43.** (a) 0.83 J; (b) 2.5 J; (c) 4.2 J; (d) 5.0 W **45.** 4.9×10^2 W **47.** (a) 1.0×10^2 J; (b) 8.4 W **49.** 7.4×10^2 W **51.** (a) 32.0 J; (b) 8.00 W; (c) 78.2° **53.** (a) 1.20 J; (b) 1.10 m/s **55.** (a) 1.8×10^5 ft · lb; (b) 0.55 hp **57.** (a) 797 N; (b) 0; (c) -1.55 kJ; (d) 0; (e) 1.55 kJ; (f) F varies during displacement **59.** (a) 1×10^5 megatons TNT; (b) 1×10^7 bombs **61.** -6 J **63.** (a) 314 J; (b) -155 J; (c) 0; (d) 158 J **65.** (a) 98 N; (b) 4.0 cm; (c) 3.9 J; (d) -3.9 J **67.** (a) 23 mm; (b) 45 N **69.** 165 kW

71. -37 J **73.** (a) 13 J; (b) 13 J **75.** 235 kW **77.** (a) 6 J; (b) 6.0 J
79. (a) 0.6 J; (b) 0; (c) -0.6 J

CHAPTER 8

CP 1. no (consider round trip on the small loop) **2.** 3, 1, 2 (see Eq. 8-6) **3.** (a) all tie; (b) all tie **4.** (a) CD, AB, BC (0) (check slope magnitudes); (b) positive direction of x **5.** all tie
Q 1. (a) 3, 2, 1; (b) 1, 2, 3 **3.** (a) 12 J; (b) -2 J **5.** (a) increasing; (b) decreasing; (c) decreasing; (d) constant in AB and BC, decreasing in CD **7.** $+30$ J **9.** 2, 1, 3
P 1. 89 N/cm **3.** (a) 167 J; (b) -167 J; (c) 196 J; (d) 29 J; (e) 167 J; (f) -167 J; (g) 296 J; (h) 129 J **5.** (a) 4.31 mJ; (b) -4.31 mJ; (c) 4.31 mJ; (d) -4.31 mJ; (e) all increase **7.** (a) 13.1 J; (b) -13.1 J; (c) 13.1 J; (d) all increase **9.** (a) 17.0 m/s; (b) 26.5 m/s; (c) 33.4 m/s; (d) 56.7 m; (e) all the same **11.** (a) 2.08 m/s; (b) 2.08 m/s; (c) increase **13.** (a) 0.98 J; (b) -0.98 J; (c) 3.1 N/cm **15.** (a) 2.6×10^2 m; (b) same; (c) decrease **17.** (a) 2.5 N; (b) 0.31 N; (c) 30 cm **19.** (a) 784 N/m; (b) 62.7 J; (c) 62.7 J; (d) 80.0 cm **21.** (a) 8.35 m/s; (b) 4.33 m/s; (c) 7.45 m/s; (d) both decrease **23.** (a) 4.85 m/s; (b) 2.42 m/s **25.** -3.2×10^2 J **27.** (a) no; (b) 9.3×10^2 N **29.** (a) 35 cm; (b) 1.7 m/s **31.** (a) 39.2 J; (b) 39.2 J; (c) 4.00 m **33.** (a) 2.40 m/s; (b) 4.19 m/s **35.** (a) 39.6 cm; (b) 3.64 cm **37.** -18 mJ **39.** (a) 2.1 m/s; (b) 10 N; (c) $+x$ direction; (d) 5.7 m; (e) 30 N; (f) $-x$ direction **41.** (a) -3.7 J; (c) 1.3 m; (d) 9.1 m; (e) 2.2 J; (f) 4.0 m; (g) $(4 - x)e^{-x/4}$; (h) 4.0 m **43.** (a) 5.6 J; (b) 3.5 J **45.** (a) 30.1 J; (b) 30.1 J; (c) 0.225 **47.** 0.53 J **49.** (a) -2.9 kJ; (b) 3.9×10^2 J; (c) 2.1×10^2 N **51.** (a) 1.5 MJ; (b) 0.51 MJ; (c) 1.0 MJ; (d) 63 m/s **53.** (a) 67 J; (b) 67 J; (c) 46 cm **55.** (a) -0.90 J; (b) 0.46 J; (c) 1.0 m/s **57.** 1.2 m **59.** (a) 19.4 m; (b) 19.0 m/s **61.** (a) 1.5×10^{-2} N; (b) $(3.8 \times 10^2)g$ **63.** (a) 7.4 m/s; (b) 90 cm; (c) 2.8 m; (d) 15 m **65.** 20 cm **67.** (a) 7.0 J; (b) 22 J **69.** 3.7 J **71.** 4.33 m/s **73.** 25 J **75.** (a) 4.9 m/s; (b) 4.5 N; (c) 71°; (d) same **77.** (a) 4.8 N; (b) $+x$ direction; (c) 1.5 m; (d) 13.5 m; (e) 3.5 m/s **79.** (a) 24 kJ; (b) 4.7×10^2 N **81.** (a) 5.00 J; (b) 9.00 J; (c) 11.0 J; (d) 3.00 J; (e) 12.0 J; (f) 2.00 J; (g) 13.0 J; (h) 1.00 J; (i) 13.0 J; (j) 1.00 J; (l) 11.0 J; (m) 10.8 m; (n) It returns to $x = 0$ and stops. **83.** (a) 6.0 kJ; (b) 6.0×10^2 W; (c) 3.0×10^2 W; (d) 9.0×10^2 W **85.** 880 MW **87.** (a) $v_0 = (2gL)^{0.5}$; (b) $5mg$; (c) $-mgL$; (d) $-2mgL$ **89.** (a) 109 J; (b) 60.3 J; (c) 68.2 J; (d) 41.0 J **91.** (a) 2.7 J; (b) 1.8 J; (c) 0.39 m **93.** (a) 10 m; (b) 49 N; (c) 4.1 m; (d) 1.2×10^2 N **95.** (a) 5.5 m/s; (b) 5.4 m; (c) same **97.** 80 mJ **99.** 24 W **101.** -12 J **103.** (a) 8.8 m/s; (b) 2.6 kJ; (c) 1.6 kW **105.** (a) 7.4×10^2 J; (b) 2.4×10^2 J **107.** 15 J **109.** (a) 2.35×10^3 J; (b) 352 J **111.** 738 m **113.** (a) -3.8 kJ; (b) 31 kN **115.** (a) 300 J; (b) 93.8 J; (c) 6.38 m **117.** (a) 5.6 J; (b) 12 J; (c) 13 J **119.** (a) 1.2 J; (b) 11 m/s; (c) no; (d) no **121.** (a) 2.1×10^6 kg; (b) $(100 + 1.5t)^{0.5}$ m/s; (c) $(1.5 \times 10^6)/(100 + 1.5t)^{0.5}$ N; (d) 6.7 km

CHAPTER 9

CP 1. (a) origin; (b) fourth quadrant; (c) on y axis below origin; (d) origin; (e) third quadrant; (f) origin **2.** (a)–(c) at the center of mass, still at the origin (their forces are internal to the system and cannot move the center of mass) **3.** (Consider slopes and Eq. 9-23.) (a) 1, 3, and then 2 and 4 tie (zero force); (b) 3 **4.** (a) unchanged; (b) unchanged (see Eq. 9-32); (c) decrease (Eq. 9-35) **5.** (a) zero; (b) positive (initial p_y down y; final p_y up y); (c) positive direction of y **6.** (No net external force; $\vec{P}$ conserved.) (a) 0; (b) no; (c) $-x$ **7.** (a) 10 kg·m/s; (b) 14 kg·m/s; (c) 6 kg·m/s **8.** (a) 4 kg·m/s; (b) 8 kg·m/s; (c) 3 J **9.** (a) 2 kg·m/s (conserve momentum along x); (b) 3 kg·m/s (conserve momentum along y)
Q 1. (a) 2 N, rightward; (b) 2 N, rightward; (c) greater than 2 N, rightward **3.** b, c, a **5.** (a) x yes, y no; (b) x yes, y no; (c) x no, y yes

7. (a) c, kinetic energy cannot be negative; d, total kinetic energy cannot increase; (b) a; (c) b **9.** (a) one was stationary; (b) 2; (c) 5; (d) equal (pool player's result) **11.** (a) C; (b) B; (c) 3
P 1. (a) -1.50 m; (b) -1.43 m **3.** (a) -6.5 cm; (b) 8.3 cm; (c) 1.4 cm **5.** (a) -0.45 cm; (b) -2.0 cm **7.** (a) 0; (b) 3.13×10^{-11} m **9.** (a) 28 cm; (b) 2.3 m/s **11.** $(-4.0 \text{ m})\hat{i} + (4.0 \text{ m})\hat{j}$ **13.** 53 m **15.** (a) $(2.35\hat{i} - 1.57\hat{j})$ m/s^2; (b) $(2.35\hat{i} - 1.57\hat{j})t$ m/s, with t in seconds; (d) straight, at downward angle 34° **17.** 4.2 m **19.** (a) 7.5×10^4 J; (b) 3.8×10^4 kg·m/s; (c) 39° south of due east **21.** (a) 5.0 kg·m/s; (b) 10 kg·m/s **23.** 1.0×10^3 to 1.2×10^3 kg·m/s **25.** (a) 42 N·s; (b) 2.1 kN **27.** (a) 67 m/s; (b) $-x$; (c) 1.2 kN; (d) $-x$ **29.** 5 N **31.** (a) 2.39×10^3 N·s; (b) 4.78×10^5 N; (c) 1.76×10^3 N·s; (d) 3.52×10^5 N **33.** (a) 5.86 kg·m/s; (b) 59.8°; (c) 2.93 kN; (d) 59.8° **35.** 9.9×10^2 N **37.** (a) 9.0 kg·m/s; (b) 3.0 kN; (c) 4.5 kN; (d) 20 m/s **39.** 3.0 mm/s **41.** (a) $-(0.15 \text{ m/s})\hat{i}$; (b) 0.18 m **43.** 55 cm **45.** (a) $(1.00\hat{i} - 0.167\hat{j})$ km/s; (b) 3.23 MJ **47.** (a) 14 m/s; (b) 45° **49.** 3.1×10^2 m/s **51.** (a) 721 m/s; (b) 937 m/s **53.** (a) 33%; (b) 23%; (c) decreases **55.** (a) $+2.0$ m/s; (b) -1.3 J; (c) $+40$ J; (d) system got energy from some source, such as a small explosion **57.** (a) 4.4 m/s; (b) 0.80 **59.** 25 cm **61.** (a) 99 g; (b) 1.9 m/s; (c) 0.93 m/s **63.** (a) 3.00 m/s; (b) 6.00 m/s **65.** (a) 1.2 kg; (b) 2.5 m/s **67.** -28 cm **69.** (a) 0.21 kg; (b) 7.2 m **71.** (a) 4.15×10^5 m/s; (b) 4.84×10^5 m/s **73.** 120° **75.** (a) 433 m/s; (b) 250 m/s **77.** (a) 46 N; (b) none **79.** (a) 1.57×10^6 N; (b) 1.35×10^5 kg; (c) 2.08 km/s **81.** (a) 7290 m/s; (b) 8200 m/s; (c) 1.271×10^{10} J; (d) 1.275×10^{10} J **83.** (a) 1.92 m; (b) 0.640 m **85.** (a) 1.78 m/s; (b) less; (c) less; (d) greater **87.** (a) 3.7 m/s; (b) 1.3 N·s; (c) 1.8×10^2 N **89.** (a) $(7.4 \times 10^3 \text{ N·s})\hat{i} - (7.4 \times 10^3 \text{ N·s})\hat{j}$; (b) $(-7.4 \times 10^3 \text{ N·s})\hat{i}$; (c) 2.3×10^3 N; (d) 2.1×10^4 N; (e) $-45°$ **91.** $+4.4$ m/s **93.** 1.18×10^4 kg **95.** (a) 1.9 m/s; (b) $-30°$; (c) elastic **97.** (a) 6.9 m/s; (b) 30°; (c) 6.9 m/s; (d) $-30°$; (e) 2.0 m/s; (f) $-180°$ **99.** (a) 25 mm; (b) 26 mm; (c) down; (d) 1.6×10^{-2} m/s^2 **101.** 29 J **103.** 2.2 kg **105.** 5.0 kg **107.** (a) 50 kg/s; (b) 1.6×10^2 kg/s **109.** (a) 4.6×10^3 km; (b) 73% **111.** 190 m/s **113.** 28.8 N **115.** (a) 0.745 mm; (b) 153°; (c) 1.67 mJ **117.** (a) $(2.67 \text{ m/s})\hat{i} + (-3.00 \text{ m/s})\hat{j}$; (b) 4.01 m/s; (c) 48.4° **119.** (a) -0.50 m; (b) -1.8 cm; (c) 0.50 m **121.** 0.22%

CHAPTER 10

CP 1. b and c **2.** (a) and (d) ($\alpha = d^2\theta/dt^2$ must be a constant) **3.** (a) yes; (b) no; (c) yes; (d) yes **4.** all tie **5.** 1, 2, 4, 3 (see Eq. 10-36) **6.** (see Eq. 10-40) 1 and 3 tie, 4, then 2 and 5 tie (zero) **7.** (a) downward in the figure ($\tau_{net} = 0$); (b) less (consider moment arms)
Q 1. (a) c, a, then b and d tie; (b) b, then a and c tie, then d **3.** all tie **5.** (a) decrease; (b) clockwise; (c) counterclockwise **7.** larger **9.** c, a, b
P 1. 14 rev **3.** (a) 4.0 rad/s; (b) 11.9 rad/s **5.** 11 rad/s **7.** (a) 4.0 m/s; (b) no **9.** (a) 3.00 s; (b) 18.9 rad **11.** (a) 30 s; (b) 1.8×10^3 rad **13.** (a) 3.4×10^2 s; (b) -4.5×10^{-3} rad/s^2; (c) 98 s **15.** 8.0 s **17.** (a) 44 rad; (b) 5.5 s; (c) 32 s; (d) -2.1 s; (e) 40 s **19.** (a) 2.50×10^{-3} rad/s; (b) 20.2 m/s^2; (c) 0 **21.** 6.9×10^{-13} rad/s **23.** (a) 20.9 rad/s; (b) 12.5 m/s; (c) 800 rev/min^2; (d) 600 rev **25.** (a) 7.3×10^{-5} rad/s; (b) 3.5×10^2 m/s; (c) 7.3×10^{-5} rad/s; (d) 4.6×10^2 m/s **27.** (a) 73 cm/s^2; (b) 0.075; (c) 0.11 **29.** (a) 3.8×10^3 rad/s; (b) 1.9×10^2 m/s **31.** (a) 40 s; (b) 2.0 rad/s^2 **33.** 12.3 kg·m^2 **35.** (a) 1.1 kJ; (b) 9.7 kJ **37.** 0.097 kg·m^2 **39.** (a) 49 MJ; (b) 1.0×10^2 min **41.** (a) 0.023 kg·m^2; (b) 1.1 mJ **43.** 4.7×10^{-4} kg·m^2 **45.** -3.85 N·m **47.** 4.6 N·m **49.** (a) 28.2 rad/s^2; (b) 338 N·m **51.** (a) 6.00 cm/s^2; (b) 4.87 N; (c) 4.54 N; (d) 1.20 rad/s^2; (e) 0.0138 kg·m^2 **53.** 0.140 N **55.** 2.51×10^{-4} kg·m^2 **57.** (a) 4.2×10^2 rad/s^2; (b) 5.0×10^2 rad/s **59.** 396 N·m **61.** (a) -19.8 kJ; (b) 1.32 kW **63.** 5.42 m/s **65.** (a) 5.32 m/s^2; (b) 8.43 m/s^2; (c) 41.8° **67.** 9.82 rad/s **69.** 6.16×10^{-5}

kg·m^2 **71.** (a) 31.4 rad/s^2; (b) 0.754 m/s^2; (c) 56.1 N; (d) 55.1 N
73. (a) 4.81 × 10^{5}N; (b) 1.12 × 10^{4}N·m; (c) 1.25 × 10^{6}J   **75.** (a) 2.3
rad/s^2; (b) 1.4 rad/s^2 **77.** (a) −67 rev/min^2; (b) 8.3 rev   **81.** 3.1 rad/s
83. (a) 1.57 m/s^2; (b) 4.55 N; (c) 4.94 N   **85.** 30 rev   **87.** 0.054 kg·m^2
89. 1.4 × 10^{2}N·m   **93.** 4.6 rad/s^2 **95.** 2.6 J   **97.** (a) 5.92 × 10^{4}m/s^2;
(b) 4.39 × 10^{4}s^{-2} **99.** (a) 0.791 kg·m^2; (b) 1.79 × 10^{-2}N·m   **101.**
(a) 1.5 × 10^{2}cm/s; (b) 15 rad/s; (c) 15 rad/s; (d) 75 cm/s; (e) 3.0 rad/s
103. (a) 7.0 kg·m^2; (b) 7.2 m/s; (c) 71°

CHAPTER 11

CP   **1.** (a) same; (b) less   **2.** less (consider the transfer of energy
from rotational kinetic energy to gravitational potential energy)   **3.**
(draw the vectors, use right-hand rule) (a) ±z; (b) +y; (c) −x   **4.**
(see Eq. 11-21) (a) 1 and 3 tie; then 2 and 4 tie, then 5 (zero); (b) 2
and 3   **5.** (see Eqs. 11-23 and 11-16) (a) 3, 1; then 2 and 4 tie (zero);
(b) 3   **6.** (a) all tie (same τ, same t, thus same ΔL); (b) sphere, disk,
hoop (reverse order of I)   **7.** (a) decreases; (b) same (τ_{net} = 0, so L
is conserved); (c) increases

Q   **1.** a, then b and c tie, then e, d (zero)   **3.** (a) spins in place; (b)
rolls toward you; (c) rolls away from you   **5.** (a) 1, 2, 3 (zero); (b) 1
and 2 tie, then 3; (c) 1 and 3 tie, then 2   **7.** (a) same; (b) increase; (c)
decrease; (d) same, decrease, increase   **9.** D, B, then A and C tie

P   **1.** (a) 0; (b) (22 m/s)$\hat{i}$; (c) (−22 m/s)$\hat{i}$; (d) 0; (e) 1.5 × 10^3 m/s^2; (f)
1.5 × 10^3 m/s^2; (g) (22 m/s) $\hat{i}$; (h) (44 m/s) $\hat{i}$; (i) 0; (j) 0; (k) 1.5 × 10^3
m/s^2; (l) 1.5 × 10^3 m/s^2 **3.** −3.15 J   **5.** 0.020   **7.** (a) 63 rad/s; (b) 4.0 m
9. 4.8 m   **11.** (a) (−4.0 N) $\hat{i}$; (b) 0.60 kg·m^2 **13.** 0.50   **15.** (a)
−(0.11 m)ω; (b) −2.1 m/s^2; (c) −47 rad/s^2; (d) 1.2 s; (e) 8.6 m; (f) 6.1

m/s   **17.** (a) 13 cm/s^2; (b) 4.4 s; (c) 55 cm/s; (d) 18 mJ; (e) 1.4 J; (f) 27
rev/s   **19.** (−2.0 N·m)$\hat{i}$ **21.** (a) (6.0 N·m)$\hat{j}$ + (8.0 N·m)$\hat{k}$; (b)
(−22 N·m)$\hat{i}$ **23.** (a) (−1.5 N·m)$\hat{i}$ − (4.0 N·m)$\hat{j}$ − (1.0 N·m)$\hat{k}$;
(b) (−1.5 N·m)$\hat{i}$ − (4.0 N·m)$\hat{j}$ − (1.0 N·m)$\hat{k}$ **25.** (a) (50 N·m)$\hat{k}$;
(b) 90°   **27.** (a) 0; (b) (8.0 N·m)$\hat{i}$ + (8.0 N·m)$\hat{k}$ **29.** (a) 9.8
kg·m^2/s; (b) +z direction   **31.** (a) 0; (b) −22.6 kg·m^2/s; (c) −7.84
N·m; (d) −7.84 N·m   **33.** (a) (−1.7 × 10^2 kg·m^2/s)$\hat{k}$; (b) (+56
N·m)$\hat{k}$; (c) (+56 kg·m^2/s^2)$\hat{k}$ **35.** (a) 48$t\hat{k}$ N·m; (b) increasing
37. (a) 4.6 × 10^{-3} kg·m^2; (b) 1.1 × 10^{-3} kg·m^2/s; (c) 3.9 × 10^{-3}
kg·m^2/s   **39.** (a) 1.47 N·m; (b) 20.4 rad; (c) −29.9 J; (d) 19.9 W
41. (a) 1.6 kg·m^2; (b) 4.0 kg·m^2/s   **43.** (a) 1.5 m; (b) 0.93 rad/s; (c) 98
J; (d) 8.4 rad/s; (e) 8.8 × 10^2 J; (f) internal energy of the skaters   **45.**
(a) 3.6 rev/s; (b) 3.0; (c) forces on the bricks from the man transferred energy from the man's internal energy to kinetic energy   **47.**
0.17 rad/s   **49.** (a) 750 rev/min; (b) 450 rev/min; (c) clockwise   **51.**
(a) 267 rev/min; (b) 0.667   **53.** 1.3 × 10^3 m/s   **55.** 3.4 rad/s   **57.** (a)
18 rad/s; (b) 0.92   **59.** 11.0 m/s   **61.** 1.5 rad/s   **63.** 0.070 rad/s   **65.** (a)
0.148 rad/s; (b) 0.0123; (c) 181°   **67.** (a) 0.180 m; (b) clockwise   **69.**
0.041 rad/s   **71.** (a) 1.6 m/s^2; (b) 16 rad/s^2; (c) (4.0 N)$\hat{i}$ **73.** (a) 0; (b)
0; (c) −30$t^3\hat{k}$ kg·m^2/s; (d) −90$t^2\hat{k}$ N·m; (e) 30$t^3\hat{k}$ kg·m^2/s; (f) 90$t^2\hat{k}$
N·m   **75.** (a) 149 kg·m^2; (b) 158 kg·m^2/s; (c) 0.744 rad/s   **77.** (a)
6.65 × 10^{-5}kg·m^2/s; (b) no; (c) 0; (d) yes   **79.** (a) 0.333; (b) 0.111
81. (a) 58.8 J; (b) 39.2 J   **83.** (a) 61.7 J; (b) 3.43 m; (c) no   **85.** (a)
$mvR/(I + MR^2)$; (b) $mvR^2/(I + MR^2)$ **87.** rotational speed would
decrease; day would be about 0.8 s longer   **89.** (a) 12.7 rad/s; (b)
clockwise   **91.** (a) 0.81 mJ; (b) 0.29; (c) 1.3 × 10^{-2}N   **93.** (a) $mR^2/2$;
(b) a solid circular cylinder

Figures are noted by page numbers in *italics*, tables are
indicated by t following the page number.

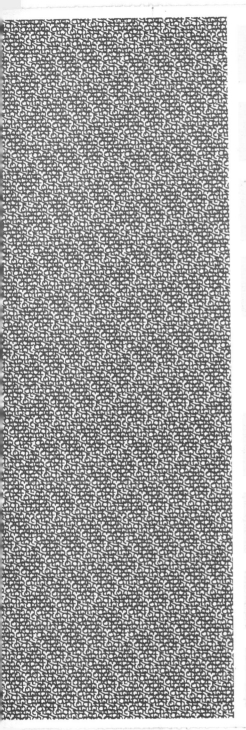

Some Physical Constants*

Speed of light	c	2.998×10^8 m/s
Gravitational constant	G	6.673×10^{-11} N $\cdot$ m^2/kg^2
Avogadro constant	N_A	6.022×10^{23} mol^{-1}
Universal gas constant	R	8.314 J/mol $\cdot$ K
Mass–energy relation	c^2	8.988×10^{16} J/kg
		931.49 MeV/u
Permittivity constant	ε_0	8.854×10^{-12} F/m
Permeability constant	μ_0	1.257×10^{-6} H/m
Planck constant	h	6.626×10^{-34} J $\cdot$ s
		4.136×10^{-15} eV·s
Boltzmann constant	k	1.381×10^{-23} J/K
		8.617×10^{-5} eV/K
Elementary charge	e	1.602×10^{-19} C
Electron mass	m_e	9.109×10^{-31} kg
Proton mass	m_p	1.673×10^{-27} kg
Neutron mass	m_n	1.675×10^{-27} kg
Deuteron mass	m_d	3.344×10^{-27} kg
Bohr radius	a	5.292×10^{-11} m
Bohr magneton	μ_B	9.274×10^{-24} J/T
		5.788×10^{-5} eV/T
Rydberg constant	R	$1.097\,373 \times 10^7$ m^{-1}

*For a more complete list, showing also the best experimental values, see Appendix B.

The Greek Alphabet

Alpha	A	α	Iota	I	ι	Rho	P	ρ
Beta	B	β	Kappa	K	κ	Sigma	Σ	σ
Gamma	Γ	γ	Lambda	Λ	λ	Tau	T	τ
Delta	Δ	δ	Mu	M	μ	Upsilon	Υ	υ
Epsilon	E	ϵ	Nu	N	ν	Phi	Φ	ϕ, φ
Zeta	Z	ζ	Xi	Ξ	ξ	Chi	X	χ
Eta	H	η	Omicron	O	o	Psi	Ψ	ψ
Theta	Θ	θ	Pi	Π	π	Omega	Ω	ω